GENETIC AND MOLECULAR ASPECTS OF SPORT PERFORMANCE

GENETIC AND MOLECULAR ASPECTS OF SPORT PERFORMANCE

VOLUME XVIII OF THE ENCYCLOPAEDIA OF SPORTS MEDICINE

AN IOC MEDICAL COMMISSION PUBLICATION

EDITED BY

CLAUDE BOUCHARD, PhD

and

ERIC P. HOFFMAN, PhD

A John Wiley & Sons, Ltd., Publication

Library of Congress Cataloging-in-Publication Data

Genetic and molecular aspects of sport performance/edited by Claude Bouchard and Eric Hoffman.
      p. ; cm.—(Encyclopaedia of sports medicine; v. 18)
   An IOC Medical Commission publication.
   Includes bibliographical references.
   ISBN 978-1-4443-3445-6
   1. Sports—Phyiological aspects.   2. Exercise—Phyiological aspects.   3. Human genetics. 4. Athletic ability.   I. Bouchard, Claude.   II. Hoffman, Eric (Eric P.)   III. IOC Medical Commission.   IV. Series: Encyclopaedia of sports medicine; v. 18.
   [DNLM: 1. Athletic Performance—physiology.   2. Genetic Phenomena.   3. Sports Medicine—methods. QT 13 E527 1988 v.18]
   RC1235.G457 2011
   612'.044—dc22

                                                                2010010778

A catalogue record for this book is available from the British Library.

This book is published in the following electronic formats: ePDF 9781444327342; Wiley Online Library 9781444327335

Set in 9/12pt Sabon by MPS Limited, a Macmillan Company, Chennai, India
Printed and bound in Malaysia by Vivar Printing Sdn Bhd

1   2011

Contents

v

List of Contributors

ANNA BAOUTINA, PhD, *National Measurement Institute, Lindfield, NSW, Australia*

DAVID BASSETT, PhD, *University of Tennessee Obesity Research Center, Knoxville, TN, USA*

GASTON P. BEUNEN, PhD, *Faculty of Kinesiology and Rehabilitation Sciences, Department of Biomedical Kinesiology, Research Centre for Exercise and Health, Katholieke Universiteit Leuven, Leuven, Belgium*

JOHN BLANGERO, PhD, *Department of Genetics, Southwest Foundation for Biomedical Research, San Antonio, TX, USA*

YOHAN BOSSÉ, PhD, *Department of Molecular Medicine, Laval University, and Institut Universitaire de Cardiologie et de Pneumologie de Québec, Québec, QC, Canada*

CLAUDE BOUCHARD, PhD, *Human Genomics Laboratory, Pennington Biomedical Research Center, Baton Rouge, LA, USA*

MOLLY S. BRAY, PhD, *Department of Epidemiology, Heflin Center for Genomic Sciences, University of Alabama at Birmingham, Birmingham, AL, USA*

STEVEN L. BRITTON, PhD, *Department of Anesthesiology, University of Michigan Medical School, Ann Arbor, MI, USA*

KRISTY J. BROWN, PhD, *Research Center for Genetic Medicine, Children's National Medical Center, Washington, DC, USA*

EUNHEE CHUNG, PhD, *Department of Molecular, Cellular, and Developmental Biology, University of Colorado, Boulder, CO, USA*

KARINE CLÉMENT, MD, PhD, *INSERM, Cordelier Research Center, Paris, France; University Pierre et Marie Curie-Paris 6, Paris, France; Endocrinology and Nutrition Department, Pitié-Salpêtrière Hospital, Paris, France*

CATHERINE A. FORMOLO, BS, *Research Center for Genetic Medicine, Children's National Medical Center, Washington, DC, USA*

JANET E. FULTON, PhD, *Division of Physical Activity, Nutrition, and Obesity, Centers for Disease Control and Prevention, Atlanta, GA, USA*

ECO J.C. DE GEUS, PhD, *Department of Biological Psychology, VU University Amsterdam, Amsterdam, The Netherlands*

MARLEEN H.M. DE MOOR, PhD, *Department of Biological Psychology, VU University Amsterdam, Amsterdam, The Netherlands*

JAMES M. HAGBERG, PhD, *Department of Kinesiology, School of Public Health, University of Maryland, College Park, MD, USA*

YETRIB HATHOUT, PhD, *Research Center for Genetic Medicine, Children's National Medical Center, Washington, DC, USA*

JOHN A. HAWLEY, PhD, *HIRi, School of Medical Sciences, RMIT University, Bundoora, VIC, Australia*

JORN W. HELGE, PhD, *Center for Healthy Ageing, Section of Systems Biology Research, Panum Institute, University of Copenhagen, Copenhagen, Denmark*

CORNELIU HENEGAR, MD, PhD, *INSERM, Cordelier Research Center, Paris, France; University Pierre et Marie Curie-Paris 6, Paris, France; Endocrinology and Nutrition Department, Pitié-Salpêtrière Hospital, Paris, France*

DUSTIN S. HITTEL, PhD, *Roger Jackson Center for Health and Wellness, Human Performance Laboratory, Faculty of Kinesiology, University of Calgary, Calgary, AB, Canada*

ERIC P. HOFFMAN, PhD, *Department of Integrative Systems Biology, George Washington University School of Medicine, Research Center for Genetic Medicine, Children's National Medical Center, Washington, DC, USA*

HENRI HOOTON, MS, *INSERM, Cordelier Research Center, Paris, France; University Pierre et Marie Curie-Paris 6, Paris, France; Endocrinology and Nutrition Department, Pitié-Salpêtrière Hospital, Paris, France*

MONICA HUBAL, PhD, *Department of Integrative Systems Biology, George Washington University School of Medicine, Research Center for Genetic Medicine, Children's National Medical Center, Washington, DC, USA*

FAWZI KADI, PhD, *School of Health and Medical Sciences, Örebro University, Örebro, Sweden*

NISHAN SUDHEERA KALUPAHANA, M PHIL, *University of Tennessee Obesity Research Center, Knoxville, TN, USA*

LEONIDAS G. KARAGOUNIS, PhD, *HIRi, School of Medical Sciences, RMIT University, Bundoora, VIC, Australia*

JACK W. KENT Jr., PhD, *Department of Genetics, Southwest Foundation for Biomedical Research, San Antonio, TX, USA*

CHAD M. KERKSICK, PhD, *Applied Biochemistry and Molecular Physiology Laboratory, Department of Health and Exercise Science, University of Oklahoma, Norman, OK, USA*

JUNG HAN KIM, PhD, *University of Tennessee Obesity Research Center, Knoxville, TN, USA; Department of Pharmacology, Physiology, and Toxicology, Marshall University, Huntington, WV, USA*

LAUREN GERARD KOCH, PhD, *Department of Anesthesiology, University of Michigan Medical School, Ann Arbor, MI, USA*

SHINYA KUNO, PhD, *Department of Sports Medicine, University of Tsukuba, Ibaraki, Japan*

LESLIE A. LEINWAND, PhD, *Department of Molecular, Cellular, and Developmental Biology, University of Colorado, Boulder, CO, USA*

HUAI LI, PhD, *Bioinformatics Unit, National Institute on Aging, National Institutes of Health, Baltimore, MD, USA*

J. TIMOTHY LIGHTFOOT, PhD, FACSM, RCEP, CES, *Department of Health and Kinesiology, Texas A&M University, College Station, TX, USA*

SIGMUND LOLAND, PhD, *Department of Social and Cultural Studies, The Norwegian School of Sport Sciences, Oslo, Norway*

DANIEL G. MACARTHUR, PhD, *Discipline of Paediatrics and Child Health, Faculty of Medicine, University of Sydney, Sydney, NSW, Australia; Human Evolution, Wellcome Trust Sanger Institute, Cambridge, UK*

ROBERT M. MALINA, PhD, *Department of Kinesiology and Health Education, University of Texas at Austin and Tarleton State University, Stephenville, TX, USA*

ANDY MIAH, PhD, *Chair of Ethics and Emerging Technologies, Faculty of Business & Creative Industries, University of the West of Scotland, Ayr Campus, UK*

DAVID J. MILLER, PhD, *Department of Electrical Engineering, Pennsylvania State University, University Park, PA, USA*

HUGH E. MONTGOMERY, MBBS, MD, *UCL Institute for Human Health and Performance, London, UK*

NAIMA MOUSTAID-MOUSSA, PhD, *University of Tennessee Obesity Research Center, Knoxville, TN, USA*

JOHN J. MULVIHILL, MD, *Department of Pediatrics, University of Oklahoma Health Sciences Center, Oklahoma City, OK, USA*

HARUKA MURAKAMI, PhD, *National Institute of Health and Nutrition, Tokyo, Japan*

JAVAD NAZARIAN, PhD, *Research Center for Genetic Medicine, Children's National Medical Center, Washington, DC, USA*

KATHRYN N. NORTH, MD, MBBS, *Discipline of Paediatrics and Child Health, Faculty of Medicine, University of Sydney, Sydney, NSW, Australia; Institute for Neuroscience and Muscle Research, Children's Hospital at Westmead, Sydney, Australia*

MAARTEN W. PEETERS, PhD, *Faculty of Kinesiology and Rehabilitation Sciences, Department of Biomedical Kinesiology, Research Centre for Exercise and Health, Katholieke Universiteit Leuven, Leuven, Belgium; Research Foundation-Flanders (FWO), Belgium*

LOUIS PÉRUSSE, PhD, *Department of Social and Preventive Medicine, Division of Kinesiology, Laval University, Québec, QC, Canada*

YANNIS P. PITSILADIS, PhD, *College of Medicine, Veterinary & Life Sciences, Institute of Cardiovascular & Medical Sciences, University of Glasgow, Glasgow, UK*

ZUDIN A. PUTHUCHEARY, MBBS, *UCL Institute for Human Health and Performance, London, UK*

TUOMO RANKINEN, PhD, *Human Genomics Laboratory, Pennington Biomedical Research Center, Baton Rouge, LA, USA*

JAI RAWAL, MBBS, *UCL Institute for Human Health and Performance, London, UK*

STEPHEN M. ROTH, PhD, *Department of Kinesiology, School of Public Health, University of Maryland, College Park, MD, USA*

MARK A. SARZYNSKI, PhD, *Human Genomics Laboratory, Pennington Biomedical Research Center, Baton Rouge, LA, USA*

JAMES R.A. SKIPWORTH, MBBS, *UCL Institute for Human Health and Performance, London, UK*

MARK A. TARNOPOLSKY, MD, PhD, *Department of Pediatrics and Medicine, McMaster University Medical Center, Hamilton, ON, Canada*

MARTINE A.I. THOMIS, PhD, *Faculty of Kinesiology and Rehabilitation Sciences, Katholieke Universiteit Leuven, and Department of Biomedical Kinesiology, Research Centre for Exercise and Health, Leuven, Belgium*

JAMES A. TIMMONS, PhD, *Lifestyle Research Group, Royal Veterinary College, University of London, London, UK*

BRYNN H. VOY, PhD, *University of Tennessee Obesity Research Center, Knoxville, TN, USA*

DOUGLAS C. WALLACE, PhD, *Center for Mitochondrial and Epigenomic Medicine and Department of Pathology and Laboratory Medicine, Children's Hospital of Philadelphia and University of Pennsylvania, Philadelphia, PA, USA*

YUE WANG, PhD, *Bradley Department of Electrical and Computer Engineering, Virginia Polytechnic Institute and State University, Arlington, VA, USA*

ZUYI WANG, PhD, *Department of Integrative Systems Biology, George Washington University School of Medicine, Research Center for Genetic Medicine, Children's National Medical Center, Washington, DC, USA*

KLAAS J. WIERENGA, MD, *Department of Pediatrics, University of Oklahoma Health Sciences Center, Oklahoma City, OK, USA*

BERND WOLFARTH, MD, *Department of Preventive and Rehabilitative Sports Medicine, Technical University Munich, Munich, Germany*

JIANHUA XUAN, PhD, *Bradley Department of Electrical and Computer Engineering, Virginia Polytechnic Institute and State University, Arlington, VA, USA*

HIROFUMI ZEMPO, MS, *Department of Sports Medicine, University of Tsukuba, Ibaraki, Japan*

Foreword

Knowledge concerning the relationship of genetics to human performance has grown markedly during the last quarter century. For the performance of sport, the potentials passed on by parents to their offspring for physical size, tissue composition, neural control, and metabolic function establish a foundation for the athlete's nutritional needs, physical and mental conditioning, and motor skill development.

Professors Bouchard and Hoffman and their contributing authors have produced a comprehensive and authoritative volume that summarizes and interprets the complex research that has recently become available. The initial chapters address the basic science of genomics and genetics and the regulation of gene expression. Additional authoritative information is provided in chapters on the genetics of complex performance phenotypes, the contributions of small animal research, family and twin studies, and ethnic comparisons. A final section addresses the issue of the contribution of specific genes and molecular markers as related to endurance, strength and power, and responsiveness to specific conditioning programs. The many fundamental advances in our understanding of the genetic and molecular basis of human physical performance have important implications for success in sport on all levels of competition.

This volume constitutes a landmark contribution to the understanding of the human organism engaged in sport and many other physical activities. We welcome its addition to the IOC Medical Commission's series, *The Encyclopaedia of Sports Medicine*.

Dr Jacques Rogge
IOC President

Preface

CLAUDE BOUCHARD[1] AND ERIC P. HOFFMAN[2]

[1]Human Genomics Laboratory, Pennington Biomedical Research Center, Baton Rouge, LA, USA
[2]Research Center for Genetic Medicine, Children's National Medical Center, Washington, DC, USA

An elite athlete trains to ensure that all key tissues, organs, and systems reach the highest degree of efficiency, specialization, and precision. For instance, the shot putter trains for efficiency in motor coordination and muscle power. The marathon runner trains for high cardiac output and muscle endurance while the sprinter for motor coordination and speed. Training is critical for success, but we all recognize that training is only one component of the equation leading to the status of an elite athlete. Among the other determinants, the palette of genetic variation imparted to the athlete at conception stands out as one of the least understood but most critical dimensions to consider.

The world-class elite athlete exemplifies successful interactions of nature versus nurture—of one's genomic and epigenomic traits and the training, dietary, and other lifestyle and environmental demands imposed on them. One could argue that the remodeling and adaptation of muscle based on patterns of use, combined with genetic predisposition, is probably the best example of the intrinsic ability of a tissue to respond to the environment and adapt to it. As such, academics have been fascinated with muscle adaptation, working toward understanding how muscle works in both health and disease, as well as skeletal muscle structures and properties have changed through evolution. Indeed, muscle is the single binding feature of all living "animals"—the ability to move toward a food source.

Advances in training regimens produce increasingly impressive athletes who are constantly pushing the boundaries of the "citius, altius, fortius."

In parallel, scientists have been similarly pursuing excellence, honing the tools of genomic and other omic sciences to understand tissue and organ structure, function, and response to training. In this regard, the decoding of the human genome in the late 1990s led to a new toolbox to support the discovery process, and genomic-based technologies have been quickly applied to understanding tissue and organ remodeling with exercise training, specially skeletal muscle, by scientists worldwide.

The first comprehensive effort at summarizing the evidence for a genetic contribution to indicators of fitness and physical performance was in the form of a volume written in the mid-1990s by C. Bouchard, R.M. Malina, and L. Perusse and published in 1997 by Human Kinetics Publisher under the title "Genetics of Fitness and Physical Performance." It described the methods and technologies available at the time for the study of the genetic basis of complex human traits, and summarized the evidence accumulated at that time on a number of endophenotypes and fitness and performance traits. The volume preceded by several years the publication of the human genome sequence, the advent of high-throughput sequencing and genotyping technologies, and the surge in genomic and other omic sciences.

This volume represents a major attempt to bring under a single publication the advances that have occurred since then. However, the reader will quickly realize that the genetics and molecular biology of performance is far from a "completed chapter in science." Even though progress has been made on several fronts, little is known about the roles of genetic differences and epigenomic events in performance in which cardiorespiratory endurance or muscle strength and power predominate. Our understanding of the effects of human genomic variation on the ability of tissues and organs to be trained, particularly cardiac and skeletal muscle, remains very scanty. Nowhere is the limitation of our current knowledge base more obvious than for the role of human genomic and epigenomic variation in motor coordination and motor learning, which are key determinants of a large number of sports performances. The contributing authors of this volume point out what is known, and emphasize what is unknown and holds promise for future research. We believe that most of the genome-enabled research tools are in hand to achieve a "complete" understanding of muscle remodeling, and these tools are described in some detail in this book. However, an enormous amount of research work is needed before we can predict the effects of an individual's genomic and perhaps epigenomic characteristics on his or her ability to be trained and to reach elite athlete status in a given sport. This volume covers the current status of the genetic and molecular aspects of sports performance, and the competent scientists and authors assembled to produce it voice their collective excitement for what is to come.

The overarching goal of this publication is to summarize the evidence from all relevant sources on the genetic and molecular basis of sports and other human physical performance. The book is divided into five parts. Part 1 deals with the basic science of genomics and genetics that are relevant to sports performance. Chapters focus on topics such as the human genome, mitochondrial DNA, genes and regulation of gene expression, extent of human DNA sequence variation, imprinting and epigenetics, technologies to identify genes influencing complex human traits, and computational biology and bioinformatic tools. Part 2 defines the evidence from genetic epidemiology studies and includes chapters on the genetics of complex performance phenotypes, physical activity level, selection experiments in rodents, family and twin studies, and ethnic differences in performance.

Part 3 addresses the issue of the contribution of specific genes and molecular markers. Key chapters deal with evidence for endurance performance, strength and power phenotypes, other types of human performance, as well as responsiveness to training. Evidence from transgenic and knock-out mouse models, and human candidate genes, linkage and genome-wide association studies is reviewed. Specific chapters are devoted to the role of *ACE*, *ACTN3*, and mitochondrial DNA to performance. Three chapters deal with the contributions of genetic differences to carbohydrate, lipid, and protein metabolism. This part of the volume is completed by chapters focused on hemodynamic traits, the drive to be active, and psychological factors. Part 4 deals with systems biology applied to exercise and training. Chapters on proteomics, mRNA profiling molecular networks, metabolomics, and other relevant topics are grouped in this section. One chapter is devoted to stem cell biology and performance enhancement. Part 5 highlights the implications of these fundamental advances for our understanding of the genetic and molecular basis of human physical performance, for the world of elite sports and society in general. Chapters on ethical issues posed by current practices and those that can be anticipated from the predicted advances in the genetics of performance are incorporated as well as on culture and policy. One chapter is devoted to talent selection. The issue of performance enhancement resulting from gene augmentation technologies (gene doping) is covered in a specific chapter. All these chapters together comprise a photograph of the situation at the time of the writing along with the rationale for future research directions.

This publication would not have been possible without the contributions of the talented authors that we were able to recruit from 48 different laboratories in 13 countries. We want to thank them for their valuable manuscripts and for responding diligently to our queries throughout the production process. This project was first proposed to Bouchard by Dr. Howard Knuttgen as part of the *Encyclopaedia of Sports Medicine* series published under the egis of the International Olympic Committee Medical Commission. We would like to thank "Skip" for this mark of confidence and for his support as the project evolved. Our heartfelt thanks to Mrs. Agnes Gaillard from the Lausanne IOC office, who was instrumental in helping us secure the participation of the contributing authors. Ms. Kate Newell, Development Editor, at Wiley-Blackwell, Oxford, was very helpful throughout the whole process and we want to express our appreciation for her assistance. Our thanks also to Ms. Cathryn Gates, Production Editor at Wiley-Blackwell, and Ms. Geetha Williams of Macmillan Publishing Solutions for their support. Finally, our deepest gratitude to Mrs. Nina Laidlaw and Ms. Allison Templet in the office of Bouchard for their strong support to both editors. They brought competence and dedication to the editing process, and the volume is clearly better because of their sustained involvement.

PART 1

THE SCIENCE OF GENOMICS AND GENETICS

Chapter 1

The Human Genome and Epigenome

JOHN J. MULVIHILL,[1] KLAAS J. WIERENGA[1] AND CHAD M. KERKSICK[2]

[1]*Department of Pediatrics, University of Oklahoma Health Sciences Center, Oklahoma City, OK, USA*
[2]*Applied Biochemistry and Molecular Physiology Laboratory, Department of Health and Exercise Science, University of Oklahoma, Norman, OK, USA*

The primary intention of this chapter is to introduce topics and concepts central to genetics. For this reason, any individual with extensive background in the area may skip it altogether. However, people who come from an applied physiology or exercise-centered focus of training will likely find value in the introductory material and how it sets the table for the remainder of this book. As such, the principles of genetics, genomics, and epigenetics will be the central focus so those interested individuals can quickly refresh any gaps in their knowledge from reading this chapter of the book. Overall, the goal of this chapter is to provide enough definitions and examples of the fundamentals of genetics to permit an understanding and context of the details in specific chapters. Similar overviews, some at greater length, are available (Bouchard et al., 1997; Gibson, 2009; Roth, 2007).

Genetics and Genomics

Genetics is the study of intrinsic causes of variation in living organisms. *Human genetics* narrows the scope to variation among human beings, and *medical genetics* emphasizes the variations that are considered diseases or pathologic conditions. *Genomics*, which entered the scientific lexicon at the outset of the Human Genome Project (HGP) (http://www.genome.gov/), was invented for the

Genetic and Molecular Aspects of Sport Performance, 1st edition.
Edited by Claude Bouchard and Eric P. Hoffman.
Published 2011 by Blackwell Publishing Ltd.

purpose of naming a new genetics journal and was coined in 1986, by one of its founding editors, Dr Thomas H. Roderick, of the Jackson Laboratory, Bar Harbor, ME. Genetics and genomics can be considered synonyms, with genomics being trendier, implying something new, modern, and big, and embracing the entire anatomy and physiology of all human genes, including the elements that control the action of genes (i.e., turning them on and off). "Genome" is traced to Hans Winkler in 1920, referring to the entire set of genes in a germ cell (sperm or ovum) (Yadav, 2007). Human genomics is the study all human genes acting together over a lifespan.

The gene is the basic physical–chemical unit of inheritance, consisting of the famous deoxyribonucleic acid (DNA), configured as a double helix of paired complementary strands, as deduced by Watson and Crick in 1953. The total DNA of a single cell is approximately 2.5 m in length (if stretched out straight) and 3 billion base pairs. Considering that an estimated 100 trillion cells make up the human body, the amount of DNA is massive. DNA encodes instructions for the development and function of the entire human being, producing the nickname, the "blueprint of life." The definition of a gene was, until the HGP, a piece of DNA that encodes the amino acid sequence of a single protein, hence the cliché, "one gene—one protein," and the estimated total of 100,000 genes, the numbers of distinct human proteins. When the entire sequence of all nucleotides found in all nuclear DNA was analyzed, the definition of a human gene has become a physical–chemical one, based on the typical structure of a human gene (Figure 1.1). The estimated number of human genes has been revised downward to 25,000, with the realization that each stretch of DNA can code for an average of three proteins.

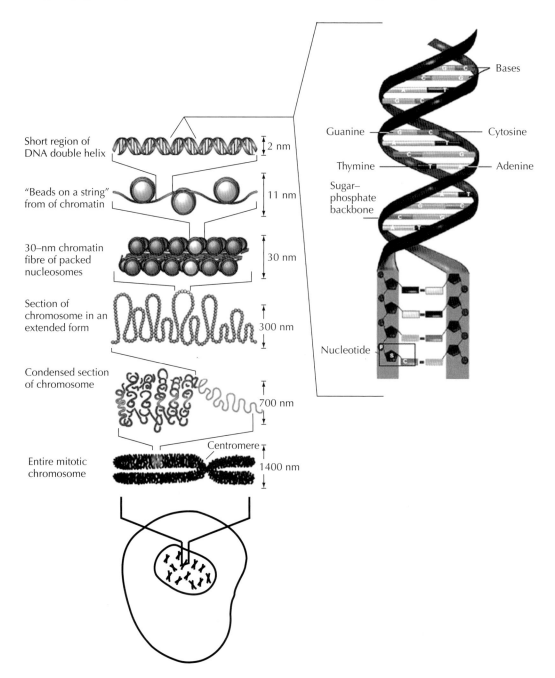

Figure 1.1 Visual outlay of genetic materials starting at the level of the cell and finishing at the level of nucleotide base pairing.

Genes are passed from parents to offspring and contain the information needed to specify physical, biochemical, and behavioral traits. Genes are arranged, one next to another in a linear fashion, on structures called chromosomes. A chromosome contains a single, long DNA molecule, only portions of which correspond to single genes. This organization results in all 25,000 human genes being arranged on 23 pairs of chromosomes in the cell nucleus, in addition to 37 genes which are located in the mitochondria in the cell cytoplasm.

Transcription and Translation

Each DNA strand is made of four chemical units, called nucleotide bases, which comprise the DNA "alphabet." The bases are adenine (A), thymine (T), guanine (G), and cytosine (C). Binding is restrictive and complementary resulting in a scenario where an adenine (A) on one strand always binds to a thymine (T) on the opposite strand. Similar binding occurs with guanine (G) and cytosine (C). The order of these nucleotide bases (e.g., the As, Ts, Cs, and Gs) determines the meaning of the information encoded in that part of the DNA molecule just as the order of letters determines the meaning of a word, except the DNA alphabet has just four letters! Located on 23 pairs of chromosomes packed into the nucleus of a human cell, genes direct the production of proteins with the assistance of enzymes and messenger molecules. Specifically, an enzyme called an RNA polymerase copies the information in a gene's DNA into a molecule called messenger ribonucleic acid (mRNA). The mRNA travels out of the nucleus and into the cell's cytoplasm, where the mRNA is read by ribosomes. The information located on mRNA and read by ribosomes directs the specific linking of various amino acids in the right order to form a specific protein.

Proteins make up body structures like organs and tissue, as well as control chemical reactions and carry signals between cells. If a cell's DNA is mutated, an abnormal protein may be produced, which can disrupt the body's usual processes and lead to a disease, such as cancer or result in a situation which may alter a particular physiological process that may impact how effectively a physiological system can respond to exercise training (i.e., sickle cell trait).

Mitochondria are highly pleiomorphic cytoplasmic organelles composed of compartments that include the outer membrane, the inter-membrane space, the inner membrane, and the cristae and matrix, resulting in four compartments: the central mitochondrial matrix, where the Krebs cycle and beta-oxidation take place; the inner membrane where oxidative phosphorylation generates adenosine triphosphate (ATP); the inner membrane space with key proteins; and, the outer membrane that allows for communication. Mitochondrial DNA (mtDNA) is a 16,569-nucleotide, closed-circular molecule, located within the mitochondrial matrix, and present in thousands of copies per cell. This mtDNA contains 37 genes: 2 ribosomal RNAs, 22 transfer RNAs (tRNAs) that can interpret the entire mtDNA genetic code (which differs slightly from the nuclear genetic code), and 13 genes that code for key polypeptides in oxidative phosphorylation. Other structural and functional proteins, estimated to be around 1500, including many that take part in oxidative phosphorylation, are nuclear-encoded, and transported into the mitochondria. While nuclear genes encoding mitochondrial functions follow Mendelian inheritance patterns, genes encoded by mtDNA do not: mtDNA is strictly maternally inherited.

The mutation rate of mtDNA is higher than in nuclear DNA. When a mutation arises in a copy of mtDNA within a mitochondrion, it creates a mixed population of mutant and normal mtDNA copies, known as heteroplasmy. When a cell divides, it is a matter of chance whether mutant DNA is partitioned over the two daughter cells or not. Over several cell divisions, the percentage of mutant mtDNA can shift to all mutant (homoplasmy) or all normal, but can also remain heteroplasmic. Since different tissues have different demands for energy generation, the effects of various degrees of heteroplasmy may initially only affect those tissues with high energy demands. Tissues and organs with increased energy demands are central nervous system (including the retina), heart and skeletal muscle, the renal system, and the liver (DiMauro & Schon, 2002; Rimoin et al., 2002; Scriver et al., 2005).

Epigenetics

A related topic to polygenic inheritance and complex conditions with genetics and environmental causations is epigenetics. *Epigenetics* is concerned with heritable changes in DNA not determined by the genetic code, but instead by such mechanisms as DNA methylation and chromatin modification. DNA in humans is packaged using histone proteins to form nucleosomes, with double-stranded DNA wrapped twice around an octameric core of four histones (H2A, H2B, H3, and H4) (Rimoin et al., 2002). Chromatin modification involves conformational changes of the DNA–histone complexes, thus enhancing or diminishing transcription. The core histones are subject to posttranslational modification, including methylation and acetylation, and, depending on the pattern of modification signals, enhance or decrease gene expression. DNA methylation, in humans restricted to CpG dinucleotides, is established and maintained in humans by DNA methyltransferases, enzymes that transfer a methyl group from *S*-adenosyl methionine to the 5′ position of cytosine bases. Generally, methylated CpG islands are associated with gene silencing, as this occurs in imprinted regions of the human genome, and in the inactivated X chromosome in females. Methylation patterns can be maintained through successive cell generations. In addition to maintaining established methylation patterns, *de novo* methylation can be added during development to establish parental imprinting (Delcuve et al., 2009).

X-inactivation

Sex determination in humans is genetically identified by the presence of either two X chromosomes for females or an X and a Y chromosome in males. This situation creates an issue with gene dosage compensation, as females would have twice the X chromosome gene dosage of males. One X chromosome in female cells is condensed, and visible as a Barr body, representing an inactivated X chromosome. X-inactivation originates from the X-inactivation center (XIC), subsequently spreading to the entire X chromosome (Brown et al., 1991; Rimoin et al., 2002). RNA produced by XIST (X-inactivating specific transcript) physically "coats" most of the inactivated X chromosomes, attracting histone-modifying enzymes followed by DNA methylation. Once an X chromosome is inactivated by this process, progeny of that particular cell maintains the inactivation of that X chromosome.

Imprinting

The great majority of autosomal genes are expressed equally from both parental chromosomes. However, a small number (estimated around 100) of genes are expressed only from one parental chromosome, as the other allele is silenced or imprinted. This epigenetic phenomenon differs from X-inactivation as it involves single genes, and relatively small clusters of genes, up to a few million nucleotides (megabase pairs, mb) in size. Imprinting centers are DNA sequences that control expression "in cis" of clusters of closely linked imprinted genes. The action of these imprinting centers results in differential methylation of the paternal and maternal chromosomes, resulting in differential regulation and transcription of these imprinted genes. Many imprinted genes generate non-coding RNAs, the function of which is not always well-understood. While representing only a fraction of genes in the human genome, abnormalities in these imprinted genes constitute an important group of disorders, such as Prader–Willi syndrome, Angelman syndrome, and Beckwith–Wiedemann syndrome (Rimoin et al., 2002).

Disease Versus Trait

An enduring metaphysical and epistemologic question is, "What is disease?" The issue is pertinent in sports performance when the human body and spirit are pushed to limits that raise the issue of a genetic predisposition to performance improved by training, in contrast to a variation that exceeds the range of normal anatomy and physiology. The genetic issues and determinants of gender are an obvious example, but rare variants that improve endurance, muscle composition, cellular function, or adaptations to regular training could be cited as well. In fact, outstanding work has been completed in the last decade in this area. The genetic contribution to various phenotype outcomes related to exercise have been exquisitely summarized and explained (Perusse et al., 2003; Rankinen et al., 2002, 2004, 2006; Wolfarth et al., 2005).

Disease can be defined as the absence of health and the World Health Organization defines health as a state of complete physical, mental, and social well-being and not merely the absence of disease or infirmity. Barton Childs, a pediatric geneticist at Johns Hopkins University and mentor of primary author John J. Mulvihill, MD, has long advocated a Garrodian view of disease versus Oslerian (Childs, 2001). Sir William Osler (1849–1919) was Professor of Medicine at McGill, University of Pennsylvania, Johns Hopkins University, and finally Regius Professor of Medicine at Oxford University. He championed the concept of "one cause—one disease," especially since it was bolstered by 19th century successes in finding single microbial agents to account for tuberculosis, syphilis, the plague, and gonorrhea. The view says that an ill person is a broken machine for the doctor to fix, that one major cause is probably to blame, and that that agent attacked the person probably close before the onset of symptoms and signs. The disease had an essential core feature, the same among all affected persons, so trainees and doctors could safely learn about the average or typical case.

Osler's successor at Oxford University was Sir Archibald Garrod (1857–1936), famous for coining the term, "inborn errors of metabolism," a heuristic or commonsense category of diseases, like pentosuria (high levels of a pentose in the urine), alkaptonuria (high levels of alkapton or homogentisic acid in the urine), and cystinuria (high cystine in urine), all of which clearly illustrate the principle of genetic variation in response to specific environmental (in this case, dietary) exposure (Garrod, 1923). Garrod wrote late in life on less obvious examples like trauma and tuberculosis (Scriver et al., 2005) and the neologism "ecogenetics" has been applied to the concept. Derived from pharmacogenetics, ecogenetics is the study of intrinsic variations in response to environmental agents and implies that each person's disease arises from an interaction of inborn differences in susceptibility and resistance to life-long environmental factors, including prenatal (transplacental), infancy, and childhood. Hence, Garrod would hold that a sick person represents a poor fit—discongruence—between a person's genome and their total integrated chemical, physical, and microbial exposures of a life time. Many causes or determinants play a role, some proximate, some remote, interacting perhaps in a specific sequence over seconds, days, or years. Hence there is no "essence" to a disease and "sickness" does not exist, except in a genetically unique person with a unique pattern of environmental exposures. There are not "typical" cases, and no two cases will be exactly the same, coincidentally a holistic view that many patients seek and appreciate.

Genetic Diversity

Due to the availability of molecular technologies to study the human genome, the genetic composition not only of individuals but also of populations can be studied more easily than ever before. The field of population genetics studies allele frequency distributions and the change in these distributions under the influence of natural selection, genetic drift, mutation, and gene flow. Briefly, an *allele* is considered to be an alternative form of a gene located at a specific position on a specific chromosome thus creating a scenario whereby different phenotypical outcomes can result from one gene or gene section. Such studies have revealed that the genetic variability in humans is extensive, at both the individual and the population levels. It has been known for a long time that there may be multiple alleles of any given gene in a certain population. Generally, an allele with a frequency over 1% in a given population is called a polymorphism. Alleles occur at different frequencies in different human populations, with a tendency of populations that are more geographically and ancestrally remote to differ more. These differences between individuals may be caused by recombination during meiosis and by various mutational events. Natural selection may confer an adaptive advantage to individuals in a specific environment if a certain allele provides a competitive advantage: such allele is then more likely to become more prevalent in a certain population, while other allele(s) decline in frequency.

Of course, alleles that confer a selective advantage are likely to become more prevalent only in those geographic regions where they confer this advantage, which contributes to the genetic variation between various population and ethnic groups. The other main cause of genetic variation is due to *genetic drift*. Genetic drift considers that changes in allele frequency are due to random events and not due to natural selection. This concept then describes the fact that certain alleles, while not conferring a selective advantage or disadvantage, may still become less or more frequent in a population. It is felt that genetic drift is more important in small populations.

In human populations, differences in allele frequencies are relevant not only for visible differences between individuals, but also for differences in disease susceptibility, as well as differences in other characteristics of human physiology (e.g., muscle strength and/or cardiorespiratory endurance). Differences in disease susceptibility may be due to simple Mendelian traits (autosomal dominant, autosomal recessive, X-linked), but also due to variations in mitochondrial DNA. Other susceptibility factors, and at population level likely more important, are disease susceptibility genes and other polygenic factors. Lastly, multifactorial conditions may be caused by interaction between various genes, including susceptibility genes, and the environment in which an individual grows up and matures.

Generally, conditions with a large genetic contribution are better understood but at the population level are relatively rare. Conditions with a relatively smaller genetic contribution require extensive population-based studies to identify susceptibility genes or alleles. However, while the genetic contribution of these alleles to the phenotype is less clear, its relevance at the population level is likely to be high. In recent years, *genome-wide association studies* (GWAS), designed to identify genetic associations with observable traits have become very popular. In such studies, if certain genetic variations (e.g., *single nucleotide polymorphisms*—SNPs) are more frequent in people with the disease, these variations are said to be associated with the

condition. The genomic locations of associated genetic variations are considered proxies to the region of the human genome where the disease-causing problem resides. Critics say the GWAS approach produces few meaningful biologic insights and identifies too many loci rather than too few (Pillai et al., 2009).

Variation in the human genome can go beyond SNPs, and may include segments ranging 1 kb to a few Mb, named *copy number variations* (CNVs) or copy number changes (CNCs), thus causing structural variation (Iafrate et al., 2004; Sebat et al., 2004). CNVs are a source of genetic diversity in humans, and can represent benign polymorphic variants present in >1% of the population. However, there are also instances that CNVs, by various molecular mechanisms including gene dosage, gene disruption, or gene fusion, result in abnormal gene dosage and cause Mendelian traits or sporadic disease, or contribute to complex disorders. While recurrent rearrangements of the human genome are thought to be due to non-allelic homologous recombination (NAHR) and non-homologous end joining (NHEJ), non-recurrent rearrangements may be due to another mechanism, named fork stalling and template switching (FoSTeS) (Lee et al., 2007). The extent to which large duplication and deletion CNVs contribute to human genetic diversity, and may or may not convey phenotypes, is still being researched (Zhang et al., 2009).

Presently, the practice of medical genetics is still mostly concerned with strong genetic effects. Traditional tools, that have proven track records, include a relatively well-understood clinical history of Mendelian disorders, an established diagnostic approach, and the demonstrated benefit of a family history. Genetic professionals condense the information obtained from the family in a visualized form, named a pedigree. Such information can provide evidence for genetic inheritance, which, together with other relevant information, may guide the physician in the work-up. Its relevance was again stated by the US Surgeon General, supporting that "A detailed family history can predict the disorders for which a person may be at increased risk, and thereby help to develop more personalized action plans." Such approach also empowers families to discuss health issues with their healthcare providers that would otherwise miss detection.

Ethics

There are a myriad of terms, not in common parlance, that must be used in considering the ethics of any area, including the genetics of sports performance. Considering recent newsmaking stories regarding genetic identification and international competition, the issue of ethics will continue to remain a central theme. The overall goals of addressing theoretical ethics in this context are twofold, both in regard to actual human behavior: knowledge-seeking (what *do* people decide and do?) and normative (what *should* people decide and do?). Applied ethics typically addresses specific concrete situations that can be resolved in more than one way, "do this" or "do not do it", or "pick option 1, 2, or 3". The area of theoretical ethics often has religion, culture, or both as its source and foundation and often leads to abstract generalization that seems completely rational. But, applied ethics must resolve a specific dilemma, with a specific person, in a specific time and place, and often has fine distinctions not apparent in the broad theoretical construct. In a survey of ethical considerations and scenarios in multiple countries in the 1980s, Wertz and Fletcher (1989) found that clinical genetics had some consensus in resolving dilemmas in clinical genetics, but often had different rationale for even similar decisions.

Principles of Ethics

Among many principles, four big moral obligations of professional ethics of the clinician or researcher interacting with individual human beings are labeled autonomy, beneficence, non-maleficence, and justice.

Autonomy is an individual's independence, free of coercion, to decide or act in a way that the professional has a duty to respect. Beneficence is the state or quality of being kind, charitable, or beneficial, in a sense opposite to non-maleficence, meaning, "Do no harm." Justice is fairness or rightness in treating individuals equally (e.g., in access to research, offering clinical care).

In no area of biomedicine have as much advancement and change occurred in the last several decades as in the area of genetics. This dynamic scenario creates an environment which challenges those professionals who seek to use existing information to improve diagnosis and treatment of genetic-related conditions. A broad understanding of genetics is an imperative foundation to establish before moving onto to other topics. In addition, the rapid advancement of knowledge has also created opportunities in which unethical interests and practices are developed and used to manipulate how the human genome continues to express its components. As science moves forward, it is likely that many areas will continue to change while putting continued pressure on those coaches, trainers athletes, counselors, and physicians who will no doubt depend on the contribution of genetics to tell them something about their performance or health in general.

References

Bouchard, C., Malina, R. & Perusse, L. (1997). *Genetics of Fitness and Physical Performance.* Human Kinetics, Champaign, IL.

Brown, C.J., Lafreniere, R.G., Powers, V.E., et al. (1991). Localization of the X inactivation center on the human X chromosome in Xq13. *Nature* 349(6304), 82–84.

Childs, B. (2001). A logic of disease. In: C.R. Scriver, A.L. Beaudet, W.S. Sly, et al. (eds.) *The Metabolic and Molecular Bases of Inherited Disease*, 8th edn, pp. 129–153. McGraw-Hill, New York, NY.

Delcuve, G.P., Rastegar, M. & Davie, J.R. (2009). Epigenetic control. *Journal of Cell Physiology* 219(2), 243–250.

DiMauro, S. & Schon, E.A. (2003). Mitochondrial respiratory-chain diseases. *New England Journal of Medicine* 348(26), 2656–2668.

Garrod, A. (1923). *Inborn Errors of Metabolism*, 2nd edn. Henry Frowde and Hodder & Stoughton, London, England.

Gibson, W. (2009). Key concepts in human genetics: Understanding the complex phenotype. *Medicine in Sport and Science* 54, 1–10.

Iafrate, A.J., Feuk, L., Rivera, M.N., et al. (2004). Detection of large-scale variation in the human genome. *Nature Genetics* 36(9), 949–951.

Lee, J.A., Carvalho, C.M. & Lupski, J.R. (2007). A DNA replication mechanism for generating nonrecurrent rearrangements associated with genomic disorders. *Cell* 131(7), 1235–1247.

Perusse, L., Rankinen, T., Rauramaa, R., Rivera, M.A., Wolfarth, B. & Bouchard, C. (2003). The human gene map for performance and health-related fitness phenotypes: The 2002 update. *Medicine & Science in Sports & Exercise* 35(8), 1248–1264.

Pillai, S.G., Ge, D., Zhu, G., et al. (2009). A genome-wide association study in chronic obstructive pulmonary disease (COPD): Identification of two major susceptibility loci. *PLoS Genetics* 5(3), e1000421.

Rankinen, T., Perusse, L., Rauramaa, R., Rivera, M.A., Wolfarth, B. & Bouchard, C. (2002). The human gene map for performance and health-related fitness phenotypes: The 2001 update. *Medicine & Science in Sports & Exercise* 34(8), 1219–1233.

Rankinen, T., Perusse, L., Rauramaa, R., Rivera, M.A., Wolfarth, B. & Bouchard, C. (2004). The human gene map for performance and health-related fitness phenotypes: The 2003 update. *Medicine & Science in Sports & Exercise* 36(9), 1451–1469.

Rankinen, T., Bray, M.S., Hagberg, J.M. et al. (2006). The human gene map for performance and health-related fitness phenotypes: The 2005 update. *Medicine & Science in Sports & Exercise* 38(11), 1863–1888.

Rimoin, D.L., Connor, J.M., Pyeritz, R.E. & Korf, B.R. (eds.) (2002). *Emery and Rimoin's Principles and Practice of Medical Genetics*, 4th edn. Churchill Livingstone, London, Philadelphia, PA.

Roth, S. (2007). *Genetics Primer for Exercise Science and Health*. Human Kinetics, Champaign, IL.

Scriver, C., Beaudet, A., Sly, W.S., et al. (2005). *Scriver's OMMBID Online Metabolic and Molecular Bases of Inherited Diseases*. McGraw-Hill, New York, NY.

Sebat, J., Lakshmi, B. Troge, J., et al. (2004). Large-scale copy number polymorphism in the human genome. *Science* 305(5683), 525–528.

Wertz, D. & Fletcher, J. (1989). *Ethics and Human Genetics: A Cross Cultural Perspective*. Springer-Verlag, Berlin.

Wolfarth, B., Bray, M.S., Hagberg, J.M., et al. (2005). The human gene map for performance and health-related fitness phenotypes: The 2004 update. *Medicine & Science in Sports & Exercise* 37(6), 881–903.

Yadav, S.P. (2007). The wholeness in suffix -omics, -omes, and the word om. *Journal of Biomolecular Techniques* 18(5), 277.

Zhang, F., Gu, W., Hurles, M.E. & Lupski, J.R. (2009). Copy number variation in human health, disease, and evolution. *Annual Review of Genomics and Human Genetics* 10, 451–481.

Appendix to Chapter 1—Genetics Resources for Research and Education on the Internet

Primary Research Resources

NATIONAL CENTER FOR BIOTECHNOLOGY INFORMATION

http://www.ncbi.nlm.nih.gov/guide/

The National Institutes of Health's (NIH's) National Center for Biotechnology Information (NCBI) advances science and health by providing access to biomedical and genomic information. It is best known in biology and medicine for PubMed, http://www.ncbi.nlm.nih.gov/pubmed—19 million references from the periodic biomedical literature since 1949. As a whole, the resource contains multiple locations and the most relevant home pages for readers of this volume are:

NCBI—ONLINE MENDELIAN INHERITANCE IN MAN

http://www.ncbi.nlm.nih.gov/omim/

Online Mendelian Inheritance in Man® (OMIM) is a comprehensive, authoritative, and timely compendium of human genes and genetic phenotypes, which is updated several times a week. The full-text, referenced overviews in OMIM contain information on all known Mendelian disorders and over 20,000 genes, proteins, genetic diseases, and traits with a focus on the relationship between phenotype and genotype. OMIM is authored and edited at the McKusick–Nathans Institute of Genetic Medicine, Johns Hopkins University School of Medicine, under the direction of Dr. Ada Hamosh. OMIM is ideally suited for use by physicians and other professionals concerned with genetic disorders, by genetics researchers, and by advanced students in science and medicine.

NCBI—ENTREZ GENE

http://www.ncbi.nlm.nih.gov/gene/

Entrez Gene is NCBI's database for gene-specific information. It does not include all known or predicted genes; instead Entrez Gene focuses on the genomes that (1) have been completely sequenced, (2) that have an active research community to contribute gene-specific information, or (3) that are scheduled for intense sequence analysis. Sequence records are assigned unique, stable, and tracked integers as identifiers. The content (nomenclature, map location, gene products and their attributes, markers, phenotypes, and links to citations, sequences, variation details, maps, expression, homologs, protein domains, and external databases) is updated as new information becomes available. This is a powerful resource for the genetics research community.

NCBI—GENETESTS

http://www.ncbi.nlm.nih.gov/sites/GeneTests/

The GeneTests web site is a publicly funded medical genetics information resource developed for physicians, other healthcare providers, and researchers, available at no cost to all interested persons. It is funded through the NIH to facilitate further integration with all databases at NCBI. The web sites work to provide current, authoritative information on genetic testing and how this information can be used in diagnosis, management, and genetic counseling. In addition, the web site provides a search function for known genetic disorders and within each disorder provides structured expert and peer-reviewed descriptions of about 500 known genetic diseases; an international directory of genetics clinics; international list of laboratories that do DNA testing for clinical and research purposes.

Educational and Clinical Resources

NATIONAL COALITION FOR HEALTH PROFESSIONAL EDUCATION IN GENETICS

http://www.nchpeg.org/

Established in 1996 by the American Medical Association, the American Nurses Association, and the National Human Genome Research Institute, the National Coalition for Health Professional Education in Genetics (NCHPEG) is an "organization of organizations" committed to a national effort to promote health professional education and access to information about advances in human genetics. The organization has established a unique interdisciplinary approach of leaders from a diverse group of 80 health professional organizations, consumer, and volunteer groups, government agencies, private industry, managed care organizations, and genetics professional societies. NCHPEG's mission is to promote health professional education and access to information about advances in human genetics to improve the health care of the nation. They work to this mission through annual meetings, integration of genetics content between health professionals and students, development of valuable educational tools and resources to facilitate this integration and further strengthening the interdisciplinary community of organizations.

HUMAN GENOME PROJECT INFORMATION

http://www.ornl.gov/sci/techresources/Human_Genome/home.shtml

The Human Genome Project (HGP) was completed in 2003 and was a 13-year project coordinated by the U.S. Department of Energy and the National Institutes of Health. Upon initiation, other groups became major partners, including the Wellcome Trust (U.K.), Japan, France, Germany and China. Sequences of all human genes and base pairs were detailed regarding human DNA and in the process transformed the information and resources available to researchers, clinicians, and other relevant biomedical organizations and companies. The web site is considered an excellent general information about many aspects of the HGP including: its background, research-related aspect to the HGP, the impact of HGP on medicine and our new grasp of genetic information, ethical, legal, and social related to the acquiring of this information and numerous media tools, publications, educational resources, related meetings, and auxiliary research programs in genomics.

LEARN. GENETICS™ GENETIC SCIENCE LEARNING CENTER—THE UNIVERSITY OF UTAH

http://learn.genetics.utah.edu/

An exceptional, dynamic, and cutting-edge display of topics centered upon all aspects of genetics. Although multiple university web pages across the globe have valuable information, this web site provides valuable information in multiple levels of learning context from text, interview, videos, and animations. Major sections of the learning page contain various information sources on topics centered upon genetic technology, laboratory techniques, and basic genetic concepts. Videos include animations on how to clone a mouse, how polymerase chain reaction (PCR) analysis is completed, and basic diagrams of fundamental concepts such as DNA, RNA, etc. In addition, this web site recently launched a section which provides teacher resources and lesson plans on various genetics topics.

COLD SPRING HARBOR LABORATORY'S DOLAN DNA LEARNING CENTER

http://www.dnalc.org/

An educational web site with several programs targeted to middle school and high school aged students, summer camps, and various training and fellowship opportunities for educators. This web site provides educational material on the fundamental aspects of genetics like many other web sites, but also places significant emphasis on ethics and medicine, an important concept for young minds and future researchers and practitioners to grasp and appreciate early. Additional online educational web sites are available for both teachers and students alike, which contain dynamic animations and streamed video with audio. Web site is an outstanding resource for those wanted to improve their own understanding of concepts or to find information to use for instruction.

NATIONAL HUMAN GENOME RESEARCH INSTITUTE

http://www.genome.gov/Education/

A comprehensive educational and professional resource of information related to the human genome. Major education-related topics include background information on the HGP, a talking glossary, professional experts speaking on various topics, and links to many other related web sites. In addition, major sections are provided for the public, patients, and practitioners as well as relevant, contemporary issues related to genetics. The web site also provides significant information for researchers including summaries of clinical research, scientific profiles, recommended books and articles, research centers affiliated with the institute, and several areas related to grant funding in areas related to genetics.

Professional Resources

AMERICAN SOCIETY OF HUMAN GENETICS

http://www.ashg.org/

Founded in 1948, the American Society of Human Genetics (ASHG) is the primary professional membership organization for human geneticists from around the world; its 8000 members include researchers, academicians, clinicians, laboratory practice professionals, genetic counselors, nurses, and others who have a special interest in the field of human genetics. It holds annual Fall meetings, support a high impact journal, *The American Journal of Human Genetics*, advocates for research support, enhances genetics education, informs the public, promotes genetic services, and supports responsible social and scientific policies.

GENETIC ALLIANCE

http://www.geneticalliance.org/

Founded in 1986 as the Alliance for Genetic Support Groups, Genetic Alliance has the mission to transform health through genetics. Its open network connects members of parent and family groups, community organizations, disease-specific advocacy organizations, professional societies, educational institutions, corporations, and government agencies to create novel partnerships. It strives to improve access to information for individuals, families, and communities, while supporting the translation of research into services.

AMERICAN COLLEGE OF MEDICAL GENETICS

http://www.acmg.net/

The American College of Medical Genetics (ACMG) provides education, resources, and a voice for the medical genetics profession. To make genetic services available to and improve the health of the public, the ACMG promotes the development and implementation of methods to diagnose, treat, and prevent genetic diseases. ACMG promotes high standards for all professionals in education, practice, and research while serving as an advocate for geneticists and providers of clinical genetic services. From a professional perspective, they have taken and continue to take great strides to develop clinical practice guidelines, laboratory services, including uniform standards, quality assurance and proficiency testing; databases, population screening guidelines, and preparing/sponsoring relevant position statements. The organization also organizes educational programs on an annual basis for healthcare providers, the public, and geneticists, including an Annual Clinical Genetics Meeting (http://www.acmgmeeting.net/acmg10/public/enter.aspx).

NATIONAL SOCIETY OF GENETICS COUNSELORS

http://www.nsgc.org/

An organization established in 1979 to promote the development and recognition of the profession within genetics. Genetics counselors possess unique knowledge and skills related to many aspects of counseling families and professionals and education on issues related to human genetics, thus becoming an integral member of the clinical genetics team. The society continues its efforts as the leading voice, authority, and advocate for the genetic counseling profession by fostering education, research, and public policy to ensure the availability of quality genetic services. The society maintains a web site devoted to the profession providing opportunity to interact with other counselors, recent research, and further development through hosting an annual scientific meeting, while also making contributions to a scientific journal titled, *Journal of Genetic Counseling*, the official journal of the society, read internationally by practitioners, researchers, instructors, and students.

Chapter 2

Mitochondrial Medicine in Health and Disease: Interface Between Athletic Performance and Therapeutics

DOUGLAS C. WALLACE

Center for Mitochondrial and Epigenomic Medicine and Department of Pathology and Laboratory Medicine, Children's Hospital of Philadelphia and University of Pennsylvania, Philadelphia, PA, USA

It is becoming increasingly clear that energy metabolism plays a central role in human health and disease. This is particularly apparent in elite athletes which strive for maximum performance. Genetic studies have identified a number of nuclear DNA (nDNA) encoded bioenergetic genes that affect athletic performance. However, the unique role of mitochondria DNA (mtDNA) variation in athletes is just beginning to be appreciated. Of particular interest is the association between ancient regional mtDNA variation and endurance. Because of the prevalence of mitochondrial diseases, progress is being made in developing both metabolic and genetic therapies to increase the mitochondrial function of patients. However, these same principles and procedures could be applied to enhancing performance of normal individuals. Germline gene therapy for mtDNA disease is rapidly approaching a reality. This procedure could also be used to change the mtDNA genotype of an embryo in hopes of influencing future athletic performance. However, the value of any individual mtDNA genotype depends on its context including

nDNA genetic variants, environmental conditions, and cultural norms. Hence, germline gene therapy to enhance performance may not only prove counterproductive, but it might also be deleterious.

Since life is the interplay between structure, energy, and information, and the mitochondria generate most of the cellular energy, it should not be surprising that mitochondria would play a key role in human performance (Ostrander et al., 2009; Wallace, 2007; Wallace & Fan, 2010; Wallace et al., 2010). Even so, the role of bioenergetics has been largely overlooked by Western medicine. This is because Western medical traditions have been strongly rooted in the anatomical description and diagnosis of disease and the Mendelian inheritance of genes. The Western emphasis on anatomy and Mendelian genetics is internally consistent since all of the genes for determining anatomy are encoded by chromosomal genes, and thus Mendelize. However, bioenergetic dysfunction can affect a variety of tissues depending on their energy requirements, and the most important mitochondrial energy genes are not located on the chromosomes but are on the maternally inherited mtDNA (Figure 2.1). Hence, until recently the role of altered bioenergetics has been overlooked in both medicine and athletics (Holt et al., 1988; Holt et al., 1990; Ostrander et al., 2009; Shoffner et al., 1990; Wallace et al., 2007).

Genetic and Molecular Aspects of Sport Performance, 1st edition. Edited by Claude Bouchard and Eric P. Hoffman. Published 2011 by Blackwell Publishing Ltd.

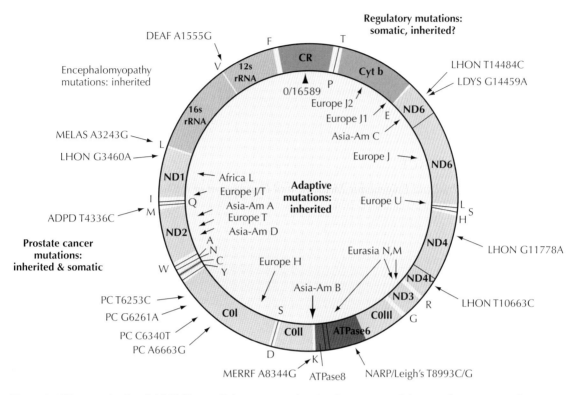

Figure 2.1 Human mitochondrial DNA map. D-loop = control region. Letters around the outside perimeter indicate cognate amino acids of the tRNA genes. Other gene symbols are defined in the text. Arrows followed by continental names and associated letters on the inside of the circle indicate the position of defining polymorphisms of selected region-specific mtDNA lineages. Arrows associated with abbreviations followed by numbers around the outside of the circle indicate representative pathogenic mutations, the number being the nucleotide position of the mutation. DEAF, deafness; MELAS, mitochondrial encephalomyopathy, lactic acidosis, and stroke-like episodes; LHON, Leber's hereditary optic neuropathy; ADPD, Alzheimer and Parkinson diseases; MERRF, myoclonic epilepsy and ragged red fiber disease; NARP, neurogenic muscle weakness, ataxia, retinitis pigmentosum; LDYS, LHON + dystonia. (Reprinted with permission from Wallace, 2005b.)

The importance of mitochondrial energetics in athletic performance has become apparent even through association studies on chromosomal genes of elite athletes. This is because several nuclear genes that have been found to be important in performance turned out to be central in the transcriptional regulation of nuclear-encoded genes involved in mitochondrial biogenesis (Ostrander et al., 2009; Wallace, 2005a).

The discovery that human mtDNA variation is highly regionally specific and the proposal that it correlates with adaptation to environmental energetic demands (Mishmar et al., 2006; Ruiz-Pesini et al., 2004; Wallace, 2007) has resulted in a rise in interest of the role of mtDNA functional variants in performance (Wallace, 2005a). It is now apparent that mtDNA variation can not only cause disease and limit elite performance, but that ancient mtDNA adaptive polymorphisms may be directly affecting maximum performance capacity. As a consequence, pharmacological and genetic interventions that could be used to treat mitochondrial disease patients could also be applied to enhance athletic performance (Ostrander et al., 2009; Wallace, 2005a).

Bioenergetic Genetics

The bioenergetic genome encompasses thousands of nDNA genes as well as the mtDNA. The nDNA genes encompass an estimated 1500 nDNA genes involved in mitochondrial bioenergetics, the nDNA genes involved in glycolysis and associated pathways, the genes for oxygen distribution and delivery to the cell, the genes involved in the application of energy to functions such as muscle contraction, the neuron-physiological genes involved in coordinating energy production and utilization, the array of signal transduction pathways that monitor energy and translate that information into altered fluxes through metabolic pathways, and all of the proteins regulated by redox metabolism (Wallace & Fan, 2009; Wallace et al., 2010). The coordination of these nDNA genes is modulated by high energy intermediates generated by the flux of environmental energy through mitochondrial oxidative phosphorylation (OXPHOS) and glycolysis. The most important high energy intermediates are adenosine triphosphate (ATP) for phosphorylation, acetyl-CoA for acetylation, and S-adenosyl methionine for methylation. These key intermediates are used to modify histones and coordinate nDNA gene expression. Thus the epigenome is the system by which nDNA gene expression is modulated by the availability of energy supply and energy demands on the organism mediated by high energy intermediates generated by the mitochondrion (Wallace 2010; Wallace & Fan, 2010).

The relationship between nDNA bioenergetic gene polymorphisms and elite athletic performance has revealed associations with the mitochondrial transcription factors nuclear regulatory factor 1 (NRF1), NRF2, and peroxisome proliferator-activated receptor γ (PPARγ) transcriptional coactivator 1α (PGC-1α). PGC-1α interacts with NRF1 and NRF2 to induce mitochondrial biogenesis (Kelly & Scarpulla, 2004; Sonoda et al., 2007). Additional relevant variants to bioenergetics include the angiotensin-converting enzyme (ACE) which is involved in modulating blood pressure, respiratory drive, tissue oxygenation, and skeletal muscle efficiency; nitric oxide synthetase 3 (NOS3) and endothelin 1 which regulate blood pressure; α-Actinin-3 (ACTN3) which in mice affects muscle energy efficiency; and erythropoietin which regulates red blood cell production and thus oxygen delivery (Ostrander et al., 2009).

However, the relevance of mtDNA variation in human performance is only beginning to be explored. This delay is because an entirely new system of genetics needed to be elaborated for the mtDNA and the role of bioenergetics in cellular physiology needed to be reexamined. Now that these principles are in place, they are being applied to an increasing array of human concerns including physical performance (Wallace, 2005a, b).

Mitochondrial Function

Muscle requires ATP for contraction and for the transport of ions across the sarcolemmal membranes. Hence, the rapidity and efficiency of generating ATP is critical for peak performance. Cellular ATP can be generated by both cytosolic glycolysis and mitochondrial OXPHOS. However, the mitochondria produce most of the cellular energy, control cellular calcium (Ca^{++}) levels, generate reactive oxygen species (ROS), regulate cellular redox status, and initiate cell death, and these five mitochondrial functions are central to both health and performance.

Mitochondrial OXPHOS generates energy by burning hydrogen derived from the carbohydrates and fats of our diet with the oxygen that we breathe to generate water (H_2O) and release the energy. The energy released can be used to generate ATP for performing work, heat for maintaining body temperature, membrane potentials for ion transport, etc. The reducing equivalents (electrons) from dietary calories are collected by the tricarboxylic acid (TCA) cycle and β-oxidation and transferred to either nicotinamide adenine dinucleotide, oxidized form (NAD^+) to generate the reduced form (NADH) or flavin adenine dinucleotide, oxidized form (FAD) to give the reduced form ($FADH_2$). The electrons are then oxidized by the electron transport chain (ETC). NADH transfers electrons to complex I (NADH dehydrogenase) and succinate from the TCA cycle transfers electrons to complex II (succinate dehydrogenase). Both complexes then transfer their electrons to ubiquinone (coenzyme Q_{10} or CoQ) to generate ubisemiquinone (CoQH) and then ubiquinol ($CoQH_2$). From $CoQH_2$ the electrons are transferred to complex III, then cytochrome c, then complex IV, and finally to ½ O_2 to give water. The energy that is released during electron transport is used to pump protons from the mitochondrial matrix out across the inner membrane to create an electro-chemical gradient ($\Delta P = \Delta \Psi + \Delta \mu^{H+}$). This biological capacitor serves as a source of potential energy to drive complex V (ATP synthase) to condense adenosine diphosphate (ADP) + P_i to give ATP. The mitochondrial ATP is then exchanged for cytosolic ADP by the adenine nucleotide translocator (ANT). The efficiency by which calories are converted to ΔP and ΔP is converted to ATP is known as the coupling efficiency. Tightly coupled OXPHOS generates the maximum ATP and minimum heat per calorie burned. By contrast, loosely coupled OXPHOS generates less ATP per calorie. Therefore, uncoupled mitochondria must burn more calories for the same amount of ATP, the extra calories being dissipated as heat (Wallace, 2005b; Wallace, 2007).

The mitochondria regulate cytosolic Ca^{++} concentrations through uptake of the Ca^{++} via mitochondrial inner membrane carriers. This Ca^{++} uptake is driven by ΔP.

The mitochondria also generate most of the endogenous ROS. When mitochondrial function becomes impaired, then excess electrons can accumulate in the ETC. These electrons have an increased tendency to be transferred from complex I and III to O_2, presumably from bound ubisemiquinone. This produces superoxide anion (O_2^-), a potent oxidizing agent. Superoxide anion from complex I is produced in the mitochondrial matrix, while the superoxide anion from complex III is produced in the intermembrane space. Matrix superoxide is detoxified by manganese superoxide dismutase (MnSOD, *Sod2*) while the intermembrane space superoxide is detoxified by Cu/ZnSOD (*Sod1*). Superoxide dismutases convert two O_2^- to a hydrogen peroxide (H_2O_2) which is relatively stable. However, in the presence of transition metals, H_2O_2 can be reduced to hydroxyl radical (OH) which is the most toxic ROS of all. H_2O_2 can be detoxified by reduction to water by glutathione peroxidase (GPx1) or conversion to O_2 and H_2O by catalase. If not neutralized, ROS can damage mitochondrial proteins, lipids, and nucleic acids; inhibit OXPHOS; and further exacerbate ROS production (Wallace, 2005b).

The flux of electrons through the mitochondrion regulates the redox state of a series of redox control nodes. These redox nodes then regulate the redox state of a wide range of redox sensitive metabolic and transcriptional control proteins (Wallace & Fan, 2010; Wallace et al., 2010). Separate but interconnected redox couples function in the mitochondrion, cytosol, endoplasmic reticulum, lysosomes, peroxisomes, and nucleus. These include the NADH/NAD$^+$, NADPH/NADP$^+$, glutathione (reduced and oxidized), cysteine and cystine, and thioredoxins (mitochondrial versus nucleus–cytosol). The redox status of these redox control nodes then regulate a broad array of cellular metabolic pathways and the activity of multiple transcription factors involved in cellular energy production, antioxidant defenses, growth, status, and death (Wallace et al., 2010).

When the mitochondria become sufficiently energetically impaired so that the cell malfunctions, the cell, with its defective mitochondria, is removed from the tissue by apoptosis. This is initiated through the activation of the mitochondrial permeability transition pore (mtPTP). The mtPTP spans the mitochondrial outer and inner membrane and interacts with the pro- and anti-apoptotic Bcl2-Bax gene family proteins and cyclophillin D. When mitochondrial energy production declines, as perceived by decreased ΔP and high energy phosphates, or if Ca^{++} or ROS increases beyond critical limits, the mtPTP is activated opening a channel across the mitochondrial membranes. This shorts ΔP and releases proteins from the mitochondrial intermembrane space into the cytosol. Among the proteins released is cytochrome c which activates cytosolic Apaf-1 to convert pro-caspases 9 to active caspase 9. Caspase 9 then initiates the caspase cascade which degrades the cellular and mitochondrial proteins. Apoptosis-initiating factor and endonuclease G are released from the mitochondrial intermembrane space and are transported to the nucleus where they degrade the nDNA (Wallace, 2005b).

The Nature of the Mitochondrial Genome

The mitochondrial genome is unique in that it is composed of 1500 nDNA-encoded mitochondrial genes plus 37 mtDNA-encoded mitochondrial genes. The expression of the genes of the mtDNA bioenergetic genome is coordinated relative to the available calories and the energetic demands of the environment by the epigenome (Wallace & Fan, 2010; Wallace et al., 2010).

The human mtDNA is a circular molecule of approximately 16,569 nucleotide pairs (nps) (Figure 2.1). It retains the mitochondrial genes for the small (12S) and large (16S) ribosomal RNAs (rRNA) and the 22 transfer RNAs (tRNAs) necessary to translate the 13 mtDNA polypeptides. The mtDNA uses a simplified and modified genetic coded. Key modifications of the mtDNA genetic code include the use of the opal stop codon (UGA) as a second tryptophan codon and the absence of a tRNA for the AGG and AGA codons making them stop codons. The 13 mtDNA proteins are all key components of OXPHOS and include seven (ND1, ND2, ND3, ND4L, ND4, ND5, and ND6) of the 45 polypeptides of OXPHOS complex I, one (cytochrome b, cytb) of the 11 proteins of complex III, three (COI, COII, and COIII) of the 13 proteins of complex IV, and two (ATP6, ATP8) of the approximately 16 proteins of complex V. All of the remaining proteins of the mitochondrial genome, including those for the DNA and RNA polymerases, ribosomal proteins, structural proteins, and enzymes, are encoded in the nucleus (Wallace & Lott, 2002).

In addition to the RNA and polypeptide genes, the mtDNA contains an approximately 1000 np control region (CR). The CR encompasses the light (L)-strand promoter (P_L) and two adjacent heavy (H)-strand promoters (P_{H1} & P_{H2}) and their associated mitochondrial transcription factor binding sites; three conserved sequence blocks (CSBI–III); the origin of H-strand replication, O_H; and the termination-associated sequences. Transcription is initiated at P_L or the two P_Hs and proceeds continuously around the mtDNA. The tRNAs that punctuate the larger genes then fold in the transcripts and are cleaved out to generate the larger transcripts which are then polyadenylated. The P_L promoter transcribes the *ND6* gene and some of the tRNAs. The P_H promoters transcribe all of the rest of the genes. DNA replication has been proposed to initiate from a 3′-OH generated by the cleavage of the L-strand transcript after conserved

sequence block 1 (CBS1). Replication then proceeds two-thirds of the way around the molecule making a new H-strand until the L-strand origin is exposed. L-strand replication is then initiated and proceeds back around the displaced H-strand (Clayton, 1982). This replication model has recently been questioned (Bowmaker et al., 2003; Holt & Jacobs, 2003).

The mitochondrial 55S ribosomes are distinct from the cytosolic 80S ribosomes. The mitochondrial ribosomes still retain certain characteristics of bacterial ribosomes including their sensitivity to chloramphenicol (Wallace, 1982).

Each cell has hundreds of mitochondria and each mitochondrion contains multiple mtDNAs. Therefore, a cell contains thousands of mtDNAs. The mtDNAs are maternally inherited and have a very high mutation rate. *De novo* mtDNA mutations result in an intracellular mixture of mutant and normal mtDNAs, a state known as heteroplasmy. When a heteroplasmic cell divides, it is a matter of chance which mitochondria and thus mtDNAs become distributed into the daughter cells. Therefore, during mitosis, the mtDNA genotype will segregate with the result that cells in different derived lineages can drift toward either more mutant or normal mtDNAs, ultimately cells are generated that are either pure mutant or wild type (homoplasmy).

For deleterious mutations, as the percentage of mutant mtDNAs increases, the energy output declines. Ultimately, the energy output becomes sufficiently diminished that it impairs the function of the most energetic tissues: the brain, heart, muscle, renal, and endocrine systems. The point where the energy deficiency results in severe malfunction is known as the bioenergetic threshold. However, this threshold is contextual. For example, the office worker can perform the job even when it is difficult for him to walk upstairs while the elite athlete will fall behind other elite athletes when performance declines by a few percent (Wallace & Fan, 2010).

Genetics of the nDNA Mitochondrial Genes and Disease Mutations

Mutations in the nDNA-encoded mitochondrial genes can result in a wide spectrum of clinical symptoms. These nDNA mutations can affect structural and assembly genes of OXPHOS, genes involved in mtDNA maintenance, and genes affecting mitochondrial fusion and mobility (Wallace et al., 2007).

Severe mutations in the structural genes of OXPHOS enzyme subunits often result in Leigh syndrome, a frequently lethal childhood disease. Examples include mutations in the structural subunits of complex I (Procaccio & Wallace, 2004; Smeitink & van den Heuvel, 1999) and in the *SURF1* assembly gene of complex IV (Tiranti et al., 1998; Wallace et al., 2007; Zhu et al., 1998).

Mutations in genes involved in mtDNA replication, substrate uptake and processing, and polymerization can result in destabilization and depletion of the mtDNA (Wallace et al., 2007). Mutations in the cytosolic thymidine phosphorylase result in mitochondrial neurogastrointestinal encephalomyopathy (Nishino et al., 1999); mutations in deoxyguanosine kinase (Mandel et al., 2001), mitochondrial thymidine kinase 2 (Saada et al., 2001), and *SUV3* (Khidr et al., 2008) result in mtDNA depletion; mutations in ANT1 (Kaukonen et al., 2000), *Twinkle* helicase (Spelbrink et al., 2001) and in mtDNA polymerase γ (Van Goethem et al., 2001, 2003) have been linked to autosomal dominant or recessive–progressive external ophthalmoplegia, while null mutations in ANT1 result in myopathy and cardiomyopathy (Palmieri et al., 2005).

Mutations in proteins involved in mitochondrial fusion can cause neurological disease. Mutations in the dynamin-related GTPase fusion protein result in dominant optic atrophy-1 (*OPA-1*) (Delettre et al., 2000). Mutations in mitofusin 2 (*MFN2*) gene result in Charcot–Marie–Tooth 2 (Wallace et al., 2007; Zuchner et al., 2004).

The Role of mtDNA Mutations in Health and Disease

Three primary classes of mtDNA variation impact human health and disease: (1) recent deleterious mutations frequently resulting in maternally inherited disease; (2) ancient polymorphisms, a subset of which permitted our ancestors to adapt to new environmental conditions; and (3) somatic mtDNAs that accumulate in post-mitotic tissue mtDNAs with age and provide the aging clock (Wallace & Lott, 2002).

Recent Pathogenic Mutations

The high mtDNA mutation rate means that new mtDNA mutations are continuously arising in the human population. Indeed, a survey of 15 known pathogenic mtDNA mutations in the cord bloods of normal newborns revealed that 1 in 200 harbored a pathogenic mtDNA mutation (Elliott et al., 2008). Based on common mtDNA mutations, it has been estimated that in the order of 1 in 5000 individuals suffer from clinically apparent mitochondrial disease (Schaefer et al., 2004, 2007).

Since all of the mtDNA genes are essential for life and relatively highly conserved, most new mutations are deleterious. Indeed, if all mtDNA mutations were introduced into the population, it is likely that the genetic load would be so great that the species would go extinct. This is avoided by there being an intra-ovarian selection system that eliminates those proto-oocytes which harbor the most deleterious mtDNA mutations (Fan et al., 2008; Stewart et al., 2008). Still, many mildly deleterious mtDNA mutations pass through the ovarian selection window. The most severe of these reduce reproductive fitness sufficiently to be eliminated from the population in one or a few generations. These present in the clinic as spontaneous or maternally inherited diseases. Pathogenic mtDNA mutations include both rearrangement mutations and base substitution mutations. Deletions frequently remove one or more tRNAs and thus impair protein synthesis. Base substitution mutations can alter proteins or protein synthesis (rRNA and tRNA) genes.

Rearrangement mutations can occur virtually anywhere within the mtDNA and can include both maternally inherited and spontaneous cases. Maternally inherited rearrangement mutations are often associated with Type II diabetes and deafness and are thought to be inherited as duplication mutations (Ballinger et al., 1992, 1994). Deletions in the mtDNA, with or without attendant insertions, are primarily spontaneous and result in multisystem disorders encompassing a continuum of phenotypes ranging from chronic progressive external ophthalmoplegia to the more severe Kearns–Sayre syndrome. The most severe mtDNA rearrangement disorder is the Pearson's marrow pancreas syndrome which frequently leads to death in child-hood from pancytopenia (Wallace et al., 2007).

Examples of pathogenic mtDNA polypeptide missense mutations include the np G11778A mutation in ND4 (R340H) which causes Leber's hereditary optic neuropathy (LHON), a form of sudden-onset, mid-life blindness (Wallace et al., 1988) and the np T8993G missense mutation in ATP6 (L158R) causing neurogenetic muscle weakness, ataxia, retinitis pigmentosum as well as the Leigh syndrome (Holt et al., 1990). Examples of a pathogenic protein synthesis mutations include the np A8344G in the tRNA^{Lys} gene which causes myoclonic epilepsy and ragged red fiber disease (MERRF) and the tRNA$^{Leu(UUR)}$ np A3243G that causes mitochondrial encephalomyopathy, lactic acidosis, and stroke-like episodes (MELAS). When present as a high percent of the mtDNAs, the A3243G mutation causes MELAS but when present as a low percentage mutant, it causes Type II diabetes (Goto et al., 1990; Wallace et al., 2007).

Because of the differential reliance of tissues on mitochondrial energy production, sensitivity to redox state and oxidative stress, predilection to Ca^{++} overload, and potential for the activation of the mtPTP, systemic mitochondrial bioenergetic defects can result in an array of tissue-specific clinical systems. Moreover, the stochastic segregation of mtDNA mutations, the potential for faulty nuclear–cytoplasmic interaction (Potluri et al., 2009), and the complexity of the pathophysiology of mitochondrial diseases results in complex genotype–phenotype correlations. However, the organs most consistently affected by energy deficiency are the brain, heart, and muscle—the organs most critical for athletic performance (Wallace, 2010; Wallace et al., 2007).

Ancient Adaptive mtDNA Mutations

While most functional mtDNA mutations disrupt mitochondrial bioenergetics and thus result in disease, a small subset of mtDNA mutations results in alterations in mitochondrial bioenergetics that can be beneficial in certain energetic environments.

Because of exclusive maternal inheritance, the mtDNA accumulates mutations sequentially along radiating maternal lineages (Ruiz-Pesini et al.,

2007). All mtDNA variation coalesces to a founding mtDNA lineage that existed in Africa about 150,000 to 200,000 years before present (YBP). From this origin, four mtDNA lineages arose that are specific to sub-Saharan Africa: L0, L1, L2, and L3. About 65,000 ago, two new lineages arose from haplogroup L3 in northeastern Africa, designated macrohaplogroups M and N. Only the descendents of these two lineages ultimately colonized the rest of the World. One branch of the out-of-Africa migration carried macrohaplogroup M through southeast Asia and then moved northward to radiate throughout Asia. The second branch carried macrohaplogroup N northward into Europe, giving rise to the European-specific lineages H, I, J, K, T, U, V, W, and X. Macrohaplogroup N also moved eastward into Asia. In Asia, both M and N radiated to generate a plethora of Asian-specific mtDNA lineages including A, B, and F from N and C, D, and G from M. Among all of the Asian mtDNA lineages, only A, C, and D successfully occupied arctic northeastern Siberia. These were the mtDNAs that were present when the Bering land bridge appeared and thus they became the first individuals to colonize the Americas about 20,000 YBP. Another migration across the arctic brought haplogroup X to the Great Lakes region of North America about 15,000 YBP. Then, a coastal migration brought haplogroup B to mix with A, C, and D in southern North America, Central America, and northern South America about 12,000–15,000 YBP. These five mtDNA lineages represent the Paleo-Indians. Subsequently, a migration from around the Sea of Okhostk about 7000–9000 YBP brought a modified A to found the Na-Dene. Finally a recent migration bringing A and D mtDNAs gave rise to the Eskimos and Aleuts (Wallace et al., 1999).

The mtDNA distribution among indigenous people is remarkable for the striking discontinuities that exist in mtDNA diversity between climatic zones. Of the extensive mtDNA variation present in Africa, only two mtDNA lineages (M and N) succeeded in colonizing all of Eurasia. Of the plethora of Asian mtDNA types that subsequently accumulated, only three haplogroups A, C, D, and, much later, G came to occupy the extreme northeastern Chukotka Peninsula of Siberia. This striking correlation between mtDNA lineages, latitude, and climate led to the hypothesis that mutations in the mtDNA that decreased the coupling efficiency increased mitochondrial heat production and permitted people to survive the cold of the more northern latitudes (Mishmar et al., 2003; Ruiz-Pesini & Wallace, 2006; Ruiz-Pesini et al., 2004). Since changes in mitochondrial physiology can affect a wide range of cellular physiological parameters, mtDNA variation may have provided the capacity of populations to adapt to a wide range of environmental parameters (Wallace & Fan, 2010; Wallace et al., 2010).

Direct evidence that the ancient regional mtDNA variation has been adaptive has come from correlations between mtDNA variation and individual predisposition to a spectrum of diseases. The first such association was the demonstration that European haplogroup J increased the penetrance of milder LHON pathogenic mutations (Brown et al., 1995, 1997, 2002; Torroni et al., 1997). Haplogroup J was subsequently correlated with longevity in Europeans (De Benedictis et al., 1999; Ivanova et al., 1998; Niemi et al., 2003; Rose et al., 2001) and D with longevity in Asians (Tanaka et al., 1998, 2000). Haplogroup H was associated with increased risk and haplogroups J and Uk with decreased risk for developing Parkinson disease (Ghezzi et al., 2005; Khusnutdinova et al., 2008; van der Walt et al., 2003) and haplogroup H sublineage nt 4336 has been associated with increased Alzheimer disease risk, while haplogroups U and T have been associated with decreased Alzheimer risk in certain contexts (Carrieri et al., 2001; Chagnon et al., 1999; Shoffner et al., 1993; van der Walt et al., 2004). Haplogroup H has been correlated with reduced risk of age-related macular degeneration while haplogroups J and U are associated with increased risk for macular degeneration (Udar et al., 2009) and increased drusen levels and retinal pigment abnormalities (Jones et al., 2007). Haplogroup J has been associated with increased risk of diabetes in certain European descent populations (Crispim et al., 2006; Mohlke et al., 2005; Saxena et al., 2006), while haplogroup N9a is protective of diabetes, metabolic syndrome, and myocardial infarction in Asians (Fuku et al., 2007; Nishigaki et al., 2007).

Haplogroup H has been associated with protection against sepsis (Baudouin et al., 2005) while haplogroup U has been associated with increased serum IgE levels (Raby et al., 2007). Certain mtDNA haplogroups have been correlated with sensitivity to AIDS-associated progression (Hendrickson et al., 2008) and predilection to lipodystrophy in response to anti-retroviral therapy (Hendrickson et al., 2009). Finally, haplogroups have also been correlated with altered risk for various cancers (Bai et al., 2007; Booker et al., 2006; Darvishi et al., 2007).

Direct demonstration that mtDNA variants affect mitochondrial function comes from the demonstration that the founding missense variants of macrohaplogroup N, ND3 np 10398 A115T and ATP6 np 8701 A59T, affect complex I activity, membrane potential, and Ca^{++} regulation (Kazuno et al., 2006). Furthermore, a CR variant associated with haplogroup J at np 295 $C > T$ has been found to increase L-strand transcription and the mtDNA copy number (Suissa et al., 2009).

Based on the thermal adaptation hypothesis for regional mtDNA lineages, the mtDNAs of indigenous Africans would be expected to be more tightly coupled than those of Europeans and Asians. This would imply that African mtDNAs might have the highest ATP production and lowest heat generation per calorie consumed. This would result in their oxidative Type I fibers being more efficient at generating ATP making indigenous Africans better endurance athletes. By contrast, temperate zone peoples would have more loosely coupled mitochondria. Hence, their Type I fibers would need to consume more oxygen and calories per ATP generated, with the excess energy being dissipated as heat. As a result, these individuals might be more reliant on Type II glycolytic fibers and thus make better athletes for events that require rapid, short-term bursts of energy. Preliminary data suggest that this may be the case. Among Finnish elite athletes, all of the haplogroup I mtDNAs were found in the endurance athletes while all of the haplogroup K and most of the J mtDNAs were found in the sprinters (Niemi & Majamaa, 2005). For Spanish athletes, haplogroup T was strongly underrepresented among endurance athletes (Castro et al., 2007). Among elite Kenyan long-distance runners, haplogroup

L0 was strongly associated with endurance performance. By contrast, haplogroup L3* mtDNAs were underrepresented among the top endurance athletes (Scott et al., 2009). While a study of Ethiopian elite athletes did not reveal a strong association between mtDNA haplogroup and athletic performance, the Ethiopian samples lacked African haplogroup L0 mtDNAs. Assuming that the greatest difference in coupling efficiency occurred between haplogroup L0, the founding mtDNA haplogroup, and haplogroups L1, L2, and L3 which are similar to each other and less tightly coupled, then there might have been little functional differences between the Ethiopian mtDNA haplogroups. Hence the Ethiopian endurance athletes did not differ significantly in the mtDNA parameter (Scott et al., 2005). Still, at international competitions, Ethiopian long-distance runners generally perform better than European long-distance runners. Based on these results and similar trends observed in relation to various clinical problems, we can propose a preliminary hierarchy for mtDNA haplogroups: L0 > (L1, L2, L3) > I > H > U > J > Uk. For endurance long-distance runners, this list is arranged from high to low, while for sprinters the list is from low to high.

Somatic mtDNA Mutations

It is well established that athletic performance peaks in the latter teens and twenties, and then declines with age. This correlates with the accumulation of somatic mtDNA mutations in post-mitotic tissues with age. These can include rearrangement, coding region base substitution, and CR mutations. Since mtDNA diseases generally have a delayed-onset and progressive course and affect the same tissues and cause the same symptoms as aging, it has been hypothesized that the accumulation of somatic mtDNA mutations results in the age-related decline in mitochondrial function ultimately resulting in degenerative diseases and senescence. For the elite athlete, this accumulation of somatic mtDNA mutations eventually limits peak performance (Wallace & Lott, 2002).

Consistent with these concepts, mtDNA deletions have been found to accumulate with age in those tissues most prone to age-related decline (Corral-Debrinski et al., 1992). Furthermore, the level of somatic mtDNA deletions is elevated in the affected tissues in Alzheimer disease (Corral-Debrinski et al., 1994) and Parkinson disease brains (Bender et al., 2006; Kraytsberg et al., 2006).

Somatic mtDNA point mutations also accumulate with age. The T414G mutation in the mtTFA binding site for P_L has been found to accumulate in the mtDNAs of skin fibroblasts from older individuals, reaching levels in excess of 60% in some older patient cell lines (Coskun et al., 2003; Michikawa et al., 1999). Furthermore, somatic point mutation levels are elevated in neurodegenerative diseases. Sixty-five percent of Alzheimer disease brains test positive for the T414G CR mutation while no controls harbored this mutation (Coskun et al., 2004).

Somatic mtDNA mutations have also been found in a variety of solid tumors including prostate, breast, colon, bladder, head, and neck tumors (Brandon et al., 2006). Evidence that mtDNA mutations contribute to tumorigenesis comes from demonstration that introduction of the pathogenic np T8993G ATP6 mutation into prostate cancer PC3 cell lines resulted in increased tumor growth (Petros et al., 2005). Furthermore, introduction of an mtDNA bearing an ND6 G13997A missense mutation (P25L) increased ROS production and tumorigenesis (Ishikawa et al., 2008).

Therapeutic Approaches to Enhance Mitochondrial Function

Mitochondrial deficiencies might be treated by either metabolic or genetic therapies. Presumably, these same strategies could be used to enhance normal performance.

Metabolic and Pharmacological Therapeutics

Metabolic and pharmacological therapeutic strategies have focused on four potential mitochondrial targets: enhancing energy generation, reducing ROS production, inhibiting mtPTP activation, and activating mitochondrial biogenesis.

A variety of metabolic strategies have been used to increase mitochondrial energy production. The most ancient and still one of the most effective approaches to treating energy deficiency is the ketogenic diet. This diet shifts the calorie distribution from carbohydrates which are metabolized through glycolysis and results in the down-regulation of OXPHOS to fatty acids and/or ketones (acetoacetate and β-hydroxybutryate) which can only be metabolized by OXPHOS and thus up-regulates OXPHOS. The mechanisms by which this shift of metabolism is achieved are just beginning to be understood. However, one aspect may be the induction of OXPHOS via the NAD^+ activation of Sirt1. During glucose catabolism by glycolysis, cytosolic NAD^+ is reduced to NADH. By contrast, during the mitochondrial oxidation of fatty acids and ketones, it is the mitochondrial NAD^+ that is reduced, the cytosolic NAD^+ remaining oxidized. The FOXO and PGC-1α transcription factors which act together to induce OXPHOS can be inactivated by acetylation. However, in the presence of cytosolic NAD^+, Sirt1 will deacetylate and activate these factors, up-regulating OXPHOS. The ketogenic diet would keep cytosolic NAD^+ oxidized, thus driving the induction of OXPHOS (Wallace et al., 2010). The value of the ketogenic diet has been documented by culturing cells that were heteroplasmic for an mtDNA deletion for 5 days in ketones. This resulted in the increase in the percentage of normal mtDNAs from 13% to 22% (Santra et al., 2004).

Another approach toward increasing energy production in patients with mitochondrial defects has involved attempts to bypass the metabolic blocks in the ETC. In one chronic progressive external ophthalmoplegia patient with respiratory failure due to a predominantly complex I defect, the patient was treated with succinate plus CoQ. The succinate was used to feed electrons into the ETC through complex II, thus bypassing the complex I block. CoQ_{10} was added to enhance the transfer of the electrons from complex II to complex III and to act as an antioxidant. This treatment successfully overcame the patient's respiratory failure (Shoffner et al., 1989). In another instance, a patient with a complex III deficiency due to a cytb mutation was treated with ascorbate and menadione to feed electrons into complex IV from complex I and ascorbate. This treatment significantly improved the patient's muscle strength and mitochondrial energy production, as assessed by ^{31}P MR spectroscopy (Argov et al., 1986; Eleff et al., 1984).

Dichloroacetate has been employed in an effort to increase electron flux through the ETC. Dichloroacetate stimulates pyruvate dehydrogenase, by blocking inhibition by the regulatory kinase, thus driving pyruvate into the mitochondrion as acetyl-CoA. While this approach has been reported to reduce lactic acidosis, in MELAS patients, it has been found to increase the risk of peripheral neuropathy (Kaufmann et al., 2006; Stacpoole et al., 2006). Perhaps the addition of electrons to the already stalled ETC simply increased ROS production.

De novo pyrimidine biosynthesis is also inhibited by blocking the respiratory chain since dihydroorotate CoQ oxidoreductase (dihydroorate dehydrogenase) uses CoQ_{10} of the ETC as an electron sink (King & Attardi, 1989). Therefore, some benefit may be obtained by providing the bioavailable form of uridine, triacetyluridine, in cases of severe OXPHOS dysfunction (Klivenyi et al., 2004).

Efforts to reduce mitochondrial ROS toxicity have employed both natural and synthetic antioxidants. Natural antioxidants in different combinations have been used to treat mitochondrial disease patients, but with varying levels of success (Wallace et al., 2010). CoQ_{10} and its analogs including Idebenone have been given to mitochondrial disease patients based on the logic that the CoQ could enhance electron flux through the ETC and act as an antioxidant (Geromel et al., 2002; Wallace et al., 2010). While the efficacy of CoQ has been disputed in certain contexts (Wallace et al., 2010), clinical trials with Idebenone have reported successful treatment of the cardiomyopathy and neurological symptoms associated with Friedreich's ataxia (Di Prospero et al., 2007; Hart et al., 2005; Rustin et al., 1999).

Vitamin E and ascorbate have also been used in antioxidant therapies. However, these compounds can act as both antioxidants and pro-oxidants which might explain their variable efficacy. Moreover, their redox status may ultimately equilibrate with the redox state of the individual, thus diminishing long-term efficacy (Wallace et al., 2010).

Efforts have been made to increase the therapeutic benefit of vitamin E and CoQ by selectively targeting them to the mitochondrion via linking them to the cationic alkyltriphenylphosphonium ion. The positive charged compound is selectively electrophoresed into the mitochondrial matrix. *In vitro* studies have shown that these hybrid drugs have increased mitochondrial antioxidant activity (James et al., 2004; Jauslin et al., 2003; Smith et al., 2003).

Synthetic catalytic superoxide dismutase and catalase mimetics have much greater therapeutic potential than natural compounds, as indicated by studies in model systems of mitochondrial disease. The mimetic MnTBAP (manganese 5-, 10-, 15-, 20-tetrakis (4-benzoic acid) porphyrin) ameliorated the neonatal cardiomyopathy associated with *Sod2* deficiency (Melov et al., 1998). Treatment of the MnSOD-deficient animals with the salen manganese complexes EUK-8, -134, -189, which can cross the blood brain barrier, protected *Sod2*-deficient mice from both cardiomyopathy and spongiform encephalomyopathy (Melov et al., 2001).

Treatment of mutant fruit flies with the metaloporphrin Mn(II) tetrakis (1,3-diethyl imidazolium-2-yl) meso-substitute porphyrin rectified the increased mitochondrial ROS production and decreased lifespan resulting from inactivation of the *Drosophila* homolog of the neurofibromatosis-1 (*Nf1*) gene. Inactivation of *Nf1* modifies mitochondrial function by inhibiting cAMP production. Overexpression of *Nf1* or treatment with cAMP analogues reduced mitochondrial ROS and increased lifespan (Tong et al., 2007).

Efforts to inhibit the activation of the mtPTP have recently met with success, in the treatment of myopathy associated with a defect in the extracellular matrix protein, collagen VI. The activation of the mtPTP can be inhibited by cyclosporine A, but this drug has immunosuppressive activity. A variety of cyclosporine A analogues have recently been developed, which modulate the mtPTP without affecting the immune system (Rasola & Bernardi, 2007). Mutations in the collagen VI gene result in Bethlem myopathy and Ullrich congenital muscular dystrophy. A mouse model of these diseases has been generated by knocking out the collagen VI gene. Treatment of collagen VI-deficient mice with the cyclosporine A analogues has been shown to eliminate the myopathy, and treatment of both the mouse and human mitochondria with the drugs

inhibits the activation of the mtPTP (Angelin et al., 2007; Irwin et al., 2003). These results suggest that pharmacological regulation of the mtPTP might have broad applications for the treatment of degenerative diseases.

Mitochondrial biogenesis has been enhanced by agonists of nuclear receptor (NR) proteins and activators of the sirtuins. The NR family is central to the transcriptional regulation of metabolism (Shulman & Mangelsdorf, 2005; Sonoda et al., 2008). NRs interact with the co-transcriptional activators of the PGC-1 family to modulate mitochondrial biogenesis (Kelly & Scarpulla, 2004; Sonoda et al., 2007). Overexpression of PPARδ in mouse skeletal muscle resulted in a shift of the skeletal muscle fiber types from the glycolytic fast twitch Type II fibers to the slow twitch oxidative Type I fibers. This resulted in a remarkable increase in exercise endurance in the mice (Wang et al., 2004). In a mouse model of mitochondrial diseases resulting from inactivation of the complex IV assembly gene, *COX10*, overexpression of PGC-1α delayed the onset of symptoms and increased the mouse life expectancy (Wenz et al., 2008). Treatment of these mice with benzafibrate, a known activator of the PPAR/PGC-1α pathway, had a similar effect in mutant mice and patient cell lines (Bastin et al., 2008; Djouadi & Bastin, 2008).

OXPHOS can be pharmacologically induced by Sirt1-mediated deacetylation of PGC-1α and the FOXO. Sirt1 can be activated by resveratrol, a molecule found in grape skins, and its more potent derivatives SRT1720, SRT2183, etc. (Milne et al., 2007). Treatment of mice with resveratrol has been shown to induce mitochondria biogenesis and increase exercise performance. It also protects against mitochondrial genetic defects and metabolic disease (Lagouge et al., 2006; Srivastava et al., 2007). Since aerobic exercise relies heavily on ATP generation by the Type I fiber mitochondria, the pharmacological induction of muscle and heart mitochondrial biogenesis could have a significant effect on athletic endurance.

Mitochondrial Gene Therapy

Mitochondrial diseases may also be treated by gene therapy. Mitochondrial gene therapy can include both somatic therapy to ameliorate symptoms and germline therapy to replace mtDNA genotypes in the maternal germ line.

Somatic Mitochondrial Gene Therapy

Somatic mitochondrial gene therapy could take three approaches: direct genetic modification of an nDNA-encoded mutant mitochondrial gene, insertion of a modified mtDNA protein gene into the nucleus and redirection of the normal polypeptide back to the mitochondrion, and direct modification of the mtDNA.

Treatment regimes for mitochondrial diseases resulting from mutations in nDNA-encoded mitochondrial genes fall into three major categories: those affecting structural genes, those affecting mtDNA maintenance and expression genes, and those regulating mitochondrial antioxidant defense genes. An OXPHOS defect in CHO cells due to a deficiency in an nDNA-encoded complex I gene has been treated by transfection with a plasmid carrying the yeast *NDI1* gene which encodes a single polypeptide NADH dehydrogenase. *NDI1* transfection restored the NADH oxidase activity in this cell line and the resulting complex I was insensitive to rotenone but sensitive to flavone, characteristics of the yeast but not mammalian NADH dehydrogenase (Seo et al., 1998). The efficacy of transfer of the yeast *NDI1* gene was greatly increased by its introduction into an adeno-associated virus (AAV) vector (Seo et al., 2000). This AAV-*NDI1* transduction system was then used to transduce the yeast *NDI1* gene into mouse and rat neuronal cells (Seo et al., 2002). This AAV-*NDI1* vector was also used to transduce the brain substantia nigra and striatum as well as the skeletal muscle of rats (Seo et al., 2004).

The AAV-transduction system has also been used to induce a LHON-like pathology. The retina and optic nerve were transduced with viruses carrying ribozymes that destroy the mRNAs for either mitochondrial MnSOD (*Sod2*) (Qi et al., 2003a) or the nDNA-encoded NDUFA1 subunit of complex I (Qi et al., 2003b). The loss of optic nerve cells in the later rodent retinal transduction model for optic neuropathy was then partially ameliorated by a secondary transduction of the retina with an AAV carrying the human *Sod2* gene. The AAV-*Sod2*

transduction was found to reduce apoptosis in the retinal ganglion cells and degeneration of the optic nerve (Qi et al., 2004).

The allotopic expression of the cyanide-insensitive alternative oxidase from the ascidian *Ciona intestinalis* has been used to confer significant cyanide-insensitive cytochrome c oxidase (COX) activity in cultured cells. Transduction of this enzyme into the nucleus was able to bypass a patient's COX defects and prevent over-reduction of the mitochondrial quinone pool responsible for ROS thus acting as an antioxidant (Hakkaart et al., 2006).

Transformation of cells with the yeast *NDI1* gene has also been used to treat mitochondrial defects resulting from mtDNA mutations. AAV-transduction or calcium phosphate transfection of the yeast *NDI1* gene was shown to restore NADH dehydrogenase activity in a human cybrid cell line that was homoplasmic for an mtDNA ND4 gene frame-shift mutation. A high expression nDNA *NDI1* transformant was found to reconstitute the NADH-dependent respiration and restore the P:O ratio on malate/glutamate substrates to two-thirds that of the wild type human cells. The reduced ATP production would be expected since the yeast NADH dehydrogenase cannot transport protons. The *NDI1* gene also restored the ability to grow on galactose (Bai et al., 2001).

Transforming mouse mtDNA-deficient (ρ°) cells with the *Saccharomyces cerevisiae* NADH dehydrogenase gene (*Ndi1*) and the *Emericella nidulans* alternative oxidase gene (*Aox*) was able to completely restore NADH/CoQ plus CoQ oxidase activity. This relieved the ρ° cell's dependence on exogenous pyruvate and uridine, but since neither protein can transport protons, ATP synthesis was not restored (Perales-Clemente et al., 2008).

Finally, AAV-mediated transduction was used to deliver the mouse *Ant1* cDNA into *Ant1*-deficiency fibroblasts, myoblasts, and skeletal muscle derived from an *Ant1* knockout mouse. AAV-*Ant1* cDNA transduction partially restored the mitochondrial ANT1 activity, mitochondrial export of ATP, and significantly reduced the muscle histopathology associated with the *Ant1* defect (Flierl et al., 2005).

An alternative strategy to treat diseases resulting from mutations in mtDNA-encoded polypeptide genes is to clone the wild type version of the mtDNA gene, use site-directed mutagenesis to convert the mitochondrial genetic code to the universal genetic code, add a mitochondrial-targeting peptide, and transform the nucleus with this construct. This strategy has been used on cell lines harboring the LHON ND4 np 11778 missense mutation. An AAV-vector containing a functional FLAG-tagged ND4 fused to the ATP1 or aldehyde dehydrogenase targeting peptide was able to restore the mitochondrial ATP production from 40% normal caused by the 1178 mutation from back to 100% and also to partially restore survival and growth in galactose medium (Guy et al., 2002). The same strategy was used to treat a cell line that was homoplasmic for the mtDNA ATP6 gene T8993G mutation, which reduces mitochondrial ATP production about 70% (Trounce et al., 1994). Transformation of ATP6 T8993G cells with a recoded FLAG-tagged *ATP6* gene fused to the *COX8*-targeting peptide resulted in the import of the nDNA ATP6 polypeptide into the mitochondrion, assembly into complex V, increased mitochondrial ATP production, and improved growth on galactose medium (Manfredi et al., 2002).

In addition to treating mtDNA defects by transduction of nDNA, direct modification of the mtDNA sequence may eventually be possible. A fluorescein-labeled oligonucleotide has been introduced into the mitochondria of cells by covalently linking the amines at the 3′ end of the oligonucleotide to those of the mitochondrial-targeting peptide from ornithine transcarbamylase using glutaraldehyde. When this construct was encapsulated into liposomes and the complex exposed to skin fibroblasts, the oligonucleotide-targeting peptide was deposited in the cytosol and subsequently taken up into the mitochondria, with the mitochondrial fluorescence persisting up to 8 days (Geromel et al., 2001). While this procedure targeted the oligonucleotide to the mitochondria, the covalent linkage of the oligonucleotide to the protein could limit its bioavailability.

As an alternative, protein nucleic acids (PNAs) have been substituted for oligonucleotides. PNAs substitute a peptide backbone for the sugar phosphate background of nucleic acids. Since peptide bonds are not charged, the PNAs bind with greater affinity to complementary nucleic acid strands than do other nucleic acid strands. In *in vitro* mtDNA replication studies, a PNA-synthesized complementary to the common 4977 np deletion breakpoint inhibited the replication of the deletion template 80% without affecting the replication of the normal template. Similarly, a PNA complementary to the *MERRF* A8344G variant inhibited replication of the mutant mtDNA by about 75% but did not inhibit the normal. PNAs have also been observed to be taken up by mammalian cells. However, no *in vivo* effects of this system have yet been demonstrated (Taylor et al., 1997a, b). This system has been further elaborated by coupling the PNA to the 4-thiobutyltriphenylphosphonium ion through a disulfide bond. This construct was rapidly taken up by the 143B osteosarcoma cells studied. However, the disulfide bond was then reduced in the cytosol, releasing the PNA. Consequently, the PNA remained in the cytosol while the 4-thiobutyltriphenylphosphonium ion continued into the mitochondrion (Filipovska et al., 2004).

The final PNA rendition has involved the coupling of a mitochondrial-targeting peptide to a PNA. A homologous-labeled oligonucleotide was then annealed to the PNA (mtTargeting peptide-PNA: homologous oligonucleotide-labeled) and the complex introduced into myoblast and myotube cells using either branched chain polyethylenimine (PEI) of Streptolysin-O permeabilization. Under these circumstances, the "targeting peptide-PNA:oligonucleotide-labeled" constructs were delivered to the cytosol and then rapidly taken up by the mitochondria. Moreover, undecagold-labeled oligonucleotides could be observed in the mitochondrial matrix by electron microscopy (Flierl et al., 2003). While all of these approaches suggest that it may be possible to modify the mtDNA via exposure of cells to exogenous DNA or PNA, it will be some time before they will be used in any meaningful clinical application.

Germline Mitochondrial Gene Therapy

While somatic mtDNA gene therapy may hold promise for treating the symptoms of patients with mtDNA disease, these approaches cannot help the woman who harbors a homoplasmic pathogenic mtDNA mutation, as she will have a probability of one of transmitting her mutant mtDNA to her children. To permit this woman to have healthy children, it will be necessary to break the maternal transmission of the mutant mtDNA. One procedure proposed in the 1980s could be to fertilize the mother's egg *in vitro* with the father's sperm and then transfer the zygote nucleus to an enucleated egg containing normal mtDNAs (Wallace, 1987). This has now been accomplished using rhesus macaques as a model system. The mother's egg was fertilized and the zygote MII spindle collected, encapsulated within a bleb of the oocyte plasma membrane. This diploid MII karyoplast was then inserted under the zona pellucida of an enucleated egg and fused to the oocyte using inactivated Sendai virus. This procedure generated apparently normal offspring with the biparental nuclei but the recipient oocyte mtDNA. Moreover, this procedure was successful at a relatively high frequency. Hence, this technique may ultimately permit germline gene therapy for mtDNA diseases in the future (Tachibana et al., 2009).

Mitochondrial Therapeutics and Performance Enhancement

It is now clear that not all mtDNA variation is deleterious. Indeed, about 25% of all ancient mtDNA variation appears to have caused functional mitochondrial changes and thus been adaptive. Those mtDNA variants that are adapted to warm climates result in more tightly coupled OXPHOS, thus maximizing ATP output and minimizing heat production. The presence of these mtDNAs permits maximum muscle performance but also predisposes sedentary individuals that consume excess calories to multiple problems including obesity, diabetes, excessive ROS production, and predilection to a variety of degenerative diseases, cancer, and premature aging. Partially uncoupled mitochondria generate more heat, but

at the expense of efficiency of ATP production. Individual's with these variants are better able to tolerate the cold, and are less prone to obesity, generate less ROS, and are more resistant to degenerative diseases and aging, but have reduced endurance.

All of these factors and numerous others are areas that influence our daily lives. Consequently, some individuals may wish to change their phenotype by altering their mtDNA genotype. As mitochondrial metabolic and genetic therapies advance for treating mitochondrial disease, they will also become available to influence the personal lives of healthy individuals. For example, parents might consider changing the mtDNAs of their children in hopes of increasing their potential for athletic performance. Nuclear transfer could be used to substitute a high performance African haplogroup L0 mtDNA (Scott et al., 2009) for a low performance European haplogroup K (Niemi & Majamaa, 2005) or T mtDNA (Castro et al., 2007). If done at conception, such a substitution of mtDNAs would be an integral part of the individual's genotype and thus undetectable by any anti-doping strategy.

While it might seem attractive to change an unborn child's mtDNA genotype to achieve the parent's aspirations for athletic prowess, it is important to keep in mind that the definition of a good or bad mtDNA is contextual. Therefore, it is possible that an mtDNA might perform very differently or even be deleterious in an alternative nDNA context (Potluri et al., 2009). Furthermore, what might be a highly desirable trait for one generation might become undesirable in subsequent generations.

Acknowledgments

The author would like to thank Ms Marie T. Lott for her assistance in assembling this document. The work has been supported by NIH grants NS21328, AG24373, DK73691, AG13154, AG16573, a CIRM Comprehensive Grant RC1-00353-1, a Doris Duke Clinical Interfaces Award 2005, and an Autism Speaks High Impact Grant.

References

Angelin, A., Tiepolo, T., Sabatelli, P., et al. (2007). Mitochondrial dysfunction in the pathogenesis of Ullrich congenital muscular dystrophy and prospective therapy with cyclosporins. *Proceedings of the National Academy of Sciences USA* 104(3), 991–996.

Argov, Z., Bank, W.J., Maris, J., et al. (1986). Treatment of mitochondrial myopathy due to complex III deficiency with vitamins K3 and C: A 31P-NMR follow-up study. *Annals of Neurology* 19(6), 598–602.

Bai, R.K., Leal, S.M., Covarrubias, D., Liu, A. & Wong, L.J. (2007). Mitochondrial genetic background modifies breast cancer risk. *Cancer Research* 67(10), 4687–4694.

Bai, Y., Hajek, P., Chomyn, A., et al. (2001). Lack of complex I activity in human cells carrying a mutation in mtDNA-encoded ND4 subunit is corrected by the *Saccharomyces cerevisiae* NADH-quinone oxidoreductase (NDI1) gene. *Journal of Biological Chemistry* 276(42), 38808–38813.

Ballinger, S.W., Shoffner, J.M., Hedaya, E.V., et al. (1992). Maternally transmitted diabetes and deafness associated with a 10.4 kb mitochondrial DNA deletion. *Nature Genetics* 1(1), 11–15.

Ballinger, S.W., Shoffner, J.M., Gebhart, S., Koontz, D.A. & Wallace, D.C. (1994). Mitochondrial diabetes revisited. *Nature Genetics* 7(4), 458–459.

Bastin, J., Aubey, F., Rotig, A., Munnich, A.& Djouadi, F. (2008). Activation of peroxisome proliferator-activated receptor pathway stimulates the mitochondrial respiratory chain and can correct deficiencies in patients' cells lacking its components. *Journal of Clinical Endocrinology and Metabolism* 93(4), 1433–1441.

Baudouin, S.V., Saunders, D., Tiangyou, W., et al. (2005). Mitochondrial DNA and survival after sepsis: A prospective study. *Lancet* 366(9503), 2118–2121.

Bender, A., Krishnan, K.J., Morris, C.M., et al. (2006). High levels of mitochondrial DNA deletions in substantia nigra neurons in aging and Parkinson disease. *Nature Genetics* 38(5), 515–517.

Booker, L.M., Habermacher, G.M., Jessie, B.C., et al. (2006). North American white mitochondrial haplogroups in prostate and renal cancer. *Journal of Urology* 175(2), 468–472.

Bowmaker, M., Yang, M.Y., Yasukawa, T., et al. (2003). Mammalian mitochondrial DNA replicates bidirectionally from an initiation zone. *Journal of Biological Chemistry* 278(51), 50961–50969.

Brandon, M., Baldi, P. & Wallace, D.C. (2006) Mitochondrial mutations in cancer. Oncogene 25(34), 4647–4662.

Brown, M.D., Torroni, A., Reckord, C.L. & Wallace, D.C. (1995). Phylogenetic analysis of Leber's hereditary optic neuropathy mitochondrial DNAs indicates multiple independent occurrences of the common mutations. *Human Mutation* 6(4), 311–325.

Brown, M.D., Sun, F. &Wallace, D.C. (1997). Clustering of Caucasian Leber hereditary optic neuropathy patients containing the 11778 or 14484 mutations on an mtDNA lineage. *American Journal of Human Genetics* 60(2), 381–387.

Brown, M.D., Starikovskaya, E., Derbeneva, O., et al. (2002). The role of mtDNA background in disease expression: A new primary LHON mutation associated with Western Eurasian haplogroup J. *Human Genetics* 110(2), 130–138.

Carrieri, G., Bonafe, M., De Luca, M., et al. (2001). Mitochondrial DNA haplogroups and APOE4 allele are non-independent variables in sporadic Alzheimer's disease. *Human Genetics* 108(3), 194–198.

Castro, M.G., Terrados, N., Reguero, J.R., Alvarez, V. & Coto, E. (2007). Mitochondrial haplogroup T is negatively associated with the status of elite endurance athlete. *Mitochondrion* 7(5), 354–357.

Chagnon, P., Gee, M., Filion, M., Robitaille, Y., Belouchi, M. & Gauvreau, D. (1999). Phylogenetic analysis of the mitochondrial genome indicates significant differences between patients with Alzheimer disease and controls in a French-Canadian founder population. *American Journal of Medical Genetics* 85(1), 20–30.

Clayton, D.A. (1982). Replication of animal mitochondrial DNA. *Cell* 28(4), 693–705.

Corral-Debrinski, M., Horton, T., Lott, M.T., Shoffner, J.M., Beal, M.F. & Wallace, D.C. (1992). Mitochondrial DNA deletions in human brain: Regional variability and increase with advanced age. *Nature Genetics* 2(4), 324–329.

Corral-Debrinski, M., Horton, T., Lott, M.T., et al. (1994). Marked changes in mitochondrial DNA deletion levels in Alzheimer brains. *Genomics* 23(2), 471–476.

Coskun, P.E., Ruiz-Pesini, E.E. & Wallace, D.C. (2003). Control region mtDNA variants: Longevity, climatic adaptation and a forensic conundrum.

Proceedings of the National Academy of Sciences USA 100(5), 2174–2176.

Coskun, P.E., Beal, M.F. & Wallace, D.C. (2004). Alzheimer's brains harbor somatic mtDNA control-region mutations that suppress mitochondrial transcription and replication. *Proceedings of the National Academy of Sciences USA* 101(29), 10726–10731.

Crispim, D., Canani, L.H., Gross, J.L., Tschiedel, B., Souto, K.E. & Roisenberg, I. (2006). The European-specific mitochondrial cluster J/T could confer an increased risk of insulin-resistance and type 2 diabetes: An analysis of the m.4216T > C and m.4917A > G variants. *Annals of Human Genetics* 70(Pt 4), 488–495.

Darvishi, K., Sharma, S., Bhat, A.K., Rai, E. & Bamezai, R.N. (2007). Mitochondrial DNA G10398A polymorphism imparts maternal haplogroup—a risk for breast and esophageal cancer. *Cancer Letters* 249(2), 249–255.

De Benedictis, G., Rose, G., Carrieri, G., et al. (1999). Mitochondrial DNA inherited variants are associated with successful aging and longevity in humans. *FASEB Journal* 13(12), 1532–1536.

Delettre, C., Lenaers, G., Griffoin, J.M., et al. (2000). Nuclear gene OPA1, encoding a mitochondrial dynamin-related protein, is mutated in dominant optic atrophy. *Nature Genetics* 26(2), 207–210.

Di Prospero, N.A., Baker, A., Jeffries, N. & Fischbeck, K.H. (2007). Neurological effects of high-dose idebenone in patients with Friedreich's ataxia: A randomised, placebo-controlled trial. *Lancet Neurology* 6(10), 878–886.

Djouadi, F. & Bastin, J. (2008). PPARs as therapeutic targets for correction of inborn mitochondrial fatty acid oxidation disorders. *Journal of Inherited Metabolic Disease* 31(2), 217–225.

Eleff, S., Kennaway, N.G., Buist, N.R., et al. (1984). 31P NMR study of improvement in oxidative phosphorylation by vitamins K3 and C in a patient with a defect in electron transport at complex III in skeletal muscle. *Proceedings of the National Academy of Sciences USA* 81(11), 3529–3533.

Elliott, H.R., Samuels, D.C., Eden, J.A., Relton, C.L. & Chinnery, P.F. (2008). Pathogenic mitochondrial DNA mutations are common in the general population. *American Journal of Human Genetics* 83(2), 254–260.

Fan, W., Waymire, K., Narula, N., et al. (2008). A mouse model of mitochondrial disease reveals germline selection against severe mtDNA mutations. *Science* 319(5865), 958–962.

Filipovska, A., Eccles, M.R., Smith, R.A. & Murphy, M.P. (2004). Delivery of antisense peptide nucleic acids (PNAs) to the cytosol by disulphide conjugation to a lipophilic cation. *FEBS Letters* 556(1–3), 180–186.

Flierl, A., Jackson, C., Cottrell, B., Murdock, D., Seibel, P. & Wallace, D.C. (2003). Targeted delivery of DNA to the mitochondrial compartment via import sequence-conjugated peptide nucleic acid. *Molecular Therapy* 7(4), 550–557.

Flierl, A., Chen, Y., Coskun, P.E., Samulski, R.J. & Wallace, D.C. (2005). Adeno-associated virus-mediated gene transfer of the heart/muscle adenine nucleotide translocator (ANT) in mouse. *Gene Therapy* 12(7), 570–578.

Fuku, N., Park, K.S., Yamada, Y., et al. (2007). Mitochondrial haplogroup N9a confers resistance against type 2 diabetes in Asians. *American Journal of Human Genetics* 80(3), 407–415.

Geromel, V., Cao, A., Briane, D., et al. (2001). Mitochondria transfection by oligonucleotides containing a signal peptide and vectorized by cationic liposomes. *Antisense and Nucleic Acid Drug Development* 11(3), 175–180.

Geromel, V., Darin, N., Chretien, D., et al. (2002). Coenzyme Q(10) and idebenone in the therapy of respiratory chain diseases: Rationale and comparative benefits. *Molecular Genetics and Metabolism* 77(1–2), 21–30.

Ghezzi, D., Marelli, C., Achilli, A., et al. (2005). Mitochondrial DNA haplogroup K is associated with a lower risk of Parkinson's disease in Italians. *European Journal of Human Genetics* 13(6), 748–752.

Goto, Y., Nonaka, I. & Horai, S. (1990). A mutation in the tRNA$^{Leu(UUR)}$ gene associated with the MELAS subgroup of mitochondrial encephalomyopathies. *Nature* 348(6302), 651–653.

Guy, J., Qi, X., Pallotti, F., et al. (2002). Rescue of a mitochondrial deficiency causing Leber hereditary optic neuropathy. *Annals of Neurology* 52(5), 534–542.

Hakkaart, G.A., Dassa, E.P., Jacobs, H.T. & Rustin, P. (2006). Allotopic expression of a mitochondrial alternative oxidase confers cyanide resistance to human cell respiration. *EMBO Reports* 7(3), 341–345.

Hart, P.E., Lodi, R., Rajagopalan, B., et al. (2005). Antioxidant treatment of patients with Friedreich ataxia: Four-year follow-up. *Archives of Neurology* 62(4), 621–626.

Hendrickson, S.L., Hutcheson, H.B., Ruiz-Pesini, E., et al. (2008). Mitochondrial DNA haplogroups influence AIDS progression. *AIDS* 22(18), 2429–2439.

Hendrickson, S.L., Kingsley, L.A., Ruiz-Pesini, E., et al. (2009). Mitochondrial DNA haplogroups influence lipoatrophy after highly active anti-retroviral therapy. *Journal of Acquired Immune Deficiency Syndromes* 51(2), 111–116.

Holt, I.J. & Jacobs, H.T. (2003). Response: The mitochondrial DNA replication bubble has not burst. *Trends in Biochemical Sciences* 28(7), 355–356.

Holt, I.J., Harding, A.E. & Morgan-Hughes, J.A. (1988). Deletions of muscle mitochondrial DNA in patients with mitochondrial myopathies. *Nature* 331(6158), 717–719.

Holt, I.J., Harding, A.E., Petty, R.K. & Morgan-Hughes, J.A. (1990). A new mitochondrial disease associated with mitochondrial DNA heteroplasmy. *American Journal of Human Genetics* 46(3), 428–433.

Irwin, W.A., Bergamin, N., Sabatelli, P., et al. (2003). Mitochondrial dysfunction and apoptosis in myopathic mice with collagen VI deficiency. *Nature Genetics* 35(4), 367–371.

Ishikawa, K., Takenaga, K., Akimoto, M., et al. (2008). ROS-generating mitochondrial DNA mutations can regulate tumor cell metastasis. *Science* 320, 661–664.

Ivanova, R., Lepage, V., Charron, D. & Schachter, F. (1998). Mitochondrial genotype associated with French Caucasian centenarians. *Gerontology* 44(6), 349.

James, A.M., Smith, R.A. & Murphy, M.P. (2004). Antioxidant and prooxidant properties of mitochondrial coenzyme Q. *Archives of Biochemistry and Biophysics* 423(1), 47–56.

Jauslin, M.L., Meier, T., Smith, R.A. & Murphy, M.P. (2003). Mitochondria-targeted antioxidants protect Friedreich ataxia fibroblasts from endogenous oxidative stress more effectively than untargeted antioxidants. *FASEB Journal* 17(13), 1972–1974.

Jones, M.M., Manwaring, N., Wang, J.J., Rochtchina, E., Mitchell, P. & Sue, C.M. (2007). Mitochondrial DNA haplogroups and age-related maculopathy. *Archives of Ophthalmology* 125(9), 1235–1240.

Kaufmann, P., Engelstad, K., Wei, Y.,et al. (2006). Dichloroacetate causes toxic neuropathy in MELAS: A randomized, controlled clinical trial. *Neurology* 66(3), 324–330.

Kaukonen, J., Juselius, J.K., Tiranti, V., et al. (2000). Role of adenine nucleotide translocator 1 in mtDNA maintenance. *Science* 289(5480), 782–785.

Kazuno, A.A., Munakata, K., Nagai, T., et al. (2006). Identification of mitochondrial DNA polymorphisms that alter mitochondrial matrix pH and intracellular calcium dynamics. *PLoS Genetics* 2(8), e128.

Kelly, D.P. & Scarpulla, R.C. (2004). Transcriptional regulatory circuits controlling mitochondrial biogenesis and function. *Genes and Development* 18(4), 357–368.

Khidr, L., Wu, G., Davila, A., Procaccio, V., Wallace, D. & Lee, W.H. (2008). Role of SUV3 helicase in maintaining mitochondrial homeostasis in human cells. *Journal of Biological Chemistry* 283(40), 27064–27073.

Khusnutdinova, E., Gilyazova, I., Ruiz-Pesini, E., et al. (2008). A mitochondrial etiology of neurodegenerative diseases: Evidence from Parkinson's disease. *Annals of the New York Academy of Sciences* 1147, 1–20.

King, M.P. & Attardi, G. (1989). Human cells lacking mtDNA: Repopulation with exogenous mitochondria by complementation. *Science* 246(4929), 500–503.

Klivenyi, P., Gardian, G., Calingasan, N.Y., et al. (2004). Neuroprotective effects of oral administration of triacetyluridine against MPTP neurotoxicity. *Neuromolecular Medicine* 6(2–3), 87–92.

Kraytsberg, Y., Kudryavtseva, E., McKee, A.C., Geula, C., Kowall, N.W. & Khrapko, K. (2006). Mitochondrial DNA deletions are abundant and cause functional impairment in aged human substantia nigra neurons. *Nature Genetics* 38(5), 518–520.

Lagouge, M., Argmann, C., Gerhart-Hines, Z., et al. (2006). Resveratrol improves mitochondrial function and protects against metabolic disease by activating SIRT1 and PGC-1alpha. *Cell* 127(6), 1109–1122.

Mandel, H., Szargel, R., Labay, V., et al. (2001). The deoxyguanosine kinase gene is mutated in individuals with depleted hepatocerebral mitochondrial DNA. *Nature Genetics* 29(3), 337–341.

Manfredi, G., Fu, J., Ojaimi, J., et al. (2002). Rescue of a deficiency in ATP synthesis by transfer of MTATP6, a mitochondrial DNA-encoded gene, to the nucleus. *Nature Genetics* 30(4), 394–399.

Melov, S., Schneider, J.A., Day, B.J., et al. (1998). A novel neurological phenotype in mice lacking mitochondrial manganese superoxide dismutase. *Nature Genetics* 18(2), 159–163.

Melov, S., Doctrow, S.R., Schneider, J.A., et al. (2001). Lifespan extension and rescue of spongiform encephalopathy in superoxide dismutase 2 nullizygous mice treated with superoxide dismutase–catalase mimetics. *Journal of Neuroscience* 21(21), 8348–8353.

Michikawa, Y., Mazzucchelli, F., Bresolin, N., Scarlato, G. & Attardi, G. (1999). Aging-dependent large accumulation of point mutations in the human mtDNA control region for replication. *Science* 286(5440), 774–779.

Milne, J.C., Lambert, P.D., Schenk, S., et al. (2007). Small molecule activators of SIRT1 as therapeutics for the treatment of type 2 diabetes. *Nature* 450(7170), 712–716.

Mishmar, D., Ruiz-Pesini, E.E., Golik, P., et al. (2003). Natural selection shaped regional mtDNA variation in humans. *Proceedings of the National Academy of Sciences USA* 100(1), 171–176.

Mishmar, D., Ruiz-Pesini, E., Mondragon-Palomino, M., Procaccio, V., Gaut, B. & Wallace, D.C. (2006). Adaptive selection of mitochondrial complex I subunits during primate radiation. *Gene* 378, 11–18.

Mohlke, K.L., Jackson, A.U., Scott, L.J., et al. (2005). Mitochondrial polymorphisms and susceptibility to type 2 diabetes-related traits in Finns. *Human Genetics* 118(2), 245–254.

Niemi, A.K. & Majamaa, K. (2005). Mitochondrial DNA and ACTN3 genotypes in Finnish elite endurance and sprint athletes. *European Journal of Human Genetics* 13(8), 965–969.

Niemi, A.K., Hervonen, A., Hurme, M., Karhunen, P.J., Jylha, M. & Majamaa, K. (2003). Mitochondrial DNA polymorphisms associated with longevity in a Finnish population. *Human Genetics* 112(1), 29–33.

Nishigaki, Y., Yamada, Y., Fuku, N., et al. (2007). Mitochondrial haplogroup N9b is protective against myocardial infarction in Japanese males. *Human Genetics* 120(6), 827–836.

Nishino, I., Spinazzola, A. & Hirano, M. (1999). Thymidine phosphorylase gene mutations in MNGIE, a human mitochondrial disorder. *Science* 283(5402), 689–692.

Ostrander, E.A., Huson, H.J. & Ostrander, G.K. (2009). Genetics of athletic performance. *Annual Review of Genomics and Human Genetics* 10, 407–429.

Palmieri, L., Alberio, S., Pisano, I., et al. (2005). Complete loss-of-function of the heart/muscle-specific adenine nucleotide translocator is associated with mitochondrial myopathy and cardiomyopathy. *Human Molecular Genetics* 14(20), 3079–3088.

Perales-Clemente, E., Bayona-Bafaluy, M.P., Perez-Martos, A., Barrientos, A., Fernandez-Silva, P. & Enriquez, J.A. (2008). Restoration of electron transport without proton pumping in mammalian mitochondria. *Proceedings of the National Academy of Sciences USA* 105(48), 18735–18739.

Petros, J.A., Baumann, A.K., Ruiz-Pesini, E., et al. (2005). mtDNA mutations increase tumorigenicity in prostate cancer. *Proceedings of the National Academy of Sciences USA* 102(3), 719–724.

Potluri, P., Davila, A., Ruiz-Pesini, E., et al. (2009). A novel NDUFA1 mutation leads to a progressive mitochondrial complex I-specific neurodegenerative disease. *Molecular Genetics and Metabolism* 96(4), 189–195.

Procaccio, V. & Wallace, D.C. (2004). Late-onset Leigh syndrome in a patient with mitochondrial complex I NDUFS8 mutations. *Neurology* 62(10), 1899–1901.

Qi, X., Lewin, A.S., Hauswirth, W.W. & Guy, J. (2003a). Optic neuropathy induced by reductions in mitochondrial superoxide dismutase. *Investigative Ophthalmology and Visual Science* 44(3), 1088–1096.

Qi, X., Lewin, A.S., Hauswirth, W.W. & Guy, J. (2003b). Suppression of complex I gene expression induces optic neuropathy. *Annals of Neurology* 53(2), 198–205.

Qi, X., Lewin, A.S., Sun, L., Hauswirth, W.W. & Guy, J. (2004). SOD2 gene transfer protects against optic neuropathy induced by deficiency of complex I. *Annals of Neurology* 56(2), 182–191.

Raby, B.A., Klanderman, B., Murphy, A., et al. (2007). A common mitochondrial haplogroup is associated with elevated total serum IgE levels. *Journal of Allergy and Clinical Immunology* 120(2), 351–358.

Rasola, A. & Bernardi, P. (2007). The mitochondrial permeability transition pore and its involvement in cell death and in disease pathogenesis. *Apoptosis* 12(5), 815–833.

Rose, G., Passarino, G., Carrieri, G., et al. (2001). Paradoxes in longevity: Sequence analysis of mtDNA haplogroup J in

centenarians. *European Journal of Human Genetics* 9(9), 701–707.

Ruiz-Pesini, E. & Wallace, D.C. (2006). Evidence for adaptive selection acting on the tRNA and rRNA genes of the human mitochondrial DNA. *Human Mutation* 27(11), 1072–1081.

Ruiz-Pesini, E., Mishmar, D., Brandon, M., Procaccio, V. & Wallace, D.C. (2004). Effects of purifying and adaptive selection on regional variation in human mtDNA. *Science* 303(5655), 223–226.

Ruiz-Pesini, E., Lott, M.T., Procaccio, V., et al. (2007). An enhanced MITOMAP with a global mtDNA mutational phylogeny. *Nucleic Acids Research* 35(Database issue), D823–D828.

Rustin, P., von Kleist-Retzow, J.C., Chantrel-Groussard, K., Sidi, D., Munnich, A. & Rotig, A. (1999). Effect of idebenone on cardiomyopathy in Friedreich's ataxia: A preliminary study. *Lancet* 354(9177), 477–479.

Saada, A., Shaag, A., Mandel, H., Nevo, Y., Eriksson, S. & Elpeleg, O. (2001). Mutant mitochondrial thymidine kinase in mitochondrial DNA depletion myopathy. *Nature Genetics* 29(3), 342–344.

Santra, S., Gilkerson, R.W., Davidson, M. & Schon, E.A. (2004). Ketogenic treatment reduces deleted mitochondrial DNAs in cultured human cells. *Annals of Neurology* 56(5), 662–669.

Saxena, R., de Bakker, P.I., Singer, K., et al. (2006). Comprehensive association testing of common mitochondrial DNA variation in metabolic disease. *American Journal of Human Genetics* 79(1), 54–61.

Schaefer, A.M., Taylor, R.W., Turnbull, D.M. & Chinnery, P.F. (2004). The epidemiology of mitochondrial disorders—past, present and future. *Biochimica et Biophysica Acta* 1659(2–3), 115–120.

Schaefer, A.M., McFarland, R., Blakely, E.L., et al. (2007). Prevalence of mitochondrial DNA disease in adults. *Annals of Neurology* 63(1), 35–39.

Scott, R.A., Wilson, R.H., Goodwin, W.H., et al. (2005). Mitochondrial DNA lineages of elite Ethiopian athletes. *Comparative Biochemistry and Physiology Part B, Biochemistry and Molecular Biology* 140(3), 497–503.

Scott, R.A., Fuku, N., Onywera, V.O., et al. (2009). Mitochondrial haplogroups associated with elite Kenyan athlete status. *Medicine and Science in Sports and Exercise* 41(1), 123–128.

Seo, B.B., Kitajima-Ihara, T., Chan, E.K., Scheffler, I.E., Matsuno-Yagi, A. & Yagi, T.

(1998). Molecular remedy of complex I defects: Rotenone-insensitive internal NADH-quinone oxidoreductase of *Saccharomyces cerevisiae* mitochondria restores the NADH oxidase activity of complex I-deficient mammalian cells. *Proceedings of the National Academy of Sciences USA* 95(16), 9167–9171.

Seo, B.B., Wang, J., Flotte, T.R., Yagi, T. & Matsuno-Yagi, A. (2000). Use of the NADH-quinone oxidoreductase (NDI1) gene of *Saccharomyces cerevisiae* as a possible cure for complex I defects in human cells. *Journal of Biological Chemistry* 275(48), 37774–37778.

Seo, B.B., Nakamaru-Ogiso, E., Flotte, T.R., Yagi, T., and Matsuno-Yagi, A. (2002). A single-subunit NADH-quinone oxidoreductase renders resistance to mammalian nerve cells against complex I inhibition. *Molecular Therapy* 6(3), 336–341.

Seo, B.B., Nakamaru-Ogiso, E., Cruz, P., Flotte, T.R., Yagi, T. & Matsuno-Yagi, A. (2004). Functional expression of the single subunit NADH dehydrogenase in mitochondria in vivo: A potential therapy for complex I deficiencies. *Human Gene Therapy* 15(9), 887–895.

Shoffner, J.M., Lott, M.T., Voljavec, A.S., Soueidan, S.A., Costigan, D.A. & Wallace, D.C. (1989). Spontaneous Kearns–Sayre/chronic external ophthalmoplegia plus syndrome associated with a mitochondrial DNA deletion: A slip-replication model and metabolic therapy. *Proceedings of the National Academy of Sciences USA* 86(20), 7952–7956.

Shoffner, J.M., Lott, M.T., Lezza, A.M., Seibel, P., Ballinger, S.W. & Wallace, D.C. (1990). Myoclonic epilepsy and ragged-red fiber disease (MERRF) is associated with a mitochondrial DNA tRNALys mutation. *Cell* 61(6), 931–937.

Shoffner, J.M., Brown, M.D., Torroni, A., et al. (1993). Mitochondrial DNA variants observed in Alzheimer disease and Parkinson disease patients. *Genomics* 17(1), 171–184.

Shulman, A.I. & Mangelsdorf, D.J. (2005). Retinoid x receptor heterodimers in the metabolic syndrome. *New England Journal of Medicine* 353(6), 604–615.

Smeitink, J. & van den Heuvel, L. (1999). Human mitochondrial complex I in health and disease. *American Journal of Human Genetics* 64(6), 1505–1510.

Smith, R.A., Porteous, C.M., Gane, A.M. & Murphy, M.P. (2003). Delivery of bioactive molecules to mitochondria in vivo. *Proceedings of the National Academy of Sciences USA* 100(9), 5407–5412.

Sonoda, J., Mehl, I.R., Chong, L.W., Nofsinger, R.R. & Evans, R.M. (2007). PGC-1beta controls mitochondrial metabolism to modulate circadian activity, adaptive thermogenesis, and hepatic steatosis. *Proceedings of the National Academy of Sciences USA* 104(12), 5223–5228.

Sonoda, J., Pei, L., Evans, R.M. (2008). Nuclear receptors: Decoding metabolic disease. *FEBS Letters* 582(1), 2–9.

Spelbrink, J.N., Li, F.Y., Tiranti, V., et al. (2001). Human mitochondrial DNA deletions associated with mutations in the gene encoding Twinkle, a phage T7 gene 4-like protein localized in mitochondria. *Nature Genetics* 28(3), 223–231.

Srivastava, S., Barrett, J.N. & Moraes, C.T. (2007). PGC-1alpha/beta upregulation is associated with improved oxidative phosphorylation in cells harboring nonsense mtDNA mutations. *Human Molecular Genetics* 16(8), 993–1005.

Stacpoole, P.W., Kerr, D.S., Barnes, C., et al. (2006). Controlled clinical trial of dichloroacetate for treatment of congenital lactic acidosis in children. *Pediatrics* 117(5), 1519–1531.

Stewart, J.B., Freyer, C., Elson, J.L., et al. (2008). Strong purifying selection in transmission of mammalian mitochondrial DNA. *PLoS Biology* 6(1), e10.

Suissa, S., Wang, Z., Poole, J., et al. (2009). Ancient mtDNA genetic variants modulate mtDNA transcription and replication. *PLoS Genetics* 5(5), e1000474.

Tachibana, M., Sparman, M., Sritanaudomchai, H., et al. (2009). Mitochondrial gene replacement in primate offspring and embryonic stem cells. *Nature* 461(7262), 367–372.

Tanaka, M., Gong, J.S., Zhang, J., Yoneda, M. & Yagi, K. (1998). Mitochondrial genotype associated with longevity. *Lancet* 351(9097), 185–186.

Tanaka, M., Gong, J., Zhang, J., Yamada, Y., Borgeld, H.J. & Yagi, K. (2000). Mitochondrial genotype associated with longevity and its inhibitory effect on mutagenesis. *Mechanisms of Ageing and Development* 116(2–3), 65–76.

Taylor, R.W., Chinnery, P.F., Clark, K.M., Lightowlers, R.N. & Turnbull, D.M. (1997a). Treatment of mitochondrial disease. *Journal of Bioenergetics and Biomembranes* 29(2), 195–205.

Taylor, R.W., Chinnery, P.F., Turnbull, D.M. & Lightowlers, R.N. (1997b).

Selective inhibition of mutant human mitochondrial DNA replication in vitro by peptide nucleic acids. *Nature Genetics* 15(2), 212–215.

Tiranti, V., Hoertnagel, K., Carrozzo, R., et al. (1998). Mutations of SURF-1 in Leigh disease associated with cytochrome c oxidase deficiency. *American Journal of Human Genetics* 63(6), 1609–1621.

Tong, J., Schriner, S.E., McCleary, D., Day, B.J. & Wallace, D.C. (2007). Life extension through neurofibromin mitochondrial regulation and antioxidant therapy for neurofibromatosis-1 in *Drosophila melanogaster*. *Nature Genetics* 39(4), 476–485.

Torroni, A., Petrozzi, M., D'Urbano, L., et al. (1997). Haplotype and phylogenetic analyses suggest that one European-specific mtDNA background plays a role in the expression of Leber hereditary optic neuropathy by increasing the penetrance of the primary mutations 11778 and 14484. *American Journal of Human Genetics* 60(5), 1107–1121.

Trounce, I., Neill, S. & Wallace, D.C. (1994). Cytoplasmic transfer of the mtDNA nt 8993 TG (ATP6) point mutation associated with Leigh syndrome into mtDNA-less cells demonstrates cosegregation with a decrease in state III respiration and ADP/O ratio. *Proceedings of the National Academy of Sciences USA* 91(18), 8334–8338.

Udar, N., Atilano, S.R., Memarzadeh, M., et al. (2009). Mitochondrial DNA haplogroups associated with age-related macular degeneration. *Investigative Ophthalmology and Visual Science* 50(6), 2966–2974.

van der Walt, J.M., Nicodemus, K.K., Martin, E.R., et al. (2003). Mitochondrial polymorphisms significantly reduce the risk of Parkinson disease. *American Journal of Human Genetics* 72(4), 804–811.

van der Walt, J.M., Dementieva, Y.A., Martin, E.R., et al. (2004). Analysis of European mitochondrial haplogroups with Alzheimer disease risk. *Neuroscience Letters* 365(1), 28–32.

Van Goethem, G., Dermaut, B., Lofgren, A., Martin, J.J. & Van Broeckhoven, C. (2001). Mutation of POLG is associated with progressive external ophthalmoplegia characterized by mtDNA deletions. *Nature Genetics* 28(3), 211–212.

Van Goethem, G., Martin, J.J. & Van Broeckhoven, C. (2003). Progressive external ophthalmoplegia characterized by multiple deletions of mitochondrial

DNA: Unraveling the pathogenesis of human mitochondrial DNA instability and the initiation of a genetic classification. *Neuromolecular Medicine* 3(3), 129–146.

Wallace, D.C. (1982). Cytoplasmic inheritance of chloramphenicol resistance in mammalian cells. In: J.W. Shay (ed.) *Techniques in Somatic Cell Genetics,* pp. 159–187. Plenum Press, New York, NY.

Wallace, D.C. (1987). Maternal genes: Mitochondrial diseases. In: V.A. McKusick, T.H. Roderick, J. Mori & M.W. Paul (eds) *Medical and Experimental Mammalian Genetics:* A Perspective, pp. 137–190. A.R. Liss, Inc, New York, NY.

Wallace, D.C. (2005a). The mitochondrial genome in human adaptive radiation and disease: On the road to therapeutics and performance enhancement. *Gene* 354, 169–180.

Wallace, D.C. (2005b). A mitochondrial paradigm of metabolic and degenerative diseases, aging, and cancer: A dawn for evolutionary medicine. *Annual Review of Genetics* 39, 359–407.

Wallace, D.C. (2007). Why do we have a maternally inherited mitochondrial DNA? Insights from Evolutionary Medicine. *Annual Review of Biochemistry* 76, 781–821.

Wallace, D.C. (2009). Mitochondria, bioenergetics, and the epigenome in eukaryotic and human evolution (in press). In: B. Stillman, D. Stewart & J. Wirtkowski (eds) *Cold Spring Harbor Symposia on Quantitative Biology LXXIV Evolution: The Molecular Landscape,* pp. 383–393. Cold Spring Harbor Laboratory Press, Cold Spring Harbor, NY.

Wallace, D.C & Fan, W. (2009). The pathophysiology of mitochondrial disease as modeled in the mouse. *Genes and Development* 23(15), 1714–1736.

Wallace, D.C. & Fan, W. (2010). Energetics, epigenetics, mitochondrial genetics. *Mitochondrion* 10(1), 12–31.

Wallace, D.C. & Lott, M.T. (2002). Mitochondrial genes in degenerative diseases, cancer and aging. In: D.L. Rimoin, J.M. Connor, R.E. Pyeritz & B.R. Korf (eds) *Emery and Rimoin's Principles and Practice of Medical Genetics,* pp. 299–409. Churchill Livingstone, London.

Wallace, D.C., Singh, G., Lott, M.T., et al. (1988). Mitochondrial DNA mutation associated with Leber's hereditary optic neuropathy. *Science* 242(4884), 1427–1430.

Wallace, D.C., Brown, M.D. & Lott, M.T. (1999). Mitochondrial DNA variation in human evolution and disease. *Gene* 238(1), 211–230.

Wallace, D.C., Lott, M.T. & Procaccio, V. (2007). Mitochondrial genes in degenerative diseases, cancer and aging. In: D.L. Rimoin, J.M. Connor, R.E. Pyeritz & B.R. Korf (eds) *Emery and Rimoin's Principles and Practice of Medical Genetics, 5th Edition,* pp. 194–298. Churchill Livingstone Elsevier, Philadelphia, PA.

Wallace, D.C., Fan, W. & Procaccio, V. (2010). Mitochondrial energetics and therapeutics. *Annual Review of Pathology* 5, 297–348.

Wang, Y.X., Zhang, C.L., Yu, R.T., et al. (2004). Regulation of muscle fiber type and running endurance by PPARdelta. *PLoS Biology* 2(10), e294.

Wenz, T., Diaz, F., Spiegelman, B.M. & Moraes, C.T. (2008). Activation of the PPAR/PGC-1alpha pathway prevents a bioenergetic deficit and effectively improves a mitochondrial myopathy phenotype. *Cell Metabolism* 8(3), 249–256.

Zhu, Z., Yao, J., Johns, T., et al. (1998). SURF1, encoding a factor involved in the biogenesis of cytochrome c oxidase, is mutated in Leigh syndrome. *Nature Genetics* 20(4), 337–343.

Zuchner, S., Mersiyanova, I.V., Muglia, M., et al. (2004). Mutations in the mitochondrial GTPase mitofusin 2 cause Charcot-Marie-Tooth neuropathy type 2A. *Nature Genetics* 36(5), 449–451.

Chapter 3

Characterizing the Extent of Human Genetic Variation for Performance-Related Traits

JOHN BLANGERO AND JACK W. KENT JR.

Department of Genetics, Southwest Foundation for Biomedical Research, San Antonio, TX, USA

In this chapter we present a general overview of approaches to describing and understanding the extent of genetic variation in traits, how such variation impacts gene localization and discovery, and the relevance to athletic performance. Distinctions between monogenic and quantitative (polygenic) traits are explained, and how the amount of genetic contribution to a complex trait (heritability) is measured. We provide examples of measures of heritability of traits such as physical activity (PA) levels, where about 25% of activity is likely genetically determined, and it becomes possible to identify the responsible genes due to the finite nature of the human genome. We describe the meaning of "functional genetic variation" and how it determines heritability, and point out that between-population genetic variation has important consequences on quantifying and identifying responsible genes. The recent application of genome-wide association studies (GWAS) to identify quantitative traits has led to seemingly enigmatic results. Taking a trait such as height (stature), known to be about 80% genetic, multiple GWAS in large cohorts found strong statistical support for over 40 gene regions (loci), yet together they explained only ~5% of the variance in stature. This apparent contradiction raises the issue of common variants (as tested by GWAS), versus rare variants (not tested by GWAS), and the relative contribution of each toward complex quantitative traits. We conclude with a section on an integrative approach to genetic analysis of performance, with recommendations regarding the scientific approach and interpretation of the data. Emerging technologies that allow whole genome or deep resequencing at a reasonable cost are likely to show the complexity of human variation, while also finding those sequence variants responsible for physical performance traits.

Human Genetic Variation

Human genetic variation is vast and manifests itself in all aspects of human phenotypic variation. For the purpose of this volume, we focus on human genetic variation in relation to variation in performance-related phenotypes. Such phenotypes belong to a large class of *quantitative traits*—traits whose natural measures vary across some continuous scale. Traits of this type have long been a primary concern of agricultural plant and animal breeders and, due to Darwin's extrapolation from artificial to natural selection, of evolutionary biologists as well. At the dawn of modern genetics, Gregor Mendel carried out controlled breeding experiments focusing on *discrete* traits, and his ingenious mathematical interpretation of his results led him to a factorial model of inheritance. However, early geneticists struggled with the apparent dilemma that Mendel's elegant biological mechanism of inheritance, well supported by experiment for *some* traits, seemed inconsistent with the complex distribution of traits

Genetic and Molecular Aspects of Sport Performance, 1st edition. Edited by Claude Bouchard and Eric P. Hoffman. Published 2011 by Blackwell Publishing Ltd.

like body size, fecundity, disease resistance, etc. that defined the differential vigor and physiological capacity of individuals.

This dilemma was resolved theoretically by scientists like Ronald Fisher (1918) and Sewall Wright (1921, 1969), as follows:

- Genetic variation is DNA sequence variation.[1] The human genome consists of ~3.4 billion base pairs of DNA. All of these sites are likely variable somewhere throughout the expanse of living humans, except for those sites whose resulting phenotypes are incompatible with life. Thus, the potential extent of human genetic variation is truly vast, although most such sequence variation is obligately rare in frequency.
- Quantitative distributions of phenotype can arise through some combination of the following (Lander & Schork, 1994):
  - The combined effects of multiple genes (oligogenic or polygenic inheritance);
  - Differential expression of genes due to gene–environment interactions;
  - Heterogeneity of causes in different subpopulations. This may reflect systematic differences in allele frequency due to founder effects or migration (population stratification) or extremes of environmental exposure that mimic genetic effects (phenocopy).

"Mendelian" or monogenic variation is often characterized by an extreme change in phenotype due to loss (or, less commonly, excess expression) of gene product. By contrast, quantitative variation is expected to represent incremental variation in the quantity or activity of functional gene products, perhaps more likely due to regulatory variation than to non-synonymous coding changes. Recognizing this distinction, several researchers have begun to measure individual variation in gene transcription (as RNA level) in tissues of interest (de Koning & Haley, 2005; Pastinen et al., 2006; Storey et al., 2007).

Such *expression phenotypes* represent novel quantitative traits that are the direct consequence of gene regulation and thus may offer strong and unambiguous signal for gene localization (Göring et al., 2007).

However the quantitative phenotype is defined, an overriding challenge, as compared to analysis of monogenic traits, is that multiple loci may be contributing to the phenotype so the effect size of any locus is likely to be relatively small. In addition, there may be multiple types of sequence variation in play, ranging from substitution of single nucleotides to rearrangements of chromosomal structure (sequence deletion, duplication, or inversion), which may not be equally detectable by any one analytical technique. Finally, even if the majority of genetic effect were confined to a single locus, this could be due either to a single variable site or multiple rare alleles segregating in the population(s) under study.

Measuring the Extent of Human Genetic Variation in Quantitative Traits

Heritability is the proportion of the total variance of a phenotype that is attributable to the additive effects of alleles.[2] Thus, heritability tells us how important a role genetic variation is likely to play in the causation of a variable human trait. In a classical variance components-based approach to quantitative genetic analysis, heritability is readily estimated by decomposing the phenotypic covariance between pairs of individuals based on their kinship:

$$Cov(i,j) = 2\phi_{ij}\sigma_g^2$$
$$Cov(i,i) = \sigma_g^2 + \sigma_e^2$$
$$h^2 = \sigma_g^2 / \sigma_p^2$$

where $Cov(i, j)$ is the covariance between different individuals i and j, $Cov(i, i)$ represents the variance for the i-th individual, ϕ_{ij} is the kinship coefficient between i and j, σ_g^2 is the additive genetic variance,

---

[1] "Epigenetic" variation—such as altered expression of DNA sequences by methylation or other chemical modification—is itself influenced by sequence (some sequences may define targets for modification; methylation may be regulated by factors encoded elsewhere in the genome, etc.).

[2] This is sometimes called "narrow-sense heritability," with "broad-sense heritability" used to describe the proportion of total variance due to *all* genetic effects including dominance and epistasis. In outbred populations, additive effects account for most of the genetic variance and are the focus of this discussion.

σ_e^2 is the error (sometimes termed environmental) variance, σ_p^2 is the total phenotypic variance of the trait given by $\sigma_p^2 = \sigma_g^2 + \sigma_e^2$, and h^2 is the additive genetic heritability. Twice the "kinship" coefficient in this context refers to the expected proportion of alleles shared *identical by descent* (IBD) by two individuals given their degree of relatedness: siblings share half their alleles IBD, half-siblings one-fourth of their alleles, and so on. Alleles shared IBD are necessarily also *identical by state* (IBS), a distinction that we discuss in more detail below. The genetic variance is a cumulative variance; it represents the summation over all additive genetic factors for the phenotype. Hence, depending upon the phenotype, it may represent the influence of a single genetic variant or that of many hundreds of genetic variants.

If variation in a phenotype were entirely due to genetic causes the heritability of the trait would, of course, be 1. Due to multifactorial causation and measurement error, a typical range of heritability for many quantitative traits is 20–80% of total variance, and estimates may differ widely from one study to another due to sampling error. The heritability is a critical measure of the importance of within-population genetic variation. This single metric conveys whether or not the search for the individual contributing genes is merited for a given phenotype.

The existing scientific literature has numerous examples of heritability measures for traits relevant to human performance. Table 3.1 compares heritability estimates from four populations for levels of PA, a behavioral phenotype. Estimation of heritabilities requires phenotypic information on family members, and the data in Table 3.1 come from study designs based either on nuclear families or extended pedigrees. Both types of study suggest only modest heritabilities for these behavioral phenotypes (measured by either questionnaires or accelerometers). Although genetic factors account generally for only about a quarter or so of the total phenotypic variation in PA levels, this number likely reflects truly causal genetic factors that feasibly can be searched for and ultimately identified due to the finite nature of the human genome. In contrast, the 75% or so residual phenotypic variance must be due to causal factors to be sought from the infinite state space of

Table 3.1 Heritability of selected performance measures in several populations

Study/Phenotype	Instrument	Population	Study design	N	Heritability (SE)
Seabra et al. (2008)					
Work physical activity (PA)	Recall questionnaire	Portuguese	Nuclear families	9500	0.06 (0.02)**
Sport PA				9500	0.19 (0.02)**
Leisure PA				9500	0.25 (0.02)**
Combined PA				9500	0.23 (0.02)**
Kent and Blangero (unpublished data)					
Work PA	Recall questionnaire	Mexican Americans	Extended families	1187	0.16 (0.05)***
Leisure PA				1189	0.19 (0.05)***
Combined PA				1186	0.10 (0.05)*
Simonen et al. (2002)					
%Time inactivity	Recall questionnaire	Quebecois	Nuclear families	696	0.25 (0.04[a])***
%Time				696	0.16 (0.02)***
Combined PA				696	0.19 (0.04)***
Williams et al. (2008)					
%Time inactivity	7-day accelerometer	Nepalese	Extended	309	0.41 (0.17)*
%Time vigorous PA				309	0.15 (0.17)[b]

[a] Standard errors approximated from published 95% confidence intervals.
[b] Not significant.
*$P<0.05$; **$P<0.001$; ***$P<0.0001$.
SE = Standard error.

(unspecified) environmental variation. The results from this small example also show that heritabilities can be population-specific, a point we will address shortly.

Functional Genetic Variation Determines Heritability

What is the biological source of heritability in humans? Ultimately, it comes from observable *functional* genetic variation at the sequence level. A functional variant is one that influences the focal phenotype via some molecular mechanism (thus, functionality is phenotype-specific). If the variant influences a quantitative trait (such as performance), we term it a *quantitative trait nucleotide* (QTN) variant. The effect of a functional variant on the phenotype can be quantified by the QTN-specific variance $\sigma_q^2 = 2p(1-p)\alpha^2$, where p is the minor allele frequency of the QTN and α is one-half the difference between phenotypic means of the two homozygotes. Biologically, we expect α to be determined by biophysical molecular properties of the QTN and relatively constant across populations. The term " $2p(1-p)$," also known as the expected heterozygosity of the underlying genotype, expresses the dependence of trait variance on the frequency of minor alleles in the diploid genotype. The relative genetic signal intensity for this QTN is given by the QTN-specific heritability $h_q^2 = \sigma_q^2 / \sigma_P^2$ where σ_P^2 is the total variance of the phenotype.

Figure 3.1 shows a hypothetical case of a *quantitative trait locus* (QTL) whose contribution to a trait is due to a single diallelic QTN. More generally, the relative genetic signal for the QTL will be determined by the sum of the QTN-specific heritabilities (although these must be corrected for possible linkage disequilibrium, LD, amongst variant sites) in the immediate region of the QTL and thus will be influenced by all of the relevant functional variants in the region. In algebraic form, the QTL-specific heritability is given as:

$$h_Q^2 = \frac{\sum 2p_i(1-p_i)\alpha_i}{\sigma_P^2} = \sum h_{qi}^2$$

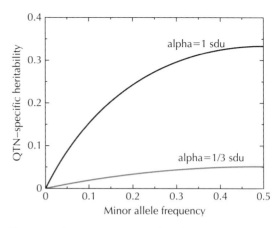

Figure 3.1 Quantitative trait nucleotide (QTN) heritability as a function of allele frequency. Two displacements α of the homozygote means are shown: either one or one-third standard deviation unit (sdu) of a hypothetical quantitative trait, representing larger or smaller genetic effect sizes, respectively. The QTN-specific heritability, a measure of genetic signal to noise, is a function of both the effect size and the frequency of the minor allele.

where the summation is over the functional variants in the regions of the QTL. Similarly, the total heritability of the phenotype is given by the sum of all of the QTL-specific heritabilities over the whole genome or $h^2 = \sum h_{Qi}^2$.

Between-Population Genetic Variation Has Important Consequences

Due to the obvious dependence of the relevant genetic signal on allele frequency, QTN-specific heritabilities will vary from population to population. Figure 3.1 shows the relationship between the QTN-specific heritability and the minor allele frequency for two different choices of α. Clearly, there is great potential for allele frequency differences across human populations to have large effects on the relative genetic signals observed for specific performance traits, and efficient gene discovery will be enhanced by information from multiple populations (Blangero et al., 2003).

The extent of genetic variation can also differ greatly across human populations. Such variation is often measured by a statistic first proposed by

Sewall Wright, termed F_{st}. For our current purpose, F_{st} may be defined as the standardized genetic variance between populations. It estimates the relative importance of between-group versus within-group variation in allele frequency. It can be written as the observed variance in allele frequency across populations divided by that which would be expected if there were a single population exhibiting an allele frequency ($\bar{p}$) equal to the mean across populations:

$$F_{st} = \frac{\sum_{i=1}^{n}(p_i - \bar{p})^2}{\bar{p}(1-\bar{p})n}$$

where the summation is over populations. For differences across major geographic groupings of human populations, average F_{st} across many loci typically ranges from 0.10 to 0.15 (Guthery et al., 2007; Li et al., 2008; Myles et al., 2008) so that about 10 of the total genetic variation is due to between-population differences. Within major geographic areas, average F_{st} is more modest (generally no more than 0.02 or so). However, F_{st} can vary even more widely for specific genetic variants. Such between-population differences in allele frequencies can lead to large differences in QTN-specific heritabilities across populations (as shown in Figure 3.1). Under the assumption of evolutionary neutrality, there is a simple relationship between between-population quantitative genetic variation and allele frequency variation given by:

$$\sigma_B^2 = 2\frac{F_{st}}{1-F_{st}}\sigma_g^2$$

However, given that performance-related phenotypes may have been under some kind of natural selection, we are unsure if this simple expected relationship between between-group and within-group genetic variation will hold for a given phenotype.

Example: A Functional Variant in *ACTN3*

As an example of the extent of human between-population variation, we examine the observed distribution of a functional coding variant in the *ACTN3* gene, which is believed to impact performance-related phenotypes (Yang et al., 2009) via its role in skeletal muscle. This variant (R577X, dbSNP rs1815739) induces a premature stop to

the alpha-actinin-3 protein at the 577th amino acid (which is usually an arginine). Individuals homozygous for the stop codon completely lack ACTN3 protein, although this deficiency is not associated with any known disease. This coding sequence variation was first identified by North and colleagues (1999), who also first hypothesized that the deficiency genotype would reduce athletic performance in sprint/power events. Since its discovery, there have been multiple papers examining the association of R577X with elite sports performance (Eynon et al., 2009; Lucia et al., 2006) and various physiological parameters related to performance (Clarkson et al., 2005; Delmonico et al., 2008; Norman et al., 2009; Walsh et al., 2008).

Using the publicly available data from Li et al. (2008) on over 1000 individuals from 50 populations, we have examined the between-population variation in the allele frequency of R577X. Figure 3.2 shows the extensive frequency variation amongst human populations for this sequence variant. It ranges from near 0 in African populations to over 0.8 in some Amerindian populations. The observed F_{st} is ~0.16, which is somewhat higher than the average of a typical variant. In fact, this high a level of population differentiation may be indicative of the effect of natural selection on this specific variant (Yngvadottir et al., 2009).

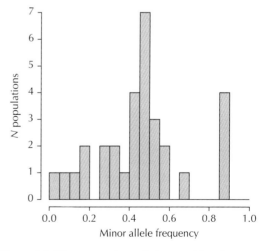

Figure 3.2 Histogram of population-specific minor allele frequencies for the *ACTN3* stop codon variant.

Understanding how this large between-group variation might impact performance-related traits requires some basic information about how *ACTN3* genotypes differ functionally. The definition of "function" is phenotype-specific; for this example, we will use a quantitative phenotype, mRNA levels, that is closest to the action of the gene. Recently, Norman et al. (2009) established that individuals with two copies of the *ACTN3* stop codon (the XX genotype) exhibited *ACTN3* mRNA levels that are more than 10-fold lower than those observed for RR individuals. Using other statistical information available in Norman et al. (2009), we determine that the estimated α for this variant is ~0.60 standardized units of mRNA level. Using this estimate and the observed between-group variation shown in Figure 3.2, we calculate that the average QTN-specific heritability for this trait is about 0.129. However, across populations, this heritability varies widely from 0 (in the African San population where the X variant is missing) to about 0.15 (in several Asian populations where heterozygosity is maximized). Thus, depending upon the population examined, this "functional" variant can appear to have a very small effect or a very large effect.

Identifying Genes Responsible for Heritability

The magnitude of both within-population and between-population genetic variation has direct implications for the discovery of the underlying genes responsible for the observed heritability of performance traits. Localization and identification of these causal genes is the main goal of any genetic study of human performance. Many recent advances in analysis of human quantitative traits have been made in the context of genetically complex diseases (Almasy & Blangero, 2009; Blangero, 2004), although most of the analytical methods are applicable to non-pathological traits. Studies of complex disease are relevant to athletic performance to the extent that they focus on normal variation in risk: just as the disease state results from the intersection of heightened susceptibility and a pathogenic environment, so exceptional physical performance results from superior capacity developed by training and practice.

In the absence of complete sequence information for all study participants (which still lies in the future in terms of technical and economic feasibility), gene localization depends either on the random effect of known genetic markers assessed via linkage, or the main effect of the markers via association. Association methods depend on alleles shared, not necessarily IBD, but IBS. If a marker is physically close enough to a functional variant, linkage between the functional and marker alleles will not be disrupted by meiotic recombination: that is, the alleles are in *LD* (Lewontin & Kojima, 1960). Association is conceptually and mathematically simple: analysis is based on *measured genotypes* (MGs) either for individual markers (often diallelic single-nucleotide polymorphisms, SNPs, which are closely spaced throughout the human genome), or on *haplotype blocks* of marker alleles in high LD. In family-based studies, association may be measured as the main effect of the MG in a mixed model along with the random effects of allele-sharing (Boerwinkle et al., 1986). In principle the main effect may be tested in unrelated individuals as well, although the test is then susceptible to confounding effects of population stratification (Devlin & Roeder, 1999).

If one is so lucky as to have an actual functional variant in hand—or a marker in perfect LD with a functional variant—MG analysis is necessarily the most powerful test of genetic effect. However, LD falls off quite sharply with physical distance (by ~250 kb), which is both a blessing and a curse. It is why association is capable of providing high-resolution positional data, but it also places a severe limitation on detectable effect. The observed effect of a marker locus $h_m^2 = h_q^2 \rho^2$, where h_q^2 and ρ^2—both unobserved—are the effect size of the functional variant and the squared correlation between the marker and functional genotypes, respectively.

Common and Rare Variants: Critical Differences

Because MG association is possible without family data, some investigators have envisioned a golden age of gene discovery via mass genotyping of thousands of unrelated individuals and subsequent genome-wide association scans (Healy, 2006;

Risch & Merikangas, 1996).Their choice of unrelated subjects was strategic: in its simplest mathematical form, MG analysis assumes each datum to be statistically independent, and very large cohorts of subjects could conceivably be drawn from routine clinical examinations without specific recruitment of their relatives.

As we have seen, this approach to gene discovery is necessarily very sensitive to the allele frequencies of both the causative sites and their associated markers. Proponents of GWAS argue that, if susceptibility to highly prevalent diseases/phenotypes is due to many genetic variants each of small effect, such variants should be shielded from strong selection and therefore (potentially) maintained at high frequency for many generations—the so-called "common disease–common variant" (CDCV) hypothesis (Chakravarti, 1999; Lander, 1996). But the assumption that common variants are the *primary* cause of human phenotypic variation is unproven and highly questionable. As Sewall Wright (1949, 1969) first proposed, in the absence of strong purifying selection—the very condition assumed by the CDCV hypothesis—allele frequencies are expected to drift toward fixation in finite populations (Figure 3.3; Pritchard, 2001; Pritchard & Cox, 2002), so that the majority of allelic variants should actually be rare. This theoretical prediction agrees with findings from deep sequencing. A reasonable conclusion is that GWAS in unrelated individuals must *necessarily* miss most functional genetic variation. To assume that the missed variation is largely irrelevant to the phenotype of interest is, at best, to bet one's entire stake on a very lucky confluence of biology and population history.

Recent studies have applied GWAS methods to stature, a canonical, non-pathological biometrical phenotype. In particular, three studies published simultaneously (Gudbjartsson et al., 2008; Lettre et al., 2008; Weedon et al., 2008) were based on extraordinarily large samples ($N = \sim13{,}000\text{–}27{,}000$), taking advantage of the routine measurement of height in clinical examinations carried out for other purposes. Together these studies identified 44 loci at genome-wide significance, but cumulatively these account for only ~5% of the genetic variance of this highly heritable ($h^2 > 0.8$) phenotype

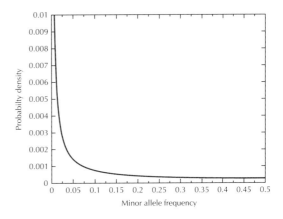

Figure 3.3 Rare allele hypothesis: more Wright than wrong? Predicted distribution of minor allele frequencies in a population at equilibrium, based on classical population genetics principles (Wright, 1949, 1969). We assume a biologically plausible balance between mutation and selection (assuming a polygenic trait, the effect sizes of most variants will be small and subject to no or weak purifying selection). The vast majority of variants are rare (minor allele frequency <0.05).

(Weedon & Frayling, 2008). These results also reflect only genomic localizations of stature-related QTLs; they do not represent true gene identifications, which have been slow to follow such localizations.

These results reflect a critical weakness of the GWAS design: complete dependence upon common sequence variations misses most of the true biological signal. Thus, these massive studies fundamentally underestimated the overwhelming importance of causal genetic factors on height, which is known from every family-based heritability study of stature that has ever been undertaken. Stature may be an extreme example, as it probably reflects variation in nearly every developmental process in the human body—but the same could be said for many measures of physical capacity.

An Integrative Approach to Genetic analysis of Performance

While CDCV-motivated GWAS has provided multiple robust findings (Amos, 2007), even its proponents have been disappointed that the discovery rate has not been greater. In particular, while genomic regions identified by linkage often contribute substantially to the phenotype of interest—QTL-specific heritabilities in the range of 10–20% are not uncommon for phenotypes acting proximate to gene action—most association findings to date account for only a few percent of phenotypic variance, leading to some anguish about "missing heritability" (Maher, 2008; Manolio et al., 2009). As the initial enthusiasm for GWAS wanes, complex disease geneticists are reconsidering their approach to gene discovery, including renewed interest in family-based methods (Schork et al., 2009).

We assert that the greatest potential to capture genetic variability for complex traits must come from a combined approach capable of assessing the cumulative effects of rare *and* common polymorphisms. The preceding discussion indicates that any investigation of the genetic basis of human variation in performance should include the following considerations.

Where possible, use families. Only pedigree information provides precise information on allele-sharing IBD, without which one cannot determine either heritability or linkage. Association is readily tested in family data; in fact, family-based designs are likely to be optimal for detecting the effects of rare, even private, genetic variants, which must account for a large proportion of the causative variation for complex phenotypes. An allele that is rare in the population is more likely to be found in multiple copies in a pedigree due to the correlated genotypes of relatives, thus increasing the power for detection and association. The expected distribution of these "extra" copies of rare variants is a function of the pedigree size and complexity and the entry point of the variant into the pedigree. Extremely large pedigree configurations will provide our best hopes for detecting such rare variants, although such designs may be difficult to pursue for many of the phenotypes classically examined in human sports performance.

Use multiple populations. As the *ACTN3* example makes clear, allele frequencies can vary dramatically across populations. In polygenic or oligogenic traits, this means that different contributing loci may contribute larger proportions of the genetic effect (and thus be detectable) in different samples.

Precise definition of the phenotype is a crucial early step in genetic analysis, although its importance is often overlooked. Quantitative geneticists have long recognized that a broadly defined physiological trait (like diabetes) may be a weaker starting point for gene discovery than a correlated but more specific phenotype (like fasting glucose). In the literature of psychiatric genetics, these more specific traits are known as *endophenotypes* (Almasy & Blangero, 2001), a term that is gaining wider acceptance (Comuzzie et al., 2001). Ideally, such phenotypes are more narrowly representative of some physiochemical pathway underlying the broader trait, and therefore closer to gene action.

The repertoire of possible endophenotypes can be expanded dramatically by inclusion of transcriptomic data. Advances in high-throughput microarray technology allow quantitative measurement of mRNA levels for nearly the entire human transcriptome, yielding thousands of novel phenotypes for genetic analysis. Expression phenotypes may be the ultimate endophenotypes, representing the quantitative activity of individual genes and their regulatory networks. Such measures are obligately proximate to gene action and thus are likely to be substantially less complex than focal performance phenotypes that are far downstream from genes. Many expression phenotypes are highly heritable, reflecting both the extent of genetic variation and the strength of the genetic signal (Göring et al., 2007). To the extent that these phenotypes, or networks of these, are correlated with "higher-order" traits, they offer a powerful tool for dissecting the underlying biology.

Gene expression also provides an intermediate level of resolution between linkage and association. Where linkage represents the observable effect of all functional variation within a relatively broad QTL, an expression *eQTL* may account for the regulatory variation within a specific gene locus. The number of possible targets (~48,000 gene or hypothetical gene products in the human transcriptome) is at least an order of magnitude less than the number of markers in a typical GWAS, simplifying the statistical problem of correcting for multiple tests.

Use the complementary forms of positional information available from linkage and LD. Identification of a linked genomic region, followed by tests for association of measured markers in the region, has provided several notable successes including identification of *TCF7L2* as a candidate gene for diabetes (Grant et al., 2006; Helgason et al., 2007) and identification of likely functional polymorphisms in the selenoprotein S gene *SEPS1* (Curran et al., 2005) and the clotting factor VII gene *F7* (Soria et al., 2005). Similarly, identification of candidate regions by linkage, gene expression, or candidate gene data, followed by QTL-targeted association, has been productively applied to performance traits in the HERITAGE (Health, Risk Factors, Exercise Training, and Genetics) Family Study (Table 3.2). Linkage conditioned on association has been used as evidence that the associated polymorphisms account for part or all of the QTL variance (Almasy & Blangero, 2004; Sun et al., 2002). Association conditioned on linkage may defend against false-positive association due to population structure (Kent et al., 2007). It follows that a strategy of combined linkage and association has the potential to maximize the information in a sample of related individuals, amplifying signal where both lines of evidence agree and reducing noise elsewhere.

We have recently implemented genome-wide joint linkage/association (JLA) in our genetics software package SOLAR (Almasy & Blangero, 1998). Each SNP in the GWA marker set is mapped not only to its physical location in base pairs but also to its genetic position based on a recombination map. Each SNP location can then be tested for the main effect of the SNP genotype and the random effect of allele-sharing in its vicinity, as well as the joint effect of these, providing maximum power for localizing the genetic effect. It has the great benefit of being able to detect the effects of both common and rare functional variants.

Example: Expression Endophenotypes Related to Physical Performance

The San Antonio Family Heart Study is an ongoing genetic investigation of normal variation in cardiovascular and other disease risk in members of about 40 extended Mexican American families (Mitchell et al., 1996). We have performed genome-wide transcriptional profiling on lymphocytes obtained from 1240 individuals (Göring et al., 2007), obtaining about 22,000 heritable gene-centric quantitative measures of mRNA levels. Although absolute levels of gene expression are expected to differ between tissues (e.g., lymphocyte vs. muscle), our approach merely requires that there be correlations across tissue types for successful exploitation. The larger the correlation between tissues within each individual, the greater the utility of using lymphocytes as a surrogate tissue.

Table 3.2 QTL-based association in the HERITAGE Family Study

Phenotype	Locus	Evidence	Genes with associated SNPs*	Reference
Response of cardiac stroke volume to routine exercise	Chr 10p11	Linkage	KIF5B	Argyropoulos et al. (2009)
Insulin/insulin sensitivity response to routine exercise	Chr Xq26	Gene expression in muscle	FHL1	Teran-Garcia et al. (2007)
Change in HDL-cholesterol levels with endurance training	Chr 19q13.2	Candidate gene	APOE	Spielmann et al. (2007)
Change in HDL-cholesterol levels with endurance training	Chr 16q21	Candidate gene	CETP	Spielmann et al. (2007)
Cardiorespiratory fitness response to endurance training	Chr 6p21.2-p21.1	Candidate gene	PPARD	Hautala et al. (2007)

HERITAGE—Health, Risk Factors, Exercise Training, and Genetics.
*Locus-wide $P<0.05$.

Table 3.3 presents selected endophenotypic measures (both mRNA levels and serum protein levels) for genes that have been shown to be important to human performance. As expected, there are substantial differences in the level of evidence for causal genetic factors acting on these proximate phenotypes. For example, about 57% of angiotensin-converting enzyme (ACE) protein level variation is likely due to genetic variation; variation in this gene is of interest as a possible link between exercise and blood pressure (Zhang et al., 2003). Although we find little evidence for genetic factors influencing *IGF1* mRNA levels in this data set, we do find evidence for genetic factors influencing levels of IGF1 protein, a performance candidate because of its numerous effects on human growth and muscular development. Several of the mRNA

phenotypes of great relevance show strong genetic components. For example, the transcription factor A (*TFAM*) gene is critically important in the biogenesis of mitochondria. The heritability for *TFAM* mRNA levels is ~52%. Similarly, transcript levels for uncoupling protein 2 (*UCP2*), a gene implicated in body temperature regulation and resting metabolic rate (Bouchard et al., 1997) appear to also be strongly genetically determined.

Example: Deep Sequencing

We emphasize again that evidence from either linkage or association is merely positional, pointing to a genomic region likely to harbor functional variants. It does not provide the biologically far more exciting result of gene discovery. The next

Table 3.3 Heritability of selected performance-related endophenotypes

Gene symbol	Gene name	Gene aliases	Location	Gene product	Heritability Standard error (SE)
ACE	Angiotensin I converting enzyme 1	*CD143, DCP*	17q23.3	protein	0.567 (0.058)***
ACE	Angiotensin I converting enzyme 1	*CD143, DCP*	17q23.3	RNA isoform 1[a]	0.159 (0.052)**
ACE	Angiotensin I converting enzyme 1	*CD143, DCP*	17q23.3	RNA isoform 2[b]	0.116 (0.047)*
IGF1	Insulin-like growth factor 1	somatomedin C	12q22-q23	protein	0.293 (0.069)***
IGF1	Insulin-like growth factor 1	somatomedin C	12q22-q23	RNA	0.032 (0.038)[c]
IGF1R	Insulin-like growth factor 1 receptor	*CD221*	15q26.3	RNA	0.101 (0.049)*
NRF1	Nuclear respiratory factor 1	*ALPHA-PAL*	7q32	RNA	0.152 (0.051)**
PPARD	Peroxisome proliferators-activated receptor delta	*FAAR, NUC1*	6p21.2-21.1	RNA	0.376 (0.062)***
PPP3R1	Protein phosphatase 3, regulatory subunit B, alpha isoform	*CALNB1, CNB*	2p15	RNA	0.421 (0.050)***
TFAM	Transcription factor A, mitochondrial	*MtTF1, TCF6*	10q21	RNA	0.520 (0.056)***
UCP2	Uncoupling protein 2	*SLC25A8, UCPH*	11q13	RNA	0.499 (0.058)***
UCP3	Uncoupling protein 3	*SLC25A9*	11q13	RNA	0.255 (0.057)***

National Center for Biotechnology Information (NCBI) reference numbers: [a] NM_000789.2; [b] NM_152830.1; [c] Not significant.
*$P < 0.01$; **$P < 0.001$; ***$P < 10^{-6}$.

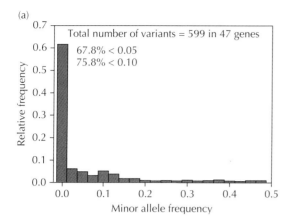

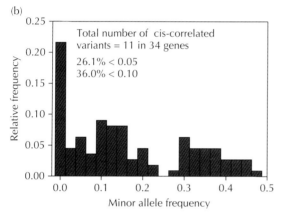

Figure 3.4 Empirical distribution of minor allele frequencies from deep sequencing. (See text for explanation.) (a) Distribution of SNP variants in 192 founder (presumed unrelated) individuals in the San Antonio Family Heart Study. The majority of variants are rare, consistent with theoretical prediction (Figure 3.2). (b) Distribution of the *cis*-associated SNPs that localize to the promoter regions of their respective genes: rare variants comprise a large proportion of variants with detectable genetic effect.

step for any finding of interest is deep sequencing: exhaustive re-sequencing of the localized region in founder individuals to identify all of the genetic variants present in the sample, followed by association tests of new variants and, finally, functional analysis (e.g., promoter binding assays; Curran et al., 2005, 2009) of associated SNPs to identify causal variants and, hence, causal genes.

Extensive sequence information provides direct evidence of the extent and type of genetic variation relevant to quantitative traits. We have recently

seq-uenced promoter regions for 47 diabetes-correlated expression phenotypes in the 192 San Antonio Family Heart Study founders, then genotyped the 599 discovered variants in all 1240 participants. The majority of variants are rare (Figure 3.4a), consistent with theory (Figure 3.2). Approximately 68% exhibited minor allele frequencies less than 0.05 and 75% less than 0.10. If these variants were functional, it would be extremely unlikely to detect them in a conventional GWAS. Not all of these variants are expected to alter gene expression; however, a large proportion of the significantly *cis*-associated SNPs (i.e., associated variants localized to putative regulatory regions of the focal gene) also display low minor allele frequencies (Figure 3.4b). It should be noted that this last analysis is biased *against* low-frequency variants, since association alone may lack power to detect their genetic effects. Therefore, it underestimates the frequency of rare functional variants. Overall, these sequence-based data strongly support the idea that many functional variants are likely to be of relatively low frequency and therefore unlikely to be detectable by association methods alone.

Summary

In summary, the full range of genetic variation responsible for human variability in athletic performance is likely to include both rare and common polymorphisms. The recent history of genetic research into the causes of complex human diseases provides reason for both caution and optimism that many such variants can and will be found— provided that we remain aware of the challenges and use methods capable of sampling all of the genetic information available from human populations.

Acknowledgments

Development of the statistical methods described in this report was supported in part by U.S. National Institute of Mental Health grant MH59490 and carried out in computing facilities funded in part by the AT&T Foundation. Acquisition of lymphocyte expression phenotypes was made possible by the generous support of the Azar/ Shepperd Family Foundation of San Antonio, Texas. The authors declare no competing interests.

References

Almasy, L. & Blangero, J. (1998). Multipoint quantitative-trait linkage analysis in general pedigrees. *American Journal of Human Genetics* 62, 1198–1211.

Almasy, L. & Blangero, J. (2001). Endophenotypes as quantitative risk factors for psychiatric disease: Rationale and study design. *American Journal of Medical Genetics (Neuropsychiatric Genetics)* 105, 42–44.

Almasy, L. & Blangero, J. (2009). Human QTL linkage mapping. *Genetica* 136, 333–340.

Almasy, L. & Blangero, J. (2004). Exploring positional candidate genes: Linkage conditional on measured genotype. *Behavioral Genetics* 34, 173–177.

Amos, C.I. (2007). Successful design and conduct of genome-wide association studies. *Human Molecular Genetics* 16, R220–R225.

Argyropoulos, G. Stütz, A.M., Ilnytska, O., et al. (2009). KIF5B gene sequence variation and response of cardiac stroke volume to regular exercise. *Physiological Genomics* 36, 79–88.

Blangero, J. (2004). Localization and identification of human quantitative trait loci: King Harvest has surely come. *Current Opinion in Genetics & Development* 14, 233–240.

Blangero, J., Williams, J.T. & Almasy, L. (2003). Novel family-based approaches to genetic risk in thrombosis. *Journal of Thrombosis and Hemostasis* 1, 1391–1397.

Boerwinkle, E., Chakraborty, R. & Sing, C.F. (1986). The use of measured genotype information in the analysis of quantitative phenotypes in man. I. Models and analytical methods. *Annals of Human Genetics*, 50, 181–194.

Bouchard, C., Perusse, L., Chagnon, Y.C., Warden, C. & Ricquier, D. (1997). Linkage between markers in the vicinity of the uncoupling protein 2 gene and resting metabolic rate in humans. *Human Molecular Genetics* 6, 1887–1889.

Chakravarti, A. (1999). Population genetics—making sense out of sequence. *Nature Genetics* 21, 56–60.

Clarkson, P.M., Devaney, J.M., Gordish-Dressman, H., et al. (2005). ACTN3 genotype is associated with increases in muscle strength in response to resistance training in women. *Journal of Applied Physiology* 99, 154–163.

Comuzzie, A.G., Funahashi, T., Sonnenberg, G., et al. (2001). The genetic basis of plasma variation in adiponectin, a global endophenotype for obesity and the metabolic syndrome. *Journal of Clinical Endocrinology and Metabolism* 86, 4321–4325.

Curran, J.E., Jowett, J.B.M., Elliott, K.S., et al. (2005). Genetic variation in selenoprotein S influences inflammatory response. *Nature Genetics* 37, 1234–1241.

Curran, J.E., Jowett, J.B., Abraham, L.J., et al. (2009). Genetic variation in *PARL* influences mitochondrial content. *Human Genetics* 127, 183–190.

Delmonico, M.J., Zmuda, J.M., Taylor, B.C., et al. (2008). Association of the ACTN3 genotype and physical functioning with age in older adults. *Journal of Gerontology A: Biological Science & Medical Science* 63, 1227–1234.

Devlin, B. & Roeder, K. (1999). Genomic control for association studies. *Biometrics* 55, 997–1004.

Eynon, N., Duarte, J.A., Oliveira, J., et al. (2009). ACTN3 R577X polymorphism and Israeli top-level athletes. *International Journal of Sports Medicine* 30, 695–698.

Fisher, R.A. (1918). The correlation between relatives on the supposition of Mendelian inheritance. *Transactions of the Royal Society of Edinburgh* 52, 399–433.

Göring, H.H.H., Curran, J.E., Johnson, M.P., et al. (2007). Discovery of expression QTLs using large-scale transcriptional profiling in human lymphocytes. *Nature Genetics* 39, 1208–1216.

Grant, S.F., Thorleifsson, G., Reynisdottir, I., et al. (2006). Variant of transcription factor-7-like 2 (*TCF7L2*) gene confers risk of type 2 diabetes. *Nature Genetics* 38, 320–323.

Gudbjartsson, D.F., Walters, G.B., Thorleifsson, G., et al. (2008). Many sequence variants affecting diversity of adult human height. *Nature Genetics* 40, 609–615.

Guthery, S.L., Salisbury, B.A., Pungliya, M.S., Stephens, J.C. & Bamshad, M. (2007). The structure of common genetic variation in United States populations. *American Journal of Human Genetics* 81, 1221–1231.

Hautala, A.J., Leon, A.S., Skinner, J.S., Rao, D.C., Bouchard, C. & Rankinen, T. (2007). Peroxisome proliferator-activated receptor-delta polymorphisms are associated with physical performance and plasma lipids: the HERITAGE Family Study. *American Journal of Physiology. Heart and Circulatory Physiology* 292, H2498–H2505.

Healy, D.G. (2006). Case-control studies in the genomic era: A clinician's guide. *Lancet Neurology* 5, 701–707.

Helgason, A., Pálsson, S., Thorleifsson, G., et al. (2007). Refining the impact of *TCF7L2* gene variants on type 2 diabetes and adaptive evolution. *Nature Genetics* 39, 218–225.

Kent, J.W. Jr., Dyer, T.D., Göring, H.H.H. & Blangero, J. (2007). Type I error rates in association versus joint linkage/association tests in related individuals. *Genetic Epidemiology*, 31, 173–177.

de Koning, D.J. & Haley, C.S. (2005). Genetical genomics in humans and model organisms. *Trends in Genetics* 21, 377–381.

Lander, E.S. & Schork, N.J. (1994). Genetic dissection of complex traits. *Science* 265, 2037–2048.

Lander, E.S. (1996). The new genetics: Global views of biology. *Science* 274, 536–539.

Lettre, G., Jackson, A.U., Gieger, C., et al. (2008). Identification of ten loci associated with height highlights new biological pathways in human growth. *Nature Genetics* 40, 584–591.

Lewontin, R.C. & Kojima, K. (1960). The evolutionary dynamics of complex polymorphisms. *Evolution* 14, 458–472.

Li, J.Z., Absher, D.M., Tang, H., et al. (2008). Worldwide human relationships inferred from genome-wide patterns of variation. *Science* 319, 1100–1104.

Lucia, A., Gómez-Gallego, F., Santiago, C., et al. (2006). *ACTN3* genotype in professional endurance cyclists. *International Journal of Sports Medicine* 27, 880–884.

Maher, B. (2008). Personal genomes: The case of the missing heritability. *Nature* 456, 18–21.

Manolio, T.A., Collins, F.S., Cox, N.J., et al. (2009). Finding the missing heritability of complex diseases. *Nature* 46, 747–753.

Mitchell, B.D., Kammerer, C.M., Blangero, J., et al. (1996). Genetic and environmental contributions to cardiovascular risk factors in Mexican Americans. The San Antonio Family Heart Study. *Circulation* 94, 2159–2170.

Myles, S., Davison, D., Barrett, J., Stoneking, M. & Timpson, N. (2008). Worldwide population differentiation at disease-associated SNPs. *BMC Medical Genomics* 1, 22.

Norman, B., Esbjörnsson, M., Rundqvist, H., Osterlund, T., von Walden, F. & Tesch, P.A. (2009). Strength, power, fiber types, and mRNA expression in trained men and women with different ACTN3 R577X genotypes. *Journal of Applied Physiology* 106, 959–965.

North, K.N., Yang, N., Wattanasirichaigoon, D., Mills, M., Easteal, S. & Beggs, A.H. (1999). A common nonsense mutation results in alpha-actinin-3 deficiency in the general population. *Nature Genetics* 21, 353–354.

Pastinen, T., Ge, B. & Hudson, T.J. (2006). Influence of human genome polymorphism on gene expression. *Human Molecular Genetics* 15, R9–R16.

Pritchard, J.K. & Cox, N.J. (2002). The allelic architecture of human disease genes: Common disease–common variant … or not? *Human Molecular Genetics* 11, 2417–2423.

Pritchard, J.K. (2001). Are rare variants responsible for susceptibility to common diseases? *American Journal of Human Genetics* 69, 124–137.

Risch, N. & Merikangas, K. (1996). The future of genetic studies of human complex disease. *Science* 273, 1516–1517.

Schork, N.J., Murray, S.S., Frazer, K.A. & Topol, E.J. (2009). Common vs. rare allele hypotheses for complex diseases. *Current Opinion in Genetics & Development* 19, 212–219.

Seabra, A.F., Mendoça D.M., Göring, H.H.H., Thomís, M.A. & Maia, J.A. (2008). Genetic and environmental factors in familial clustering in physical activity. *European Journal of Epidemiology* 23, 205–211.

Simonen, R.L., Perusse, L., Rankinen, T., Rice, T., Rao, D.C. & Bouchard, C. (2002). Familial aggregation of physical activity levels in the Québec Family Study.

Medicine & Science in Sports & Exercise 34, 1137–1142.

Soria, J.M., Almasy, L., Souto, J.C., Sabater-Lleal, M., Fontcuberta, J. & Blangero, J. (2005). The *F7* gene and clotting factor VII levels: Dissection of a human quantitative trait locus. *Human Biology* 77, 561–575.

Spielmann, N., Leon, A.S., Rao, D.C., et al. (2007). CETP genotypes and HDL-cholesterol phenotypes in the HERITAGE Family Study. *Physiological Genomics* 31, 25–31.

Storey, J.D., Madeoy, J., Strout, J.L., Wurfel, M., Ronald, J. & Akey, J.M. (2007). Gene-expression variation within and among human populations. *American Journal of Human Genetics* 80, 502–509.

Sun, L., Cox, N.J. & McPeek, M.S. (2002). A statistical method for identification of polymorphisms that explain a linkage result. *American Journal of Human Genetics*, 70, 399–411.

Teran-Garcia, M., Rankinen, T., Rice, T., et al. (2007). Variations in the four and a half LIM domains 1 gene (FHL1) are associated with fasting insulin and insulin sensitivity responses to regular exercise. *Diabetologia* 50, 1858–1866.

Walsh, S., Liu, D., Metter, E.J., Ferrucci, L. & Roth, S.M. (2008). ACTN3 genotype is associated with muscle phenotypes in women across the adult age span. *Journal of Applied Physiology* 105, 1486–1491.

Weedon, M.N. & Frayling, T.M. (2008). Reaching new heights: Insights into the genetics of human stature. *Trends in Genetics* 24, 595–603.

Weedon, M.N., Lango, H., Lindgren, C.M., et al. (2008). Genome-wide association analysis identifies 20 loci that influence adult height. *Nature Genetics* 40, 575–583.

Williams, K.D., Czerwinski, S.A., Sibedi, J., Jha, B., Blangero, J., Williams-Blangero, S., Towne B. (2008). Physical activity assessment of children from the Jirel ethnic group in eastern Nepal. American Journal of Physical Anthropology, Supp. 46, 222–223.

Wright, S. (1921). Systems of mating. I. The biometric relations between parent and offspring. *Genetics* 6, 111–123.

Wright, S. (1949). Adaptation and selection. In: G.I. Jepsen, G.G. Simpson & E. Mayr (eds.) *Genetics, Palaeontology and Evolution*, pp. 365–389. Princeton University Press, Princeton, NJ.

Wright, S. (1969). *Evolution and the Genetics of Populations. Vol. 2: The Theory of Gene Frequencies*, pp. 345–392. The University of Chicago Press, Chicago, IL.

Yang, N., Garton, F. & North, K. (2009). Alpha-actinin-3 and performance. *Medicine & Sport Science* 5, 88–101.

Yngvadottir, B., Xue, Y., Searle, S., et al. (2009). A genome-wide survey of the prevalence and evolutionary forces acting on human nonsense SNPs. *American Journal of Human Genetics* 84, 224–234.

Zhang, B., Tanaka, H., Shono, N., et al. (2003). The I allele of the angiotensin-converting enzyme gene is associated with an increased percentage of slow-twitch type I fibers in human skeletal muscle. *Clinical Genetics* 63, 139–144.

Chapter 4

Current Proteome Profiling Methods

YETRIB HATHOUT, CATHERINE A. FORMOLO, KRISTY J. BROWN
AND JAVAD NAZARIAN

Research Center for Genetic Medicine, Children's National Medical Center, Washington, DC, USA

Proteins are the most diversified components of a living cell and participate in almost every process within the cell. The term "proteome" or "proteomics" refers to the global study of proteins expressed in a given cell, tissue, or organism under a defined condition and at a particular time. Since its introduction by Wilkins in a conference meeting (Wasinger et al., 1995), proteomics has rapidly expanded and become an integral part of many biochemical and biological studies. Almost all proteomic studies involve the identification of proteins using mass spectrometry. Thanks to major advancements in mass spectrometry instrumentation and bioinformatics tools that occurred during the last decade, proteomics has emerged as an indispensable research tool.

Proteomics is a versatile technology and includes a number of analysis options ranging from simple quantitative analysis (proteome profiling) to more complex studies such as post-translational modifications, protein–protein interactions and pathway analysis. Depending on the goals of a study, complex and sophisticated analytical strategies are needed to generate data. In this chapter, we will focus mainly on current proteome profiling methods with some basics on mass spectrometry and protein separation techniques. To detect and quantify a protein in a complex mixture usually requires that the absolute quantity of the protein be present in a detectable level and its relative abundance in the mixture to fall within the dynamic range covered by the technique used. Proteome coverage has been and remains a major challenge due to the inherent complexity and dynamic range of protein mixtures in biological extracts. With advances in protein separation techniques and mass spectrometry instrumentation, proteome profiling had become a fairly standardized technology but still laborious and slow compared to gene expression profiling (e.g., cDNA microarray).

Proteomics in Muscle Biology

Muscle is becoming an attractive target for proteomics studies owing to its exquisite physiological adaptation to exercise and its association with a number of metabolic disorders and other diseases such as muscular dystrophies and age-related sarcopenia. About 416 articles dealing with muscle proteomics have been published to date. Chapter 29 (Hittel) later in this volume describes a number of proteomic technologies applied to exercise biology. In addition to these, a wealth of information continues to be uncovered in the muscle proteome and is helping elucidate the molecular mechanisms underlying muscle function in health and disease (Flueck, 2009; Guelfi et al., 2006; Hittel et al., 2007; Kavazis et al., 2009; Lanza et al., 2008).

Muscle fibers are remarkably dynamic cells and resist tremendous amounts of stress and tears without collapsing, making muscle an interesting tissue to study (Hoffman, 2003). However, the study of muscle proteome and especially that of skeletal muscle has been challenging due to the wide range

Genetic and Molecular Aspects of Sport Performance, 1st edition.
Edited by Claude Bouchard and Eric P. Hoffman.
Published 2011 by Blackwell Publishing Ltd.

in abundances of contractile proteins and other functional proteins. Therefore, such studies often employ subcellular fractionation and protein separation techniques to reduce the sample complexity and narrow the focus of the investigation.

Protein Separation Methods

A typical proteomic experiment requires the extraction of proteins from control and experimental samples followed by processing for mass spectrometry analysis. Each sample usually contains thousands of proteins of different size and abundance therefore

requiring extensive fractionation and separation before mass spectrometry analysis. Two methods are commonly used in proteomics to separate complex protein mixtures. The two-dimensional gel electrophoresis (2-DE) based approach and the shotgun approach (Figure 4.1). In both the methods, mass spectrometry is required for protein identification and quantitation.

Introduction to Mass Spectrometry

Mass spectrometry is a technique that measures masses of molecules and macromolecules. A mass

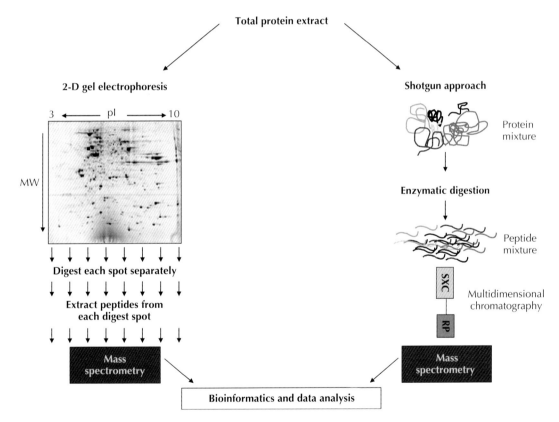

Figure 4.1 Commonly used separation methods in proteomics. In 2-D gel electrophoresis total proteins are first separated according to their isoelectric points (pI's) using an immobilized pH gradient strip (pH 3–10). The strip is then placed at the top of an SDS-PAGE gel to separate proteins based in their molecular masses (MW). The proteins are then visualized by a staining solution resulting in a pattern of spots on the gel. Each spot is then excised, digested with trypsin and the resulting peptides are extracted and analyzed by mass spectrometry. In the shotgun approach the whole protein mixture is digested by trypsin resulting in an even more complex peptide mixture. A thousand proteins will generate about 20,000–30,000 peptides. This complex peptide mixture is then fractionated by liquid chromatography (LC) and continuously introduced into the mass spectrometer for analysis.

spectrometer is composed of three components linked in tandem: an ion source, a mass analyzer, and a detector (Figure 4.2). In order for mass spectrometers to detect proteins or fragments of proteins (peptides), the peptide must be given a charge and then fly through an oppositely charged vacuum chamber for detection. In this manner, the mass spectrometer determines the mass-to-charge ratio (*m/z*) of an ion, and derives the mass of the peptide from this ratio. Ionization of molecules occurs in the ion source by either subtracting or adding electrons, protons, cations, anions, or a chemical moiety (for more details, see the book by Gary Siuzdak, 2006). In the case of peptide and protein analysis, the ionization method of choice is often protonation (e.g., addition of protons).

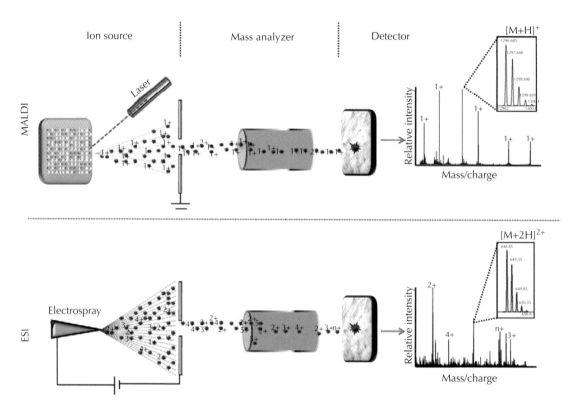

Figure 4.2 Basics of matrix-assisted laser desorption ionization (MALDI) and electrospray ionization (ESI) techniques. In MALDI, peptide solution (few picomoles) is mixed with a saturated matrix solution, 50 mM (a UV-absorbing component such alpha cyano hydroxicinnamic acid or dihydroxybenzoic acid) and deposited on a MALDI plate to form thin crystal layer. A laser beam is then directed toward these crystals resulting in the formation of a plum of ionized peptides. MALDI generates mainly singly charged ions $[M+H]^+$ which are then ejected into the mass analyzer via −20 kV differential potential. Often a time-of-flight (TOF) is used to resolve ions generated by MALDI. Light ions will travel faster than heavy ions and will reach the detector first. The instrument uses the TOF of ions to deduce their exact masses and a spectrum is recorded at each laser pulse. In ESI, peptides are desorbed from a liquid solution (usually acidic solution composed of organic compounds and water). The peptide solution is infused through a tiny capillary needle. A 2 to 4 kV potential is applied to the needle resulting in the formation of a spray with a very fine droplets containing charged peptide molecules. The water around peptide molecule evaporates leaving multiply protonated peptide molecules in the gaze phase. The charged peptides are then introduced into the mass analyzer and resolved based on their masses and charges. With ESI often other kind of mass analyzers are used (Table 4.1).

There are two main ionization techniques used in this case: matrix-assisted laser desorption ionization (MALDI) and electrospray ionization (ESI) (Figure 4.2). The invention of these ionization techniques has revolutionized the field of mass spectrometry by allowing analysis of large molecules for the first time and by contributing to major progress in the field of protein chemistry. Hence, two Nobel Prizes in chemistry were awarded in 2002 (one to Dr Koichi Tanaka for MALDI invention and one to Dr John B. Fenn for ESI invention). While MALDI generates mostly singly charged ions, ESI often generates multiply charged ions. Thus a peptide with neutral molecular mass of 1295.6 Da will be detected at m/z of 1296.6 if it is singly protonated, at m/z of 648.8 if it is doubly protonated, and at m/z of 432.8 if it is triply protonated. Ions are introduced into the mass analyzer via a differential potential and resolved depending on their kinetic energy, size, and charge.

There are several mass analyzers that can be used in conjunction with these ion sources, and these are listed in Table 4.1 with their specifications and characteristics. A good mass analyzer should provide high resolution and high accuracy of measurement. Resolution is defined by the ability of an instrument to detect the different ionic m/z values as separate peaks, and accuracy is defined by the difference between the measured mass and the theoretical mass. For example, angiotensin I can be detected as single, broad peak with a low-resolution mass spectrometer and as a series of peaks separated by one mass unit with high-resolution instruments (Figure 4.3). Currently most modern instruments provide resolution and accuracy sufficient for proteomic studies. An instrument with 10,000 resolution will be able to resolve an ion with m/z of 1000 from an ion with m/z of 1000.1 Da.

Besides measuring molecular masses of peptides, some mass spectrometers can also perform fragmentation of the peptides and further characterize their sequences. This technique is called tandem mass spectrometry or MS/MS. Individual peptides are selected among others using off-resonance electrical and/or magnetic fields and are fragmented in a collision cell, using an inert gas such as helium, argon, or simply air. The resulting fragments are then detected by the detector and recorded. Peptides tend to fragment along the amide backbone resulting in spectral peaks that differ by the mass of an amino acid (Figure 4.4). This sequence data with the mass of the peptide is used to determine the identity of the parent protein.

Quantitative Proteomics and Proteome Profiling

The main goal of many proteomics studies is to measure protein abundances in samples representing different biological states or conditions. Quantitation using the 2-DE approach is quite straightforward

Table 4.1 Commonly used mass analyzers

Ion source	Mass analyzer	Resolution	Accuracy (ppm)
MALDI	Linear time-of-flight (TOF)	300	1000
MALDI	Reflectron TOF	7000–14,000	50–100
ESI	Ion trap	100–200	500
ESI or MALDI	Quadrupole time-of-flight (QTOF)	up to 17,000	10–50
ESI	Ion trap-Orbitrap	up to 60,000	>10
ESI	Fourier transform ion cyclotron resonance (FTICR)	up to 500,000	>3

MALDI, matrix-assisted laser desorption ionization; ESI, electrospray ionization; ppm, part per million.

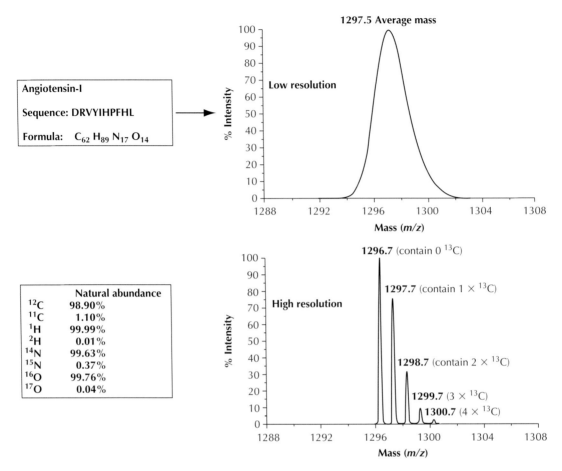

Figure 4.3 Low-resolution and high-resolution mass spectra of angiotensin-I peptide. With a low-resolution mass spectrometer this peptide is detected as a single broad peak while with a high-resolution mass spectrometer the peptide is detected as series of five peaks "isotopic distribution" reflecting the abundance of natural isotopes in the peptide. The monoisotopic peak, usually the first peak from the left, compromises only ^{12}C, but no ^{13}C, while the adjacent peaks contain 1, 2, 3, and 4 ^{13}C, respectively.

and utilizes densitometer measurements of spot intensities. With the shotgun approach, it becomes more complex because of the large number of generated peptides and less mapping to the same parent protein; shotgun quantitative proteomics thus requires sophisticated strategies. In 2-DE, thousands of proteins can be displayed in one gel slab, providing a simplistic way to simultaneously compare relative protein quantities between two or more samples using computer-assisted image analysis software of the 2-D gel images (Rabilloud, 2002; Zhan & Desiderio, 2003). Measurement of a

differentially expressed protein across samples is based on the quantitation of the relative intensity of each spot. Proteins of interest are then identified by a combination of in-gel digestion, mass spectrometry analysis, and database searching (Jensen et al., 1999). A protein spot is excised from the gel, digested with an enzyme (e.g., trypsin), and the resulting peptides are analyzed by mass spectrometry often using MALDI time-of-flight (TOF). Each protein should provide a specific peptide mass fingerprint that can be searched against a protein database using search engine tools such as Mascot, PepMapper, Profound,

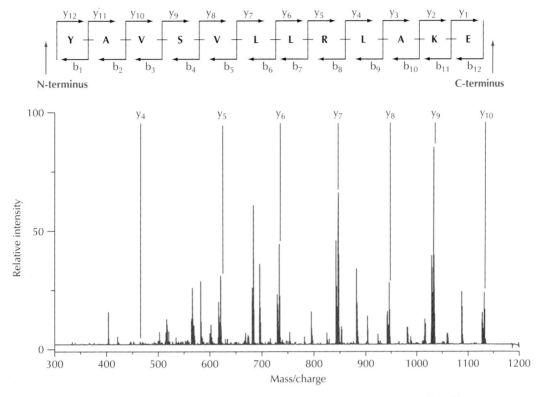

Figure 4.4 A tandem mass spectrum of a single peptide showing fragmentation along the amide backbone. Fragment ions containing the N-terminus are termed "b-ions" while C-terminus containing ions are termed "y-ions." The mass difference between consecutive b- or y-ions corresponds to an amino acid.

or ProteinProspector (see http://au.expasy.org/). Usually peptide masses without MS/MS are enough to identify a protein because of the specificity of the mass fingerprint. While 2-DE can provide quick information about the protein pattern in a given sample, some proteins (e.g., low-abundance proteins, proteins that show a very high net positive or negative charge (pI), or particularly large or small molecular weight (masses)) are not represented. As a result, considerable efforts have been devoted to develop new technical strategies for comprehensive and deeper proteome profiling.

Shotgun Proteome Profiling

In shotgun proteomics, the whole protein mixture is digested by trypsin, resulting in a complex peptide mixture of many thousands of different peptides derived from the different proteins in the original mixture. The peptide mixture is fractionated or separated by chromatography and continuously introduced into the mass spectrometer via ESI. The mass spectrometer takes an MS survey of the eluting peptides followed by MS/MS fragmentation of individual peptides. The cycle can be repeated throughout the chromatographic run. Modern mass spectrometers, such as quadruple TOF and linear ion traps, can generate 10,000–20,000 MS/MS spectra per hour, which is compatible with the chromatographic timescale resulting in comprehensive proteome coverage. In this case, each peptide is identified by its correct mass and partial sequence information (Figure 4.4).

Usually two or more peptides are sufficient to identify a protein in a mixture. However, quantitation of proteins using this approach is not straightforward and has proven to be challenging for several reasons. First, mass spectrometry is inherently not quantitative meaning that the intensity of a peptide ion depends more on its amino acid content and length rather than its absolute quantity. Second, lack of reproducibility between liquid chromatography tandem mass spectrometry (LC-MS/MS) runs renders comparative analysis between samples difficult. Indeed, there is often poor overlap in identified proteins between replicate samples and even between replicate LC-MS/MS runs of the same sample. This is especially the case with low-abundance proteins. These proteins are sometimes detected and sometimes not. To circumvent these obstacles, samples to be compared are usually paired for analysis within the same LC-MS/MS run using differential stable isotope-labeling techniques (Bantscheff et al., 2007; Julka & Regnier, 2004). In this approach, proteins or peptides in control and experimental pools are labeled with light and heavy isotope tags respectively then mixed together in one LC-MS/MS run. The light and heavy peptide pairs co-elute from the chromatographic column but their masses are resolved by the mass spectrometer thus allowing measurement of their respective intensities (and therefore their quantities) in the control versus experimental samples. Labeling with light and heavy tags can be achieved at the peptide or protein level via enzymatic or chemical reaction or at the cellular level via metabolic labeling (Aebersold & Mann, 2003).

Chemical Tagging

The overall concept behind this strategy is to label the control sample with a "light" tag and the experimental sample with a "heavy" tag or vice versa (Figure 4.5). The light and heavy tags have similar chemical structure but contain light and heavy stable isotopes respectively such as ^{12}C and ^{13}C, ^{14}N and ^{15}N, and 16O and ^{18}O. Thus, homologous peptides in control sample and experimental samples labeled with light and heavy tag respectively will co-elute in the LC and will be detected as pairs of heavy and light peptides in the mass spectrometer. The differential mass between light and heavy peptides depends on the number and mass difference of isotopes in the tag. The intensity ratio of heavy-to-light peptide will reflect the amount of the corresponding protein in experimental versus control samples.

Several tags have been designed and developed in the last decade (Julka & Regnier, 2004). The tags are usually targeted against a specific amino acid or reactive group in the protein or peptide.

Isotope-coded affinity tag or ICAT (Gygi et al., 1999) consists of a thiol-reactive group linked to a biotin moiety. The ICAT reagent specifically reacts with cysteine residues in the proteins. Typically, control and experimental samples are labeled with the light and heavy ICAT reagent, respectively. The samples are then mixed at 1:1 ratios and digested with trypsin, and the resulting peptides passed through an avidin affinity column to pull-down ICAT derivitized peptides. In the original ICAT reagent, the linker between the thiol-reactive group and the biotin moiety contained either eight hydrogen atoms for the light form or eight deuterium atoms for the heavy form resulting in a mass difference of 8 Da. The use of deuterium atoms as a labeling isotope resulted in a slight shift in the elution between the peptide pairs during chromatography and thus affected accurate quantitation by mass spectrometry (Zhang et al., 2002). Hence, deuterium-labeled ICAT reagent has been replaced by a new generation ICAT reagent in which the linker is labeled with ^{12}C and ^{13}C instead of hydrogen and deuterium. In this case, the heavy and light peptides co-elute perfectly and resolve in the mass spectrometer. The peptides are then identified by MS/MS, and their intensity ratio is used to determine the relative amount of the corresponding protein in the control sample versus experimental sample.

Isobaric tag for relative and absolute quantitation (iTRAQ) is another widely used strategy in proteome profiling (Zieske, 2006). This strategy is based on the labeling of tryptic peptides with tags of equal masses that generate specific reporter ions (e.g., specific masses) when fragmented in the mass spectrometer (e.g., MS/MS analysis). The tag compromises an amine-reactive group that is connected to the reporter group via a linker or balance.

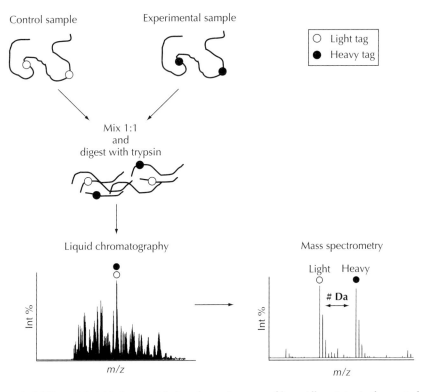

Figure 4.5 Concept of differential stable isotope labeling for proteome profiling. All proteins in the control sample are labeled with a light tag while all proteins in the experimental samples are labeled with a heavy tag. The samples are then mixed at 1:1 ratio and processed for LC-MS/MS analysis. Corresponding peptides in the two biological samples will co-elute in the LC because they have similar amino acid sequence but will be resolved by the mass spectrometer due to the light and heavy isotopic tags. The intensities of peptide pairs will reflect their relative amount in the control and experimental sample.

There are currently two sets of iTRAQ reagents: the 4-plex and the 8-plex reagents constituted of four and eight tags, respectively. While each reporter ion has a unique mass, the balance group counters this mass such that all iTRAQ tags will have similar masses. For example, in the 4-plex reagent, the four tags have reporter masses of 114, 115, 116, and 117 Da linked to the amino-reactive group with linker balances of 31, 30, 29, and 28 Da, respectively. Thus a peptide belonging to the same protein across 4 different samples will be detected as a single ion but when this ion is fragmented, the reporter masses fall off the peptide and are detected in low mass range. The relative intensity of these reporter ions reflects the relative amount of the peptide in each of the four samples. One advantage of iTRAQ strategy is that it can be used to compare up to eight samples at once thus reducing analysis time while allowing high-throughput analysis. However, this kind of analysis requires careful sample preparation and a compatible mass analyzer that can scan lower masses in MS/MS mode.

Stable Isotope Labeling via Enzymatic Reaction

This strategy takes advantage of trypsin hydrolysis to incorporate heavy isotopes at the C-termini of peptides. One of the samples is digested with trypsin in ^{18}O-enriched water ($H_2{}^{18}O$, 99%) while the comparable sample is digested in regular water ($H_2{}^{16}O$). During enzymatic digestion, trypsin uses molecule of water to hydrolyze the amide bonds resulting in the incorporation of oxygen atoms at the C-termini of tryptic peptides. Homologous peptides generated from the same proteins in sample digested in the presence of $H_2^{16}O$ or $H_2^{18}O$ will have mass differences of 4 Da due to the incorporation of two atoms of ^{16}O or ^{18}O at the C-terminal end of each tryptic peptide. These pairs of peptides will again co-elute in the LC but their masses will be resolved by the mass spectrometer, allowing quantitative analysis. The method was first used to compare the proteome of two serotypes of adenovirus (Yao et al., 2001) and quickly become popular and is currently implemented in a number of basic and clinical studies (Fenselau & Yao, 2009; Nazarian et al., 2008).

Metabolic Labeling

Metabolic labeling by stable isotopes is the most systemic way to uniformly label all the proteins within a cell. Microorganisms or cells use exogenous carbon, nitrogen, or amino acids to synthesize new proteins. If these nutrients are replaced with ^{13}C- and/or ^{15}N-containing nutrients, then these isotopes will be incorporated into the newly synthesized proteins resulting in a shift of their molecular masses. Mass spectrometry analysis of a mixture of protein extracts from labeled and unlabeled cells will result in the detection of pairs of light and heavy peptides, thus facilitating measurement of protein abundance between pairs of samples. The metabolic labeling strategy was first tested on yeast using ^{15}N uniformly labeled glucose (Oda et al., 1999), then extended to mammalian cells using stable isotope-labeled amino acids (Ong et al., 2002) and even to whole organisms such as *Caenorhabaditis elegans, Drosophila melanogaster*, and mouse (Krijgsveld et al., 2003; Kruger et al., 2008). Metabolic labeling with stable isotopes is currently the most accurate strategy in quantitative proteome profiling because labeled and unlabeled cells can be mixed before protein extraction and processing; thus variations that would result from sample handling are minimized (Figure 4.6).

Label-free Proteome Profiling

Although differential stable isotope-labeling strategies mentioned above are the method of choice to generate accurate data, most of these strategies are costly and require complex experimental procedures. Tremendous efforts continue to develop and troubleshoot label-free proteome profiling methods. Because of the complexity of samples in shotgun proteomics and the large variability in LC-MS/MS runs, matching homologous peptides across individual sample runs and monitoring their retention times and intensities has been challenging. However, with the development of reproducible nanoscale chromatography and sensitive mass spectrometers, the reproducibility and stability of LC-MS/MS runs has increased. Thus, homologous peptides across individual runs can be more easily assigned and compared.

Two main approaches were and are still being evaluated for label-free proteomics: spectral counting (SC) and mass spectrometry–based chromatogram alignments. SC calculates protein abundances by counting the number of MS/MS spectra assigned to each protein, assuming that the number of observed peptides directly correlates with the abundance of the assigned protein (Old et al., 2005). In mass spectrometry–based chromatogram alignments, homologous peptides are matched across multiple samples based on their retention time, their molecular masses, and their charges. Once matched, the intensities of these peptides are used for quantification.

There are several advantages to using label-free methods. First, there is a monetary incentive, as label-free assays tend to cost much less compared to stable isotope-labeling techniques. Secondly, and more importantly, while stable isotope-labeling strategies are capable of analyzing a limited number of samples, the label-free method permits analysis and comparison of an unlimited number of samples. The third advantage of the label-free method is its capability to measure both relative and absolute abundances of different proteins in different runs (Lu et al., 2007; Old et al., 2005).

The downfalls of the label-free method are also twofold. First, a slight variability between runs that goes undetected (due to sample preparation or technical issues) can have profound effects on experimental outcome. To prevent this, a known amount of specific exogenous peptides or proteins are spiked into samples and used for data normalization.

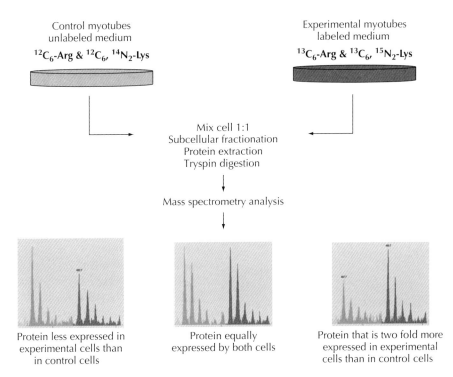

Control myotubes
unlabeled medium
$^{12}C_6$-Arg & $^{12}C_6$, $^{14}N_2$-Lys

Experimental myotubes
labeled medium
$^{13}C_6$-Arg & $^{13}C_6$, $^{15}N_2$-Lys

Mix cell 1:1
Subcellular fractionation
Protein extraction
Tryspin digestion

Mass spectrometry analysis

Protein less expressed in
experimental cells than
in control cells

Protein equally
expressed by both cells

Protein that is two fold more
expressed in experimental
cells than in control cells

Figure 4.6 Stable isotope labeling by amino acid in cell culture (SILAC). Control cells are grown in regular medium containing light $^{12}C_6$-Arg and $^{12}C_6$-Lys while experimental cells are grown in medium containing heavy $^{13}C_6$-Arg and $^{13}C_6$-Lys. The heavy and light amino acids are incorporated into all the proteins within the cells thus resulting in differential mass of 6 Da between homologous peptides generated from control and experimental samples. Peptides will be detected as pairs of heavy and light ions, and the ratio of intensities will reflect the relative quantity in the two samples. Individual peptides are then fragmented in the mass spectrometer to identify the corresponding protein.

The second disadvantage is that the true limit to label-free profiling depends on software and hardware availability and capability. Because the label-free method accounts for every peptide in each LC-MS/MS run, accumulation of multiple runs causes an exponential increase in the number of peptides to be analyzed. As such, a handful of runs will generate millions of peptides. Analyses of their expression values and comparison to the expression values of other peptides require sophisticated software and high-power hardware. At the current time, only few software programs exist for label-free protein profiling.

Bioinformatics in Proteomics

Bioinformatics has been and still is an indispensable tool in proteomics. It provides support at every step, from identification of proteins to quantitation and pathway analysis. Shotgun proteomics as described above can generate large files of raw mass spectrometry data (150 MB to 250 MB per LC-MS/MS run) resulting in gigabytes of data for any given experiment. The raw data is then searched against protein databases using specialized search engines and goes through several computational steps before obtaining a list of identified proteins. These steps include signal-to-noise assessment, peak detection, *in silico* simulation, and mass spectra data correlation. Three major databases are often used in proteomics: SwissProt/UniProt database, the National Center for Biotechnology and Information (NCBI) database, and the International Protein Index (IPI) database. All contain well-annotated protein sequences with valuable information, such as subcellular localization, post-translational modifications, and potential function. Protein identification has become a highly automated process using integrated software packages that contain useful tools for data filtration and visualization. The only required input from the user is to set certain parameters before starting the search, such as taxonomy of the sample, type of database, type of enzyme used to process sample, and, most importantly, the mass tolerance or mass accuracy of the instrument. The search engine uses all these parameters to search the database and generate a list of identified proteins ranked from the most to the least probable proteins. The users then set the filtration criteria such as peptide score, peptide probability–based score, and number of peptides per protein that must be identified to generate a confident list of proteins. Usually peptide scores are enough to filter the data. These scores indicate how closely the experimental mass spectrum and theoretical mass spectrum are correlated.

While software for protein identification has been well developed, computational tools for quantitative proteomics are still evolving. In high-throughput proteome profiling, the use of stable isotope-labeling or label-free methods generates thousands and thousands of peaks that need to be assigned and matched to each other. Usually this is done through sophisticated software that merges the list of identified peptides and the corresponding raw data. Each identified peptide can be traced back to the raw data using its retention time and scan number. This step is often the bottleneck in quantitative proteomic studies, but much improvement has been made recently, and true high-throughput processing is attainable in the near future.

Summary

About two decades ago, protein identification by mass spectrometry required laborious and time-consuming experiments to achieve. Nowadays, with the improvement of protein separation methods and mass spectrometry instrumentation, proteomics has become a fairly routine and accessible technology to a large category of researchers. Unfortunately, even though major advances have been made in the mass spectrometry instrumentation in terms of speed and sensitivity, sample preparation and separation still remain a laborious step in proteomics. There is no single standardized technique in proteome profiling. Depending on the goals of the study, specialized sample preparation in conjunction with mass spectrometry is often required.

Proteome profiling using 2-DE mass spectrometry–based strategy is still the gold standard in many studies and works very well with soluble proteins (Rabilloud, 2002). However, the technique has limited use in respect to proteome coverage, detection of low-abundance proteins, and separation of membrane proteins. As a result, alternative approaches combining shotgun proteomics, stable isotope-labeling, and label-free strategies have been developed. With shotgun proteomics, thousands of proteins are identified and quantified compared to a few hundred with 2-DE. However, unlike gene expression profiling (e.g., cDNA microarray) which is a mature and well-standardized technique, proteomics is still an emerging technology with much room for improvement and standardization. Perhaps one day, a universal and simple method for proteome profiling will be accessible to every laboratory. Until then, methods and instruments continue to make great progress in the proteomics field. A number of studies have already started using proteome profiling to understand the biology of human physical performance. With the development of new cutting-edge proteome profiling techniques and easy to use mass spectrometry instruments, the number of such studies will likely grow in the next years and thus increase our understanding of the exercise physiology. Defining molecular markers and pathways associated with physical performance will be valuable tool to accurately measure and monitor functional outcome of physical exercise in health and disease conditions.

Chapter 29 describes the application of some of the above proteomics techniques toward human physical performance. Proteomics continues to require fairly large amounts of tissue (often grams), and this precludes use of human clinical biopsies in training interventions or elite athletes.

Also, the most accurate quantitative proteomics requires cell culture models (e.g., SILAC), and *in vitro* models of muscle training do not yet exist. Despite these shortcomings, proteomic studies of rodent models of training are emerging, and human muscle culture studies of cell metabolism are shedding light on metabolic syndrome (Chapter 29).

References

Aebersold, R. & Mann, M. (2003). Mass spectrometry-based proteomics. *Nature* 422, 198–207.

Bantscheff, M., Schirle, M., Sweetman, G., Rick, J. & Kuster, B. (2007). Quantitative mass spectrometry in proteomics: A critical review. *Analytical and Bioanalytical Chemistry* 389, 1017–1031.

Fenselau, C. & Yao, X. (2009). $^{18}O_2$-labeling in quantitative proteomic strategies: A status report. *Journal of Proteome Research* 8, 2140–2143.

Flueck, M. (2009). Plasticity of the muscle proteome to exercise at altitude. *High Altitude Medicine & Biology* 10, 183–193.

Guelfi, K.J., Casey, T.M., Giles, J.J., Fournier, P.A. & Arthur, P.G. (2006). A proteomic analysis of the acute effects of high-intensity exercise on skeletal muscle proteins in fasted rats. *Clinical and Experimental Pharmacology & Physiology* 33, 952–957.

Gygi, S.P., Rist, B., Gerber, S.A., Turecek, F., Gelb, M.H. & Aebersold, R. (1999). Quantitative analysis of complex protein mixtures using isotope-coded affinity tags. *Nature Biotechnology* 17, 994–999.

Hittel, D.S., Hathout, Y. & Hoffman, E.P. (2007). Proteomics and systems biology in exercise and sport sciences research. *Exercise and Sport Science Reviews* 35, 5–11.

Hoffman, E.P. (2003). Desminopathies: Good stuff lost, garbage gained, or the trashman misdirected? *Muscle Nerve* 27, 643–645.

Jensen, O.N., Wilm, M., Shevchenko, A. & Mann, M. (1999). Sample preparation methods for mass spectrometric peptide mapping directly from 2-DE gels. *Methods in Molecular Biology* 12, 513–530.

Julka, S. & Regnier, F. (2004). Quantification in proteomics through stable isotope coding: A review. *Journal of Proteome Research* 3, 350–363.

Kavazis, A.N., Alvarez, S., Talbert, E., Lee, Y. & Powers, S.K. (2009). Exercise training induces a cardioprotective phenotype and alterations in cardiac subsarcolemmal and intermyofibrillar mitochondrial proteins. *American Journal of Physiology. Heart and Circulatory Physiology* 297, H144–H152.

Krijgsveld, J., Ketting, R.F., Mahmoudi, T., et al. (2003). Metabolic labeling of *C. elegans* and *D. melanogaster* for quantitative proteomics. *Nature Biotechnology* 21, 927–931.

Kruger, M., Moser, M., Ussar, S., et al. (2008). SILAC mouse for quantitative proteomics uncovers kindlin-3 as an essential factor for red blood cell function. *Cell* 134, 353–364.

Lanza, I.R., Short, D.K., Short, K.R., et al. (2008). Endurance exercise as a countermeasure for aging. *Diabetes* 57, 2933–2942.

Lu, P., Vogel, C., Wang, R., Yao, X. & Marcotte, E.M. (2007). Absolute protein expression profiling estimates the relative contributions of transcriptional and translational regulation. *Nature Biotechnology* 25, 117–124.

Nazarian, J., Santi, M., Hathout, Y. & MacDonald, T.J. (2008). Protein profiling of formalin fixed paraffin embedded tissue: Identification of potential biomarkers for pediatric brainstem glioma. *Proteomics—Clinical Applications* 2, 915–924.

Oda, Y., Huang, K., Cross, F.R., Cowburn, D. & Chait, B.T. (1999). Accurate quantitation of protein expression and site-specific phosphorylation. *Proceedings of the National Academy of Sciences U S A* 96, 6591–6596.

Old, W.M., Meyer-Arendt, K., Aveline-Wolf, L., et al. (2005). Comparison of label-free methods for quantifying human proteins by shotgun proteomics. *Molecular Cell Proteomics* 4, 1487–1502.

Ong, S.E., Blagoev, B., Kratchmarova, I., et al. (2002). Stable isotope labeling by amino acids in cell culture, SILAC, as a simple and accurate approach to expression proteomics. *Molecular Cell Proteomics* 1, 376–386.

Rabilloud, T. (2002). Two-dimensional gel electrophoresis in proteomics: Old, old fashioned, but it still climbs up the mountains. *Proteomics* 2, 3–10.

Siuzdak, G (2006). Ionization and the mass spectrometer. In: G. Siuzdak (ed.) *The Expanding Role of Mass Spectrometry in Biotechnology*, 2nd edn, MCC Press, San Diego, CA.

Wasinger, V.C., Cordwell, S.J., Cerpa-Poljak, A., et al. (1995). Progress with gene-product mapping of the Mollicutes: *Mycoplasma genitalium*. *Electrophoresis* 16, 1090–1094.

Yao, X., Freas, A., Ramirez, J., et al. (2001). Proteolytic ^{18}O labeling for comparative proteomics: Model studies with two serotypes of adenovirus. *Analytical Chemistry* 73, 2836–2842.

Zhan, X. & Desiderio, D.M. (2003). Spot volume vs. amount of protein loaded onto a gel: A detailed, statistical comparison of two gel electrophoresis systems. *Electrophoresis* 24, 1818–1833.

Zhang, R., Sioma, C.S., Thompson, R.A., Xiong, L. & Regnier, F.E. (2002). Controlling deuterium isotope effects in comparative proteomics. *Analytical Chemistry* 74, 3662–3669.

Zieske, L.R. (2006). A perspective on the use of iTRAQ reagent technology for protein complex and profiling studies. *Journal of Experimental Botany* 57, 1501–1508.

Chapter 5

Bioinformatics and Public Access Resources

YUE WANG,[1] HUAI LI,[2] DAVID J. MILLER[3] AND JIANHUA XUAN[1]

[1]Bradley Department of Electrical and Computer Engineering, Virginia Polytechnic Institute and State University, Arlington, VA, USA
[2]Bioinformatics Unit, National Institute on Aging, National Institutes of Health, Baltimore, MD, USA
[3]Department of Electrical Engineering, Pennsylvania State University, University Park, PA, USA

High-throughput molecular profiling technologies provide powerful methods through which systems biology can be used to address important issues in biology and medicine. These technologies are often used to identify genomic and epigenetic markers that may have a functional role in specific phenotypes. These technologies present scientists with the task of extracting meaningful statistical and biological information from high-dimensional data spaces, wherein each sample is defined by a very large number of measurements, usually concurrently obtained. The emerging field of bioinformatics applies principles of information sciences and technologies to make the vast, diverse, and complex life sciences data more understandable and useful. In this chapter, we briefly review the high-throughput experimental platforms commonly used for generating highly parallel DNA, mRNA, and protein data sets. We then describe some of the key computational steps for sensitive and reliable interpretation of the data (e.g., signal processing), and subsequent integration of the processed data into biological context (networks and pathways). We review the development and application of computational tools and approaches for expanding the use of biological, medical, behavioral, or health data. Finally, we describe current efforts to bring large data sets into the public domain, for access and analysis by scientists worldwide.

High-Throughput Molecular Profiling and Platforms

Gene Expression Microarrays

A gene expression microarray is a glass or silicon chip attached with an arrayed series of thousands of microscopic spots of DNA oligonucleotides, each containing picomoles of a specific DNA sequence. After hybridization with fluorophore-labeled complementary DNA (cDNA) reversely transcribed from mRNA, the intensity of fluorescent light reflects the abundance of cDNA sequences and thus indicates the abundance of mRNA sequences in the original mRNA sample. Gene expression microarrays simultaneously measure the expression levels of thousands of genes and produce a vast amount of data. Gene expression data contains significant noise that exists in both the biological system to be studied and the different stages for assaying gene expression. Gene expression microarrays can be used to produce expression levels of both static populations and time-course samples.

Single-Nucleotide Polymorphisms

Single-nucleotide polymorphisms (SNPs) are DNA sequence variations that occur when a single nucleotide in the genome sequence is altered.

Genetic and Molecular Aspects of Sport Performance, 1st edition.
Edited by Claude Bouchard and Eric P. Hoffman.
Published 2011 by Blackwell Publishing Ltd.

A high-throughput SNP array measures the hybridization of sample DNA to the templates of two different alleles for each SNP, and the relative hybridization strength is used to call genotypes. Heterozygous SNPs have balanced hybridization for each allele, while homozygous SNPs tend to hybridize to one allele type. Several commercial SNP chips, including Affymetrix GeneChips™ and Illumina BeadChips, can genotype over one million SNPs. Rare variants that can be detected by deep sequencing have relatively lower frequency compared to SNPs; both SNPs and the summation of rare variants may contribute to significant associations with common traits.

Copy Number Variation

The same experimental technology utilized to generate genotype data for SNPs can also output gene dosage (copy number) across the entire genome in a single experiment. DNA copy number variations (CNVs) refer to copy number changes of DNA segments, compared to a reference genome. CNVs are a highly prevalent form of structural variation in human genomes, and can cover from a few thousand to several million nucleotide bases (McCarroll & Altshuler, 2007). The copy numbers of DNA segments can be measured using high-throughput SNP arrays and oligonucleotide-based Comparative Genomic Hybridization (CGH), where a DNA sample is digested to snippets that are amplified and hybridized to a large number of probes, corresponding to different genome loci. The hybridization intensities, proportional to copy numbers, are arranged in a one-dimensional signal profile according to the genomic locations of the probes.

ChIP-on-chip

ChIP-chip (or ChIP-on-chip) is a technology for measuring *in vivo* the direct physical interaction between specific DNA binding proteins (e.g., transcriptional factors) and their target genes. ChIP-on-chip combines chromatin immunoprecipitation (*ChIP*), which is a method for isolating DNA fragments that are bound by specific DNA

binding proteins, with DNA microarray technology (*chip*) for measuring the concentrations of these DNA fragments. Affymetrix offers whole-genome tiled arrays for human, mouse, yeast, etc., while recent ChIP-sequencing uses chromatin immunoprecipitation to investigate interactions between proteins and DNA and more accurate higher-throughput sequencing to localize interaction points.

Other Platforms (Proteomics, Epigenetics)

Other molecular profiling platforms include proteomics and epigenetics analysis. The most common way to study protein modifications is via two-dimensional (2D)-gel electrophoresis. Mass spectrometry (MS) coupled with liquid chromatography (LC) and multiple reaction monitoring (MRM) has been successfully used to characterize and quantify proteins in complex mixtures. The two most common epigenetic events are DNA methylation and chromatin remodeling. Various experimental techniques have been developed for genome-wide mapping of epigenetic information, the most widely used being ChIP-on-chip, ChIP-seq, and bisulfite sequencing.

Signal and Data Pre-processing

All highly parallel assays of nucleic acids (DNA, mRNA) use hybridization to microarrays. This method relies on the highly specific binding of the experimental DNA or RNA in solution, to its match synthesized on a solid substrate (microarray). To increase the accuracy and sensitivity of microarrays, most experimental systems query the target molecule many times, often with overlapping sequences. Thus, a single target gene or mRNA is queried by multiple features in a "probe set." Hybridization is an imperfect science, and there are key computation steps used to distill many signals deriving from multiple microarray features into a single number representation of the gene or mRNA under study. The following subsections describe some of these computational and bioinformatic steps.

Probe-Set Algorithms

In the widely used Affymetrix GeneChips™, the level of gene expression is summarized from a probe set designed to span a target region based on a UniGene cluster. Summarized expression measurements for a probe set can be derived using several algorithms, including MAS5.0, model-based-expression indices (MBEI; Li & Wong, 2001), robust multi-chip-average (RMA; Irizarry et al., 2003), and probe logarithmic intensity error (PLIER; Affymetrix, 2005). Each algorithm consists of (1) background correction to remove background noise from signal intensities, (2) normalization to even out unwanted non-biological variation across arrays, and (3) summarization to calculate the expression index. PLIER is the current default method provided by Affymetrix, which produces an improved signal by accounting for experimentally observed patterns of signal behavior and by handling error appropriately at both low and high abundance. Although there is still some debate, PLIER appears to be superior to other algorithms in its ability to avoid false positives for poorly performing probe sets. Seo and Hoffman (2006) pointed out that "background" is an enormously complex variable that can only be vaguely quantified, and thus the "best" probe-set algorithm may be application-dependent.

Normalization

Normalization is a prerequisite for almost all follow-up steps in genomic data analysis. Accurate normalization across different experiments and phenotypes assures a common base for comparative yet quantitative studies. Normalization aims to correct the differences across different arrays mainly in probe labeling, probe concentration, and hybridization efficiency. Popular normalization methods include (1) global normalization or linear regression (LR), (2) Loess normalization (Yang et al., 2002), (3) quantile normalization (Bolstad et al., 2003), and (4) iterative nonlinear regression (INR) (Wang et al., 2002). Regardless of their large technical differences, the basic steps in all

these methods involve selection of reference genes for normalization and choice of a linear or nonlinear regression function for normalization. Normalization has been found to have a profound impact on accuracy in detecting differentially expressed genes and in classification of phenotypes. Hoffmann et al. (2002) employed four different normalization methods and all possible combinations with three different statistical algorithms for detection of differentially expressed genes. They found that the choice of normalization method has a much greater influence on performance than the choice of the subsequent statistical analysis procedure. Hua et al. (2006) used a model-based approach to generate synthetic gene expression values for studying the normalization procedure's impact on classification performance. Their experimental results demonstrated that use of suitable normalization can give a significant benefit in classification accuracy under difficult experimental conditions.

Missing Values

Typically, to apply clustering and classification algorithms, it is required that gene expression patterns be *complete*, with no missing values. However, missing values may occur for various reasons, including dust on slides, image spotting, improper probe-set calibration, and human error. Ouyang et al. (2004) have estimated that more than 90% of genes may have missing values in a given data set. It is thus often necessary to impute missing expressions prior to data analysis. A variety of such methods have been proposed, including mean value based, nearest neighbor, and least squares techniques. Some methods also embed missing value imputation as an algorithm step directly within clustering or classification.

SNP allele values may also be missing due to technical artifacts. However, SNP imputation is also often applied to estimate alleles for SNPs that are not even genotyped. This can be achieved by exploiting both the strong correlation (linkage disequilibrium, LD) between genotyped (tag) SNPs and missing SNPs and the available, large reference data sets such as HapMap data that give LD structure on a genome-wide scale. Imputed SNPs are tested for trait association in the same way as genotyped SNPs. In a paper by Spencer et al. (2009), genome-wide SNP imputation was shown to significantly boost power (probability of rejecting a false null hypothesis) compared to association testing using only tag SNPs. SNP imputations can also greatly assist the pooling of multiple data sets created using different SNP chips (with different tag SNP sets). Some standard approaches for SNP imputation are based on hidden Markov model (HMM) inference.

Sample Stratification

The role of sample stratification is well-illustrated by the confounding problem of *hidden* population structure (admixtures) in genome-wide association studies (GWAS) case–control studies; there may be subgroups with very different ancestry in a given sample, and with the subgroup affiliations unknown. For example, the population of Great Britain is quite heterogeneous due to multiple immigration waves (Wellcome Trust Group, 2007). Different groups may carry wholly different trait propensities. This further means a group with high trait risk is likely *overrepresented* in the cases. Stratification corrects such sampling biases by first partitioning the sample into subgroups (e.g., via clustering), then randomly sampling from each group so as to make a "more representative" population. For GWAS, this may also involve choosing matched case–control pairs (agreeing at a large number of SNP loci). One disadvantage of stratification is that it reduces the effective sample size.

An excellent example of problems posed by stratification (or hidden stratification) is provided in Chapter 19. Here, the authors describe attempts to associate mitochondrial DNA polymorphisms with elite athletic performance (e.g., endurance), yet the population stratification due to distinct maternal lineages giving rise to mitochondria results in great difficulties in comparing data from different ethnic subgroups.

Task-Focused Data Modeling and Analysis

Genetic Association Studies

Genetic association studies tie certain genetic polymorphisms with specific human traits. For example, it is well documented that a polymorphism in the Factor V gene imparts different rates of blood clotting to individuals, dependent on their genotype. There are two experimental approaches used for SNP association studies: candidate genes, and GWAS. Candidate gene approaches rely on implicit assumptions or hypotheses regarding a causal connection between a specific protein and a specific phenotype. For example, clotting Factor V is a well-characterized and important blood coagulation protein, and when polymorphisms were found in the Factor V gene these became excellent candidates for modulation of blood clotting times. Candidate gene studies have been applied to complex traits unrelated to disease, including athletic ability (Gordon et al., 2005; Yang et al., 2003) and related performance measures, for example $\dot{V}O_2$max performance (Henderson et al., 2005).

GWAS are not hypothesis driven, but instead canvass the entire genome in a single experiment in an unbiased manner, through genotyping of a million SNPs in a single sample. GWAS studies hold great potential for yielding innovative analysis of complex traits and diseases, based on their unbiased evaluation of association at hundreds of thousands of loci, sampled across the genome. These studies typically involve SNP genotyping for thousands of individuals, both those possessing the trait (cases) and those that do not (controls), with statistical testing then applied to detect the loci with significant associations. Common variants (e.g., minor allele frequency (MAF) > 1%) and rare variants (MAF << 1%) together may account for a substantial proportion of trait risk. Using detected SNPs as features, one can build a statistical classifier to predict whether a given individual will manifest the trait. The proximity of SNP markers to unexpected genes may also help researchers formulate novel pathway hypotheses.

In addition to individual SNPs, for some traits there are SNP *interactions*, involving multiple SNPs "acting together," which confer either increased or decreased risk (Musani et al., 2007). An interaction is *epistatic* if the association is present only when *all* involved SNPs are jointly considered. It is generally believed that complex interactions must be identified if one is to accurately characterize the genetic architectures of complex traits (Musani et al., 2007). Thus, it is important to devise methods for detecting interactions involving multiple (e.g., up to as many as 5) SNPs.

Association testing can be characterized by *lack of power* and by false discoveries, particularly when very large number of SNPs are tested in GWAS studies, leading to challenges in statistical corrections for multiple testing. Power is a function of the allele (or interaction) frequency, penetrance, sample size, the significance level, and the number of SNPs/interactions tested. By chance false discoveries increase with the number of tests, which itself grows both with the number of SNPs and the order of SNP interactions (pairs, triples, etc.) tested. Thus, to have adequate power and at the same time limit false discoveries, a large sample size may be needed. This can be mitigated by SNP pre-screening, which discards "uninformative" SNPs prior to testing, and thus reduces the number of tests. One can also increase sample size by pooling data from multiple studies, that is, GWAS "meta-analysis." To assess statistical significance of associations, it may be possible to directly evaluate exact *P*-values. Alternatively, permutation testing is a valid approach.

Population admixtures based on unknown ancestries or other confounding variables can dilute effects and thus reduce power; they may also increase false discoveries by overrepresenting a subgroup among cases. Care must also be taken in defining the phenotype. For example, for GWAS meta-analysis, subtle differences in phenotype definition in the different studies may reduce detection power. Mislabeling samples will also decrease power.

Even exhaustive evaluation only of all candidate *pairwise* SNP interactions appears to be computationally infeasible for hundreds of thousands of SNPs and, moreover, demands very large sample sizes in order to make statistically significant detections. Thus, alternatives to standard testing have been sought for detecting SNP interactions and SNP subsets that contain interacting SNPs. Inherent difficulties include the following: interactions may be epistatic, penetrance function forms are unknown, or the orders of interactions are unknown, as are the number of interactions. One approach is to apply *wrapper-based* feature selection from machine learning, which "wraps" feature selection around the training of a statistical classifier. This is suitable for identifying an SNP marker subset. However, this may not give information about the specific interactions between SNPs belonging to this subset. Moreover, the accuracy of this approach is limited by the sensitivity of the wrapper's criterion function to detecting complex interaction effects. The criterion function usually chosen is classification error rate or a related objective function with limited sensitivity. One recent, promising work devised a more sensitive criterion function for detecting complex interactions, of variable order (Yu et al., 2008). Another recent, promising variant of the wrapper approach is a model-based wrapper that explicitly adds interactions to the model, applies a heuristic, variable-order interaction search, and uses model order selection criteria to determine the interactions, their orders, and their number (Miller et al., 2009).

There are several difficulties in assessing significance of detected markers and interactions. First, standard multiple testing ignores LD structure, assuming all tests are statistically independent, which gives conservative significance levels. There are also dependencies between tests of association for candidate interactions that involve *common* SNPs, for example (S1=0, S37=1, C="phenotype"), (S1=0, S17=0, S42=1, C="phenotype"), and (S1=1, C="phenotype") are all correlated events as they all involve SNP S1. It may also be difficult to determine whether the true effect is a weak main effect (S1=1, C="phenotype"), a weak interaction (S1=0, S37=1, C="phenotype"), or whether both are genuine effects. Finally, for a machine learning-based approach, permutation testing is needed for significance evaluation of detected interactions.

However, this entails numerous wrapper-based model retrainings, which may be computationally prohibitive.

Differential Analysis

Differential analysis is a common task that complements association studies and is often applied to identify reproducible gene markers with downstream effects using expression data. For predictive classification, only a subset of differentially expressed genes may be needed to achieve predictive accuracy that generalizes well to unseen (potentially noisy) test data. In unsupervised clustering, differential analysis/feature selection may be essential for helping to discern the underlying grouping tendency that may be "buried" in a much lower-dimensional subspace—in the presence of many structurally irrelevant genes, clustering algorithms are likely to identify false group structure. The most common differential analysis method is gene filtering that applies knowledge of the phenotypic labels to evaluate the differential expression of either individual genes or collections of genes, based on criteria such as signal-to-noise ratio, correlation measures, and mutual information. A recent study found that for small sample sizes, univariate methods fared comparably with multivariate methods, whose performance may be affected by overfitting. Most filtering methods for multicategory phenotypes are straightforward extensions of binary differential analysis and are theoretically suboptimal because they overemphasize large between-phenotype distances, that is, these methods, which choose gene subsets that preserve the distances of well-separated phenotypes, are prone to missing critical genes needed to differentiate between neighboring (molecularly similar) phenotypes. A good alternative is to select genes that are highly expressed in one phenotype relative to each of the remaining phenotypes.

Clustering

The mRNA expression profiling can assess all gene transcription changes as a function of intervention (such as a bout of muscle training), and it is generally assumed that the changes are not independent

of each other. In other words, changes may occur in clusters or group. For example, aerobic training will increase oxidative capacity of muscle, and thus many or most genes involved in oxidative metabolism may be coordinately upregulated. Data clustering groups data points into clusters so that data points within a cluster are similar while data points in different clusters are dissimilar, with cluster exemplars ("centers") providing a high-level overview of the whole data set. Data clustering has been widely applied to gene expression data to identify both co-expressed genes and samples, where co-expressed genes may be co-regulated by certain genomic regulation mechanisms and thus have similar genomic functions, and where co-expressed samples may represent a phenotypic population.

Clustering approaches are often broadly categorized into supervised and unsupervised. Supervised clustering involves one or more assumptions regarding the underlying variables that are driving changes observed in the data set. For example, in an exercise intervention where muscle biopsies are taken before and after training, the analysis can be run assuming that specific genes are responding to the training in each person in a similar manner and those genes can be sought. An unsupervised approach of the same data set would simply ask, "How do these samples all differ from each other." The latter might identify highly significant genes between two groups of biopsies, but these differences may or may not be related to the exercise intervention. For example, it is likely that transcripts on the Y chromosome may be found, but these simply reflect both male and female subjects.

Unsupervised clustering must be cautiously applied and may be unnecessary when samples come with appropriate and reliable supervising labels. Clearly, in the previous example, the investigator conducting the training intervention was not out to discover the molecular basis for males versus females. However, unsupervised clustering constitutes an important tool for discovering underlying subtypes or gene modules, and may suggest possible refinement to established phenotypic or functional categories corresponding to distinct biological pathways involving subtype-specific markers. Popular clustering methods include hierarchical clustering, k-means clustering, self-organizing maps, fuzzy clustering, finite mixture modeling, affinity propagation clustering, biclustering, and their various combinations such as consensus clustering. Unsupervised sample clustering is extremely challenging in high dimensions with few samples. Standard methods evaluate distances between data points using all equally weighted genes. Thus, many noisy/irrelevant genes will dominate the much smaller set of relevant genes in determining how data points are partitioned. Rather than clustering samples using all genes, a practical alternative is to embed gene selection within an iterative unsupervised clustering algorithm. Removal of noisy genes improves clustering accuracy which, in turn, guides a more accurate round of gene selection. Another major challenge for clustering in high dimensions is estimating the number of clusters. Model-based methods choose cluster number by best fitting the data while incurring least model complexity, but widely used model selection criteria fail when there are many irrelevant dimensions. A possible strategy for identifying cluster structure is top–down divisive clustering that explores and generates hierarchical mixtures in nested subspaces. Stability analysis is another method for estimating cluster number by examining alternative models in the presence of data perturbation.

Classification

Genomic expression data provide new opportunities for molecular classification of heterogeneous phenotypes. Phenotype identification and intervention responsiveness prediction are both supervised classification problems, where the sample classes are defined in advance, independently of the genomic expression data. There are three main components in developing a predictive classifier. The first component is determining which genes to include in the predictor. This can be addressed via differential analysis, which has already been discussed in this chapter. The second component is complete specification of the mathematical function that will provide a predicted class label

for any given expression vector. Popular classifiers include Regression Models, k-Nearest Neighbor Rule, Artificial Neural Network, Classification Tree, Support Vector Machine, Naïve Bayes Classifier, and Weighted Votes. Many comparative studies show that support vector machine-based classifiers outperform other methods on most benchmark microarray data sets. The third component is parameter estimation. Most predictors have parameters that must be assigned values before the predictor is fully specified. These parameters are akin to the regression coefficients in ordinary linear regression. The machine-learning literature calls the process of specifying the parameters "learning from the data." Completely specifying the predictor means specifying all of these aspects of the predictor: the type of predictor, the genes included, and the predictor's parameter values. Performance of a predictive model depends on the interrelationship between sample size, data dimensionality, and model complexity. The accuracy of learned models tends to deteriorate in high dimensions, a phenomenon called the "curse of dimensionality" or "bias/variance dilemma," where simple models may be biased but will have low variance. More complex models have greater representation power (low bias) but overfit to the particular training set (high variance). Clearly, classification accuracy must be assessed on labeled samples "unseen" during training. However, single batch held-out test data are often precluded in microarray studies, as there will be insufficient samples for both accurate classifier training and accurate validation. The alternative is a sound cross-validation procedure, wherein all the data are used for both training and testing, but not at the same time; that is, some samples are held out from the training in a testing fold that emulates the set of future samples for which class labels are to be predicted. These held-out test samples cannot be used in any way for the development of the prediction model. The cross-validated error rate is an estimate of the error rate to be expected when using this model for prediction on future samples. This assumes that the joint distribution of classes and expression profiles is the same for future samples as for the currently available samples. Specialized to two classes, one can use a similar approach to obtain cross-validated estimates of the sensitivity, specificity, and receiver operating characteristic curves.

Visualization

Visual exploration has proven to be a powerful tool for data mining and knowledge discovery. Two major visualization modes are data visualization and information visualization. Most data visualization algorithms aim to find a projection from the data space down to a visually perceivable rendering space. Popular projection methods include principal component analysis, multidimensional scaling, projection pursuit, locality preserved projection, and locally linear embedding. Various information visualization methods aim to interactively exploit unique characteristics of heterogeneous data sources and high dimensionality. In hierarchical visualization with tree-maps, the heterogeneous data elements related to a subject are intuitively organized in an ontology tree, equipped with various summary statistics for each abstract (ontology tree) level and adjustable details. Multilevel semantic zooming enables users to browse different contents depending on the current zoom level not constrained to offspring elements; see the summary of one data element to multiple subjects (ontology trees) and additional intermediate contents when available.

Introduction to Public Access Resources

Bioinformatics resources include databases and computational tools. Bioinformatics relies on these resources to help guide biomedical research by analyzing, integrating, and modeling data. Originally, databases and computational tools were created and managed by individual research groups conducting research in related areas. Nowadays, there are many resources that have become publicly available. These resources provide easy access to various types of tools and a wealth of biological primary data and information, which biologists require to obtain more reliable analysis results and to help effectively answer research questions.

The data repositories contain both original data and interpreted informatics data. A database of curated data is also called a "knowledge base." These data and information can be analyzed and conclusions can be drawn via sophisticated search and computational tools. Currently many of the database hosts provide various tools that allow users to browse, retrieve, and analyze data. The National Center for Biotechnology Information (NCBI) site is the best example of a public resource containing databases and associated tools. For example, through the Entrez retrieval system, users can access linked resources, including DNA and protein sequence data, assembled genomes, genome maps, sequence variations, gene expression data, protein structures and domains, a sequence-based taxonomy, and biomedical literature (Rapp & Wheeler, 2005).

In this section, we will describe briefly some of the major and popular databases and tools, as summarized in Table 5.1. There are numerous web lists providing links to complete bioinformatics resources sorted into different categories. These are also included in Table 5.1.

Public Access Data Resources

Nucleotide Sequence Database

There are three major DNA sequence databases: GenBank, the European Molecular Biology Laboratory (EMBL), and the DNA Data Bank of Japan (DDBJ). GenBank is the NIH genetic sequence database, an annotated collection of all publicly available DNA sequences. There are ~106,533,156,756 bases in 108,431,692 sequence records in the GenBank divisions as of August 2009. The EMBL Bank is the European equivalent of the US's GenBank, and constitutes Europe's primary nucleotide sequence resource. DDBJ is the sole nucleotide sequence data bank in Asia, based in Japan's National Institute of Genetics, which is officially certified to collect nucleotide sequences from researchers and to issue the accession number to researchers in any country. Since DDBJ exchanges collected data with EMBL Bank and GenBank on a daily basis, the three data banks share virtually the same data at any given time.

Table 5.1 Resources for database, knowledge base, and tools

Resource	Web URL	Summary
Selected database sites		
GenBank	http://www.ncbi.nlm.nih.gov/Genbank/	DNA sequence database
EMBL Bank	http://www.ebi.ac.uk/embl/	DNA sequence database
DDBJ	http://www.ddbj.nig.ac.jp/intro-e.html	DNA sequence database
UniProt	http://www.uniprot.org/	Protein sequence database
PIR	http://pir.georgetown.edu/	Protein sequence database
Swiss-Prot	http://www.ebi.ac.uk/uniprot/	Protein sequence database
PDB	http://www.rcsb.org/pdb/	Protein 3D structure database
dbSNP	http://www.ncbi.nlm.nih.gov/SNP/	SNP database
GEO	http://www.ncbi.nlm.nih.gov/geo/	Microarray database
SMD	http://genome-www5.stanford.edu/	Microarray database
DGV	http://projects.tcag.ca/variation/	Database of genomic variants
Selected knowledge base sites		
KEGG	http://www.genome.jp/kegg/	Pathways
BioCarta	http://www.biocarta.com/genes/index.asp	Pathways
PID	http://pid.nci.nih.gov/	Pathways
Ingenuity Pathways Knowledge Base	http://www.ingenuity.com/	Pathways
BioModels	http://www.ebi.ac.uk/biomodels	Biological models
TRANSFAC	http://www.gene-regulation.com/	Transcription factor database
DIP	http://dip.doe-mbi.ucla.edu/	Protein interacting database
GO	http://www.geneontology.org	Gene ontology database
BIND	http://bond.unleashedinformatics.com/	Biomolecular network database
GeneNet	http://wwwmgs.bionet.nsc.ru/mgs/gnw/genenet/	Genetic networks
PubMed	http://www.ncbi.nlm.nih.gov/pubmed/	Database of bibliographic information
Selected tool sites		
Bioconductor	http://www.bioconductor.org/	Open source tools for the analysis and comprehension of genomic data
Cytoscape	http://cytoscape.org/	An open source tool for visualizing biological networks
The National Center for Biotechnology Information	http://www.ncbi.nlm.nih.gov/	DNA/protein sequence and structure analysis; genome and gene expression analysis; text mining
BioPortal	http://bioportal.weizmann.ac.il/	Advancing biological research using computational and mathematical tools
Bioinformatics Organization	http://bioinformatics.org/	Bioinformatics developer repository
European Bioinformatics Institute	http://www.ebi.ac.uk/	Sequence similarity and analysis; pattern and motif searches; structure analysis; text mining
Cancer Biomedical Informatics Grid	https://cabig.nci.nih.gov/tools/	Data analysis and statistical tools in the areas of imaging, genome annotation, proteomics, microarrays, pathways, etc.

Table 5.1 (Continued)

Resource	Web URL	Summary
Selected web lists		
Bioinformatics Links Directory	http://www.bioinformatics.ca/	Curated links to molecular resources, tools, and databases.
Pathguide	http://www.pathguide.org/	Complete listing of all pathway resources
Guide to Bioinformatics Internet Resources	http://www.istl.org/02-winter/internet.html	Links to Bioinformatics Internet resources

Protein Sequence Database

There are three major protein sequence databases: Universal Protein Resource (UniProt), Protein Information Resource (PIR), and Swiss-Prot. UniProt is a single database that combines the information of the major international databases from the European Bioinformatics Institute (EBI), Georgetown University Medical Center and National Biomedical Research Foundation, and Swiss Institute of Bioinformatics (SIB). PIR is a resource to assist researchers in identification and interpretation of protein sequence information. It contains functionally annotated protein sequences and has been incorporated into an integrated knowledge base system of value-added databases and analytical tools. Swiss-Prot is the major European curated protein sequence database, which strives to provide a high level of annotations such as the description of the function of a protein, its domain structure, post-translational modifications, and variants.

SNPs Database

Because SNPs are expected to facilitate large-scale association genetics studies, there has recently been great interest in SNP discovery and detection. dbSNP is a central repository for both single base nucleotide substitutions and short deletion and insertion polymorphisms. It was established by NCBI in collaboration with the National Human Genome Research Institute.

Genomic Variant Database

The database of genomic variants provides a comprehensive summary of structural variation in the human genome. Structural variation is defined as genomic alterations that involve segments of DNA that are larger than 1 kb. More than 20 k CNVs have been reported to the database of human genomic variants.

Microarray Database

Stanford Microarray Database (SMD) and Genome Expression Omnibus (GEO) are the two most famous and abundant gene expression databases in the world. GEO is an archival gene expression and hybridization array data repository, as well as an online resource for browsing and retrieval of gene expression data from any organism or artificial source. SMD stores raw and normalized data from microarray experiments, and provides web interfaces for researchers to retrieve, analyze, and visualize their data.

Public Access Knowledge Resources

Here we provide an overview of some popular web-accessible knowledge bases. These include databases on metabolic pathways, signaling pathways, transcription factor targets, gene regulatory networks, genetic interactions, protein–compound interactions, and protein–protein interactions (Bader et al., 2006).

Kyoto Encyclopedia of Genes and Genomes (KEGG) is a large database of metabolic pathways from multiple organisms. Reference pathways are available that describe a general pathway, which is then projected onto a genome based on the presence of pathway enzymes in that genome.

KEGG is composed of a number of databases: PATHWAY, GENES, and LIGAND. BioCarta is another knowledge base that describes how genes interact in dynamic graphical models. It also catalogs and summarizes important resources providing information for over 120,000 genes from multiple species. The Pathway Interaction Database (PID) from NCBI contains a highly structured, curated collection of information about known biomolecular interactions and key cellular processes assembled into signaling pathways. Users can query the database by pathway name and by molecule name or accession. Molecular details include protein post-translational modifications and cellular location. Annotations on interactions include literature citations and evidence codes. Pathways Knowledge Base from Ingenuity is a commercial repository of molecular interactions, regulatory events, gene-to-phenotype associations, and chemical knowledge that provides building blocks for pathway construction. It is a large curated database of biological networks created from millions of individually modeled relationships between proteins, genes, complexes, cells, tissues, drugs, and diseases.

BioModels database provides access to published, peer-reviewed, quantitative models of biochemical and cellular systems. Some of these systems are very simple, containing just a few processes or reactions; others contain hundreds. The models are checked to verify that they correspond to the reference publication. Human curators annotate and cross-link components of the models to other relevant data resources. This allows users to identify precisely the components of models, and helps them to retrieve appropriate models, which they can then visualize and run using any SBML-compatible software.

The Transcription Factor Database (TRANSFAC) Suite is an internationally unique knowledge base containing data on transcription factors, their experimentally proven binding sites, and regulated genes. Its broad compilation of binding sites allows the derivation of positional weight matrices, which is a commonly used representation of motifs (patterns) in biological sequences. Database of Interacting Proteins (DIP) catalogs experimentally determined interactions between proteins.

It combines information from a variety of sources to create a single, consistent set of protein–protein interactions. The data stored within the DIP database were curated both manually by expert curators and also automatically using computational approaches that utilize the knowledge about the protein–protein interaction networks extracted from the most reliable, core subset of the DIP data.

Gene Ontology (GO) provides ontology of pathways and constituent subpathways. In addition, GO adds links between biological process and molecular function which provide terminology for all of the individual reaction steps and pathway terms, and the relationships among them. Biomolecular Interaction Network Database (BIND) is a database designed to store full descriptions of interactions, molecular complexes, and pathways. An interaction record is based on the interaction between two objects. An object can be a protein, DNA, RNA, ligand, or molecular complex.

PubMed is a database of bibliographic information drawn primarily from more than 19 million citations for biomedical articles from MEDLINE and life science journals, dated back to 1948. PubMed contains links to full-text articles at participating publishers' web sites as well as links to other third-party sites such as libraries and sequencing centers. PubMed also provides access and links to the integrated molecular biology and chemistry databases maintained by NCBI.

Public Access Software Resources

Various searching and analysis tools have been developed over the past years. Their functionalities cover sequence similarity and analysis, pattern and motif searches, structure analysis, text mining, discovery of biomarkers, inference of biological networks, clustering, and classification. Some of these tools are open source software that may be copied and redistributed at no charge (Chiang, 2009; Rapp & Wheeler, 2005).

Two of the most widely used sequence matching software programs are the FASTA and the Basic Local Alignment Search Tool (BLAST). FASTA is a DNA and protein sequence alignment software package that finds exact matches between short substrings of two sequences. BLAST is another algorithm based on a similar idea. Both FASTA and BLAST searches can be carried out interactively through the links from the NCBI and EBI home pages.

Entrez is a powerful search and text mining tool provided by NCBI for searching the comment fields of the sequence databases, including PubMed, Nucleotide and Protein Sequences, Protein Structures, Complete Genomes, Taxonomy, and more. Entrez provides access to all databases simultaneously with a single query string and user interface. Entrez can efficiently retrieve related sequences, structures, and references. Entrez also generates views of gene and protein sequences and chromosome maps.

Bioconductor is an open source and open development software project to provide tools for the analysis and comprehension of genomic data.

Because it is based primarily on the R programming language, users have many data analysis resources, such as gene expression analysis, sequence and genome analysis, structural analysis, statistical analysis, and computational modeling.

Cytoscape is an open source bioinformatics software platform for visualizing molecular interaction networks and biological pathways and integrating these networks with annotations, gene expression profiles, and other state data. Cytoscape core distribution provides a basic set of features for data integration and visualization. Additional features are available as plug-ins. Plug-ins are available for network and molecular profiling analyses, new layouts, additional file format support, scripting, and connection with databases.

References

Affymetrix (2005). *Guide to Probe Logarithmic Intensity Error (PLIER) Estimation*. Affymetrix, Santa Clara, CA.

Bader, G.D., Cary, M.P. & Sander, C. (2006). Pathguide: A pathway resource list. *Nucleic Acids Research* 34, D504–D506.

Bolstad, B.M., Irizarry, R.A., Astrand, M. & Speed, T.P. (2003). A comparison of normalization methods for high density oligonucleotide array data based on variance and bias. *Bioinformatics* 19, 185–193.

Chiang, J.H. (2009). Tech-Ware: Bioinformatics and computational biology resources. *IEEE Signal Processing Magazine* 269, 153–158.

Gordon, E.S., Gordish-Dressman, H.A., Devaney, J., et al. (2005). Nondisease genetic testing: Reporting of muscle SNPs shows effects on self-concept and health orientation scales. *European Journal of Human Genetics* 13, 1047–1054.

Henderson, J., Withford-Cave, J.M., Duffy, D.L., et al. (2005). The EPAS1 gene influences the aerobic–anaerobic contribution in elite endurance athletes. *Human Genetics* 118, 416–423.

Hoffmann, R., Seidl, T. & Dugas, M. (2002). Profound effect of normalization on detection of differentially expressed genes in oligonucleotide microarray data analysis. *Genome Biology* 3, research0033.1–research0033.11.

Hua, J., Balagurunathan, Y., Chen, Y., et al. (2006). Normalization benefits microarray-based classification.

EURASIP Journal on Bioinformatics and Systems Biology 43056, 1–13.

Irizarry, R.A., Hobbs, B., Collin, F., et al. (2003). Exploration, normalization, and summaries of high density oligonucleotide array probe level data. *Biostatistics* 4, 249–264.

Li, C. & Wong, W.H. (2001). Model-based analysis of oligonucleotide arrays: Expression index computation and outlier detection. *Proceedings of the National Academy of Sciences U S A* 98, 31–36.

McCarroll, S.A. & Altshuler, D.M. (2007). Copy-number variation and association studies of human disease. *Nature Genetics* 39(Suppl. 7), S37–S42.

Miller, D.J., Zhang, Y., Yu, G., et al. (2009). An algorithm for learning maximum entropy probability models of disease risk that efficiently searches and sparingly encodes multilocus genomic interactions. *Bioinformatics* 25(19), 2478–2485.

Musani, S., Shriner, D., Liu, N., et al., (2007). Detection of gene x gene interactions in genome-wide association studies of human population data. *Human Heredity* 63, 67–84.

Ouyang, M., Welsh W.J., & Georgopoulos, P. (2004). Gaussian mixture clustering and imputation of microarray data. *Bioinformatics* 20, 917–923.

Rapp, B.A. & Wheeler, D.L. (2005). Bioinformatics resources from the National Center for Biotechnology Information: An integrated foundation for discovery. *Journal of the American*

Society for Information Science and Technology 56, 538–550.

Seo, J. & Hoffman, E.P. (2006). Probe set algorithms: Is there a rational best bet? *BMC Bioinformatics* 7, 395.

Spencer, C.C, Su, Z., Donnelly, P. & Marchini, J. (2009). Designing genome-wide association studies: Sample size, power, imputation, and the choice of genotyping chip. *PloS Genetics* 5, 1–13.

Wang, Y., Lu, J., Lee, R., Gu, Z. & Clarke, R. (2002). Iterative normalization of cDNA microarray data. *IEEE Transactions on Information Technology in Biomedicine* 6, 29–37.

The Wellcome Trust Case Control Consortium (2007). Genome-wide association study of 14,000 cases of seven common diseases and 3,000 shared controls. *Nature* 447(7145), 661–677.

Yang, Y.H., Dudoit, S., Luu, P., et al. (2002). Normalization for cDNA microarray data: A robust composite method addressing single and multiple slide systematic variation. *Nucleic Acids Research* 30, e15.

Yang, N., MacArthur, D.G., Gulbin, J.P., et al. (2003). *ACTN3* genotype is associated with human elite athletic performance. *American Journal of Human Genetics* 73, 627–631.

Yu, G., Herrington, D., Langefeld, C. & Wang, Y. (2008). Detection of complex interactions of multi-locus SNPs. *IEEE Workshop on Machine Learning for Signal Processing* Oct. 16–19, 85–90.

Chapter 6

Search for Genes Influencing Complex Traits

CORNELIU HENEGAR, HENRI HOOTON AND KARINE CLÉMENT

INSERM, Cordelier Research Center, Paris France; University Pierre et Marie Curie-Paris 6, Paris France; Endocrinology and Nutrition Department, Pitié-Salpêtrière Hospital, Paris, France

With the advent of ever better versions of the human genome map and the continuous development of "omic" technologies, greater opportunities to identify causative genetic factors for complex diseases and traits continue to emerge. Researchers devote substantial effort to isolate genetic markers associated with these conditions. These strategies have rapidly progressed and led to completely new strategies for exploring genetic determinants of complex traits and multifactorial diseases. One example is the genome-wide association studies (GWAS), which seek to identify and mass genotype molecular variants, in particular single-nucleotide polymorphisms (SNPs). The scale of such genetic studies has also changed from several hundreds of subjects studied in the late 1990s to tens of thousands in recent years. In parallel with the development of GWAS, functional "omic" approaches, inspired by the systems biology concept, were devised to explore the complex interactions that characterize molecular signatures of various traits and diseases. These approaches were applied to identify potential gene candidates and quantify the importance of their roles. More recently, a number of studies have successfully combined genetic variant association approaches with large-scale gene expression analyses. Such investigations can advance research but also face numerous methodological challenges. This chapter presents a brief overview

of the ongoing research and reviews the current approaches, techniques, and challenges in exploring genetic determinants of complex traits and diseases.

Complex traits, such as obesity, muscle strength, and fitness, are known to have very strong genetic contribution, with 30–70% of all phenotypic variation typically attributed to genes. GWAS promised to identify these genes, and many obesity-related loci have been identified. However, it has been surprising that the relative contribution of each gene is quite small, where the 10 strongest loci add up to only 1% of the variation. This paradox is discussed with regard to study design, sensitivity and specificity of tools used to measure specific phenotypes, and data interpretation.

From Candidate Gene Approaches to Genome-wide Studies

The identification of genes associated with complex human traits or diseases (such as obesity, type II diabetes, high blood pressure, and other common diseases) has proved to be a challenge but recent successes have been obtained in parallel with new methodological developments and with increased knowledge of the human genome map. Before 2000, the vast majority of genetic studies have been conducted on candidate genes identified either through association studies or familial linkage studies based on previous biochemical knowledge and animal studies. While association studies compare the allelic frequencies of genetic variants among different unrelated populations (i.e., typically

Genetic and Molecular Aspects of Sport Performance, 1st edition.
Edited by Claude Bouchard and Eric P. Hoffman.
Published 2011 by Blackwell Publishing Ltd.

cases versus controls), linkage studies focus on the relationship of a given phenotype and each chromosomal region by considering the way they are transmitted from parents to offspring.

Linkage studies rely on a large cohort of families who are recruited to explore relationships between a given phenotype and each chromosomal region either by considering markers that can specifically identify the origin of each chromosomal fragment (i.e., whether it was transmitted by the father or the mother to the offspring) or whether an increased allelic sharing is observed between related subjects (typically siblings). When co-transmission between a particular chromosomal region and the phenotype of interest is observed, the region is considered linked with the considered phenotype.

Genetic studies using microsatellite markers have successfully identified a relatively large number of genes involved in mostly rare forms of monogenic diseases, but have been relatively poor in identifying gene candidates for common diseases or complex traits. The genome-wide resolution that can be reasonably achieved using microsatellites is relatively low, ranging between 10 cM and 5 cM. Higher resolutions can be achieved using high-density SNP chips for linkage mapping (Matsuzaki et al., 2004; Middleton et al., 2004). The genome-wide scan task is performed without preconceptions about the functions of the genes and aims to identify known or unknown genes predisposing to a target trait or disease. Then molecular tools enable the newly identified genes to be positioned and eventually cloned. In the first studies, it was usually arduous to find the exact causative gene when genomic-linked regions encompass thousand of bases and sometimes hundred of genes.

Marker-based systems in genome-wide linkage studies suffer from limited power without the availability of complete pedigree information. Complex diseases do not follow a Mendelian inheritance pattern, so non-parametric linkage methods have been applied (Blackwelder & Elston, 1985; Haseman & Elston, 1972). Most of these studies relied on the estimation of the sharing of genomic fragments that are identical by descent (IBD) between two related individuals. The estimated IBD status is then compared to the expected value under the assumption of no linkage with the disease. A significant deviation from the expected IBD is evidence for a disease gene at the linked locus. Where only one affected relative pair is available from each family, it may not be possible to determine IBD unambiguously. In cases where no extended pedigree genotype information is available (e.g., for late-onset diseases), this can result in a significant decrease in the power to detect linkage (Evans & Cardon, 2004). These limitations have been reported for both microsatellite- and SNP-based methods (Evans & Cardon, 2004; Hoh & Ott, 2000; Matsuzaki et al., 2004). The major factor leading to this decrease in power is ambiguity in correct IBD prediction in a single relative pair with limited allele segregation data (Kruglyak et al., 1996).

An advantage of linkage and IBD studies is the application of robust statistical analyses, and the potential for highly statistically significant associations between a phenotype and a specific region of the genome. A disadvantage is their spatial resolution in the genome; often the area of linkage can be narrowed to only 10 million base pairs or even more. This large amount of "genomic real estate" contains dozens or hundreds of genes and thousands of SNPs, and finding causality between specific genes, genotypes and phenotypes can be difficult. GWAS require many more subjects to overcome limitations imposed by multiple testing, and are subject to false positive results. However, their spatial resolution in the genome is excellent, typically narrowing associations to a single gene or haplotype.

Testing Genotypic and Allelic Effects in Unrelated Populations

Genetic association studies aim to correlate disease or trait levels with genetic variant allele or genotype frequency. The measurement scale of the phenotypic variable (nominal, ordinal, or continuous) under consideration influences the manner in which the association test is conducted. The allelic formulation is valid (in terms of significance testing) if Hardy–Weinberg equilibrium (HWE) holds in the combined case–control population (Sasieni, 1997) and the two methods are asymptotically equivalent in terms of effect size estimation if the mode of inheritance at the locus under study is additive. The Hardy–Weinberg principle states that both alleles and genotype frequencies in a population remain constant; in other words, they are in equilibrium from generation to generation unless specific confounding influences are introduced. The genotypic formulation of the case–control test is more flexible in testing for non-additive effects (e.g., dominant, recessive, or overdominant effects) and does not require the assumption of HWE.

Family-Based Associations

Family-based tests of associations methods (Allison, 1997; Spielman et al., 1993), such as the transmission disequilibrium test (TDT) (Spielman et al., 1993), are effective at detecting association between a marker locus and a phenotypic trait. The TDT, described initially for binary traits, uses information on the parent-to-offspring transmission status of the associated allele at a marker locus to assess linkage or association in the presence of the other. The basis of the TDT is that alleles associated with disease have a higher probability of being found to have been transmitted to affected offspring. It is used primarily for dichotomous traits, but Allison extended it to quantitative traits (Allison, 1997) and genotypic tests (Allison, 1997; Fallin et al., 2002). However, one of the main challenges in conducting family-based association tests is the cost of recruiting parents and their affected children. The test is severely limited when parental data are unavailable. Additionally, if confounding from population stratification (see below) is suspected, the TDT will require a very large number of subjects to achieve the level of power of a non-family-based association study (McGinnis et al., 2002; Witte et al., 1999).

Population Stratification

Population stratification can be defined as the presence of subgroups within a population with different allele frequencies that are attributable to differences in ancestry. An often cited cause of population stratification is the admixture process, the event of two populations with different allele

frequencies producing an admixed population. If the prevalence of disease or the mean phenotypic values differ between the subgroups, an association between a genetic marker and a phenotype may exist and yet the marker may not cause changes in the phenotype nor be linked to a gene that influences the phenotype. Such an association should be considered as a spurious association. Although the case–control study is a widely applied strategy for conducting genetic association studies, it is well known that it is susceptible to producing spurious associations due to population substructure.

Genomic control provides a way to test for population stratification and adjust association tests accordingly. The concept of genomic control was introduced by Devlin and Roeder, who noticed that the impact of population stratification on association studies can be quantified by a parameter they denoted λ (a variance inflation factor relative to trend test) which is larger when the degree of population stratification is larger (Devlin & Roeder, 1999). Their genomic control method corrects for spurious associations by applying Armitage's trend test to evaluate the association between genotype and case–control status for each of K randomly selected markers. The resulting vector of K Armitage test statistics is used to compute a correction factor ($\hat{\lambda}$) that is subsequently used to adjust the test statistic for each of the candidate loci.

Structured association testing (SAT), on the other hand, incorporates population structure or stratification information into the association test (Pritchard & Donnelly, 2001). Although there are several forms of SAT, they typically have two steps. Rather than just estimating a quantified inflation factor based on population structure or stratification, the SAT framework first estimates each individual's admixture proportion (the proportion of an individual's alleles that originated from a specific ancestral population). Once the admixture proportion is estimated for each individual, these estimates are incorporated into the model for the association test. SAT models heavily rely on the precision of the inferred individual estimates (Hoggart et al., 2003). If the estimates are inaccurate, then the SAT tests may be invalid.

Characterizing Regions IBD

Methods measuring direct IBD mapping using an enzymatic process have been proposed to enrich IBD regions and to theoretically overcome the limitations described above, as the method does not rely on parental genetic information to determine IBD. The fact that non-IBD fragments will show sequence differences between two individuals is used to physically remove the non-IBD DNA. Casna and colleagues, Sanda and Ford introduced concepts for physical enrichment of IBD DNA between related individuals (Casna et al., 1986; Sanda & Ford, 1986). Nelson and colleagues proposed a technological solution termed Genomic Mismatch Scanning (GMS) (Nelson et al., 1993), which relies on the fact that genomic fragments from non-IBD regions will be polymorphic with respect to the sequences of the homologous region between two individuals. When DNA from both individuals is mixed to form hybrid DNA, these sites will have base pair mismatches. These mismatches can be detected by specific DNA repair enzymes, which introduce nicks into the mismatched DNA making it susceptible to digestion by an exonuclease. The non-IBD DNA can be removed physically from the solution. GMS held the promise to overcome the two major problems of marker-based linkage mapping, namely lack of resolution and ambiguity of IBD determination. However, because of the development of high resolution SNP arrays for genotyping, the method never achieved widespread application for several reasons. The protocols described for the GMS method suffer from the complexity of its many steps and the absence of sufficient control within the process. Technological improvements for direct IBD mapping were proposed in the late 1990s with simpler methods in application and more transparency in scoring the efficacy and success of each process step (e.g., genome Hip technology). However, this type of approach is less utilized because of the rapid development of other genome-wide screening technologies.

Genome-Wide Association Studies

The fact that the human genome has been completely sequenced is one of the major causes of progresses made in high throughput genome exploration in relation to complex multifactorial traits and diseases. The International HapMap Consortium (www.hapmap.org) published a detailed haplotype map of the human genome, providing the possibility of exploring the SNP variations in human genes to identify the genotypes that contribute to complex phenotypes (The International HapMap Consortium, 2007). GWAS are unbiased approach used in order to identify loci implicated in genetic disorders or complex traits, which relies on a common disorder–common variant hypothesis. The basic idea is to test for association between a given phenotype and SNPs throughout all the genome in a very large amount of individuals. The first step is to select a population for the study (a cohort, a group of cases and another of controls or a group of trios), typically including thousands of subjects. For case–control studies, an equivalent number of cases and controls should be available.

These individuals are then genotyped for a very large amount of SNPs (usually several hundreds of thousands and up to one million). Microarray-based SNP genotyping methods have increased the throughput as well as reducing the cost of SNP large-scale genotyping. Many companies, such as Affymetrix and Illumina, provide tools and high throughput SNP genotyping platforms for rapid screening. Once the genotyping stage is over, the next step is to test for association between each SNP that has been genotyped and the studied disease or trait in order to identify loci that could be implicated in the phenotype of interest.

GWAS are of several types—one stage, multistage, and meta-analysis. The one-stage study represents the prototype GWAS. The multistage study is a one-stage GWAS, with extra replication stages, in which only the SNPs that reached statistical significance are replicated in independent populations. Finally, the meta-analysis reanalyzes several independent GWAS through a combined assessment, in order to increase the power of the study and the number of potentially significant results. The meta-analysis procedures can also include a second stage of replication of findings in one or several independent populations.

Multistage Association Studies

The large-scale GWAS require a sizable genotyping effort which is limited by the financial resources that are available. On the other hand, the high number of markers to be tested for association with the phenotype of interest raises methodological issues related to multiple testing. The overall cost of a multistage study depends upon the total number of markers genotyped and the total number of individuals sampled. Strategies to maximize power while minimizing costs can be classified in two categories: group sequential methods and stepwise focusing methods.

Group sequential method—Sequential test procedures were designed to reduce the sampling cost of different type of experiments. Compared to fixed sampling methods, they offer investigators greater control over both type I error and power. Sobell and colleagues proposed a sequential procedure that starts by genotyping all markers in a subset of the total sample of individuals (Sobell et al., 1993). Using this group of individuals, they further proposed dividing the markers to be tested into three groups based on the *P*-values corresponding to the tests of association with the phenotype. The first group contains markers that are apparently unrelated to the phenotype under consideration. Markers that are significantly associated with the phenotype are placed in the second category. The third group contains all the "ambiguous markers" for which a clear-cut decision about the association between the marker and the phenotype could not be made. In this latter situation, more individuals are genotyped and used as a second independent sample to further test the ambiguous markers for association. A third sample of genotyped individuals may be necessary in order to test markers that remained ambiguous at the end of the second stage of testing.

Group sequential methods constitute an improvement over sequential designs since they do not require the evaluation of the test statistic for each unit of information collected. A group sequential association test works by dividing the overall sample into different subsets of individuals. A test for association between marker and phenotype is then conducted on the first group. The study stops if the test rejects the null hypothesis or shows signs of futility; otherwise the study continues to another stage. Within the second stage, all observations from the first and second stage are used to evaluate the hypothesis. The process continues through several stages until the null hypothesis of no association is either rejected, or it becomes obvious that this hypothesis will never be rejected or finally, when the entire sample is depleted.

Stepwise focusing method—In addition to these, there are also two possible formulations of the stepwise approach to GWAS. They correspond to the case where the investigator seeks to maximize the power of an association study under a budget constraint that limits the number of markers to be genotyped in the study (Satagopan et al., 2002). The optimum allocation in these situations consists in spending most of the available resource on markers that are more susceptible to be associated with the phenotype under study. For studies that intend to identify only one true gene among a set of unlinked markers, Satogopan et al. (2002) recommend using about 75% of the budget at the first stage to select the top 10% of markers more likely to be associated with the phenotype of interest. This ratio varies slightly, however, when the study is conducted using linked markers. For mildly correlated markers, it is better to use about 25% of markers at the first stage and then 75% of the available resource at the second stage to follow up the top 10% of markers.

The two-stage design can also be optimal when the objective is to identify not just a subset of markers but specific haplotype groups associated with a specific phenotype (Thomas et al., 2004). Other multistage approaches focus more on controlling the overall type I error rate. Measures like the false discovery rate (FDR) are used to control the number of markers selected for further analysis at end of each stage. It has been shown that a k-stage selection procedure that uses high FDR at the first $k-1$ stages and a stringent FDR at the last stage may result in a 50–70% reduction in genotyping cost (van den Oord & Sullivan, 2003). The FDR produces weaker control than the family-wise error rate (FWER) used in methods described in Elston et al. (1996), Satagopan & Elston (2003), and Satagopan et al. (2002). In general, the designs that control for the FDR are more powerful than those that use FWER to determine the number of markers that are retained from one stage to another (Kraft, 2006). This improvement comes with the added cost that the number of false positives will be higher than what is observed with the methods that use FWER.

Epistasis, Attributable Risk, and Replication Issues

Epistasis or interaction among gene loci is a basic concept in genetics. A common approach for exploring epistasis is to search for loci which have moderate to large effect and then examine possible interactions among all such loci. However, interactions with small effect loci would be missed in this approach, and such interactions among loci of small effect could be the rule rather than the exception for complex traits (Frankel & Schork, 1996). On the other hand, testing all possible interactions in GWAS with thousands of SNP markers leads to a large number of tests, a situation referred to as the "curse of dimensionality." Numerous methods have been proposed to resolve the issue of dimensionality and reduce the computational burden. Some authors have recommended using data mining and neural networks (Hoh & Ott, 2003; Moore & Ritchie, 2004) to overcome dimensionality problems. These methods to detect epistasis do not require a pre-specified statistical hypothesis and, in this respect, are better suited to search for trends or patterns in high-dimensional data sets. However, data mining methods are prone to chance patterns in the data and may thereby result in false positives. Furthermore, models of neural networks are difficult to interpret, and the results are not intuitive.

Attributable risk (AR) can be defined as the excess risk of disease or the occurrence of a particular trait for an individual with a risk factor relative to an individual without that risk factor; that is, the risk difference between individuals exposed to a risk factor and individuals not exposed to a risk factor. The rationale behind AR calculations is to provide an alternative to simple relative risk measures in order to describe the rates in absolute terms instead of relative (Walter, 1976). As with

all investigations of risk factors, the design of the study must be considered. The risk ratio is directly estimable using prospective designs; however, in a retrospective trial, in which individuals are selected based upon disease states, odds ratios would be substituted for relative risks (Walter, 1976). Within genetic studies of complex phenotypes, it is anticipated that the ARs due to specific alleles or genotypes will be small, given the multifactorial nature of the trait.

Lack of replication and contradictory results are two of the major concerns with association studies (Lyon & Hirschhorn, 2005; Redden & Allison, 2003). Several reasons have been proposed to account for the lack of reproducibility. These include population heterogeneity (genetic stratification and great environmental diversity), publication bias, epistasis, imprecise phenotypic measures, lack of type I error control, and lack of statistical power to detect small effects (Redden & Allison, 2003).

Failure to replicate may result if the initial finding is in fact either a type I error (false positive result) or a spurious association attributed to population stratification. Inadequate adjustment for multiple hypothesis tests increases the probability of at least one type I error which may lead to numerous false positive results being published. Some debate exists as to whether population stratification is a likely cause of non-replication. Investigators have pointed out that although population stratification is the most common explanation given for spurious associations, there are few actual examples to support this assumption, indicating that the problem may be overemphasized. Other reasons for non-replication may be over-interpretation of marginal findings and under-emphasis of publication bias for positive results, which is difficult to quantify.

Two other issues likely affecting the observed non-replication of association studies are low statistical power within some of the follow-up studies, as well as population heterogeneity in diverse environmental settings. If replication of a genetic finding is attempted in a readily available sample from a study of a different phenotype or marker, the replication study may have insufficient power for several reasons including the variance of the phenotype or the magnitude of the effect size associated with the marker. If genes–environment interactions are occurring, then investigation of separate populations in non-identical environments would explain variability in results.

Meta-Analysis Approaches and Phenotypic Challenges

As GWAS accumulate, the opportunity to seek greater power, precision, and clarity via meta-analysis increases. The term "meta-analysis" refers in this context to any activity that utilizes the data (whether raw data or published summary statistics) from two or more studies in a formal and quantitative manner to estimate quantities and/or test hypotheses. Meta-analysis can be used to seek clarity when there are many studies that upon narrative review seem to give an equivocal answer about a particular question (e.g., whether a particular gene is associated with a particular phenotype). Investigators use this approach with a single study of the association of a polymorphism with a phenotype to put their latest result in the context of the existing research (Swarbrick et al., 2005). Meta-analysis can also be used to refine estimates of an effect or association.

Modern GWAS aim at precisely dissecting individual sub-phenotypes in common diseases, including the evaluation of perturbed functions at the multitissue level (e.g., innovative imagery coupled with metabolic investigation) and environmental profiles in population-based studies. The individual complexity in humans relates indeed to the influence of common disease alleles in different genetic backgrounds. Taking into consideration various genetic combinations, possibly influenced by different epigenetic or environmental factors during an individual's lifetime—including the role of *in utero* environmental determinants—represents a major challenge in relation to the current efforts aimed at deepening our understanding of the genetic origin of common multifactorial traits and disorders. The precise degree to which particular genes contribute to these entities remains to be determined, while the importance of subtle environmental factors and their diversity may not have been well appreciated. In particular, the

recruitment of very large numbers of subjects (sometimes tens of thousands of individuals) to fulfill power challenges required by GWAS has been detrimental to the quality and precision of endo-phenotypes associated with complex polygenic traits. An example can be provided by the recent discovery of SNPs associated with BMI variations, replicated in large-scale populations, whose usefulness is greatly reduced by the fact that the BMI represents a very crude mean for characterizing the adiposity phenotype. The challenge is now to take into account precisely the endo-phenotypes and diverse environmental factors in heterogeneous populations.

Systems Biology Approaches for Exploring Determinants of Complex Traits and Diseases

The interactional nature of cellular processes, understood as associations of molecules whose relations to each other are instrumental in carrying out a particular function (Alon, 2003; Barabasi & Oltvai, 2004; Hartwell et al., 1999), underlines the key role of molecular interaction pattern analysis in untangling the functional architecture of cellular environments. Among various types of omic interactions, the study of transcriptional co-expression networks, built by relating transcripts (i.e., the nodes of the network) displaying similar expression profiles, has contributed to the characterization of key properties of these interactions. Among them the hierarchical architecture of molecular networks, made of modules of functionally related components (i.e., genes, enzymes, and metabolites) (Barabasi & Oltvai, 2004; Ravasz et al., 2002), and the presence of various types of omic "hubs" (i.e., highly connected or central nodes) (Guimera & Nunes Amaral, 2005), which modulate key interactions at distinct levels in the cell are of particular importance for exploring of genetic determinants of multifactorial traits and disorders.

Indeed large-scale studies (Allocco et al., 2004; Bergmann et al., 2004; Balaji et al., 2006; Carlson et al., 2006; Guimera & Nunes Amaral 2005; Jeong et al., 2001), integrating gene expression patterns with information on phylogenetic variability, the sharing of common transcriptional binding sites, results from mutagenesis experiments, and knowledge of transcripts' biological roles, have demonstrated strong correlations of various network centrality indices (i.e., nodes' connectivity or betweenness within the network) with experimental indicators of biological relevance, including lethal knockout phenotypes (Carter et al., 2004; Jeong et al., 2001) and specific patterns of phylogenetic variability (Balaji et al., 2006; Guimera & Nunes Amaral, 2005).

Such attempts are limited, however, by a recurrent difficulty: that is, how to distinguish noisy interactions from biologically meaningful co-expression patterns. To improve the biological relevance of the identified interaction patterns Henegar and colleagues proposed an integrative exploratory approach that relies on knowledge of transcripts' biological roles, extracted from genomic databases (Gene Ontology, KEGG, etc.) together with gene expression measurements (Henegar et al., 2008; Prifti et al., 2008). The core of this approach is represented by a two-layer network model generated by mapping significantly overrepresented biological themes, identified among those annotating transcripts roles, onto a conventional transcriptional co-expression network.

Conclusion

In this chapter, we attempted to give an overall image of two conceptually different families of approaches that are developed and used for exploring genetic determinants of complex traits and diseases. Although technologies employed for GWAS and molecular interactions analysis are based on fundamentally different principles, there is both a great potential and a major challenge of integrating them to increase their power and deepen our understanding of the intricate mechanisms underlying the expression of such common multifactorial conditions.

References

Allison, D.B. (1997). Transmission-disequilibrium tests for quantitative traits. *American Journal of Human Genetics* 60(3), 676–690.

Allocco, D.J., Kohane, I.S. & Butte, A.J. (2004). Quantifying the relationship between co-expression, co-regulation and gene function. *BMC Bioinformatics* 5, 18.

Alon, U. (2003). Biological networks: The tinkerer as an engineer. *Science* 301(5641), 1866–1867.

Balaji, S., Babu, M.M., Iyer, L.M., Luscombe, N.M. & Aravind, L. (2006). Comprehensive analysis of combinatorial regulation using the transcriptional regulatory network of yeast. *Journal of Moleciar Biology* 360(1), 213–227.

Barabasi, A.L. & Oltvai, Z.N. (2004). Network biology: Understanding the cell's functional organization. *Nature Reviews Genetics* 5(2), 101–113.

Bergmann, S., Ihmels, J. & Barkai, N. (2004). Similarities and differences in genome-wide expression data of six organisms. *PLoS Biol*, 2(1), E9.

Blackwelder, W.C. & Elston, R.C. (1985). A comparison of sib-pair linkage tests for disease susceptibility loci. *Genetic Epidemiology* 2(1), 85–97.

Carlson, M.R., Zhang, B., Fang, Z., et al. (2006). Gene connectivity, function, and sequence conservation: Predictions from modular yeast co-expression networks. *BMC Genomics*, 7, 40.

Carter, S.L., Brechbuhler, C.M., Griffin, M. & Bond, A.T. (2004). Gene co-expression network topology provides a framework for molecular characterization of cellular state. *Bioinformatics* 20(14), 2242–2250.

Casna, N.J., Novack, D.F., Hsu, M.T. & Ford, J.P. (1986). Genomic analysis II: Isolation of high molecular weight heteroduplex DNA following differential methylase protection and Formamide-PERT hybridization. *Nucleic Acids Research* 14(18), 7285–7303.

Devlin, B. & Roeder, K. (1999). Genomic control for association studies. *Biometrics* 55(4), 997–1004.

Elston, R.C., Guo, X. & Williams, L.V. (1996). Two-stage global search designs for linkage analysis using pairs of affected relatives. *Genetic Epidemiology* 13(6), 535–558.

Evans, D.M. & Cardon, L.R. (2004). Guidelines for genotyping in genomewide linkage studies: Single-nucleotide-polymorphism maps versus microsatellite maps. *American Journal of Human Genetics* 75(4), 687–692.

Fallin, D., Beaty, T., Liang, K.Y. & Chen, W. (2002). Power comparisons for genotypic vs. allelic TDT methods with >2 alleles. *Genetic Epidemiology* 23(4), 458–461.

Frankel, W.N. & Schork, N.J. (1996). Who's afraid of epistasis? *Nature Genetics* 14(4), 371–373.

Guimera, R. & Nunes Amaral, L.A. (2005). Functional cartography of complex metabolic networks. *Nature* 433(7028), 895–900.

Hartwell, L.H., Hopfield, J.J., Leibler, S. & Murray, A.W. (1999). From molecular to modular cell biology. *Nature* 402(Suppl. 6761), C47–C52.

Haseman, J.K. & Elston, R.C. (1972). The investigation of linkage between a quantitative trait and a marker locus. *Behaviour Genetics* 2(1), 3–19.

Henegar, C., Tordjman, J., Achard, V., et al. (2008). Adipose tissue transcriptomic signature highlights the pathological relevance of extracellular matrix in human obesity. *Genome Biology* 9(1), R14.

Hoggart, C.J., Parra, E.J., Shriver, M.D., et al. (2003). Control of confounding of genetic associations in stratified populations. *American Journal of Human Genetics* 72(6), 1492–1504.

Hoh, J. & Ott, J. (2000). Scan statistics to scan markers for susceptibility genes. *Proceedings of the National Academy of Sciences U S A*, 97(17), 9615–9617.

Hoh, J. & Ott, J. (2003). Mathematical multi-locus approaches to localizing complex human trait genes. *Nature Reviews Genetics* 4(9), 701–709.

Jeong, H., Mason, S.P., Barabasi, A.L. & Oltvai, Z.N. (2001). Lethality and centrality in protein networks. *Nature* 411(6833), 41–42.

Kraft, P. (2006). Efficient two-stage genome-wide association designs based on false positive report probabilities. *Pacific Symposium on Biocomputing* 523–534.

Kruglyak, L., Daly, M.J., Reeve-Daly, M.P. & Lander, E.S. (1996). Parametric and nonparametric linkage analysis: A unified multipoint approach. *American Journal of Human Genetics* 58(6), 1347–1363.

Lyon, H.N. & Hirschhorn, J.N. (2005). Genetics of common forms of obesity: A brief overview. *American Journal of Clinical Nutrition* 82(Suppl. 1), 215S–217S.

Matsuzaki, H., Loi, H., Dong, S., et al. (2004). Parallel genotyping of over 10,000 SNPs using a one-primer assay on a high-density oligonucleotide array. *Genome Research*, 14(3), 414–425.

McGinnis, R., Shifman, S. & Darvasi, A. (2002). Power and efficiency of the TDT and case-control design for association scans. *Behaviour Genetics* 32(2), 135–144.

Middleton, F.A., Pato, M.T., Gentile, K.L., et al. (2004). Genomewide linkage analysis of bipolar disorder by use of a high-density single-nucleotide-polymorphism (SNP) genotyping assay: A comparison with microsatellite marker assays and finding of significant linkage

to chromosome 6q22. *American Journal of Human Genetics* 74(5), 886–897.

Moore, J.H. & Ritchie, M.D. (2004). STUDENTJAMA. The challenges of whole-genome approaches to common diseases. *JAMA* 291(13), 1642–1643.

Nelson, S.F., McCusker, J.H., Sander, M.A., et al. (1993). Genomic mismatch scanning: A new approach to genetic linkage mapping. *Nature Genetics* 4(1), 11–18.

Prifti, E., Zucker, J.D., Clement, K. & Henegar, C. (2008). FunNet: An integrative tool for exploring transcriptional interactions. *Bioinformatics* 24(22), 2636–2638.

Pritchard, J.K. & Donnelly, P. (2001). Case-control studies of association in structured or admixed populations. *Theoretical Population Biology* 60(3), 227–237.

Ravasz, E., Somera, A.L., Mongru, D.A., Oltvai, Z.N. & Barabasi, A.L. (2002). Hierarchical organization of modularity in metabolic networks. *Science* 297(5586), 1551–1555.

Redden, D.T. & Allison, D.B. (2003). Nonreplication in genetic association studies of obesity and diabetes research. *Journal of Nutrition* 133(11), 3323–3326.

Sanda, A.I. & Ford, J.P. (1986). Genomic analysis I: Inheritance units and genetic selection in the rapid discovery of locus linked DNA markers. *Nucleic Acids Research* 14(18), 7265–7283.

Sasieni, P.D. (1997). From genotypes to genes: Doubling the sample size. *Biometrics* 53(4), 1253–1261.

Satagopan, J.M. & Elston, R.C. (2003). Optimal two-stage genotyping in population-based association studies. *Genetic Epidemiology* 25(2), 149–157.

Satagopan, J.M., Verbel, D.A., Venkatraman, E.S., Offit, K.E. &

Begg, C.B. (2002). Two-stage designs for gene-disease association studies. *Biometrics* 58(1), 163–170.

Sobell, J.L., Heston, L.L. & Sommer, S.S. (1993). Novel association approach for determining the genetic predisposition to schizophrenia: Case-control resource and testing of a candidate gene. *American Journal of Medical Genetics* 48(1), 28–35.

Spielman, R.S., McGinnis, R.E. & Ewens, W.J. (1993). Transmission test for linkage disequilibrium: The insulin gene region and insulin-dependent diabetes mellitus (IDDM). *American Journal of Human Genetics* 52(3), 506–516.

Swarbrick, M.M., Waldenmaier, B., Pennacchio, L.A., et al. (2005). Lack of support for the association between GAD2 polymorphisms and severe human obesity. *PLoS Biology* 3(9), e315.

The International HapMap Consortium (2007). A second generation human haplotype map of over 3.1 million SNPs. *Nature* 449(7164), 851–861.

Thomas, D., Xie, R. & Gebregziabher, M. (2004). Two-stage sampling designs for gene association studies. *Genetic Epidemiology* 27(4), 401–414.

van den Oord, E.J. & Sullivan, P.F. (2003). A framework for controlling false discovery rates and minimizing the amount of genotyping in the search for disease mutations. *Human Heredity* 56(4), 188–199.

Walter, S.D. (1976). The estimation and interpretation of attributable risk in health research. *Biometrics* 32(4), 829–849.

Witte, J.S., Gauderman, W.J. & Thomas, D.C. (1999). Asymptotic bias and efficiency in case-control studies of candidate genes and gene-environment interactions: Basic family designs. *American Journal of Epidemiology* 149(8), 693–705.

PART 2

EVIDENCE FROM GENETIC EPIDEMIOLOGY STUDIES

Chapter 7

Genetic Epidemiology, Physical Activity, and Inactivity

MOLLY S. BRAY,[1] JANET E. FULTON,[2] NISHAN SUDHEERA KALUPAHANA[3] AND J. TIMOTHY LIGHTFOOT[4]

[1]*Department of Epidemiology, Heflin Center for Genomic Sciences, University of Alabama at Birmingham, Birmingham, AL, USA*
[2]*Division of Physical Activity, Nutrition, and Obesity, Centers for Disease Control and Prevention, Atlanta, GA, USA*
[3]*University of Tennessee Obesity Research Center, Knoxville, TN, USA*
[4]*Department of Health and Kinesiology, Texas A&M University, College Station, TX, USA*

It has long been recognized that the propensity for voluntary exercise can be bred into rodents and other animals and varies widely with type of rodent strain (Lerman et al., 2002; Lightfoot et al., 2004, 2008). Through extended breeding studies, Garland and colleagues have developed high-active and low-active rodent colonies in which mice from high-active lines consistently and stably run ~2.5–3.0 times as many revolutions/day as controls, due primarily to differences in running speed (versus time spent on the running wheel) (Koteja et al., 1999; Swallow et al., 1999, 2001). Differences in physiologic measures, such as cardiac contractility, cardiac hypertrophy, and other measures, explain some of the differences between mouse strains in forced and voluntary wheel-running performance (Lerman et al., 2002). In addition, psychological parameters such as the appetitive value of running or enhanced frustration, anger, anxiety, or stress when running is denied have also been shown to be strong driving forces in the animals selectively bred for high physical activity (Rhodes et al., 2003). These studies provide strong evidence for a genetic basis for physical activity and suggest that a host of physiological and psychological factors that drive physical activity may also have substantial genetic underpinnings.

For complex traits with a genetic basis (e.g., body composition, heart rate, fasting glucose, and so on), it is generally accepted that several to many genes likely act cumulatively and/or concurrently to influence the trait, with each gene individually explaining a small portion of the variance in the trait. For complex human behaviors, such as physical activity, the model becomes even more complicated, since learning, modeling, the physical environment, and other factors may have a profound influence on the effect of any single gene encoding a predisposition to be physically active or inactive. These observations make the identification of specific genes that influence physical activity levels in humans a particularly daunting task. The most recent version of the Human Gene Map for Performance and Health-Related Fitness Phenotypes, developed in 2000 by Bouchard and colleagues, includes a total of 6 genes and 10 linkage regions putatively associated with physical activity among a wide variety of subjects (Bray et al., 2009). The ability to identify genes for physical

Genetic and Molecular Aspects of Sport Performance, 1st edition.
Edited by Claude Bouchard and Eric P. Hoffman.
Published 2011 by Blackwell Publishing Ltd.

activity depends critically on how precisely and accurately the behavior is characterized, how well the environmental factors are accounted for, and how strongly the gene(s) being studied influence the physical activity behavior. This chapter will provide an overview of assessment tools used to measure physical activity, non-genetic factors recognized to influence physical activity levels, basic concepts of genetic epidemiology, and the genetic basis of physical activity at the population level.

Physical Activity Assessment

Physical activity was defined many years ago as "any voluntary movement produced by skeletal muscles that results in energy expenditure" and can take place in many domains (or contexts) and for a variety of purposes (Caspersen et al., 1985). There are many methods available to assess physical activity behavior, and ultimately, the research question will guide the selection of the most appropriate physical activity assessment method. The accuracy of physical activity assessment is influenced directly by the type of instrument used, and inaccurate or "noisy" physical activity assessment may reduce the ability to detect associations between physical activity and both the genetic and non-genetic factors that may mediate (come between) or moderate (differentially affect) the relationship.

Physiologic Assessment of Energy Expenditure

Direct estimation of energy expenditure can be derived from *direct or indirect calorimetry*. Both methods provide an estimate of the amount of heat produced which is then converted to an estimate of energy expenditure (kcals). *Direct calorimetry* directly measures the heat produced by an individual and requires placement inside a sealed, insulated room calorimeter chamber (Lee et al., 2009). Room calorimeters provide highly accurate ($<1\%$ error) estimates of energy expenditure but are restrictive to many types of free-living physical activity. *Indirect calorimetry* calculates heat production indirectly through measurement of the consumption of oxygen and the production of carbon dioxide to estimate energy expenditure (Dishman et al., 2004).

Indirect calorimetric methods based on heart rate as a proxy for carbon dioxide production can also be used to estimate energy expenditure. *Chemical assessment* of physical activity involves the ingestion and metabolic analysis of water labeled with two different stable isotopes, ^2H and ^{18}O (hence, the method is called the doubly labeled water method), and the rate of elimination through respiration and urination of these isotopes can be used to estimate total energy expenditure in free-living environments (Schoeller et al., 1986). Although objective and highly accurate for assessing total energy expenditure from short to longer periods of time, direct physiologic measures of physical activity are generally expensive and impractical for use in large epidemiological studies.

Objective Monitoring of Physical Activity

Direct observation is often used when assessing the physical activity of children, for whom other types of measures are either not available and/or not valid. In addition to providing an objective measure of physical activity, direct observation can also include notations regarding the physical environment and the context in which the subject is moving or not moving (McIver et al., 2009; Pate, 1993). *Devices* designed to measure the quantity and/or quality of physical activity include pedometers, accelerometers, heart rate monitors, and combinations of these instruments. Pedometers record movement in the vertical plane (Schneider et al., 2004), while accelerometers are able to detect movement in multiple planes (Plasqui & Westerterp, 2007). Nevertheless, neither pedometers nor accelerometers are useful for measuring non-moving energy expenditure, and the accuracy of accelerometer measurements compared with energy expenditure measured using doubly labeled water varies widely by manufacturer (Plasqui & Westerterp, 2007). *Heart rate monitors* are designed to detect changes in heart rate associated with physical activity and movement rather than directly detecting the physical movement itself. Heart rate is highly correlated to energy expenditure for moderate to vigorous physical activity but less so for low levels of physical activity, in which confounding by stress, smoking, and other factors can lessen the accuracy of heart rate for characterizing physical activity (Livingstone, 1997). Newer types of accelerometers combine both heart rate monitoring and accelerometry to provide a more accurate and objective field assessment of energy expenditure and physical activity (Brage et al., 2005).

Questionnaires

Participant recall of physical activity or sedentary behavior can be completed using a wide range of reporting instruments, which have been designed for (a) specific populations and age groups, (b) specific lengths of physical activity or inactivity, and (c) specific types of physical activities (e.g., watching TV, outside play, and so on). Reporting of physical activity can be made through diaries or other self-report, proxy reporting (e.g., parental reporting

of children's activity), or via interviewer assistance. All types of subjective reporting require the subject or proxy to accurately remember and report past physical activity. As such, even detailed instruments, such as physical activity diaries, are susceptible to recall error (Neilson et al., 2008). Nevertheless, these types of instruments are widely used in physical activity research because they are inexpensive and easily administered in almost any setting.

Table 7.1 summarizes the characteristics of the methods outlined above. Genetic studies of physical activity must balance the need for large sample sizes that have sufficient power to detect small gene effects with the use of assessment instruments that are both cost-effective and efficient. Thus, subjective questionnaires are most often used in genetic studies, which ultimately may lessen the power to detect

Table 7.1 Physical activity assessment methods and their associated characteristics

Method	Physical activity prescription				Volume (metric)[a]	Validation standard	Inactivity	Context	Burden	
	Frequency	Intensity	Duration	Type					Investigator	Participant
Doubly labeled water	No	No	No	No	Yes (Kcal)	Yes	No	No	High	Low
Direct calorimetry	No	No	No	No	Yes (Kcal)	Yes	No	No	High	High
Indirect calorimetry	No	No	No	No	Yes (Kcal)	Yes	No	No	Medium	Medium
Direct observation	No	Yes	Yes	Yes	Yes (Time)	Yes	Yes	Yes	High	Low
Heart rate monitor	Yes	Yes	Yes	No	Yes (Time at intensity level)	Yes	Yes	No	Medium	Medium
Accelerometer	Yes	Yes	Yes	No	No (Acceleration counts)	Yes	Yes	No	High	Medium
Pedometer	No	No	No	No	No (Steps)	No	No	No	Low	Medium
Diary/log	Yes	Yes	Yes	Yes	Yes (Variety of metrics—e.g., kcals, MET-time)	Yes	Yes	Yes	Medium	High
Questionnaire	Yes	Yes	Yes	Yes	Yes (Variety of metrics—e.g., kcals, MET-time)	No	No	Yes	Low	Low

[a] MET refers to metabolic equivalent, and 1 MET is the rate of energy expenditure while sitting at rest. It is taken by convention to be an oxygen uptake of 3.5 ml/kg of body weight per minute. Physical activities frequently are classified by their intensity using the MET as a reference.

gene–physical activity relationships. In addition, temporal and/or generational differences in physical activity levels may make it difficult to identify concordance in these behaviors among heterogeneous populations or even among family members.

Non-genetic Predictors of Physical Activity

While accurate assessment of physical activity can undoubtedly influence the ability to identify genes that may influence such behavior, it is also evident that genes do not act independently of other non-genetic factors that may mediate and/or moderate the propensity to be physical activity. Many demographic, socioeconomic, psychological, and environmental factors have been examined for association to physical activity levels in humans. Among demographic and physical factors, the strongest positive correlates of physical activity are income/socioeconomic status, education, and male gender, while the strongest negative correlates remain age, obesity/overweight, and non-white race/ethnicity (Trost et al., 2002). Psychological and social factors such as enjoyment, perceived health benefits, self-efficacy, self-motivation, and social support from physicians, friends, and family have consistently been positively associated with physical activity, while perceived barriers, lack of time, and mood disturbances are negatively associated with physical activity (Trost et al., 2002). There is little evidence that knowledge of health and fitness, attitudes, normative beliefs, or susceptibility to illness is predictive of physical activity levels. In a recent systematic review, Wendel-Vos and colleagues (2007) showed that there is limited supportive evidence for the role of presumed environmental factors in determining the level of physical activity in adults. Climate is the only strong environmental predictor of physical activity, and other components of the physical environment (e.g., scenery, crime, safety, sidewalks, access to facilities, terrain, traffic) are only weakly associated with physical activity (Trost et al., 2002).

In assessing the effects of genetic variation on physical activity behavior, the most robust analyses would also take into account the physical, social, and environmental factors that are known to strongly influence physical activity behavior. While many genetic analyses of physical activity level include body mass index (BMI), age, gender, or other physical factors, social and environmental factors are rarely considered.

Basic Concepts of Genetic Epidemiology

The field of epidemiology is focused on the identification of factors that affect the health and disease of populations, while genetics is the study of inheritance and heritable traits and disease. Genetic epidemiology combines elements of both epidemiology and genetics and involves the study of genetic factors that interact with environmental factors to influence the occurrence of disease and related traits in populations (Khoury et al., 1993). Many different study designs and analytical approaches can be used to address research questions in genetic epidemiology, from family-based studies to the traditional cohort and case–control designs used in classic epidemiology.

Family-based studies can be used to estimate the amount of variation in a trait that can be accounted for by genetic variation in the population, termed heritability. While it is generally assumed that all physiologic traits have a genetic component, segregation analysis compares the joint occurrence of a trait among family members, and estimates the portion of the variability in the trait that is due to genetic variation, controlling for shared family environment, and other nongenetic contributors. It is important to emphasize that heritability is a population concept, not an individual one; in other words, if the heritability of BMI is 0.50 (or 50%), that means that 50% of the variation we observe in BMI values *in the population* can be explained by the fact that there is variation in DNA sequence among the population that regulates or influences BMI.

Genetic linkage analyses utilize family-based designs and are designed to interrogate and compare the sharing of DNA sequence variation across the genome among family members that do or do not share a phenotype or disease of interest. By taking into account the relatedness of relative pairs, for example, specialized genetic statistics can be used to determine the expected probability of genotype sharing for the specific type of relative pair, which can then be compared to the observed phenotype sharing for the pair.

Genetic association analysis may be conducted in cohorts of unrelated individuals, as well as in samples of cases and controls (for dichotomous or disease-based outcomes). Traditional statistical approaches can often be used to determine genetic association in these study designs. Genotype determined in unrelated individuals can be examined for association to a dichotomous outcome, such as disease status, or to a quantitative trait, such as BMI, blood pressure, and so on. Specialized statistical methods for analyzing association across the entire genome have also been developed that take into account issues of multiple testing as well as the joint effects of multiple variant sites in the genome.

Genetic Epidemiology of Physical Activity and Inactivity

Several researchers have shown that there is a familial aggregation of physical activity. Moore and colleagues (1991) in a study of 100 children from the Framingham Children Study showed that children with physically active parents are 5.8 times more likely to be physically active than children with inactive parents, suggesting a genetic contribution to the level of physical activity. Nevertheless, the contribution of other factors like parental influence, role modeling, and other environmental factors are difficult to separate from genetic factors (Guo, 2000). In contrast, twin and family studies have provided substantial evidence that the propensity to engage in physical activity is influenced by genes and provide more specific evidence on the heritability of a trait (MacGregor et al., 2000).

Estimates of the heritability of physical activity and exercise participation are often highly variable, likely due to differences in the population from which subjects were ascertained, differences in the type of physical activity or inactivity examined, and differences in the instruments used to perform the physical activity assessment (Cai et al., 2006;

Kaprio et al., 1981; Seabra et al., 2008; Simonen et al., 2002). In most studies thus far, the estimation of physical activity has been based upon subjective assessments, such as leisure-time physical activity, exercise participation, sports participation index, and self-perceived athletic ability, while only a few have used more objective measures such as activity-induced energy expenditure (using doubly labeled water) and the use of accelerometers (Table 7.2).

Using a family-based design in one of the first studies to estimate the genetic contribution to physical activity, Perusse and colleagues (1989) examined concordance of physical activity determined from a 3-day activity diary in 1610 subjects from 375 families who lived in the greater Québec city area. Pairs of biologic and non-biologic relatives were used to calculate familial correlations, controlling for age, sex, physical fitness, BMI, and socioeconomic status. While genetic factors accounted for a significant portion of the variability in habitual physical activity and exercise participation (heritability = 29% and 12%, respectively), non-transmissible environmental factors also contributed strongly to these two physical activity indicators in this population (Perusse et al. 1989). Simonen and colleagues examined heritabilities of a range of physical activity behaviors among French Canadian families from the Heritage Study. Physical activity in this study was examined using a 3-day activity diary, which included 2 week days and 1 weekend day of reporting. Subjects in the study recorded the dominant activity for each 15-min interval using a list of activities categorized into nine levels of intensity (Simonen et al., 2002). The category scores were summed to create an overall activity score, as well as scores for inactivity and for moderate to vigorous physical activity. In addition, a separate questionnaire was used to determine the number of physical activity sessions performed per week, along with the average frequency and duration of each session for the physical activity most often engaged in, which were extracted and used to compute the time spent on the most common physical activity (hours per week) during the previous year (Simonen et al., 2003). These researchers reported the highest heritability ($h^2 = 0.25$) for physical inactivity,

Table 7.2 Summary of twin studies with heritability estimates for physical activity phenotypes

Study	Number of subjects	Phenotype	Genetic factors/ heritability	Common environmental factors	Unique environmental factors
Stubbe et al. (2006)	85,198	Exercise participation (>60 min/week, >4 METs)	0.27–0.71	0–0.37	0.30–0.52
Duncan et al. (2008)	1389	Exercise participation (>60 min/week, >4 METs)	0.45	–	0.55
		Exercise participation (>150 min/week, >4 METs)	–	0.28	0.72
Joosen et al. (2005)	40	Activity-induced energy expenditure (using doubly labeled water)	0.72	–	0.28
		Physical activity (using accelerometers)	0.78	–	0.22
Eriksson et al. (2006)	2044	Physical activity (Baecke questionnaire)	0.49	–	0.51
Carlsson et al. (2006)	26,724	Leisure-time physical activity (questionnaire)	Male: 0.57 Female: 0.50	0.03 0.06	0.40 0.44
Franks et al. (2005)	200	Physical activity energy expenditure (respiratory gas exchange and doubly labeled water)	0.41	0.35	0.24
		Adjusted for weight	–	0.69	0.31
Maia et al. (2002)	411	Sports participation index	Male: 0.684 Female: 0.398	0.20 0.284	0.116 0.318
		Leisure-time physical activity (Baecke questionnaire)	Males: 0.63 Female: 0.32	– 0.38	0.37 0.30
Stubbe et al. (2005)	5256	Leisure-time physical activity (4 METs or more >60 min/week) 13–16 years	–	0.78–0.84	0.22–0.16
		17–18 years	0.36	0.47	0.17
		19–20 years	0.85	–	0.15
Kaprio et al. (1981)	9188	Daily physical activity	0.62	–	0.38

MET, metabolic equivalent.

followed by heritabilities of 19%, 17%, and 16% for total daily activity, time spent on a primary activity, and time spent in moderate/strenuous activity, respectively (Simonen et al., 2002). In Mexican–American families from the San Antonio Heart Study, shared genetic factors accounted for a significant but fairly small portion (9%) of the variance in physical activity, as measured by a 7-day recall questionnaire, with environmental factors again contributing strongly to these traits (Mitchell et al., 2003).

Twin studies generally provide the most liberal estimates of heritability, with estimates ranging from 0.27 to 0.78, and among the highest concordance reported between twins is for vigorous physical activity (Beunen & Thomis, 1999; De Moor et al., 2007; Eriksson et al., 2006; Maia et al., 2002; Pittaluga et al., 2004). In the largest twin study of physical activity to date, Stubbe et al. (2006) examined the concordance of self-reported physical activity behavior among 13,676 monozygotic and 23,375 dizygotic twin pairs from six European countries and Australia. These researchers reported a median heritability for exercise participation (equivalent to performing 60 or more minutes of exercise per week) of 62%, with heritability ranging from 27% to 70% and varying by gender and country. The variability of these estimates may be due, at least in part, to the different precision of the physical activity assessment instruments used in each country (Stubbe et al., 2006). The range of heritability reported by Stubbe et al. is similar to that of other twin studies (32–85%), which are summarized in Table 7.2.

Several studies have investigated the heritability of physical activity by using different exercise volumes as cutoffs. Researchers report that while there is a clear heritability for physical activity levels up to 60 min of moderate intensity physical activity per week, this effect is attenuated at a volume of 150 min/week (Duncan et al., 2008). This finding suggests that while genetic influences may determine the habitual level of physical activity, non-genetic factors may strongly contribute to a higher level of physical activity such as the volume recommended for health benefits. Interestingly, Joosen and colleagues (2005) examined the heritability of activity-related energy expenditure (AEE) and physical activity in 20 monozygotic and dizygotic twin pairs, and reported heritabilities of 72% and 78% for free-living physical activity measured with doubly labeled water and accelerometers but found no significant contribution of genetic factors to 24-h AEE measured in a respiration chamber in the same subject cohort. The findings of Joosen et al. suggest that short-term energy expenditure, even when measured very accurately in the respiration chamber, may be a less stable trait compared to habitual activity measured over longer time periods, resulting in decreased concordance between twins for the short-term trait. While most studies have shown that the heritability of physical activity was similar between males and females, twin cohorts from Norway (Stubbe et al., 2006) and Portugal (Maia et al., 2002) have shown gender-dependent differences.

Limited studies have estimated the heritability of physical activity specifically in children. Butte et al. (2006) determined heritability for percent of time spent in various types of activity measured via accelerometers in a large family study of Hispanic children aged 4–19 years. The lowest heritability observed among these children was 32% for vigorous activity, and the highest heritability was 66% for time spent in sedentary activity (Butte et al., 2006). In 100 monozygotic and dizygotic twin pairs aged 4–10 years, Franks et al. (2005) reported heritabilities of 79% for resting metabolic rate and 41% for physical activity energy expenditure after adjustment for age, sex, ethnicity, study date, and season; nevertheless, after further adjustment for body weight, additive genetic factors explained none of the variance in these traits. Estimates of the heritability of physical activity in children are generally within the range of values reported for adults, suggesting that genetic factors influence physical activity behavior across the life span. While several studies using children as subjects have shown that there is heritability of physical activity even in childhood, one study shows that it becomes more apparent in adulthood (Stubbe et al., 2005). Whether this pattern is similar in elderly adults is yet to be explored.

Conclusion

The determination of heritability is useful for estimating the amount of variance in a trait or behavior that can be expected to be accounted for by a given gene or set of genes. The studies described above provide strong evidence for the contribution of genetic factors to physical activity behavior. Nevertheless, it is important to note that, in most studies, shared and non-shared environmental factors also account for a substantial part of the variability observed in physical activity behaviors.

References

Beunen, G. & Thomis, M. (1999). Genetic determinants of sports participation and daily physical activity. *International Journal of Obesity and Related Metabolic Disorders* 23(Suppl. 3), S55–S63.

Brage, S., Brage, N., Franks, P.W., Ekelund, U. & Wareham, N.J. (2005). Reliability and validity of the combined heart rate and movement sensor Actiheart. *European Journal of Clinical Nutrition* 59, 561–570.

Bray, M.S., Hagberg, J.M., Perusse, L., et al. (2009). The human gene map for performance and health-related fitness phenotypes: The 2006–2007 update. *Medicine & Science in Sports & Exercise* 41, 35–73.

Butte, N.F., Cai, G., Cole, S.A. & Comuzzie, A.G. (2006). Viva la Familia Study: Genetic and environmental contributions to childhood obesity and its comorbidities in the Hispanic population. *The American Journal of Clinical Nutrition* 84, 646–654.

Cai, G., Cole, S.A., Butte, N., et al. (2006). A quantitative trait locus on chromosome 18q for physical activity and dietary intake in Hispanic children. *Obesity (Silver Spring)* 14, 1596–1604.

Carlsson, S., Andersson, T., Lichtenstein, P., Michaelsson, K. & Ahlbom, A. (2006). Genetic effects on physical activity: Results from the Swedish Twin Registry. *Medicine & Science in Sports & Exercise* 38, 1396–1401.

Caspersen, C.J., Powell, K.E. & Christenson, G.M. (1985). Physical activity, exercise, and physical fitness: Definitions and distinctions for health-related research. *Public Health Reports* 100, 126–131.

De Moor, M.H., Stubbe, J.H., Boomsma, D. I. & De Geus, E.J. (2007). Exercise participation and self-rated health: Do common genes explain the association? *European Journal of Epidemiology* 22, 27–32.

Dishman, R., Washburn, R. & Heath, G. (eds.) (2004). *Physical Activity Epidemiology*. Human Kinetics, Champaign, IL.

Duncan, G.E., Goldberg, J., Noonan, C., Moudon, A.V., Hurvitz, P. & Buchwald, D. (2008). Unique environmental effects on physical activity participation: A twin study. *PLoS One* 3, e2019.

Eriksson, M., Rasmussen, F. & Tynelius, P. (2006). Genetic factors in physical activity and the equal environment assumption—the Swedish young male twins study. *Behavior Genetics* 36, 238–247.

Franks, P.W., Ravussin, E., Hanson, R.L., et al. (2005). Habitual physical activity in children: The role of genes and the environment. *American Journal of Clinical Nutrition* 82, 901–908.

Guo, S. W. (2000). Familial aggregation of environmental risk factors and familial aggregation of disease. *American Journal of Epidemiology* 151, 1121–1131.

Joosen, A.M., Gielen, M., Vlietinck, R. & Westerterp, K.R. (2005). Genetic analysis of physical activity in twins *American Journal of Clinical Nutrition* 82, 1253–1259.

Kaprio, J., Koskenvuo, M. & Sarna, S. (1981). Cigarette smoking, use of alcohol and leisure-time activity among same-sexed adult male twins. In: L. Gedda, P. Parisi & W. Nance (eds) *Progress in Clinical and Biological Research*, pp. 37–46. Alan R. Liss, New York, NY.

Khoury, M.J., Beaty, T.H. & Cohen, B.H. (1993). *Fundamentals of Genetic Epidemiology*. Oxford University Press, New York, NY.

Koteja, P., Garland, T., Jr., Sax, J.K., Swallow, J.G. & Carter, P.A. (1999). Behaviour of house mice artificially selected for high levels of voluntary wheel running. *Animal Behaviour* 58, 1307–1318.

Lee, I.-M., Blair, S., Manson, J. & Paffenbarger, R. (eds) (2009). *Epidemiologic Methods in Physical Activity Studies*. Oxford University Press, New York, NY.

Lerman, I., Harrison, B.C., Freeman, K., et al. (2002). Genetic variability in forced and voluntary endurance exercise performance in seven inbred mouse strains. *Journal of Applied Physiology* 92, 2245–2255.

Lightfoot, J.T., Turner, M.J., Daves, M., Vordermark, A. & Kleeberger, S.R. (2004). Genetic influence on daily wheel running activity level. *Physiological Genomics* 19, 270–276.

Lightfoot, J.T., Turner, M.J., Pomp, D., Kleeberger, S.R. & Leamy, L.J. (2008). Quantitative trait loci for physical activity traits in mice. *Physiological Genomics* 32, 401–408.

Livingstone, M.B. (1997). Heart-rate monitoring: The answer for assessing energy expenditure and physical activity in population studies? *British Journal of Nutrition* 78, 869–871.

MacGregor, A.J., Snieder, H., Schork, N.J. & Spector, T.D. (2000). Twins. Novel uses to study complex traits and genetic diseases. *Trends in Genetics* 16, 131–134.

Maia, J.A., Thomis, M. & Beunen, G. (2002). Genetic factors in physical activity levels: A twin study. *American Journal of Preventive Medicine* 23, 87–91.

McIver, K.L., Brown, W.H., Pfeiffer, K.A., Dowda, M. & Pate, R.R. (2009). Assessing children's physical activity in their homes: The observational system for recording physical activity in children-home. *Journal of Applied Behavior Analysis* 42, 1–16.

Mitchell, B.D., Rainwater, D.L., Hsueh, W.C., Kennedy, A.J., Stern, M.P. & Maccluer, J.W. (2003). Familial aggregation of nutrient intake and physical activity: Results from the San Antonio Family Heart Study. *Annals of Epidemiology* 13, 128–135.

Moore, L.L., Lombardi, D.A., White, M. J., Campbell, J.L., Oliveria, S.A. & Ellison, R.C. (1991). Influence of parents' physical activity levels on activity levels of young children. *Journal of Pediatrics* 118, 215–219.

Neilson, H.K., Robson, P.J., Friedenreich, C.M. & Csizmadi, I. (2008). Estimating activity energy expenditure: How valid are physical activity questionnaires? *American Journal of Clinical Nutrition* 87, 279–291.

Pate, R.R. (1993). Physical activity assessment in children and adolescents. *Critical Reviews in Food Science and Nutrition* 33, 321–326.

Perusse, L., Tremblay, A., Leblanc, C. & Bouchard, C. (1989). Genetic and environmental influences on level of habitual physical activity and exercise participation. *American Journal of Epidemiology* 129, 1012–1022.

Pittaluga, M., Casini, B. & Parisi, P. (2004). Physical activity and genetic influences in risk factors and aging: A study on twins. *International Journal of Sports Medicine* 25, 345–350.

Plasqui, G. & Westerterp, K.R. (2007). Physical activity assessment with accelerometers: An evaluation against doubly labeled water. *Obesity (Silver Spring)* 15, 2371–2379.

Rhodes, J.S., Garland, T., Jr., & Gammie, S.C. (2003). Patterns of brain activity associated with variation in voluntary wheel-running behavior. *Behavioral Neuroscience* 117, 1243–1256.

Schneider, P.L., Crouter, S.E. & Bassett, D.R. (2004). Pedometer measures of free-living physical activity: Comparison of

13 models. *Medicine & Science in Sports & Exercise* 36, 331–335.

Schoeller, D.A., Ravussin, E., Schutz, Y., Acheson, K.J., Baertschi, P. & Jequier, E. (1986). Energy expenditure by doubly labeled water: Validation in humans and proposed calculation. *American Journal of Physiology* 250, R823–R830.

Seabra, A.F., Mendonca, D.M., Goring, H.H., Thomis, M.A. & Maia, J.A. (2008). Genetic and environmental factors in familial clustering in physical activity. *European Journal of Epidemiology* 23, 205–211.

Simonen, R., Perusse, L., Rankinen, T., Rice, T., Rao, D. & Bouchard, C. (2002). Familial aggregation of physical activity levels in the Quebec family study. *Medicine & Science in Sports & Exercise* 34, 1137–1142.

Simonen, R.L., Rankinen, T., Perusse, L., et al. (2003). Genome-wide linkage scan for physical activity levels in the Quebec Family study. *Medicine & Science in Sports & Exercise* 35, 1355–1359.

Stubbe, J.H., Boomsma, D.I., Vink, J.M., et al. (2006). Genetic influences on exercise participation in 37,051 twin pairs from seven countries. *PLoS One* 1, e22.

Stubbe, J.H., Boomsma, D.I. & De Geus, E.J. (2005). Sports participation during adolescence: A shift from environmental to genetic factors. *Medicine & Science in Sports & Exercise* 37, 563–570.

Swallow, J.G., Koteja, P., Carter, P.A. & Garland, T. (1999). Artificial selection for increased wheel-running activity in house mice results in decreased body mass at maturity. *Journal of Experimental Biology* 202, 2513–2520.

Swallow, J.G., Koteja, P., Carter, P.A. & Garland, T., Jr. (2001). Food consumption and body composition in mice selected for high wheel-running activity. *Journal of Comparative Physiology B* 171, 651–659.

Trost, S.G., Owen, N., Bauman, A.E., Sallis, J.F. & Brown, W. (2002). Correlates of adults' participation in physical activity: Review and update. *Medicine & Science in Sports & Exercise* 34, 1996–2001.

Wendel-Vos, W., Droomers, M., Kremers, S., Brug, J. & van Lenthe, F. (2007). Potential environmental determinants of physical activity in adults: A systematic review. *Obesity Reviews* 8, 425–440.

Chapter 8

Role of Genetics Factors in Sport Performance: Evidence from Family Studies

LOUIS PÉRUSSE

Department of Social and Preventive Medicine, Division of Kinesiology, Laval University, Québec, QC, Canada

This chapter presents a review of the contribution of genetic factors for phenotypes related to sports performance from a genetic epidemiology perspective. The objective is to review the extent of familial resemblance and provide estimates of heritability for various phenotypes related to sports performance, including aerobic and anaerobic performance, muscular endurance and strength, motor performance, and some determinants of performance such as morphological, cardiac and skeletal muscle characteristics. Two major approaches have been used to assess the contribution of genetic factors in sports performance–related traits in humans: twin studies and family studies. Here, we discuss family studies, while Chapter 9 reviews twin studies. Specifically, we focus on three family studies that have contributed the bulk of evidence regarding the role of genetic factors for sports performance phenotypes: the Quebec Family Study (QFS), the 1981 Canada Fitness Survey (CFS) and the HERITAGE Family Study. Many additional family cohorts are described that have contributed to our knowledge of genetic factors in fitness, motor performance, and morphological features.

The results reviewed in this chapter clearly indicate the presence of familial resemblance for the major morphological and physiological determinants

of sports performance. The evidence for a role of genetic factors varies considerably depending of the performance phenotype. For some of them, such as anaerobic performance or skeletal muscle characteristics, there is a paucity of data. Nevertheless, from the available evidence, we can conclude that genetic factors account for ~40–60% of the variation in aerobic performance and cardiac function, 50–90% of the variation in anaerobic performance, 30–70% of the variation in muscular fitness, and 20–30% of the variation in cardiac dimensions.

The QFS, the 1981 CFS, and the HERITAGE Family Study

The QFS was initiated in 1979 (Phase 1) with the aim of determining the genetic basis of physical performance and physical fitness. Over a period of about 4 years, data were gathered on 1630 subjects coming from 375 families of French descent including various types of relatives by descent and adoption. Phase 2 of QFS began in 1992 with the aim of investigating the genetic factors of obesity and its related metabolic complications. A total of 385 subjects from Phase 1 families agreed to be measured a second time, and 372 new subjects from 74 families with one or more obese members were recruited. Phase 3 of the study started in 1998 with 5-year follow-up measurements of families tested in Phase 2 and with recruitment of 194 additional subjects. Over a period of about 25 years, a total of

Genetic and Molecular Aspects of Sport Performance, 1st edition.
Edited by Claude Bouchard and Eric P. Hoffman.
Published 2011 by Blackwell Publishing Ltd.

2196 subjects from 493 families were tested for a variety of health-related fitness traits. Most of the QFS studies pertaining to the genetics of fitness and physical performance were based on Phase 1 data. The CFS was initiated in 1981 in a representative sample of the Canadian population to assess lifestyle and physical fitness characteristics of the population. The sample consisted of 13,440 urban and rural households from all across the country and 23,400 subjects from 11,884 households agreed to participate. For the majority of these households, the nature of the family relationship among household members was available, which allowed the creation of a family database in which a total of 18,073 subjects were linked according to their biological relationships (Perusse et al., 1988). These subjects underwent a battery of physical fitness tests to assess aerobic performance, muscular endurance and strength, and flexibility. The most important study in the field is undoubtedly the HERITAGE (Health, RIsk factors, exercise Training and Genetics) Family Study documenting the role of genetic factors in the cardiovascular, metabolic and hormonal responses to aerobic exercise training (Bouchard et al., 1995). The study was carried out in a total of 483 subjects from 99 Caucasian families and 259 subjects from 105 African-American families who were tested, before and after a supervised 20-week endurance training program on cycle ergometer, for several measures of aerobic performance and physiological fitness. The study is unique because it allows the investigation of the contribution of genetic factors in key phenotypes related to sports performance both in baseline and in response to exercise training. Only baseline phenotypes will be considered in the present review.

The reader is referred to several comprehensive reviews that have addressed the contribution of genetic factors in physical fitness and sport performance from both genetic and molecular perspectives (Beunen & Thomis, 2004; Bouchard & Malina, 1983; Bouchard et al., 1992, 1997; Brutsaert & Parra, 2006; Peeters et al., 2009; Perusse, 2001; Rupert, 2003). In the present chapter, only the evidence from family studies will be reviewed.

Genetics of Aerobic and Anaerobic Performances

Aerobic Performance

Table 8.1 presents an overview of heritability estimates for both submaximal and maximal aerobic performance phenotypes. The spousal correlations are also presented to highlight the fact that common familial environment contribute to the familial aggregation of aerobic performance and that heritability estimates derived from family studies have to be considered as maximal heritabilities. In the QFS, submaximal aerobic performance was measured on a cycle ergometer as the power output at a heart rate (HR) of 150 beats per minute and expressed per kilogram of body weight (PWC_{150}/kg) (Bouchard et al., 1984). In the CFS, PWC_{150}/kg was estimated from a progressive step test (Perusse et al., 1988). Studies performed in both populations found that submaximal power output was characterized by significant familial resemblance (Bouchard et al., 1984; Perusse et al., 1987a, b, 1988), but resemblance among family members was apparent in both biological relatives and spouses or relatives by adoption, suggesting the contribution of both genetic and common familial environment. This was confirmed by a study based on data from CFS which showed that familial risk ratios for high values (exceeding the 95th percentile of the distribution) of PWC_{150} reached 1.63 and 1.81 for spouses and first-degree relatives, respectively (Katzmarzyk et al., 2000b). Similar estimates of heritability were found in both populations with values of 22% in QFS (Perusse et al., 1987b) and 28% in CFS (Perusse et al., 1988).

Early studies based on QFS data also revealed significant familial resemblance for predicted (Lortie et al., 1982) and measured (Bouchard et al., 1986a; Lesage et al., 1985) values of maximal oxygen uptake ($\dot{V}O_2max$). In the study of Bouchard et al. (1986a), $\dot{V}O_2max$ and endurance performance, measured as the total work output per kilogram of body weight during a 90-min maximal test on ergocycle, were assessed in 27 male sibling pairs and 172 twin pairs of both sexes (106 pairs of monozygotic, MZ, and 66 pairs

Table 8.1 Spouse correlations and heritability estimates for aerobic performance phenotypes derived from family studies

	Spouse correlation	h^2 (%)	Reference
Submaximal aerobic performance			
PWC_{150}/kg	0.19	NA	Bouchard et al. (1984)
	0.21	22	Perusse et al. (1987)
	0.17	28	Perusse et al. (1988)
$\dot{V}O_2$ 50W	0.45	70	Perusse et al. (2000)
$\dot{V}O_2$ 60%	0.17	30	
$\dot{V}O_2$ 80%	0.12	29	
$\dot{V}O_2$ at ventilatory threshold	0.34	58	Gaskill et al. (2001)
Maximal aerobic performance			
$\dot{V}O_2$max adjusted for body mass	0.18	NA	Montoye and Gale (1978)
	0.22	<25	Lesage et al. (1985)
	0.33	35–40	Lortie et al. (1982)
	NA	20–47	Szopa and Cempla (1985)
	NA	38–47	Bouchard et al. (1986a)
	0.35	NA	Maes et al. (1996)
	0.17	52	Bouchard et al. (1998)
Endurance performance			
	NA	66–72	Bouchard et al. (1986a)

of dizygotic, DZ, twins). For $\dot{V}O_2$max expressed per kilogram of body weight, estimates of heritability range from 38% to 47% (depending on the equation used to calculate h^2), while for endurance performance the heritability was higher ranging from 66% to 72%.

Familial aggregation of maximal and submaximal aerobic performance was also investigated in the HERITAGE Family Study (Bouchard et al., 1998; Gaskill et al., 2001; Perusse et al., 2001). At baseline, all subjects performed maximal as well as submaximal exercise tests on cycle ergometers with monitoring of gas exchange variables (VO_2, VCO_2, VE, and RER), which allowed to obtain direct measures of oxygen consumption. Besides $\dot{V}O_2$max, the submaximal aerobic performance phenotypes measured in the subjects were $\dot{V}O_2$ at an absolute power output of 50 watts ($\dot{V}O_2$50W), $\dot{V}O_2$ at relative power outputs equivalent to 60% and 80% of subjects $\dot{V}O_2$max and $\dot{V}O_2$ at the ventilatory threshold ($\dot{V}O_2$vt). The heritability estimates were 52% for $\dot{V}O_2$max, 29–70% for submaximal aerobic performance, and 58% for $\dot{V}O_2$vt.

Anaerobic Performance

Anaerobic performance refers to the ability to perform maximal exercise in tasks of short duration (generally less than 90 s). Compared to aerobic performance, relatively little is known about the heritability of anaerobic performance, and family data are limited. In one study, total work output per kilogram of body weight during a 10-s maximal exercise test was measured in biological sibships and adopted sibships as well as in DZ and MZ twins (Simoneau et al., 1986). The correlations were 0.80 in MZ twins, 0.58 in DZ twins, and 0.46 in biological siblings, but were not significantly different from zero in adopted siblings. Based on these correlations, the heritability of this anaerobic performance trait would range from 0.44 (twice the difference between the MZ and DZ correlations) to 0.92 (twice biological sibling correlation). These estimates are typically in the range of more recent estimates obtained from twin data (Calvo et al., 2002), and suggest that the heritability of anaerobic performance is probably in the range of 50–90%.

Genetics of Muscular Fitness

The role of genetic factors in muscular endurance, strength, and power and flexibility has been recently reviewed (Beunen & Thomis, 2003, 2004; Bouchard et al., 1997; Peeters et al., 2009). The most comprehensive family data on muscular fitness come from the CFS where muscular fitness was measured in a sample of 13,804 subjects (Perusse et al., 1988). In the CFS, muscular endurance was assessed by measuring the maximum number of sit-ups performed in 60 s and the number of push-ups completed without time limit. Muscular strength was assessed by measuring handgrip strength, and trunk flexibility was assessed using a sit-and-reach test. The familial risk ratios for first-degree relatives of individuals exceeding the 95th percentile of the distribution were 3.98 for muscular endurance, 3.16 for muscular strength, and 3.56 for flexibility, while the corresponding values for spouses were 2.63, 2.38, and 2.59, respectively (Katzmarzyk et al., 2000b). The higher risk ratios for first-degree relatives compared to spouses suggest the contribution of genetic factors, which was confirmed by Perusse and colleagues (1988) who reported heritability estimates of 37% for sit-ups, 44% for push-ups, 37% for grip strength, and 48% for flexibility. The 7-year changes in these muscular fitness phenotypes were also investigated using data from the 1981 CFS and a follow-up of that survey conducted in 1988 (Katzmarzyk et al., 2001). Results of the analysis of variance revealed that between 54% and 63% of the variance in the 7-year changes could be accounted for by family lines. Under the best-fitting model, the heritabilities were 41% for sit-ups, 52% for push-ups, 32% for grip strength, and 48% for flexibility (Katzmarzyk et al., 2001).

The familial aggregation of muscular endurance and muscular strength was also investigated in QFS (Perusse et al., 1987a, b). In QFS, muscular endurance was assessed using a sit-up test, while muscular strength was assessed using a knee extension test measuring the maximal isometric contraction of the quadriceps. Significant familial resemblance was observed for both indicators of muscular fitness (Perusse et al., 1987a) and heritability estimates (genetic transmission) were 21% for muscular

endurance and 30% for muscular strength, whereas cultural transmission accounted for an additional 33% and 31% of the variance, respectively (Perusse et al., 1987b). These estimates (combined genetic and cultural transmission) are slightly lower than those obtained in a study of 10-year-old twins (105 twin pairs) and their parents (97 mothers and 84 fathers) in which the heritabilities of various strength measures ranged between 63% and 77% (Maes et al., 1996). For flexibility (sit-and-reach test) there were sex differences in the heritability, and genetic factors accounted for 51% of the variance in females and 72% in males (Maes et al., 1996).

More recent studies also reported significant heritabilities for muscular strength. In a sample of 748 male Caucasian siblings from 335 Belgian families, maximal heritabilities of 63–87% were reported for concentric strength measures of knee, trunk, and elbow, while estimates of heritability for the corresponding isometric strength measures ranged between 82% and 96% (Huygens et al., 2004). In another study based on 115 sibpairs from 51 families, age- and sex-adjusted heritability of knee extension strength was 42% and increased to 60% after adjustment for height and weight (Zhai et al., 2004). Measures of elbow flexion, hand grip, and knee extension strength obtained in a large sample of siblings (154,970 sibpairs), DZ (1864 pairs) and MZ (1582 pairs) twins revealed that 50–60% of the variability in the strength measures could be accounted for by additive genetic factors showed (Silventoinen et al., 2008).

Muscular power or explosive strength, in which maximal force is released in the shortest possible time, is often measured by tests of jumping (vertical jump or standing long jump). Most of the evidence regarding the heritability of muscular power has been derived from twin studies, and family data are rather limited. In Polish families, parent–offspring correlations for the vertical jump were found to range from 0.17 to 0.54 depending on the offspring age groups (Szopa, 1982). In their study of 10-year-old twins and their parents, Maes et al. (1996) reported a heritability estimate of 65% for vertical long jump.

Table 8.2 Range of heritability estimates for muscular fitness phenotypes

	Maximal heritability (%)	Heritability (genetic transmission) (%)
Muscular endurance		
Sit-ups/push-ups	37–59	21
Muscular strength		
Grip strength	32–60	NA
Isometric strength	61–96	30
Concentric strength	50–87	NA
Muscular power		
Vertical jump	65	NA
Flexibility		
Sit and reach	48–72	NA

NA, not available.

Table 8.2 presents an overview of the range of the heritability estimates for the various muscular fitness phenotypes described in the above-reviewed studies.

Genetics of Motor Performance

Family studies of motor performance are limited. An early study compared the performances of 24 college-age men with that of their fathers 34 years earlier when the latter were also of college age and found similar performances between fathers and their sons with father–son correlations of 0.86 for the running long jump and 0.59 for the 100-yard dash (Cratty, 1960). Familial resemblance for neuromuscular performance traits assessed using various hand tasks was investigated in a small sample of 559 men and women from 76 Mennonite families (Devor & Crawford, 1984). Maximal heritabilities were 7% for hand steadiness, 16% for hand–eye coordination, 11% for movement time, and 24% for reaction time. Similar estimates of heritability (genetic and cultural transmission) were found in QFS for movement time (18%) and reaction time (27%), but additive genetic variance was significant only for reaction time ($h^2 = 20\%$) (Perusse et al., 1987b). Speed of limb movement (plate tapping) and balance were also measured in Belgium twin pairs and their parents and

heritabilities of 23% and 41%, respectively, were found for these two indicators of motor performance (Maes et al., 1996).

Genetics of Other Performance-Related Phenotypes

Morphological Characteristics

Athletic performance is influenced by several morphological characteristics, including body size (stature and weight) and body build (or physique). The role of genetic factors in body size and physique has been reviewed in detail (Bouchard et al., 1997). A review of family studies showed that sibling correlations for body height range from about 0.30 to 0.50, which suggest that about 80% of the variation in adult stature could be accounted for by genetic factors (Silventoinen, 2003). Familial correlations for body weight are less consistent than those observed for stature and suggest a lower heritability (Mueller, 1985). In the study of more than one million Swedish men, 81% of the variation in body height was attributable to additive genetic factors, 12% to common environmental factors and 7% to environmental factors unique to the individual (Silventoinen et al., 2008). For body weight, the corresponding values were 64%, 10%, and 26%. Several studies have investigated the familial resemblance of weight for stature ratios, especially the body mass index (BMI = weight in kilograms per height in meters squared), but will not be reviewed here.

Differences in physique among ethnic groups are evident and suggest that differences in body proportions, often assessed by measuring the length of different body segments (e.g., arm length, leg length, sitting height), are under genetic control. Populations of Black ancestry, for example, have relatively longer extremities compared to Caucasians, and these differences have implications in sports performance. Data from the QFS reviewed by Bouchard et al. (1997) showed that midparent–child correlations for various measures of skeletal lengths and breadths were about twice as large for individuals sharing genes by descent and the home environment (0.39–0.62) than for those sharing only the familial environment (0.01–0.30), suggesting a strong genetic influence for body proportions. These results are consistent with those observed in a study of 844 individuals from 261 Spanish families, which showed significant parent–offspring and sibling correlations for several measures of skeletal breadths and lengths (Sanchez-Andres & Mesa, 1994). One of the most widely used methods to classify individuals based on the basis of their physique is the somatotype. The somatotype quantifies an individual's physique in three components: (1) endomorphy, which is characterized by roundness and softness of contours throughout the body; (2) mesomorphy, which is characterized by the predominance of muscle, bone, and connective tissues; and (3) ectomorphy, which is characterized by linearity and a predominance of surface area over body mass. Results from early family studies reviewed in Bouchard et al. (1997) suggest a significant genetic influence for physique. In the QFS, significant sibling and parent–offspring correlations were observed in somatotype components and the correlations were higher than those observed in spouses (Song et al., 1993). These results are consistent with those found in the CFS which showed that 36%, 45%, and 42% of the variance in endomorphy, mesomorphy, and ectomorphy, respectively, could be attributed to genetic and cultural transmission (Perusse et al., 1988). A recent study conducted in 329 subjects aged 7–17 years from 132 African families

reported similar results with heritabilities of 40% for endomorphy, 30% for mesomorphy, and 31% for ectomorphy (Saranga et al., 2008). Two other studies investigated the familial resemblance of a general somatotype derived from a principal component analysis of the individual somatotype components. One of these studies based on 103 Canadian families from Northern Ontario reported maximal heritabilities of 56%, 68%, 56%, and 64% for endomorphy, mesomorphy, ectomorphy, and the general somatotype, respectively (Katzmarzyk et al., 2000a). In a larger sample of more than 3000 individuals from 1330 Biscay families (Basque Country, Spain), the maximal heritabilities were slightly lower with values of 55% for endomorphy, 52% for mesomorphy, 46% for ectomorphy, and 52% for the general somatotype (Rebato et al., 2007).

Cardiac and Skeletal Muscle Characteristics

Cardiac Structures and Functions

Cardiac structures and functions are critical limiting factors of endurance performance. The role of genetic factors on cardiac structures and dimensions has been considered in a few family studies. In the QFS, familial correlations among various kind of relatives by descent and adoption were computed for various echocardiographic measurements of cardiac dimensions and for most cardiac dimensions, correlations were significant in both biological relatives and relatives by adoption, suggesting a contribution of both genetic and environmental factors in the determination of those traits (Thériault et al., 1986). The genetic contribution of left ventricular mass (LVM) (Palatini et al., 2001) and left atrial size (Palatini et al., 2002) was examined in 290 parents and 251 children from the Tecumseh population in Michigan, USA. The parent–child correlations were 0.28 for LVM and 0.25 for left atrial size. In the Framingham Heart Study, familial correlations in first-degree relatives (1486 pairs of parent–child; 662 pairs of siblings), second-degree relatives (369 pairs of uncle/aunt–nephew/niece), and unrelated spouses (855 pairs) were computed to assess the heritability of LVM (Post et al., 1997). The correlations were higher in first-degree relatives (0.15 and 0.16), than in second-degree relatives (0.06) and spouses (0.05), and the heritability of LVM adjusted for age, height, weight, and systolic blood was between 0.24 and 0.32. Sibling correlations of LVM and left ventricular relative wall thickness (RWT), a measure of left ventricular geometry, were computed in 1664 African-American and White hypertensive siblings aged 23–87 years (Arnett et al., 2001). In African-Americans, the sibling correlations for LVM range from 0.29 to 0.44. In Whites, the corresponding correlations were lower (0.05–0.22). For RWT, the sibling correlations were lower in African-Americans (0.04–0.19) than in Whites (0.19–0.28). The heritability of left ventricular dimensions and mass was also evaluated in 1373 subjects from 445 American Indian families (Bella et al., 2004). The age- and sex-adjusted heritabilities were 0.27 for LVM, 0.22 for RWT, 0.19 for the left ventricular posterior wall thickness, and 0.26 for interventricular septal wall thickness. The results of these family studies suggest a small but significant contribution of genetic factors of about 20–30% for cardiac dimensions.

The phenotypes generally used to assess cardiac function in the context of sports performance are resting and exercise HR, blood pressure response to exercise, stroke volume, and cardiac output. The heritability of these phenotypes has been studied in the HERITAGE Family Study. Familial resemblance for resting HR, systolic blood pressure (SBP), and diastolic blood pressure (DBP) measured in the sedentary state before exercise training was investigated in 528 individuals from 98 Caucasian families (An et al., 1999). Significant familial resemblance was noted for all traits, and maximal heritability estimates were 32% for resting HR, 54% for SBP, and 41% for DBP. After further adjustment for BMI, the corresponding estimates were 34%, 51%, and 42%, respectively. The estimate of 34% for the heritability of resting HR is consistent with estimates derived from other family studies (Li et al., 2009; Perusse et al., 1988; Singh et al., 1999). Familial aggregation of cardiac output (Qc), measured using the CO_2 re-breathing technique, and stroke volume (SV = Qc/HR) during submaximal exercise was also investigated in the HERITAGE Family Study

(An et al., 2000). After adjustment for age, sex, and body surface area, about 40% of the variance in Qc and SV at 50W and at 60% of $\dot{V}O_2$max could be accounted for by family lines. Maximal heritabilities of Qc at 50W and 60% of $\dot{V}O_2$max were 42% and 46%, respectively. Similar heritability estimates were found for SV with values of 41% and 46%, respectively.

Electrocardiographic (ECG) parameters can also be used to assess cardiac function. The ECG QT interval, which represents the period of ventricular depolarization–repolarization, and the heart rate variability (HRV), which provides a quantitative measure of the cardiac autonomic function, are of critical importance in the regulation of HR, blood pressure, and for the overall stability of the cardiovascular system. The heritability of these parameters has been investigated in many family studies (Friedlander et al., 1999; Hong et al., 2001; Im et al., 2009; Li et al., 2009; Newton-Cheh et al., 2005; Singh et al., 1999; Sinnreich et al., 1998, 1999; Kupper et al., 2004; Wang et al., 2005, 2009) and results suggest that both the QT interval and HRV are influenced by genetic factors with heritabilities in the range of 30–40% for the QTc interval and 0.13–71% for HRV.

Skeletal Muscle

Skeletal muscle represents the largest tissue mass in human and is the main energy-consuming and work-producing tissue of the body, but relatively little is known about the genetic factors influencing skeletal muscle characteristics. Results from a few family studies suggest that genetic factors are important determinants of muscle size and mass. Early studies based on radiographic measurements of calf muscle diameter or anthropometric indicators (arm and calf circumferences corrected for the thickness of the triceps and calf skinfolds, respectively) suggested that muscle size is significantly influenced by genetic factors (Bouchard et al., 1997). Cross-sectional areas of the upper arm and midthigh corrected for skinfolds thicknesses were assessed in 748 male siblings from 335 Belgian families and were used to estimate muscle volume of the arm and thigh, respectively (Huygens et al., 2004).

Cross-sectional areas and volumes of the arm and thigh were highly heritable with maximal heritabilities above 90%. Other methods such as underwater weighing, CT-scan, or DEXA can be used to measure lean body mass or muscle mass or lean mass specific body segments, usually arms and legs. As muscle tissue is the major component of fat-free mass (FFM), lean body mass is often used as a proxy for muscle mass. In QFS and HERITAGE Family Study, FFM was estimated from body density measurements obtained by underwater weighing and maximal heritabilities reached 40% (30% genetic transmission and 10% cultural transmission) in QFS (Bouchard et al., 1988) and 65% in HERITAGE (Rice et al., 1997). These results agree with those obtained from other family studies that reported heritability estimates ranging from about 45% to 60% for FFM (Abney et al., 2001; Hsu et al., 2005; Pan et al., 2007; Treuth et al., 2001). Sex-specific heritabilities for FFM were reported in one of these studies (Pan et al. 2007) and higher heritability was observed in males (56%) compared to females (35%).

Arm, leg, and appendicular (combined arm and leg) lean tissue mass and calf muscle cross-sectional areas were measured by DEXA and CT-scan in 444 men and women from eight large extended Afro-Caribbean families (Prior et al., 2007). The heritabilities of arm lean mass, leg lean mass, appendicular lean mass (ALM), and calf muscle area were 23%, 18%, 18%, and 23%, respectively. Potential ethnic differences appear to exist for these limb-specific heritability estimates. Indeed, on study based on 1346 adults from 327 extended pedigrees of the Framingham study found a heritability of 42% for leg lean mass (Karasik et al., 2009). Another study based on 492 White and 49 African-American sibling pairs reported heritabilities of 49% for lean body mass and 51% for ALM, but these estimates increased to 73% and 80%, respectively, after adjustment for ethnicity (Hsu et al., 2005). More recently, a lower-leg peripheral quantitative computed tomography scan performed in Afro-Caribbean families revealed that 35% of the variance in muscle density, which is inversely related to the lipid content of the muscle, could be attributed to genetic factors (Miljkovic-Gacic et al., 2008).

Only a few family studies have investigated the familial aggregation of skeletal muscle fiber type distribution and enzymatic activity profile. In a study based on 32 pairs of brothers, 26 pairs of DZ twins, and 35 pairs of MZ twins in whom muscle biopsies of the *vastus lateralis* were obtained, Bouchard et al. (1986b) reported correlations of 0.55, 0.52, and 0.33 for the proportion of type I fibers in MZ twins, DZ twins, and brothers, respectively. These results suggest a significant familial resemblance, but the contribution of genetic factors could be considered either low if we compute the heritability as twice the difference between MZ and DZ twins correlation ($h^2 = 6\%$) or high if we compute the heritability directly from the correlation among MZ twins ($h^2 = 55\%$) or from twice the sibling correlation ($h^2 = 66\%$). Placing these results in the context of data from animal studies and considering that a fraction of the variance in the proportion of type I fibers in human skeletal muscle could be explained by a residual error component related to muscle sampling and technical variance, Simoneau and Bouchard (1995) proposed that about 45% of the variance in the proportion of type I muscle fibers in humans could be explained by genetic factors. In the HERITAGE Family Study, muscle biopsies of the *vastus lateralis* were obtained in a sample of 78 subjects from 19 families, and no significant evidence of familial aggregation was found for the fiber type proportions (Rankinen et al., 2005; Rico-Sanz et al., 2003). However, a moderate evidence of familial aggregation ($F = 2.4$; $P = 0.007$) was found for type I fiber area. From these results, we can conclude that the contribution of genetic factors in determining fiber type proportions in human skeletal muscle is probably low, but considering the relatively high technical variance associated with the biopsy technique and the small sample size of the available studies, studies with larger sample size are needed to obtain a more valid estimate of the heritability of human muscle fiber type proportions.

A clearer picture emerges when one looks at the enzymatic activity profile of the skeletal muscle. A highly significant familial aggregation was found for maximal activities of selected enzymes of the phosphagenic, glycolytic, and oxidative pathways in the HERITAGE Family Study (Rankinen et al., 2005; Rico-Sanz et al., 2003). The familial resemblance appears to be stronger for glycolytic enzymes than for oxidative metabolism enzymes. For the regulatory enzymes of the phosphagenic (creatine kinase), glycolytic (phosphorylase) and oxidative (3-beta-hydroxy-acyl CoA dehydrogenase) pathways, the F ratios were 6.3, 6.9, and 3.8, respectively. A significant familial resemblance was also observed for the number of capillaries around type I and type IIA fibers ($F = 1.9$; $P = 0.04$) and type I ($F = 2.3$; $P = 0.02$) and IIA ($F = 1.9$; $P = 0.04$) fiber areas per capillary, suggesting that genetic factors play a significant role in determining the development of capillary to the oxidative skeletal muscle fibers.

Conclusion

Family studies have demonstrated familial resemblance for the major morphological and physiological determinants of sports performance. Family studies will tend to overestimate genetic factors as the environment and social structures are typically shared between family members, making it more challenging to distinguish between genetic and environmental contributions. Twin studies (comparisons of MZ versus DZ twins) are better able to control for environmental effects. The evidence for a role of genetic factors accounts for ~40–60% of the variation in aerobic performance and cardiac function, 50–90% of the variation in anaerobic performance, 30–70% of the variation in muscular fitness, and 20–30% of the variation in cardiac dimensions.

References

Abney, M., McPeek, M.S. & Ober, C. (2001). Broad and narrow heritabilities of quantitative traits in a founder population. *American Journal of Human Genetics* 68, 1302–1307.

An, P., Rice, T., Gagnon, J., et al. (1999). Familial aggregation of resting blood pressure and heart rate in a sedentary population: The HERITAGE Family Study. Health, Risk Factors, Exercise Training, and Genetics. *American Journal of Hypertension* 12, 264–270.

An, P., Rice, T., Gagnon, J., et al. (2000). Familial aggregation of stroke volume and cardiac output during submaximal exercise: The HERITAGE Family Study. *International Journal of Sports Medicine* 21, 566–572.

Arnett, D.K., Hong, Y., Bella, J.N., et al. (2001). Sibling correlation of left ventricular mass and geometry in hypertensive African Americans and whites: The HyperGEN study. Hypertension Genetic Epidemiology Network. *American Journal of Hypertension* 14, 1226–1230.

Bella, J.N., MacCluer, J.W., Roman, M.J., et al. (2004). Heritability of left ventricular dimensions and mass in American Indians: The Strong Heart Study. *Journal of Hypertension* 22, 281–286.

Beunen, G. & Thomis, M. (2003). Genetics of strength and power characteristics in children and adolescents. *Pediatric Exercise Science* 15, 128–138.

Beunen, G. & Thomis, M. (2004). Gene powered? Where to go from heritability (h2) in muscle strength and power? *Exercise and Sport Sciences Reviews* 32, 148–154.

Bouchard, C., Daw, E.W., Rice, T., et al. (1998). Familial resemblance for V̇O$_2$max in the sedentary state: The HERITAGE Family Study. *Medicine & Science in Sports & Exercise* 30, 252–258.

Bouchard, C., Dionne, F.T., Simoneau, J.A. & Boulay, M.R. (1992). Genetics of aerobic and anaerobic performances. *Exercise and Sport Sciences Reviews* 20, 27–58.

Bouchard, C., Leon, A.S., Rao, D.C., Skinner, J.S., Wilmore, J.H. & Gagnon, J. (1995). The HERITAGE Family Study. Aims, design, and measurement protocol. *Medicine & Science in Sports & Exercise* 27, 721–729.

Bouchard, C., Lesage, R., Lortie, G., et al. (1986a). Aerobic performance in brothers, dizygotic and monozygotic twins. *Medicine & Science in Sports & Exercise* 18, 639–646.

Bouchard, C., Lortie, G., Simoneau, J.A., Leblanc, C., Theriault, G. & Tremblay, A. (1984). Submaximal power output in adopted and biological siblings. *Annals of Human Biology* 11, 303–309.

Bouchard, C. & Malina, R.M. (1983). Genetics of physiological fitness and motor performance. *Exercise and Sport Sciences Reviews* 11, 306–339.

Bouchard, C., Malina, R.M. & Perusse, L. (1997). *Genetics of Fitness and Physical Performance*. Human Kinetics, Champaign, IL.

Bouchard, C., Perusse, L., Leblanc, C., Tremblay, A. & Theriault, G. (1988). Inheritance of the amount and distribution of human body fat. *International Journal of Obesity* 12, 205–215.

Bouchard, C., Simoneau, J.A., Lortie, G., Boulay, M.R., Marcotte, M. & Thibault, M.C. (1986b). Genetic effects in human skeletal muscle fiber type distribution and enzyme activities. *Canadian Journal of Physiology and Pharmacology* 64, 1245–1251.

Brutsaert, T.D. & Parra, E.J. (2006). What makes a champion? Explaining variation in human athletic performance. *Respiratory Physiology & Neurobiology* 151, 109–123.

Calvo, M., Rodas, G., Vallejo, M., et al. (2002). Heritability of explosive power and anaerobic capacity in humans. *European Journal of Applied Physiology* 86, 218–225.

Cratty, B.J. (1960). A comparison of fathers and sons in physical ability. *Research Quarterly* 31, 12–15.

Devor, E.J. & Crawford, M.H. (1984). Family resemblance for neuromuscular performance in a Kansas Mennonite community. *American Journal of Physical Anthropology* 64, 289–296.

Friedlander, Y., Lapidos, T., Sinnreich, R. & Kark, J.D. (1999). Genetic and environmental sources of QT interval variability in Israeli families: The Kibbutz Settlements Family Study. *Clinical Genetics* 56, 200–209.

Gaskill, S.E., Rice, T., Bouchard, C., et al. (2001). Familial resemblance in ventilatory threshold: The HERITAGE Family Study. *Medicine & Science in Sports & Exercise* 33, 1832–1840.

Hong, Y., Rautaharju, P.M., Hopkins, P.N., et al. (2001). Familial aggregation of QT-interval variability in a general population: Results from the NHLBI Family Heart Study. *Clinical Genetics* 59, 171–177.

Hsu, F.C., Lenchik, L., Nicklas, B.J., et al. (2005). Heritability of body composition measured by DXA in the Diabetes Heart Study. *Obesity Research* 13, 312–319.

Huygens, W., Thomis, M.A., Peeters, M.W., Vlietinck, R.F. & Beunen, G.P. (2004). Determinants and upper-limit heritabilities of skeletal muscle mass and strength. *Canadian Journal of Applied Physiology* 29, 186–200.

Im, S.W., Lee, M.K., Lee, H.J., et al. (2009). Analysis of genetic and non-genetic factors that affect the QTc interval in a Mongolian population: The GENDISCAN study. *Experimental & Molecular Medicine* 41, 841–848.

Karasik, D., Zhou, Y., Cupples, L.A., Hannan, M.T., Kiel, D.P. & Demissie, S. (2009). Bivariate genome-wide linkage analysis of femoral bone traits and leg lean mass: Framingham study. *Journal of Bone and Mineral Research* 24, 710–718.

Katzmarzyk, P.T., Gledhill, N., Perusse, L. & Bouchard, C. (2001). Familial aggregation of 7-year changes in musculoskeletal fitness. *The Journals of Gerontology. Series A, Biological Sciences and Medical Sciences* 56, B497–B502.

Katzmarzyk, P.T., Malina, R.M., Perusse, L., et al. (2000a). Familial resemblance for physique: Heritabilities for somatotype components. *Annals of Human Biology* 27, 467–477.

Katzmarzyk, P.T., Perusse, L., Rao, D.C. & Bouchard, C. (2000b). Familial risk ratios for high and low physical fitness levels in the Canadian population. *Medicine & Science in Sports & Exercise* 32, 614–619.

Kupper, N.H., Willemsen, G., Van den Berg, M., et al. (2004). Heritability of ambulatory heart rate variability. *Circulation* 110, 2792–2796.

Lesage, R., Simoneau, J.A., Jobin, J., Leblanc, J. & Bouchard, C. (1985). Familial resemblance in maximal heart rate, blood lactate and aerobic power. *Human Heredity* 35, 182–189.

Li, J., Huo, Y., Zhang, Y., et al. (2009). Familial aggregation and heritability of electrocardiographic intervals and heart rate in a rural Chinese population. *Annals of Noninvasive Electrocardiology* 14, 147–152.

Lortie, G., Bouchard, C., Leblanc, C., et al. (1982). Familial similarity in aerobic power. *Human Biology* 54, 801–812.

Maes, H.H., Beunen, G.P., Vlietinck, R.F., et al. (1996). Inheritance of physical fitness in 10-yr-old twins and their parents. *Medicine & Science in Sports & Exercise* 28, 1479–1491.

Miljkovic-Gacic, I., Wang, X., Kammerer, C.M., et al. (2008). Fat infiltration in muscle: New evidence for familial clustering and associations with diabetes. *Obesity (Silver Spring)* 16, 1854–1860.

Montoye, H.J. & Gayle, R. (1978). Familial relationships in maximal oxygen uptake. *Human Biology* 50, 241–249.

Mueller, W.H. (1985). The genetics of size and shape in children and adults. In: F. Falkner & J.M. Tanner (eds) *Human Growth: a Comprehensive Treatise*. Plenum Press, New York, NY.

Newton-Cheh, C., Larson, M.G., Corey, D.C., et al. (2005). QT interval is a heritable quantitative trait with evidence of linkage to chromosome 3 in a genome-wide linkage analysis: The Framingham Heart Study. *Heart Rhythm* 2, 277–284.

Palatini, P., Amerena, J., Nesbitt, S., et al. (2002). Heritability of left atrial size in the Tecumseh population. *European Journal of Clinical Investigation* 32, 467–471.

Palatini, P., Krause, L., Amerena, J., et al. (2001). Genetic contribution to the variance in left ventricular mass: The Tecumseh Offspring Study. *Journal of Hypertension* 19, 1217–1222.

Pan, L., Ober, C. & Abney, M. (2007). Heritability estimation of sex-specific effects on human quantitative traits. *Genetic Epidemiology* 31, 338–347.

Peeters, M.W., Thomis, M.A., Beunen, G.P. & Malina, R.M. (2009). Genetics and sports: An overview of the pre-molecular biology era. *Medicine and Sport Science* 54, 28–42.

Perusse, L. (2001). Les bases génétiques et moléculaires de la performance et de l'adaptation à l'exercice physique. *Science & Sports* 16, 186–195.

Perusse, L., Gagnon, J., Province, M.A., et al. (2001). Familial aggregation of submaximal aerobic performance in the HERITAGE Family study. *Medicine & Science in Sports & Exercise* 33, 597–604.

Perusse, L., Leblanc, C. & Bouchard, C. (1988). Inter-generation transmission of physical fitness in the Canadian population. *Canadian Journal of Sport Sciences* 13, 8–14.

Perusse, L., Leblanc, C., Tremblay, A., et al. (1987a). Familial aggregation in physical fitness, coronary heart disease risk factors, and pulmonary function measurements. *Preventive Medicine* 16, 607–615.

Perusse, L., Lortie, G., Leblanc, C., Tremblay, A., Theriault, G. & Bouchard, C. (1987b). Genetic and environmental sources of variation in physical fitness. *Annals of Human Biology* 14, 425–434.

Perusse, L., Rice, T., Province, M.A., et al. (2000). Familial aggregation of amount and distribution of subcutaneous fat and their responses to exercise training in the HERITAGE family study. *Obesity Research* 8, 140–150.

Post, W.S., Larson, M.G., Myers, R.H., Galderisi, M. & Levy, D. (1997). Heritability of left ventricular mass: The Framingham Heart Study. *Hypertension* 30, 1025–1028.

Prior, S.J., Roth, S.M., Wang, X., et al. (2007). Genetic and environmental influences on skeletal muscle phenotypes as a function of age and sex in large, multigenerational families of African heritage. *Journal of Applied Physiology* 103, 1121–1127.

Rankinen, T., Bouchard, C. & Rao, D.C. (2005). Familial resemblance for muscle phenotypes in the HERITAGE Family Study. *Medicine & Science in Sports & Exercise* 37, 2017.

Rebato, E., Jelenkovic, A. & Salces, I. (2007). Heritability of the somatotype components in Biscay families. *Homo: internationale Zeitschrift für die vergleichende Forschung am Menschen* 58, 199–210.

Rice, T., Daw, E.W., Gagnon, J., et al. (1997). Familial resemblance for body composition measures: The HERITAGE Family Study. *Obesity Research* 5, 557–562.

Rico-Sanz, J., Rankinen, T., Joanisse, D.R., et al. (2003). Familial resemblance for muscle phenotypes in the HERITAGE Family Study. *Medicine & Science in Sports & Exercise* 35, 1360–1366.

Rupert, J.L. (2003). The search for genotypes that underlie human performance phenotypes. *Comparative Biochemistry and Physiology. Part A, Molecular & Integrative Physiology* 136, 191–203.

Sanchez-Andres, A. & Mesa, M.S. (1994). Heritabilities of morphological and body composition characteristics in a Spanish population. *Anthropologischer Anzeiger; Bericht über die biologisch-anthropologische Literatur* 52, 341–349.

Saranga, S.P., Prista, A., Nhantumbo, L., et al. (2008). Heritabilities of somatotype components in a population from rural Mozambique. *American Journal of Human Biology* 20, 642–646.

Silventoinen, K. (2003). Determinants of variation in adult body height. *Journal of Biosocial Science* 35, 263–285.

Silventoinen, K., Magnusson, P.K., Tynelius, P., Kaprio, J. & Rasmussen, F. (2008). Heritability of body size and muscle strength in young adulthood: A study of one million Swedish men. *Genetic Epidemiology* 32, 341–349.

Simoneau, J.A. & Bouchard, C. (1995). Genetic determinism of fiber type proportion in human skeletal muscle. *FASEB Journal* 9, 1091–1095.

Simoneau, J.A., Lortie, G., Leblanc, C. & Bouchard, C. (1986). Anaerobic work capacity in adopted and biological siblings. In: R.M. Malina & C. Bouchard (eds) *Sport and Human Genetics*. Human Kinetics, Champaign, IL.

Singh, J.P., Larson, M.G., O'Donnell, C.J., Tsuji, H., Evans, J.C. & Levy, D. (1999). Heritability of heart rate variability: The Framingham Heart Study. *Circulation* 99, 2251–2254.

Sinnreich, R., Friedlander, Y., Luria, M.H., Sapoznikov, D. & Kark, J.D. (1999). Inheritance of heart rate variability: The Kibbutzim Family Study. *Human Genetics* 105, 654–661.

Sinnreich, R., Friedlander, Y., Sapoznikov, D. & Kark, J.D. (1998). Familial aggregation of heart rate variability based on short recordings—the Kibbutzim Family Study. *Human Genetics* 103, 34–40.

Song, T.M.K., Malina, R.M. & Bouchard, C. (1993). Familial resemblance in somatotype. *American Journal of Human Biology* 5, 265–272.

Szopa, J. (1982). Familial studies on genetic determination of some manifestations of muscular strength in man. *Genetica Polonica* 23, 65–79.

Szopa, J. & Cempla, J. (1985). Inheritance of aerobic work capacity in man: Results of population study on family resemblances. *Genetica Polonica* 26, 267–276.

Thériault, G., Diano, R., Leblanc, C., Perusse, L., Landry, F. & Bouchard, C. (1986). The role of heredity in cardiac size: An echocardiographic study on twins, brothers and sisters. *Medicine & Science in Sports & Exercise* 18, S1.

Treuth, M.S., Butte, N.F., Ellis, K.J., Martin, L.J. & Comuzzie, A.G. (2001). Familial resemblance of body composition in prepubertal girls and their biological parents. *American Journal of Clinical Nutrition* 74, 529–533.

Wang, X., Ding, X., Su, S., et al. (2009). Genetic influences on heart rate variability at rest and during stress. *Psychophysiology* 46, 458–465.

Wang, X., Thayer, J.F., Treiber, F. & Snieder, H. (2005). Ethnic differences and heritability of heart rate variability in African- and European American youth. *American Journal of Cardiology* 96, 1166–1172.

Zhai, G., Stankovich, J., Ding, C., Scott, F., Cicuttini, F. & Jones, G. (2004). The genetic contribution to muscle strength, knee pain, cartilage volume, bone size, and radiographic osteoarthritis: A sibpair study. *Arthritis Rheumatism* 50, 805–810.

Chapter 9

Twin Studies in Sport Performance

GASTON P. BEUNEN,[1] MAARTEN W. PEETERS[1,2] AND ROBERT M. MALINA[3]

[1]*Faculty of Kinesiology and Rehabilitation Sciences, Department of Biomedical Kinesiology, Research Centre for Exercise and Health, Katholieke Universiteit Leuven, Leuven, Belgium*
[2]*Research Foundation-Flanders (FWO), Belgium*
[3]*Department of Kinesiology and Health Education, University of Texas at Austin and Tarleton State University, Stephenville, Texas, TX, USA*

Sport performance at a high level is not simply a matter of nature versus nurture. A combination of genetic factors interacting with appropriate environmental conditions is essential: gene–gene interactions also enter the mix.

Contemporary genetic studies of sport performance often focus on potential candidate genes in elite athletes, especially genes associated with aerobic metabolic pathways. However, inferences on the genetics of sport performance are derived largely from samples of non-athletes, using standardized measurements and tests. Focus is ordinarily on outcome in contrast to process of performance. It is thus difficult to make inferences about genetics of actual sport performances which are carried out under dynamic conditions far removed from those under which performance phenotypes are measured in the field or in the laboratory.

Allowing for this caveat, this chapter considers estimated genetic contributions to individual differences in physical performance based on twin studies. This chapter briefly reviews the research methodology and estimates of genetic and environmental contributions to variation in performance-related phenotypes and then considers results of twin studies dealing with performance-related

phenotypes, aerobic performance, muscular strength and power, and motor performance.

Historical Note

Discussions of genetics and sport performance are not recent. The topic was discussed in several papers in the 1950s and 1960s, while studies of sport-related phenotypes received increasing attention in the 1970s and 1980s (Bouchard & Malina, 1983; Malina & Bouchard, 1989). Studies of twins dominated much of the discussion, but data for athletes who are twins are limited to case studies. Limited information (self-reported) is available in a twin registry of Italian athletes (Parisi et al., 2001), while athlete status (non-athlete; local level, county or national level athlete) had moderate heritability ($a^2 = 0.655$) in a large sample of British female twins (De Moor et al., 2007).

Methodological Considerations

Physical abilities and related factors that underlie sport performance are evaluated on continuous scales. Distributions of such variables in the general population are Gaussian or skewed, which is typical for quantitative, multifactorial phenotypes that are influenced by multiple genes (polygenic) and environmental factors. Estimates of the contribution of genetic and environmental factors,

Genetic and Molecular Aspects of Sport Performance, 1st edition.
Edited by Claude Bouchard and Eric P. Hoffman.
Published 2011 by Blackwell Publishing Ltd.

their covariation and interaction, to sport-related performance phenotypes require separate analytical techniques (Neale & Cardon, 1992).

Two basic approaches have been used to study the genetic basis of performance phenotypes and related characteristics: (1) the unmeasured genotype approach (top–down), the focus of this and other chapters in Part 2 of this volume, and (2) the measured genotype approach (bottom–up), the focus of other chapters in Part 3. In the top–down approach, inferences about genetic influences on a phenotype are based on statistical analyses of the distributions of measures in related individuals and families based on the theoretical framework of biometrical genetics (Neale & Cardon, 1992). The *classical twin design* uses monozygotic (MZ) and dizygotic (DZ) twins and has five zygosity groups: male MZ, female MZ, male DZ, female DZ, and opposite sex DZ if both sexes are represented. With twin data, genetic and environmental factors unique to the individual and shared within families (common environmental factors) can be estimated and under certain assumptions dominant genetic effects can also be identified. The determination of zygosity, preferably with genetic markers, is essential in twin studies. Earlier studies used anthropological characteristics to determine zygosity and results do not diverge substantially from zygosity determination using genetic markers (Vogel & Motulsky, 1986).

The classical twin design can be extended with several methods: by including spouses and offspring of adult twins (twin-family design); by comparing twins discordant for exposure to an environmental factor such as systematic training in sport; by studying MZ twins reared apart who did not share common environmental factors; and by incorporating other family members, especially biological parents and siblings. None of these extensions have been used with sport-related performance phenotypes except for the design incorporating parents and siblings. This design permits use of more complex models and provides more accurate estimates of environmental and genetic contributions to performance variance (Maes et al., 1996; Neale & Cardon, 1992).

For a given phenotype, total variation (V_{tot}) includes genetic (V_G), common environmental (V_C), and individual-specific environmental (V_E) components, or $V_{tot} = V_G + V_C + V_E$. Heritability ($h^2$) is the proportion of the total variation attributed to genetic effects (V_G/V_{tot}). It is generally assumed that the effects of different genes are additive (a^2) meaning that the genetic effect of the heterozygotes on the phenotype falls exactly between the genetic effects of both homozygotes. Dominance refers to the interaction between alleles at the same locus (the heterozygote effect does not fall exactly between the two homozygote effects), and epistasis describes the interaction between alleles at different loci. The contribution of environmental factors shared by family members (common environmental factors, $c^2 = V_C/V_{tot}$) and the proportion of environmental factors unique to the individual can also be estimated ($e^2 = V_E/V_{tot}$).

Several assumptions underlie the additive model: no interaction between gene action and environment (different genotypes all react equally to similar environmental factors), no gene–environment correlation (similar exposure of environments for different genotypes), no gene–gene interaction, and no assortative mating for the trait studied. It is likely that not all assumptions are met in studies of sport-related performance phenotypes. Gene action and environmental influences probably vary at different ages and perhaps with sex, ethnicity, and socioeconomic status (SES); in some countries, low SES and ethnicity are essentially coterminous. Longitudinal (transmission) models are needed to evaluate age-specific influences on the decomposition of inter-age correlations within MZ and DZ pairs into genetically and environmentally transmitted or time-specific sources of variance (Peeters et al., 2005a, 2007a, b)

Heritabilities were generally calculated using several formulae under different assumptions (Vogel & Motulsky, 1986). Most estimates used intra-class correlations based on analysis of variance (ANOVA). Before the significance of the *F*-ratios is tested, equality of means and variances of the twin types must be verified. *F*-ratios test whether the genetic variance contributes significantly to the total variance. Subsequently, two types of genetic variance can be calculated depending on the significance of the respective *F*-ratios, among and within twin pair variance.

In more recent biometrical genetics approach using structural equation modeling (SEM) or path-analysis, equations are derived from theoretical considerations and subsequently tested for goodness-of-fit. Different models are ordinarily tested and the model with the best fit and least number of parameters (most parsimonious) is held to be the causal explanation of the observed variation in a phenotype (Neale & Cardon, 1992). Common models include (1) the AE model in which both additive (a^2) genetic and unique (to each individual twin) environmental factors (e^2) explain significant proportions of the total variance, and (2) ACE model in which additive genetic factors (a^2) and both unique (e^2) and common (to both members of twin pairs) environmental factors (c^2) all explain significant proportions of the total variance. If the ACE model is retained, it must show a better fit than the more parsimonious A and AE models. Software has been developed for continuous characteristics and for nominal and ordinal measurements (Neale et al., 2003).

Related Somatic Phenotypes

Sport performances are the outcomes of a complex combination and interaction of individual characteristics—neuromuscular, somatic, physiologic and psychological, interacting with sport-specific circumstances. Issues related to optimal environmental conditions for the development of sport talent are beyond the scope of this chapter which considers several biological characteristics related to performance.

Body size, physique, proportions, and composition are associated with performance in the general population and elite athletes. Among youth, individual differences in the timing and tempo of biological maturation are an additional source of variation (Malina et al., 2004). Although considerable advances have been made in behavioral genetics, behavioral phenotypes potentially associated with sport success are not ordinarily considered in genetic studies of performance-related phenotypes.

Variability in skeletal lengths, skeletal breadths, limb circumferences, and bone mass is genetically determined (Bouchard et al., 1997). Heritabilities for stature in twin samples ranged from 0.69 to 0.96, with a mean and standard deviation of 0.85 ± 0.07. Corresponding estimates for segment lengths were similar, 0.84 ± 0.10, while those for skeletal breadths and limb circumferences were more variable and lower.

The estimated genetic determination of body mass is less than stature and when corrected for stature, heritability is about 0.40. Heritabilities of the BMI (wt/ht^2) in MZ twins reared apart ranged between 0.40 and 0.70 (Katzmarzyk & Bouchard, 2005). Other studies, however, have reported higher genetic contributions to indicators of body fatness, for example, 0.70–0.94 for BMI, waist hip ratio, sum of skinfolds and fat mass (impedance) derived from SEM in a large twin cohort (Souren et al., 2007). Estimates from several Canadian studies indicated a maximal heritability of 0.40 for several indicators of fat distribution (Bouchard et al., 1997), while estimates for a trunk-extremity skinfold ratio ranged from 0.78 to 0.88 in a longitudinal study of twins (Peeters et al., 2007a). Variation in heritabilities is due largely to adjustments made or not made in the respective analyses; estimates tend to be lower when indicators of fatness are adjusted for size and when indicators of fat distribution are adjusted for size and fat mass.

Estimated heritabilities for distribution of skeletal muscle fiber types are variable. Results of a comprehensive analysis partition total variance in the proportion of type I muscle fibers as follows: 45% genetic, 40% environmental, 15 % error associated with tissue sampling and technical variation (Simoneau & Bouchard, 1995).

Physique is most often assessed with the Heath-Carter anthropometric somatotype protocol. Estimated heritabilities from twin studies, based mainly on adolescents, are generally >0.70 for individual somatotype components (Peeters et al., 2003). In young adult twins, heritabilities in males were 0.28, 0.86, and 0.76 for endomorphy, mesomorphy, and ectomorphy, respectively; corresponding heritabilities for the three components in females were, respectively, 0.32, 0.82, and 0.70 (Peeters et al., 2007b). Multivariate analysis indicated that more than 70% of total variation in somatotype components was explained by sources of variance shared among the three components in adults. The results highlight the need to analyze somatotype with a multivariate approach (Peeters et al., 2007b).

Variability in biological maturation in youth has a significant genetic component. Heritabilities of components of the adolescent growth spurt (ages at take-off and peak height velocity [PHV], height and height velocity at these ages) range between 0.89 and 0.96 in both sexes (Beunen et al., 2000). Other indicators of maturity status—stages of sexual maturity, age at menarche, skeletal age, and percentage of adult stature attained at a given age—all show a relatively strong genetic influence (Malina et al., 2004).

Muscular Strength and Power

Genetic contributions to muscular strength and power have been recently summarized (Beunen & Thomis, 2004; Peeters et al., 2009). Estimated heritabilities for static strength of different muscle groups in twins ranged from 0.14 to 0.83 (Table 9.1). Many of the samples spanned broad age ranges without correction for age while potential sex differences were not tested. It is possible that age-corrected heritabilities may overlook the fact that genetic and environmental influences vary with age *per se* and the timing and tempo of maturation as genes can be switched on or off during childhood and adolescence (Peeters et al.,

Table 9.1 Heritabilities of muscular strength and power derived from twin studies

Static strength	0.14–0.83 (20 studies)[a]
Dynamic strength (isokinetic concentric and excentric)	0.29–0.90 (3 studies)[b]
Explosive strength or power (jump tests, Wingate test)	0.34–0.97 (7 recent studies)[c]

Heritabilities include both commonly used indexes and estimates based on structural equation modeling.
[a] Beunen et al. (2003); Beunen and Thomis (2004); De Mars et al. (2007); Okuda et al. (2005); Peeters et al. (2005a); Ropponen et al. (2004); Tiainen et al. (2005).
[b] Thomis et al. (1998); Maridaki (2006); Ropponen et al. (2004).
[c] Beunen et al. (2003); Calvo et al. (2002); Chatterjee and Das (1995); Maes et al. (1996); Maridaki (2006); Okuda et al. (2005); Peeters et al. (2005b); Tiainen et al. (2005).

2005a). Adults also differ in the rate of aging and are exposed to different environmental influences during the life cycle.

More recent twin studies based on SEM are consistent in showing no evidence of a shared (familial) environment effect for static strength in adolescents of both sexes (Okuda et al., 2005; Peeters et al., 2005a), young adult men (De Mars et al., 2007; Silventoinen et al., 2008), and elderly women (Tiainen et al., 2005). Of course, multivariate analyses increase the power to detect shared environmental influences. Heritabilities in Belgian adolescents ranged from 0.52 to 0.82 for boys (from 1 year before PHV to 3 years after PHV) and were somewhat lower in girls, 0.22 to 0.75 with one outlier of 0.07 (Peeters et al., 2005a). The heritabilities at 3 years after PHV, which approximates young adulthood, were 0.52 in boys and 0.48 in girls, and consistent with estimates for young adult males, 0.50 and 0.60 (Silventoinen et al., 2008) and 0.70 (De Mars et al., 2007), and elderly women, 0.49 (Tiainen et al., 2005). Systematic analyses of sex differences in static strength in adulthood are limited. Among Danish twins aged 45–96 years no sex differences in the relative contribution of genes and environment was noted with a heritability of 0.52 (Frederiksen et al., 2002). There also was no evidence for familial environmental effects.

Corresponding data for maximal dynamic strength (isokinetic dynamometry) are limited since the protocol is rather impractical in large field-studies (Table 9.1). Heritabilities of eccentric elbow flexor strength in young adult male twins ranged from 0.62 to 0.82 in contrast to estimates for concentric flexion, 0.29 to 0.65 (Thomis et al., 1998). Heritabilities also tended to be lower at higher contraction velocities. More recently, an opposite trend was noted in female adolescent twins selected for similar environmental backgrounds and levels of habitual physical activity. Heritability estimates increased at higher speeds of concentric knee extension, from 0.54 at 60°/s to 0.90 at 180°/s (Maridaki, 2006). Allowing for sample size (15 pairs) and homogeneity of environmental background, the results should be interpreted with caution.

Estimated heritabilities for dynamic strength tasks in older adults are reasonably similar, 0.60

for back extension in isokinetic lifting among 35- to 69-year-old male twins (Ropponen et al., 2004) and 0.46 for dynamic leg extension in postmenopausal female twins (Arden & Spector, 1997). The latter was increased after adjustment for age, height, and weight. Given the limited data, conclusions on potential age- or sex-associated variation in heritabilities of dynamic strength are not possible.

Tests of muscular power (explosive strength) as measured by jumping tasks or the Wingate protocol provide additional insights into the heritability of dynamic strength (Table 9.1), although analytical strategies vary. Estimated heritabilities were 0.67 for the squat jump, 0.45 for the counter-movement jump, and 0.74 for maximal power developed in 5 s in a Wingate test in 16 young adult male twins pairs (Calvo et al., 2002). Age-adjusted heritability for the vertical jump was 0.75 in 54 male and female adolescent twin pairs but possible sex differences were not considered (Chatterjee & Das, 1995), while heritabilities for peak power on a Wingate test were 0.97 and 0.85 in 15 pairs of preadolescent and 15 pairs of adolescent female twins, respectively (Maridaki, 2006).

More recent studies applied SEM to data for children and adolescents and included five zygosity groups which permitted testing for sex differences. In a sample of 158 twin pairs 10–15 years of age, heritability for the standing long jump was 0.66 under an AE model with no evidence for sex differences, although age differences were not taken into account (Okuda et al., 2005). In a sample of 105 male and female twin pairs of the same age, univariate cross-sectional and longitudinal models were used in genetic analyses of vertical jump (Beunen et al., 2003; Peeters et al., 2005b). Girls showed generally higher heritabilities than boys at most measurement occasions, 0.74–0.91 compared to 0.47–0.85, respectively, with one estimate of 0.0 at 16 years (Beunen et al., 2003). With data aligned on age at PHV (i.e., adjusting for sex differences in the timing of the growth spurt), sex differences were less apparent and reached significance only 3 years after PHV, 0.89 in girls and 0.61 in boys (Peeters et al., 2005b). Limiting the analysis to twins at 10 years of age and including data for parents suggested some genetic dominance was

included in the heritability of 0.65, which was slightly lower than a heritability of 0.72 when only twins were included in the analyses (Maes et al., 1996). In summary, data for late childhood, the transition into adolescence and adolescence *per se* do not provide evidence for clear age or sex differences in the heritability of jumping performance, although a divergence in heritabilities is suggested in later adolescence (Peeters et al., 2005b).

Corresponding data for explosive strength in adults are lacking. A heritability estimate of 0.32 was observed in a bivariate analysis of leg extensor power in elderly female twins; however, in a univariate analysis with an AE model which did not fit the data significantly worse than an ACE model, heritability was 0.61 and more comparable to studies in youth (Tiainen et al., 2005).

Motor Performance

Estimated heritabilities for tests of sprinting, jumping, and throwing from earlier studies of school age twins ranged from 0.60 to 0.90 with no indication of a sex difference (Bouchard et al., 1997; Malina & Bouchard, 1989). More recent studies of twins utilized SEM to partition genetic and environmental sources of variation in several performance tasks and a full complement of twins: MZ boys and girls, same sex DZ twins and opposite sex DZ twins. Results for the standing long and vertical jumps were discussed above. Running tasks are considered subsequently; corresponding data for throwing tasks are lacking.

The ACE model provided the best fit for a shuttle run (agility, running speed) in 10-year-old twins. The total variance was partitioned as follows in males: genetic 23%, common environment 33%, and specific environment 44%; estimates were as follows for females: genetic 33%, common environment 46%, and specific environment 21% (Maes et al., 1996). The side step, a measure of agility without a running component, was also best fit by the ACE model in twins at 10–15 years of age, but there was no sex difference. Genetic factors accounted for 32% of the variance, while shared and non-shared environmental factors accounted for 39% and 28% of the variance, respectively (Okuda et al., 2005).

The 50 m dash was best fit with a sex-specific CE model in 10- to 15-year-old twins (Okuda et al., 2005). Shared and non-shared environmental components accounted for 76% and 24% of the variance respectively in boys, but contributed about equally (53% and 47%, respectively) to the variance in girls.

Little is known of genetic and environmental sources of variation in movement patterns underlying running, jumping, and throwing. Two studies in twins of both sexes, 11–15 years of age (Sklad, 1972) and 8–14 years of age (Goya et al., 1991, 1993) indicated significant genotypic contributions to variation in the kinematic structure of running a dash. Intrapair differences were smaller in MZ compared to DZ males than in MZ compared to DZ females, suggesting that the sprinting performance of girls may be more amenable to environmental influences (Sklad, 1972). In contrast to the dash, intrapair differences in the kinetic structure of a throw for distance were similar in MZ and DZ twins (Goya et al., 1991, 1993), suggesting an important role of environmental influences.

Estimated heritabilities for speed of upper and lower limb movements (plate and one foot tapping, respectively) in twins aged 8–15 years varied between 0.79 and 0.90 in males and from 0.62 to 0.81 in females (Sklad, 1973). Among 10-year-old twins, speed of limb movement (plate tapping) was best fit with the AE model with equal parameters for males and females. Genes were thought to account for about 56% of the variance in speed of limb movement with no sex difference (Maes et al., 1996).

Any number of motor tasks including an array of gross and fine motor components can be tested. Many of the fine motor tasks are often indicated as perceptual-motor or psychomotor skills and include reaction time, movement time, manipulative dexterity, maze tracing, and so on. Estimated heritabilities in studies of twins range from low to moderately high and vary among tasks, between the preferred and non-preferred hands, and also among samples (Bouchard et al., 1997).

Balance is a skill that requires a combination of gross and fine motor control in the maintenance of equilibrium in static and dynamic tasks. Heritability estimates for various balance tests in samples of twins aged 8–18 years range from 0.27 to 0.86 (Bouchard et al., 1997). Estimates for a dynamic balance (ladder climb, stabilometer) are lower than those for static balance (standing on a rail without moving). A test comprising a series of stunts which include major balance and agility components had moderately high heritabilities in twins aged 8–15 years, 0.80 in boys and 0.74 in girls (Sklad, 1973). In 10-year-old twins, the AE model provided the best fit for the flamingo stand (static balance), accounting for 50% of the variance in both boys and girls (Maes et al., 1996).

Flexibility is thought to be an important trait for an array of performance phenotypes. Heritability of lower back flexibility was 0.69 in male twins aged 11–15 years but only 0.18 in twins of both sexes aged 10–27 years; the latter increased to 0.50 when age and anthropometric characteristics were controlled (Bouchard et al., 1997). Among twins at 10 years of age, lower back flexibility (sit and reach) was best fit with an ACE model (Maes et al., 1996). Genetic factors accounted for 50% and 38% of the variance in girls and boys, respectively. Shared and non-shared environmental sources of variance contributed equally to flexibility in boys, while shared environmental factors accounted for most of the remaining variance in girls. Among twins aged 10–15 years, on the other hand, the sit and reach was best fit with the AE model. Additive genetic factors accounted for 55% of the variance with no sex difference (Okuda et al., 2005). Although flexibility is joint-specific, data for other joints are limited. Among twins of both sexes aged 12–17 years, heritabilities of trunk, hip, and shoulder flexibility were, respectively, 0.84, 0.70, and 0.91 (Bouchard et al., 1997; Malina & Bouchard, 1989).

The potential role of genotype in motor skill acquisition needs attention. Several relatively dated experimental studies have considered the learning motor skills in adolescent twins (Bouchard et al., 1997). Except for a stabilometer task, which places a premium on dynamic balance and coordination, tasks used in the experimental studies were largely fine motor skills that stressed manual dexterity and precision of movement. Results suggested similar rates of learning motor skills in MZ compared to DZ twins, but estimates of the genetic contribution varied from task to task and during the time course of learning, that is, over practice trials or training sessions.

Detailed genetic analyses of learning curves for motor skills are limited. Parameters of the learning curves of four tasks—plate tapping with the hand, tapping with one foot, mirror tracing, and a ball toss for accuracy—in twins at 9–13 years of age showed variable results (Sklad, 1975). Two parameters, level and rate of learning, were more similar in MZ than in DZ twins, and intrapair correlations tended to be higher in male than in female MZ twins. The third parameter, the final level of skill attained, was quite variable between twin types and sexes and among the four tasks. Heritability of rotary pursuit performance in adult MZ and DZ twins reared apart was moderate for initial trials, 0.66, but increased with practice, 0.74 (Fox et al., 1996). Though limited, the evidence also suggested that change in performance over trials (rate of learning) was also heritable. Overall, the results suggest genetic contributions to the skill learning process.

Aerobic Performance

Aerobic performance tests are both maximal and submaximal. Maximal aerobic power (maximal oxygen uptake) is measured under standardized laboratory conditions on a cycle ergometer or a treadmill and maximal or peak oxygen uptake is often expressed per unit body mass. Submaximal aerobic performance is usually measured as power output at a heart rate of 150 or 170 beats per minute and is expressed per kilogram body mass (W/kg). Available evidence for genetic contributions to the variation in maximal and submaximal aerobic performance has been previously summarized (Bouchard et al., 1997). Correlations for maximal oxygen uptake, adjusted for body mass, varied between 0.60 and 0.91 in MZ twins and between 0.04 and 0.53 in DZ twins; estimated heritabilities varied from 0.40 to 0.94 (Table 9.2).

SEM has been more recently applied to aerobic performance. Genetic estimates varied between 0.69 and 0.87 for adjusted (body mass, lean body mass and/or fat mass) maximal aerobic performance

(Table 9.2). Since sample sizes were rather small and subjects differed in chronological age and included males and females, studies with adequate power and appropriate control for age and gender were selected for discussion. To differentiate between a false E model (variance in the phenotype explained by unique environmental factors) and a true AE model (variance explained by additive genetic and unique environmental factors) with $a^2 = .50$ and $e^2 = .50$, a realistic assumption based on the available evidence, and for a power of 80% with an alpha-level of .05, 23 twin pairs are needed in each zygosity group (power calculations based on Martin et al., 1978). When these criteria were applied, only four studies had sufficient power. In these studies, heritabilities for adjusted aerobic performance ranged from 0.40 to 0.70 in males; heritability was 0.87 in 10-year-old girls. Only one study reported sex-specific heritabilities (Maes et al., 1996). For submaximal aerobic performance, heritabilities varied between 0.38 and 0.55. Spouse correlations for maximal and submaximal tests vary between 0.17 and 0.42 (Bouchard et al., 1997), so that if the effect of positive assortative mating is not considered in twin models, heritability estimates are theoretically biased downward. Positive assortative mating increases both the MZ and DZ twin correlations which may result in an inflated estimated shared environment which, in turn, may

Table 9.2 Estimated heritabilities of aerobic performance in studies of twins

Commonly used heritability indices[a]	
Maximal aerobic performance (adjusted for body mass)	0.40–0.94
Submaximal aerobic performance (PWC150/kg body mass)	0.38
Genetic determination estimated with SEM	
Maximal aerobic performance: Adjusted for body mass, lean mass or fat mass indicators	0.69–0.87[b]
Submaximal aerobic performance (PWC110/kg mass)	0.55[c]

[a] See Bouchard et al. (1997), table 15.4, p. 331, for specific references.
[b] Fagard et al. (1991); Maes et al. (1996).
[c] Bielen et al. (1991).

bias the estimated genetic variance downwards (Neale & Cardon, 1992).

Apart from studies conducted before the 1980s, most of the information on aerobic performance stems from twin studies in Canada and Belgium. Information to verify gender-specific or age-associated genetic contributions is limited. Among Belgian twins, age- and sex-specific analyses revealed that the heritabilities for relative peak $\dot{V}O_2$, adjusted for body mass, were above 0.65 at all age levels with the exception of 10 and 12 years in boys and 12 and 16 years in girls. There was no clear age trend between 10 and 18 years and at 11, 14, and 15 years estimates were higher in girls compared to boys (Thomis et al., 2003).

Adjustments for body mass, lean body mass, and indicators of fatness may not be the most appropriate procedures to estimate the genetics of relative or normalized aerobic performance. Since these phenotypes are correlated it is likely that they share common environmental and genetic factors that could be identified using a multivariate approach. Although this has not yet been done for aerobic performance and correlated phenotypes (body mass, lean body mass, and indicators of fatness), it has been demonstrated that muscle cross-sectional area and isometric, eccentric, and concentric strength were under the control of shared genetic and environmental factors

(De Mars et al., 2007). Shared genetic factors explained significant proportions of the variance in muscle cross-sectional area (43%) and strength: eccentric (47%), isometric (58%), and concentric (32%). The results indicate that performance and anthropometric phenotypes are in part under the control of the same set of genes so that adjusting for body mass or other dimensions could bias estimated genetic contributions.

Summary

Heritability estimates spanning a wide range have been reported for several sport-related performance phenotypes. This is explained, in part, by differences in sample characteristics and analytical strategies (correction for covariates, classical heritability estimates, model fitting). Estimated heritabilities are derived largely from well-nourished samples, most of European (White) and few of Japanese ancestry. It is possible that the genetic contribution to phenotypic variance differs among populations (Malina et al., 2004). Conclusive data are lacking for the significance of effects of age, sex, maturation, and ethnicity on heritability estimates for many performance phenotypes. Data are also limited for sufficiently large samples using the best available analytical techniques.

References

Arden, N.K. & Spector, T.D. (1997). Genetic influences on muscle strength, lean body mass, and bone mineral density: A twin study. *Journal of Bone and Mineral Research* 12, 2076–2081.

Beunen, G. & Thomis, M. (2004). Gene powered? Where to go from heritability (h²) in muscle strength and power? *Exercise and Sport Sciences Reviews* 32, 148–154.

Beunen, G., Thomis, M., Maes, H.H., et al. (2000). Genetic variance of adolescent growth in stature. *Annals of Human Biology* 27, 173–186.

Beunen, G., Thomis, M., Peeters, M., Maes, H.H., Claessens, A.L. & Vlietinck, R. (2003). Genetics of strength and power characteristics in children and adolescents. *Pediatric Exercise Science* 15, 128–138.

Bielen, E.C., Fagard, R.H. & Amery, A. (1991). Inheritance of acute cardiac changes during bicycle exercise: an echocardiographic study in twins. *Medicine & Science in sports & Exercise* 23, 1254–1259.

Bouchard, C. & Malina, R.M. (1983). Genetics of physiological fitness and motor performance. *Exercise and Sport Sciences Reviews* 11, 306–339.

Bouchard, C., Malina, R.M. & Pérusse, L. (1997). *Genetics of fitness and physical performance.* Human Kinetics, Champaign, IL.

Calvo, M., Rodas, G., Vallejo, M., et al. (2002). Heritability of explosive power and anaerobic capacity in humans. *European Journal of Applied Physiology* 86, 218–225.

Chatterjee, S. & Das, N. (1995). Physical and motor fitness in twins. *Japanese Journal of Physiology* 45, 519–534.

De Mars, G., Thomis, M.A., Windelinckx, A., et al. (2007). Covariance of isometric and dynamic arm contractions: Multivariate genetic analysis. *Twin Research and Human Genetics* 10, 180–190.

De Moor, M.H., Spector, T.D., Cherkas, L.F., et al. (2007). Genome-wide linkage scan for athlete status in 700 British female DZ twin pairs. *Twin Research and Human Genetics* 10, 812–820.

Fagard, R., Bielen, E. & Amery, A. (1991). Heritability of aerobic power and anaerobic energy generation during exercise. *Journal of Applied Physiology* 70, 357–362.

Fox, P.W., Hershberger, S.L. & Bouchard, T.J. Jr. (1996). Genetic and environmental contributions to the acquisition of a motor skill. *Nature* 384, 356–358.

Frederiksen, H., Gaist, D., Petersen, H.C., et al. (2002). Hand grip strength: A phenotype suitable for identifying genetic variants affecting mid- and late-life physical functioning. *Genetic Epidemiology* 23, 110–122.

Goya, T., Amano, Y., Hoshikawa, T. & Matsui, H. (1993). Longitudinal study on the variation and development of selected sports performance in twins: Case study for one pair of female monozygous (MZ) and dizygous (DZ) twins. *Sport Science* 14, 151–168.

Goya, T., Hoshikawa, T. & Matsui, H. (1991). *Longitudinal Study on Selected Sports Performance Related with the Physical Growth and Development of Twins*, pp. 139–141. University of Western Australia, Perth.

Katzmarzyk, P. & Bouchard, C. (2005). Genetic influences on human body composition. In: S.B. Heymsfield, T.G. Lohman, Z. Wang & S.B. Going (eds.) *Human Body Composition*, pp. 243–257. Human Kinetics, Champaign, IL.

Maes, H.H., Beunen, G.P., Vlietinck, R.F., et al. (1996). Inheritance of physical fitness in 10-yr-old twins and their parents. *Medicine & Science in Sports & Exercise* 28, 1479–1491.

Malina, R.M. & Bouchard, C. (1989). Genetic considerations in physical fitness. In: T.F. Drury (ed.) *Assessing Physical Fitness and Physical Activity in Population-based Surveys*, pp. 453–473. US Government Printing Office, Washington, DC.

Malina, R.M., Bar-Or, O. & Bouchard, C. (2004). *Growth, Maturation, and Physical Activity*, 2nd edn. Human Kinetics, Champaign, IL.

Maridaki, M. (2006). Heritability of neuromuscular performance and anaerobic power in preadolescent and adolescent girls. *Journal of Sports Medicine and Physical Fitness* 46, 540–547.

Martin, N.G., Eaves, L.J., Kearsey, M.J. & Davies, P. (1978). The power of the classical twin study. *Heredity* 40, 97–116.

Neale, M.C. & Cardon, L.R. (1992). *Methodology for Genetic Studies of Twins and Families*, 1st edn. Kluwer, Dordrecht.

Neale, M.C., Boker, S.M., Xie, G., et al. (2003). *MX: Statistical Modeling*, 6th edn. Virginia Commonwealth University, Richmond, VA.

Okuda, E., Horii, D. & Kano, T. (2005). Genetic and environmental effects on physical fitness and motor performance. *International Journal of Sport and Health Science* 3, 1–9.

Parisi, P., Casini, B., Salvo, V.D., et al. (2001). The registry of Italian twin athletes (RITA): Background, design, and procedures, and twin data analysis on sport participation. *European Journal of Sport Science* 1, 1–12.

Peeters, M.W., Thomis, M.A., Claessens, A.L., et al. (2003). Heritability of somatotype components from early adolescence into young adulthood: A multivariate analysis on a longitudinal twin study. *Annals of Human Biology* 30, 402–418.

Peeters, M.W., Thomis, M.A., Maes, H.H., et al. (2005a). Genetic and environmental determination of tracking in static strength during adolescence. *Journal of Applied Physiology* 99, 1317–1326.

Peeters, M.W., Thomis, M.A., Maes, H.H., et al. (2005b). Genetic and environmental causes of tracking in explosive strength during adolescence. *Behavior Genetics* 35, 551–563.

Peeters, M.W., Beunen, G.P., Maes, H.H., et al. (2007a). Genetic and environmental determination of tracking in subcutaneous fat distribution during adolescence. *American Journal of Clinical Nutrition* 86, 652–660.

Peeters, M.W., Thomis, M.A., Loos, R.J., et al. (2007b). Heritability of somatotype components: A multivariate analysis. *International Journal of Obesity (London)* 31, 1295–1301.

Peeters, M.W., Thomis, M.A.I., Beunen, G.P. & Malina, R.M. (2009). Genetics and sports: An overview of the pre-molecular era. In: M. Collins (ed.) *Genetics and Sports*, pp. 28–42. Karger, Basel.

Ropponen, A., Levalahti, E., Videman, T., Kaprio, J. & Battie, M.C. (2004). The role of genetics and environment in lifting force and isometric trunk extensor endurance. *Physical Therapy* 84, 608–621.

Silventoinen, K., Magnusson, P.K., Tynelius, P., Kaprio, J. & Rasmussen, F. (2008). Heritability of body size and muscle strength in young adulthood: A study of one million Swedish men. *Genetic Epidemiology* 32, 341–349.

Simoneau, J.A. & Bouchard, C. (1995). Genetic determinism of fiber type proportion in human skeletal muscle. *FASEB Journal* 9, 1091–1095.

Sklad, M. (1972). Similarity of movement in twins. *Wychowanie Fizyczne i Sport* 16, 119–141.

Sklad, M. (1975). The genetic determination of the rate of learning of motor skills. *Studies in Physical Anthropology* 1, 3–19.

Sklad, M. (1973). Rozwoj fizyczny i motorycznosc blizniat. *Materialy i Prace Antropologiczne* 85, 3–102.

Souren, N.Y., Paulussen, A.D., Loos, R.J., et al. (2007). Anthropometry, carbohydrate and lipid metabolism in the East Flanders Prospective Twin Survey: Heritabilities. *Diabetologia* 50, 2107–2116.

Thomis, M.A., Beunen, G.P., Van Leemputte, M., et al. (1998). Inheritance of static and dynamic arm strength and some of its determinants. *Acta Physiologica Scandinavica* 163, 59–71.

Thomis, M.A.I., Vanden Eynde, B., Maes, H.H., et al. (2003). Heritabilities of peak $\dot{V}O_2$ in adolescence. *Revista Portuguesa de ciências do desporto* 3, 73–74.

Tiainen, K., Sipila, S., Alen, M., et al. (2005). Shared genetic and environmental effects on strength and power in older female twins. *Medicine & Science in Sports & Exercise* 37, 72–78.

Vogel, F. & Motulsky, A.G. (1986). *Human Genetics: Problems and Approaches*, 2nd edn. Springer, Berlin.

Chapter 10

Twin and Family Studies of Training Responses

MARK A. SARZYNSKI, TUOMO RANKINEN AND CLAUDE BOUCHARD

Human Genomics Laboratory, Pennington Biomedical Research Center, Baton Rouge, LA, USA

The favorable effects of endurance training on physical performance and on several risk factors for chronic diseases have been well documented. However, it is also clear that there are marked interindividual differences in the adaptation to regular exercise. For instance, several exercise training studies have shown that there are marked interindividual differences in the trainability of cardiorespiratory endurance phenotypes after exposure to a fully standardized and monitored training program (Bouchard, 1983). Variability among individuals is common for nearly all traits, both in the untrained state and in response to training, and is thought to be influenced by genetic factors. Both animal and human studies strongly support the hypothesis that there is a significant genetic component to physical performance phenotypes. The genetic component is thought to affect both the intrinsic capacity in the untrained state and responsiveness to training.

As such, it is believed that the main cause of the heterogeneity in training response is related to as yet undetermined genetic characteristics. To test this hypothesis, multiple training studies have been performed in pairs of monozygotic (MZ) twins and in one cohort of nuclear families, the rationale being that the response pattern will be more similar for individuals having the same genotype (within twin pairs or families) or sharing about 50% of their genes by descent (parent–offspring, siblings), than for subjects with different genotypes (between twin pairs or families). Results from twin studies and data from the HERITAGE Family Study suggest that the individuality in trainability of phenotypes governing endurance performance is highly familial and genetically determined. The purpose of this chapter is to provide an overview of twin and family studies of response to exercise training. First, early studies showing individual variation in response to exercise training will be introduced. Then, the key findings from twin and family training studies will be summarized, including results showing the genetic component of trainability. Lastly, the implications of these findings on sports performance will be discussed.

Human Variation in Training Response

The notion that there were biologically meaningful interindividual differences in response to exercise programs was first introduced in the early 1980s (Bouchard, 1983). In a series of controlled exercise training experiments conducted with young, healthy adults, it was demonstrated that human variation in training-induced changes in performance and fitness phenotypes was extensive, with the range between low and high responders reaching several folds (Bouchard, 1983, 1995; Bouchard et al., 1992; Lortie et al., 1984; Simoneau et al., 1986). Table 10.1 summarizes the findings of several studies showing the presence of such individual differences in response to training in aerobic and anaerobic performance traits.

The HERITAGE Family Study has provided the most extensive data to date on individual differences in trainability. The study was based on

Genetic and Molecular Aspects of Sport Performance, 1st edition.
Edited by Claude Bouchard and Eric P. Hoffman.
Published 2011 by Blackwell Publishing Ltd.

Table 10.1 Performance phenotypes before training and the response to training in sedentary young adults from several studies

	Pre-training phenotype		Changes in phenotype				References
	Mean	SD	Mean	SD	min	max	
$\dot{V}O_2$max (l/min)							
Quebec	2.9	0.42	0.63	0.25	0.13	1.03	Lortie et al. (1984), Dionne et al. (1991)
Arizona	3.4	0.57	0.42	0.22	0.06	0.95	Dionne et al. (1991)
$\dot{V}O_2$max (l/min)	2.3	0.6	30%	15%	7%	87%	Change values are given in percent of pre-training value (Lortie et al., 1984)
Max. work output (MJ) in 90 min	567	140	213	155	42	516	Lortie et al. (1984)
Max. work output (kJ) in 90 s	25.7	4.1	7.5	2.6	2.5	13.0	Simoneau et al. (1986)
Max. work output (kJ) in 10 s	6.7	0.8	1.4	0.7	0.3	2.7	Simoneau et al. (1986)

Adapted from Bouchard et al. (1992).

healthy but sedentary subjects who were exposed to a standardized, fully monitored, laboratory-based aerobic training program consisting of 60 exercise sessions over 20 weeks. There were 742 completers from 204 families. Subjects trained three times per week at the heart rate (HR) associated with 55% of their initial $\dot{V}O_2$max for 30 min/day and gradually progressed to the HR associated with 75% of their initial $\dot{V}O_2$max for 50 min/day at the end of 14 weeks. The latter was maintained for the last 6 weeks of the program (Bouchard et al., 1995). The average increase in $\dot{V}O_2$max was 384 mL O_2/min with an SD of 202. The training responses varied from no change to increases of more than 1000 ml O_2/min (Bouchard & Rankinen, 2001; Bouchard et al., 1999). The response distribution for $\dot{V}O_2$max in HERITAGE subjects is depicted in Figure 10.1.

The high degree of heterogeneity in responsiveness to a fully standardized exercise program in the HERITAGE Family Study was not accounted for by baseline $\dot{V}O_2$max level, age, gender, or ethnic differences. For example, the same heterogeneity in response levels can be found in those who began the program with a low $\dot{V}O_2$max and in those who were initially above the $\dot{V}O_2$max median (Bouchard & Rankinen, 2001). A similar pattern of interindividual

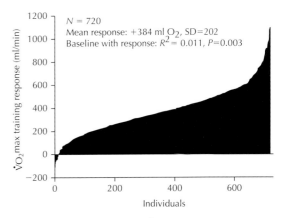

Figure 10.1 Heterogeneity of $\dot{V}O_2$max training response in the HERITAGE Family Study. For both baseline and post-training $\dot{V}O_2$max values, the mean of two separate $\dot{V}O_2$max tests was used if there was less than 5% difference between them; otherwise the highest $\dot{V}O_2$max was used. The $\dot{V}O_2$max training response was defined as the absolute difference (ml O_2/min) between post-training $\dot{V}O_2$max and baseline $\dot{V}O_2$max. From Bouchard (2001).

variation in training responses was observed for other phenotypes in HERITAGE, including stroke volume (SV), cardiac output (Q), insulin and plasma lipid levels, submaximal exercise HR, and blood pressure (Boule et al., 2005; Leon et al., 2000; Wilmore, 2001a, b), as revealed by the magnitude of the standard deviations for the training-induced changes and as illustrated in Figure 10.2 for submaximal measures of HR, systolic blood pressure (SBP), and high-density lipoprotein cholesterol (HDL-C).

In addition, previous studies have shown considerable variation in the range of individual training responses for strength and power phenotypes including one-repetition maximum (1RM, 0–250%), peak muscular torque (12–36%), maximum voluntary contraction (−32% to +149%), and eccentric strength (coefficient of variation = 504%) (Hubal et al., 2005; Thibault et al., 1986; Thomis et al., 1998). These data underline the notion that the effects of endurance and strength training on cardiovascular and other relevant traits should be evaluated not only in terms of mean changes but also in terms of response heterogeneity. Furthermore, these observations support the notion that individual variability is a normal biological phenomenon, which may largely reflect genetic diversity (Bouchard, 1995; Bouchard & Rankinen, 2001).

Twin Studies of Trainability

It is believed that the main cause of the heterogeneity in training response is to be found in as-of-yet undetermined genetic characteristics. Rodent models have provided clear evidence of large differences in running performance responses to exercise training between high- and low-responding strains of mice or rats (Koch et al., 2005; Massett & Berk, 2005; Troxell et al., 2003), suggesting that genetic background contributes significantly to adaptation to exercise. To test the role of the genotype in the response to training in humans, multiple training studies have been performed in pairs of MZ twins.

Cardiorespiratory Fitness Phenotypes

A wide range of individual training gains in absolute $\dot{V}O_2max$ was found in 10 pairs of MZ twins subjected to 20 weeks of endurance training (Prud'homme et al., 1984). However, these differences were not distributed randomly among the twins. The similarity of the training response among

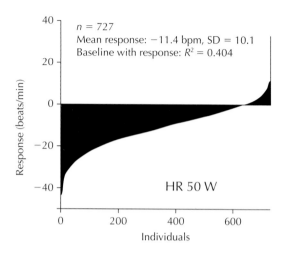

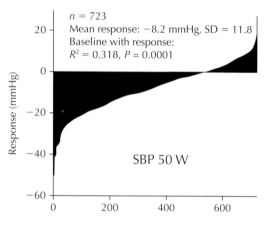

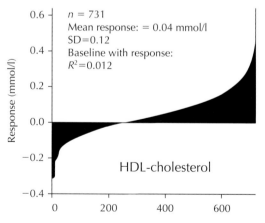

Figure 10.2 Heterogeneity of heart rate (HR 50 W), systolic blood pressure (SBP 50 W), and plasma HDL-cholesterol training responses in the HERITAGE Family Study. Adapted from Rankinen and Bouchard (2007).

members of the same MZ pair is illustrated in Figure 10.3. The intraclass correlation in response of $\dot{V}O_2$max (l/min) was 0.77, indicating that members of the same twin pair responded similarly to training. There was almost eight times (F ratio = 7.8) more variance between pairs of twins than within pairs of twins for the response to training in absolute $\dot{V}O_2$max (Prud'homme et al., 1984).

These results were replicated in two other studies of endurance training in MZ twins (Boulay et al., 1986; Hamel et al., 1986), which showed intraclass correlations in response of $\dot{V}O_2$max (l/min) of 0.44 and 0.65. Pooled results of the trainability of $\dot{V}O_2$max in MZ twins from the above three studies show there is about nine times more variance between genotypes than within genotypes in the response of $\dot{V}O_2$max to standardized training protocols (Table 10.2). Furthermore, a significant intrapair resemblance of 0.60 was observed for $\Delta\dot{V}O_2$max (l/min) in the most stringent study of exercise training in MZ twins (Table 10.2). This study involved seven male MZ twin pairs subjected to a negative energy balance protocol over a period of 3 months (Bouchard et al., 1994). The daily energy deficit was induced entirely by exercise training performed twice daily on a cycle ergometer, 9 out of 10 days, over a period of 93 days while subjects were kept on a constant daily energy and nutrient intake.

The previous results focused on $\dot{V}O_2$max. A related measure of aerobic performance is total work output during prolonged exercise. In two stud-

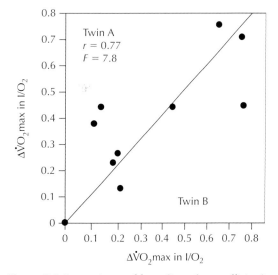

Figure 10.3 Intrapair resemblance (intraclass coefficient) in 10 pairs of MZ twins for training changes in $\dot{V}O_2$max after 20 weeks of endurance training. F represents the ratio of variance between MZ twin pairs and within MZ twin pairs for the response to training in $\dot{V}O_2$max. r is the MZ twin resemblance (correlation) in response to training for $\dot{V}O_2$max. From Bouchard et al. (1992), computed from the data in Prud'homme et al. (1984).

Table 10.2 Effects of training on performance-related traits in MZ twins and twin resemblance in training response

Trait	# twin pairs	Training response (% or Mean ± SD)	F ratio*	Intraclass coefficient*	References
Total work output in 90 min (kJ)	6	32 ± 16%	11.0	0.83	Boulay et al. (1986), Hamel et al. (1986)
	14	17 ± 12%	5.5	0.69	
$\dot{V}O_2$max (l/min)	10	16%	7.8	0.77	Bouchard et al. (1990, 1994), Boulay et al. (1986), Hamel et al. (1986), Prud'homme et al. (1984)
	6	17 ± 11%	4.6	0.65	
	14	22 ± 15%	2.6	0.44	
Pooled results of above three studies	26	0.4 ± 0.04	9.4	0.81	
	7		3.9	0.60	

*All values significant at the $P < 0.05$ level. F ratio represents the ratio of variance between pairs of twins and within pairs of twins for the response to training in the selected trait. Intraclass coefficient represents the MZ twin resemblance (correlation) in response to training for the selected trait.

ies of MZ twins (Boulay et al., 1986; Hamel et al., 1986), total power output during a 90-min maximal cycle ergometer test showed a highly significant within-pair resemblance in training response, with the ratio of between-pairs to within-pairs variances ranging from 5.5 to 11, and the intraclass coefficient for twin resemblance in response ranging from 0.69 to 0.83. The training response of anaerobic performance is also apparently determined by inherited factors. A study of 14 pairs of MZ twins found that the training response of short-term (10-s work output) and long-term (90-s work output) anaerobic performance was characterized by a high MZ twin resemblance (Simoneau et al., 1986) (Figure 10.4). For example, the trainability of the 90-s work output test was characterized by a between-pairs variance component amounting to about 65% of the training response.

The previously mentioned training studies in MZ twins were also associated with a decrease in the cardiovascular and metabolic response at a given submaximal power output. For instance, when exercising in a relative steady state at 50 watts (W), there were decreases in HR, $\dot{V}O_2$, pulmonary ventilation, ventilatory equivalent of oxygen ($V_E/\dot{V}O_2$) and in respiratory quotient (RQ) with an increase in the oxygen pulse ($\dot{V}O_2/HR$). These various metabolic improvements were, however, all characterized by a significant within-pair resemblance, with the ratio of between-pairs to within-pairs variances reaching at least 3, and the intraclass coefficients for twin resemblance in response reaching 0.53 and better (Bouchard et al., 1992). In seven male MZ twin pairs who completed a negative energy balance protocol, including twice daily exercise periods, there were significant within-pair resemblances in training response for $\dot{V}O_2$ (r = 0.87 and 0.76 at 50 and 150 W), RQ (r = 0.86 and 0.87 at 100 and 150 W), and HR (r = 0.70 and 0.63 at 50 and 150 W) phenotypes measured at submaximal intensities (Bouchard et al., 1994).

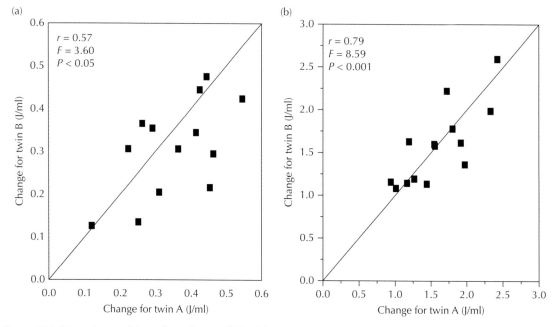

Figure 10.4 Intrapair resemblance (intraclass coefficient) for training changes in 10 s (a) and 90 s (b) cycle performance, expressed in Joules per milliliter of thigh volume, in 13 pairs of MZ twins following 15 weeks of high-intensity intermittent training. F represents the ratio of variance between MZ twin pairs and within MZ twin pairs for the response to training in $\dot{V}O_2$max. r is the MZ twin resemblance (correlation) in response to training for $\dot{V}O_2$max. From Simoneau and Bouchard (1998), based on data from Simoneau et al. (1986).

Strength and Power Phenotypes

The role of genetic factors in the response to strength training has not been examined extensively. The results from a study of five pairs of MZ twins submitted to a 10-week strength training program suggested that the response to strength training was independent of the genotype (Thibault et al., 1986). However, significant genotype by training interactions were found for dynamic strength phenotypes in 25 pairs of MZ twins subjected to a 10-week resistance training program for the elbow flexors (Thomis et al., 1998). The F ratio was 3.5 for 1RM increase with training, and 1.83 for maximal isometric contraction at 110° response, indicating twin pairs responded similarly to strength training. Model-fitting procedures indicated that about 20% of the variation in post-training 1RM, isometric strength at 110°, and concentric strength at 120°/s was explained by training-specific genetic factors that were independent from genetic factors that explained variation in the pre-training phenotype (Thomis et al., 1998).

Skeletal Muscle Phenotypes

Several experimental studies of the responses of young adult MZ twins to standardized training programs provide some insights into potential genotypic contributions to the skeletal muscle training response (Bouchard et al., 1994; Hamel et al., 1986; Simoneau et al., 1986; Thibault et al., 1986). These studies included male and female MZ twins, and there were no sex differences in training-related gains. A significant intrapair resemblance of 0.76 resulted for the training-induced changes in oxoglutarate dehydrogenase (OGDH) among five pairs of MZ twins submitted to a 10-week isokinetic strength training program (Thibault et al., 1986). After 15 weeks of intermittent endurance training in 12 pairs of MZ twins, about 50–60% of the responses of hexokinase (HK), lactate dehydrogenase (LDH), malate dehydrogenase (MDH), OGDH, and the phosphofructokinase (PFK)/OGDH ratio to training were genotype-associated, while about 80% of the creatine kinase (CK) training response appeared to be determined by the genotype (Simoneau et al., 1986).

Genotype–training interactions were evaluated for several enzymes after 7 and 15 weeks of endurance training in six pairs of MZ twins, and changes in skeletal muscle enzyme activities during the first half of training were only weakly related to genotype (Hamel et al., 1986). However, changes in the activities of PFK, MDH, 3-hydroxyacyl CoA dehydrogenase (HADH), and OGDH during the second half of training were characterized by significant within-pair resemblance (Hamel et al., 1986). This suggests that early in the program, adaptation to endurance training may be under less-stringent genetic control; however, as training continues and perhaps nears maximal trainability, the response becomes more genotype dependent.

Body Composition Phenotypes

The genotype effects for the responses of body composition phenotypes to endurance training were examined in seven pairs of young male MZ twins who completed a 3-month negative energy balance protocol as previously described (Bouchard et al., 1994). Significant intrapair resemblance was observed for the changes in body weight, fat mass (FM), percent body fat, body energy content, sum of 10 skinfolds, and abdominal visceral fat (AVF), with F ratios and intraclass correlations ranging from 6.8 to 14.1 and 0.74 to 0.87, respectively (Table 10.3).

Taken together, these studies indicate that members of the same MZ pair are significantly more alike than unrelated individuals in their physiological responses to a standardized training program. Thus, there are substantial genetic components involved in training-induced changes in performance-srelated phenotypes.

Family Studies of Trainability

The maximal heritability of a trait, that is, the combined effect of genes and shared environment across generations, can be estimated using data from family and twin studies. The early studies performed solely with identical twins do not allow for quantification of the magnitude of the heritability of trainability, although they provide rough indicators of its upper band value. However, the heritability of

Table 10.3 Effect of exercise training–induced negative energy balance in MZ twins on body composition phenotypes and intrapair resemblance in response (Mean SEM)

	Training response (Mean ± SD)	F ratio[a]	Intraclass coefficient[a]
Body weight (kg)	5.0 ± 0.6***	6.8*	0.74*
Fat mass (kg)	4.9 ± 0.6***	14.1**	0.87**
Fat-free mass (kg)	0.1 ± 0.3	8.2**	0.78**
Percent fat (%)	4.8 ± 0.6***	9.0**	0.80**
Body energy (MJ)	191 ± 24***	12.9**	0.86**
Sum of 10 skin folds (mm)	52 ± 7**	11.8**	0.84**
Abdominal vis ceral fat (cm²)	29 ± 3***	11.7**	0.84**

Adapted from Bouchard et al. (1994).
[a] F ratio represents the ratio of variance between pairs of twins and within pairs of twins for the response to training in the selected trait. Intraclass coefficient represents the MZ twin resemblance (correlation) in response to training for the selected trait.
*P ≤ 0.05; **P ≤ 0.01; ***P ≤ 0.0001.

training response in numerous performance-related phenotypes can be estimated from data of nuclear families. Family studies allow for the examination of familial aggregation, the occurrence of a trait in more members of a family than can be readily accounted for by chance. The only family study on the individual differences in trainability available at this time is the HERITAGE Family Study.

In HERITAGE, the increase in $\dot{V}O_2$max in 481 individuals from 99 two-generation families of Caucasian descent showed 2.5 times more variance between families than within families (Bouchard et al., 1999). Thus, the extraordinary heterogeneity observed for the gains in $\dot{V}O_2$max among adults is not random and is characterized by a strong familial aggregation (Figure 10.5). Figure 10.5 depicts the distribution of the age- and sex-adjusted $\dot{V}O_2$max response within and between families and illustrates the extent of familial resemblance in the trainability of $\dot{V}O_2$max. The figure also

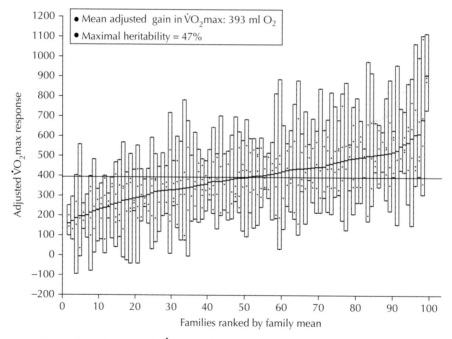

Figure 10.5 Age- and sex-adjusted response in $\dot{V}O_2$max phenotype in Caucasians of the HERITAGE Family Study plotted against family rank (i.e., families ranked by family mean). Adjusted $\dot{V}O_2$max response value for each individual was calculated as the residual from regression model plus group mean. Each family is enclosed within a bar: solid dot, individual data points; solid horizontal dash, family mean. Horizontal reference line is group mean. From Bouchard et al. (1999).

shows that there are families with a predominantly low-response phenotype and others with large concentrations of high responders.

A model-fitting procedure found the most parsimonious models yielded a maximal heritability estimate of 47% for absolute $\dot{V}O_2$max response (Bouchard et al., 1999). Adjusting the $\dot{V}O_2$max response data for baseline $\dot{V}O_2$max did not modify this estimate, indicating that the familial and genetic factors underlying these phenotypes appear to be different. Additionally, data from studies in rodents have provided corroborating evidence regarding the heritability of aerobic capacity. For example, heritability for treadmill running capacity response to 24 days of treadmill exercise was 43% after one generation of selective breeding in rats for adaptational response to exercise (low and high responders) (Troxell et al., 2003).

In addition to $\dot{V}O_2$max, familial aggregation and the heritability of training-induced changes in several other phenotypes, including submaximal aerobic performance (Perusse et al., 2001); resting and submaximal exercise blood pressure, SV, Q, and HR (An et al., 2000a, b, 2003; Rice et al., 2002a); plasma lipid, lipoprotein, and apolipoprotein levels (Rice et al., 2002b); body composition and body fat distribution (Perusse et al., 2000; Rice et al., 1999); and skeletal muscle characteristics (Rankinen et al., 2005; Rico-Sanz et al., 2003) were investigated in the HERITAGE Family Study. The maximal heritabilities for these traits ranged from 25% to 55%, further confirming the contribution of familial factors to the person-to-person variation in responsiveness to endurance training (Rankinen & Bouchard, 2008). Table 10.4 shows the maximal heritabilities of training response in several performance phenotypes, as derived from the HERITAGE Family Study.

The response to exercise training of all measured submaximal aerobic performance phenotypes showed significant familial resemblance in Caucasian families in HERITAGE (Perusse et al., 2001). Maximal heritabilities ranged from 23% to 57% for $\Delta\dot{V}O_2$ at 60% and 80% of $\dot{V}$Omax, $\Delta\dot{V}O_2$ 50 W, and ΔPower output 60% and 80%. Submaximal exercise SV and Q were also characterized by a significant familial

aggregation in response to endurance training in HERITAGE. The between-family variation in age, sex, body surface area, and baseline phenotype level-adjusted SV and Q training responses at 50 W were 1.5–2.2 times greater than the within-family variation (An et al., 2000a). In Caucasian families, maximal heritability estimates were 29% and 38% for SV and Q training responses at 50 W, and 24% and 30% for the training-induced changes in SV and Q at 60% of $\dot{V}O_2$max.

In the whole HERITAGE cohort, there was 1.8 times more variance between families than within families for the HR during submaximal exercise at 50 W response (Bouchard & Rankinen, 2001). Maximal heritabilities reached 34% and 29% for HR training responses at 50 W and 60% of $\dot{V}O_2$max, respectively, in Caucasian families (An et al. 2003). Additionally, there was evidence of familial aggregation in training response of SBP at 50 W, with an F ratio ranging from 1.2 to 1.4 depending on the adjustments used ($P < 0.02$) (An et al., 2003; Bouchard & Rankinen, 2001). Heritability was 22% for SBP training response at 50 W, but negligible at 60% of $\dot{V}O_2$max in Caucasian families. Furthermore, no significant heritabilities were found for diastolic blood pressure responses at either 50 W or 60% of $\dot{V}O_2$max (An et al., 2003). The heritabilities for exercise HR and blood pressure response to training were lower than heritability estimates for their baseline response levels, which reached 50%. The results for response to exercise training were

Table 10.4 Maximal heritability estimates of training response in various phenotypes from the HERITAGE Family Study

Trait	Maximal heritability (%)	Reference
$\dot{V}O_2$max (l/min)	47	Bouchard et al. (1999)
$\dot{V}O_2$ 60% (L/min)	23	Perusse et al. (2001)
SV 60% (ml/beat)	24	An et al. (2000a)
Q 60% (l/min)	30	An et al. (2000a)
HR 60% (beats/min)	29	An et al. (2003)
TC and TG	29–32	Rice et al. (2002b)
HDL	26–29	Rice et al. (2002b)

TC, total cholesterol; TG, triglyceride.

controlled for baseline phenotype levels, which accounted for up to 40% of the variance in training response and may explain the lower heritability estimates of response phenotypes.

Familial aggregation of the response of blood lipids to regular exercise was found in HERITAGE (Rice et al., 2002b). For example, the F ratio of the between-family variance to the within-family variance for the HDL-C changes in response to training reached 1.8 ($P < 0.0001$) (Bouchard & Rankinen, 2001). Heritability estimates for blood lipids response to training were significant for most phenotypes, as familial factors accounted for ~30% of the variation of adjusted training responses, with little evidence for ethnic differences (Rice et al., 2002b). Maximal heritabilities ranged from 25% to 38%, with the exceptions being no familial influence on Δlow-density lipoprotein cholesterol (LDL-C) in blacks and higher estimates approaching 60% for Δapolipoprotein B (ApoB) in blacks and 64% for ΔHDL-C subfraction 2 in whites. Significant heritability estimates for the amount and distribution of subcutaneous fat response to exercise training were also found in HERITAGE (Perusse et al., 2000). The heritabilities were 15% for the sum of skinfold thickness measured at eight sites and the sum of four extremity skinfolds, 21% for the sum of four trunk skinfolds, and 14% for the trunk-to-extremity skinfolds ratio. Additionally, the familial etiology of response of total FM and AVF to exercise training was found to be due to putative major genes in HERITAGE (Rice et al., 1999). Segregation analyses detected a putative recessive locus accounting for 18% of the variance for age-adjusted ΔAVF, while a putative dominant locus accounting for 31% of the variance was found for age-adjusted ΔFM. Lastly, training-induced changes in muscle enzyme activities were investigated and the results confirmed that there was consistent familial resemblance for the training responses in activities of marker enzymes of the glycolytic and oxidative pathways (Rico-Sanz et al., 2003; Rankinen et al., 2005).

Implications for Sports Performance

With the continuous expansion of the pool of young participants and competitors in sports, the level of competition has increased to the point that only those individuals who are highly gifted can expect to reach elite status. The elite athlete is probably an individual with a favorable profile in terms of the morphological, physiological, metabolic, motor, perceptual, biomechanical, and personality determinants of the relevant sport. There is evidence for a role of genes in modulating the status of numerous performance determinants in the untrained state. Furthermore, the elite athlete needs to be an obligatory, highly responsive individual to regular training and practice. Thus, two avenues exist for the identification of potential elite athletes: selection based on performance in the untrained state and on individual responsiveness to exercise training. The first avenue could simply involve measuring baseline performance, such as $\dot{V}O_2max$. The second, and possibly more potent, avenue would be more difficult to measure and determine prior to the actual exercise training. A panel of genetic markers for training response would be needed in order to determine which individuals are more likely to be high responders to training and thus likely to exhibit greater performance over time compared to low responders. The optimal strategy would be to select an individual with optimal performance attributes in the untrained state who is also a high responder to exercise training.

The heterogeneity of response to regular exercise has been examined, and the evidence clearly indicates considerable individual differences in the capacity to adapt and benefit from an exercise training program. There are people who experience little, if any, improvements in a given performance-related trait after exercise training, and others who experience great gains and benefits in these traits. Additionally, the heterogeneity in trainability is a trait-specific phenomenon. In other words, an individual who is a low responder for one phenotype may be an average or even high responder for another trait. These observations emphasize that even if an individual cannot improve cardiorespiratory fitness with exercise training, it is likely that he or she will still achieve other significant benefits from regular exercise. This has important public health and sports performance implications.

This chapter has provided ample evidence of the inherited differences in trainability for numerous physiological and metabolic determinants of sports performance. Individuals with the same genotype respond more similarly to training than those with different genotypes. Thus, the evidence suggests that there are separate genetic components affecting performance-related phenotypes at baseline in the untrained state and in response to regular exercise. However, in a few cases, the initial trait level accounts for a significant proportion of the variance in its response to training (Bouchard et al., 1997). From a conceptual point of view, the variance in training response represents one of the most striking examples of a genotype–environmental effect, in this case a genotype–training interaction effect. The search for genetic markers of trainability is a fascinating area of research, but so is the investigation of molecular markers of performance phenotypes in the untrained state.

It should eventually be possible to identify the pool of gifted and highly trainable individuals, especially if genes with large effects are implicated. Regular exercise is recommended to all people for its numerous health benefits; however, it should be recognized that exercise may have only marginal benefits for a fraction of the population. Thus, understanding the genomic determinants of the response of various phenotypes to regular exercise would allow for individualized approaches to exercise recommendations in a public health and rehabilitation context. It would also make it possible to augment the training protocol of the talented and trainable individuals whose goal is to perform at the highest possible level.

Conclusions

Interindividual variability in response to exercise training is a normal biological phenomenon, which is not limited to the phenotypes discussed in this chapter. It is ubiquitous and commonly observed. Furthermore, human heterogeneity in response to regular exercise is not randomly distributed, as it is characterized by familial aggregation. Significant heritability estimates have been reported for training-induced changes in several phenotypes, confirming that familial factors account for some of the person-to-person variation in responsiveness to exercise training. Results from rodent models have provided a convergence of data including evidence for a genetic effect in training responses. Although the research on the genetics of physical performance and health-related fitness is a young field of science, understanding the genetic components of interindividual differences in responsiveness to acute and regular exercise holds great promise. Such data would help to improve sports performance training for aspiring as well as elite athletes. It would also provide a stronger foundation regarding the use of physical activity in the prevention and treatment of chronic diseases.

References

An, P., Perusse, L., Rankinen, T., et al. (2003). Familial aggregation of exercise heart rate and blood pressure in response to 20 weeks of endurance training: The HERITAGE Family Study. *International Journal of Sports Medicine* 24, 57–62.

An, P., Rice, T., Gagnon, J., et al. (2000a). Familial aggregation of stroke volume and cardiac output during submaximal exercise: The HERITAGE Family Study. *International Journal of Sports Medicine* 21, 566–572.

An, P., Rice, T., Perusse, L., et al. (2000b). Complex segregation analysis of blood pressure and heart rate measured before and after a 20-week endurance exercise training program: The HERITAGE Family Study. *American Journal of Hypertension* 13, 488–497.

Bouchard, C. (1983). Human adaptability may have a genetic basis. In: F. Landry (ed.) *Risk Reduction and Health Promotion Proceedings of the 18th Annual Meeting of the Society of Prospective Medicine*, pp. 463–476. Canadian Public Health Association, Ottawa.

Bouchard, C. (1995). Individual differences in the response to regular exercise. *International Journal of Obesity and Related Metabolic Disorders* 19(Suppl. 4), S5–S8.

Bouchard, C. & Rankinen, T. (2001) Individual differences in response to regular physical activity. *Medicine and Science in Sports and Exercise* 33, S446–451.

Bouchard, C., Boulay, M., Dionne, F., Perusse, L., Thibault, M. & Simoneau, J. (1990). Genotype, aerobic performance and response to training. In: G. Beunen, J. Ghesquiere, T. Reybrouck & A. Claessens (eds.) *Children and Exercise*, pp. 124–135. Ferdinand Enke Verlag, Stuttgart.

Bouchard, C., Dionne, F.T., Simoneau, J.A. & Boulay, M.R. (1992). Genetics of aerobic and anaerobic performances. *Exercise and Sport Sciences Reviews* 20, 27–58.

Bouchard, C., Tremblay, A., Despres, J.P., et al. (1994). The response to exercise with constant energy intake in identical twins. *Obesity Research* 2, 400–410.

Bouchard, C., Leon, A.S., Rao, D.C., Skinner, J.S., Wilmore, J.H. & Gagnon, J. (1995). The HERITAGE Family Study. Aims, design, and measurement protocol. *Medicine and Science in Sports and Exercise* 27, 721–729.

Bouchard, C., Malina, R.M. & Perusse, L. (1997). *Genetics of Fitness and Physical Performance*. Human Kinetics, Champaign, IL.

Bouchard, C., An, P., Rice, T., et al. (1999). Familial aggregation of $\dot{V}O_2$max response to exercise training: Results from the HERITAGE Family Study. *Journal of Applied Physiology* 87, 1003–1008.

Boulay, M.R., Lortie, G., Simoneau, J.A. & Bouchard, C. (1986). Sensitivity of maximal aerobic power and capacity to anaerobic training is partly genotype dependent. In: R. Malina & C. Bouchard (eds.) *Sport and Human Genetics*, pp. 173–181. Human Kinetics, Champaign, IL.

Boule, N.G., Weisnagel, S.J., Lakka, T.A., et al. (2005). Effects of exercise training on glucose homeostasis: The HERITAGE Family Study. *Diabetes Care* 28, 108–114.

Dionne, F.T., Turcotte, L., Thibault, M.C., Boulay, M.R., Skinner, J.S. & Bouchard, C. (1991). Mitochondrial DNA sequence polymorphism, $\dot{V}O_2$max, and response to endurance training. *Medicine and Science in Sports and Exercise* 23, 177–185.

Hamel, P., Simoneau, J.A., Lortie, G., Boulay, M.R & Bouchard, C. (1986). Heredity and muscle adaptation to endurance training. *Medicine and Science in Sports and Exercise* 18, 690–696.

Hubal, M.J., Gordish-Dressman, H., Thompson, P.D., et al. (2005). Variability in muscle size and strength gain after unilateral resistance training. *Medicine and Science in Sports and Exercise* 37, 964–972.

Koch, L.G., Green, C.L., Lee, A.D., Hornyak, J.E., Cicila, G.T. & Britton, S.L. (2005). Test of the principle of initial value in rat genetic models of exercise capacity. *American Journal of Physiology—Regulatory, Integrative and Comparative Physiology*, R466–R472.

Leon, A.S., Rice, T., Mandel, S., et al. (2000). Blood lipid response to 20 weeks of supervised exercise in a large biracial population: The HERITAGE Family Study. *Metabolism* 49, 513–520.

Lortie, G., Simoneau, J.A., Hamel, P., Boulay, M.R., Landry, F. & Bouchard, C.

(1984). Responses of maximal aerobic power and capacity to aerobic training. *International Journal of Sports Medicine* 5, 232–236.

Massett, M.P. & Berk, B.C. (2005). Strain-dependent differences in responses to exercise training in inbred and hybrid mice. *American Journal of Physiology—Regulatory, Integrative and Comparative Physiology*, R1006–1013.

Perusse, L., Gagnon, J., Province, M.A., et al. (2001). Familial aggregation of submaximal aerobic performance in the HERITAGE Family Study. *Medicine and Science in Sports and Exercise* 33, 597–604.

Perusse, L., Rice, T., Province, M.A., et al. (2000). Familial aggregation of amount and distribution of subcutaneous fat and their responses to exercise training in the HERITAGE Family Study. *Obesity Research* 8, 140–150.

Prud'homme, D., Bouchard, C., Leblanc, C., Landry, F. & Fontaine, E. (1984). Sensitivity of maximal aerobic power to training is genotype-dependent. *Medicine and Science in Sports and Exercise* 16, 489–493.

Rankinen, T. & Bouchard, C. (2007). Genetic differences in the relationships among physical activity, fitness, and health. In: C. Bouchard, S. Blair & W. Haskell (eds.) *Physical Activity and Health*, pp. 337–358. Human Kinetics, Champaign, IL.

Rankinen, T. & Bouchard, C. (2008). Gene-physical activity interactions: Overview of human studies. *Obesity (Silver Spring)* 16(Suppl. 3), S47–S50.

Rankinen, T., Bouchard, C. & Rao, D.C. (2005). Corrigendum—Familial resemblance for muscle phenotypes: The HERITAGE Family Study. *Medicine and Science in Sports and Exercise* 37, 2017.

Rice, T., Hong, Y., Perusse, L., et al. (1999). Total body fat and abdominal visceral fat response to exercise training in the HERITAGE Family Study: Evidence for major locus but no multifactorial effects. *Metabolism* 48, 1278–1286.

Rice, T., An, P., Gagnon, J., et al. (2002a). Heritability of HR and BP response to exercise training in the HERITAGE

Family Study. *Medicine and Science in Sports and Exercise* 34, 972–979.

Rice, T., Despres, J.P., Perusse, L., et al. (2002b). Familial aggregation of blood lipid response to exercise training in the health, risk factors, exercise training, and genetics (HERITAGE) Family Study. *Circulation* 105, 1904–1908.

Rico-Sanz, J., Rankinen, T., Joanisse, D.R., et al. (2003). Familial resemblance for muscle phenotypes in the HERITAGE Family Study. *Medicine and Science in Sports and Exercise* 35, 1360–1363.

Simoneau, J.A. & Bouchard, C. (1998). The effects of genetic variation on anaerobic performance. In: E.V. Praagh (ed.) *Pediatric Anaerobic Performance*, pp. 5–21. Human Kinetics, Champaign, IL.

Simoneau, J.A., Lortie, G., Boulay, M.R., Marcotte, M., Thibault, M.C. & Bouchard, C. (1986). Inheritance of human skeletal muscle and anaerobic capacity adaptation to high-intensity intermittent training. *International Journal of Sports Medicine* 7, 167–171.

Thibault, M.C., Simoneau, J.A., Cote, C., et al. (1986). Inheritance of human muscle enzyme adaptation to isokinetic strength training. *Human Heredity* 36, 341–347.

Thomis, M.A., Beunen, G.P., Maes, H.H., et al. (1998). Strength training: Importance of genetic factors. *Medicine and Science in Sports and Exercise* 30, 724–731.

Troxell, M.L., Britton, S.L. & Koch, L.G. (2003). Selected contribution: Variation and heritability for the adaptational response to exercise in genetically heterogeneous rats. *Journal of Applied Physiology* 94, 674–681.

Wilmore, J.H., Stanforth, P.R., Gagnon, J., et al. (2001a). Cardiac output and stroke volume changes with endurance training: The HERITAGE Family Study. *Medicine and Science in Sports and Exercise* 33, 99–106.

Wilmore, J.H., Stanforth, P.R., Gagnon, J., et al. (2001b). Heart rate and blood pressure changes with endurance training: The HERITAGE Family Study. *Medicine and Science in Sports and Exercise* 33, 107–116.

Chapter 11

Ethnic Differences in Sport Performance

YANNIS P. PITSILADIS

College of Medicine, Veterinary & Life Sciences, Institute of Cardiovascular & Medical Sciences, University of Glasgow, Glasgow, UK

The athletic events at the XXIX Olympiad in Beijing will be remembered most for the achievements of distance runners from Kenya and Ethiopia and sprinters from Jamaica. Collectively, these three nations won a quarter of all possible track-and-field medals. The sporting achievements of these nations are even more impressive when only the track events are considered; Ethiopia, Kenya, and Jamaica won 36% of all track medals for men and women.

The results of the Beijing Olympics have sparked interest from scientists, the media, and the general public to identify the putative biological mechanisms responsible for the selective dominance of various ethnic groups on the running track. The limited studies reviewed in this chapter would constitute the only available studies on the genetics of the sprint and endurance running phenomena and demonstrate that these athletes, although arising from distinct regions of east Africa or Jamaica (the latter with earlier ancestry from west Africa), do not arise from a limited genetic isolate. Ethiopians and Kenyans do not share similar genetic ancestry but they do share a similar environment: namely moderate altitude as well as high levels of relevant physical activity. Furthermore, the implications of the Jamaican motto "out of many, one people" that is seemingly overlooked by those arguing a genetic explanation for the Jamaican sprint success, is that these islanders are potentially of even greater

genetic diversity than either Kenyans or Ethiopians. It is unlikely, therefore, that these remarkable athletes from east Africa or Jamaica possess unique genotypes that cannot be matched in other areas of the world, but more likely that athletes from these areas with an advantageous genotype realize their biological/genetic potential.

Ethnicity and Athletics

The success of east African athletes at distance running and Jamaican athletes at sprinting will have undoubtedly enhanced the concept that certain ethnic groups possess some inherent genetic advantage predisposing them to superior athletic performance the idea of "black athletic supremacy." Such thinking is not new and has emerged from simplistic interpretations of performances such as those in Table 11.1, combined with the belief that similar skin color indicates similar genetics. This stereotype has been strengthened, albeit inadvertently, by scientists investigating ethnic differences in sports performance. One approach compared the skeletal muscle characteristics of sedentary subjects from different ethnic/racial backgrounds and revealed that African students from Cameroon, Senegal, Zaire, Ivory Coast, and Burundi had 8% lower type I and 6.7% higher type IIa fiber proportions than whites (Ama et al., 1986). In addition, the Africans had 30–40% higher enzyme activities of the phosphagenic (e.g., creatine kinase) and glycolytic (e.g., hexokinase, phosphofructokinase, and lactate dehydrogenase) metabolic pathways. These

Genetic and Molecular Aspects of Sport Performance, 1st edition. Edited by Claude Bouchard and Eric P. Hoffman. Published 2011 by Blackwell Publishing Ltd.

Table 11.1 Male world records from 100 m to marathon

Distance	Athlete	Time	Ancestral origin[a]
100 m	Usain Bolt (JAM)	9.58	West Africa
110 m hurdles	Dayron Robles (CUB)	12.87	West Africa
200 m	Usain Bolt (JAM)	19.19	West Africa
400 m	Michael Johnson (USA)	43.18	West Africa
400 m hurdles	Kevin Young (USA)	46.78	West Africa
800 m	Wilson Kipketer (KEN)	1:41.11	East Africa
1000 m	Noah Ngeny (KEN)	2:11.96	East Africa
1500 m	Hicham El Guerrouj (MOR)	3:26.00	North Africa
Mile	Hicham El Guerrouj (MOR)	3:43.13	North Africa
3000 m	Daniel Komen(KEN)	7:20.67	East Africa
5000 m	Kenenisa Bekele (ETH)	12:37.35	East Africa
10,000 m	Kenenisa Bekele (ETH)	26:17.53	East Africa
Marathon	Haile Gebrselassie (ETH)	2:03:59	East Africa

JAM, Jamaica; CUB, Cuba; USA, United States of America; KEN, Kenya; MOR, Morocco; ETH, Ethiopia.
[a] Ancestral origin is derived from geographical and ethnic status.

authors concluded, "The racial differences observed between Africans and Caucasians in fiber type proportion and enzyme activities of the phosphagenic and glycolytic metabolic pathways may well result from inherited variation. These data suggest that sedentary male Black individuals are, in terms of skeletal muscle characteristics, well endowed for sport events of short duration." Notably, this conclusion was based on small sample sizes; $n = 23$ black and $n = 23$ white subjects (Ama et al., 1986). Another approach compared the work capacities of sedentary subjects from different ethnic/racial backgrounds. In a review of performance studies from as early as 1941, it was concluded that there were no differences between racial groups in maximal aerobic power, while small differences reported in submaximal work efficiency and endurance performance were attributed to differences in mechanical efficiency owing to test mode and/or level of habituation to the ergometers used for testing (Boulay et al., 1988). These authors concluded, "Thus there does not appear to be valid and reliable evidence to support the concept of clear racial differences in work capacities and powers."

Despite almost no evidence for ethnic/racial differences in performance among sedentary individuals (Boulay et al., 1988), the phenomenal sporting success of east African runners and black sprinters encouraged greater interest among scientists wishing to discover the putative biological factors responsible for this success. A number of studies compared the physiological characteristics of black and white athletes. Studies compared such characteristics as skeletal muscle, maximal aerobic capacity ($\dot{V}O_2$max), lactate accumulation and running economy between groups of black and white athletes. For example, black South African athletes were found to have lower lactate levels than white athletes for the given exercise intensities (Bosch, 1990; Coetzer et al., 1993; Weston et al., 1999). Black athletes also had better running economy (Weston et al., 2000) and higher fractional utilization of $\dot{V}O_2$max at race pace (Bosch, 1990; Coetzer et al., 1993; Weston et al., 2000). It was suggested that "if the physiological characteristics of sub-elite black African distance runners are present in elite African runners, this may help to explain the success of this racial group in distance running" (Weston et al., 2000). However, this assertion is difficult to reconcile with the earlier studies concluding that their findings were compatible with the idea that "Black male individuals are well endowed to perform in sport events of short duration" (Ama et al., 1986). This area of science is clearly confusing when studies comparing subjects of different skin color can conclude on the one hand that the results can explain the success of this racial group in distance running (Weston et al., 2000), and on the other hand that results are compatible with black athletes being suited to events of short duration (Ama et al., 1986). Such contradictions highlight the problem associated with grouping athletes based on skin color and making conclusions about racial/ethnic characteristics on the basis of such small subject numbers.

Endurance Versus Sprint Performance

The fact that many of the world's best distance runners originate from distinct regions of Ethiopia and Kenya, rather than being evenly distributed throughout their respective countries (Onywera et al., 2006; Scott et al., 2003), appears to further sustain the idea that certain ethnic groups possess some inherent genetic advantage predisposing them to superior athletic performance; a typical explanation for the success of Ethiopian and Kenyan distance runners. Notably, a similar phenomenon is also observed in Jamaica where the majority of successful sprinters trace their origins to the northwest region of Jamaica in the parish of Trelawny (Robinson, 2007). Geographical disparities in athlete production have been proposed to reflect a genetic similarity among those populating these regions for an athletic genotype and phenotype (Manners, 1997). In isolated populations, genetic drift can cause certain alleles to increase or decrease in frequency and if the variants are beneficial to sprint or endurance ability, may predispose the population to that type of performance. Alternatively, there may be selection for a particular phenotype such as sprint or endurance, if it offers a selective advantage in that environment. Indeed, some believe that the Nandi tribe in Kenya has been self-selected over centuries for endurance performance through cultural practices such as cattle raiding (Manners, 1997). It is not surprising, therefore, that there are assertions in the literature that Kenyans have the "proper genes" for distance running (Larsen, 2003). Similarly, others have suggested that the superior sprint performances of African-Americans of primarily west African origin

were due to a favorable biology (e.g., muscle fiber characteristics, metabolic pathways, and pulmonary physiology) hypothesized to have been concentrated by natural selection over three centuries in the Afrocentric peoples displaced from west Africa to the New World during the slave trade (Morrison & Cooper, 2006). Morrison and Cooper (2006) proposed that biochemical differences between west African and west African-descended populations and all other groups, including other black Africans, began but did not end with the sickling of the hemoglobin molecule. The authors advocate that individuals with the sickle-cell trait possessed a significant selective advantage in the uniquely lethal west African malarial environment that triggered a series of physiological adjustments and compensatory mechanisms that had favorable consequences for sprinting (Morrison & Cooper, 2006). While this hypothesis remains to be tested, another untested hypothesis that also warrants investigation is that the favorable west African biology referred to by Morrison and Cooper (2006) and others may have been concentrated as a consequence of the displacement process and the harsh living conditions where mainly the "fittest" slaves survived. During the whole slave trade period, ~10 million west Africans were enslaved and some 10 million more died during the process of capture and transportation. The transportation across the Atlantic was extremely brutal, lasting many weeks and at least 1 in 4 people who were transported from Africa died before reaching their destination. Despite these untested hypotheses having potential theoretical underpinnings, it is unjustified at present to regard the phenomenal athletic success of Jamaicans, or indeed east Africans, as genetically mediated; to justify doing so one must identify the genes that are responsible for that success. However, it is a justified hypothesis and is eminently amenable to experimental verification or disproof. Scientists advocating a biological/genetic explanation typically ignore the socioeconomic and cultural factors that appear to better explain ethnic differences in performance (Scott & Pitsiladis, 2007). This chapter aims to address these factors and other beliefs in the context of the limited scientific evidence available.

Genetic Explanations for Ethnic Differences in Sports Performance

There is more genetic variation among Africans than between Africans and Eurasians (Yu et al., 2002), especially bearing in mind that Eurasians are descended from Africans. The genetics of race is controversial and gives rise to a number of contrasting viewpoints with particular emphasis on the use of race as a tool in the diagnosis and treatment of disease. Some argue that there is a role for the consideration of race in biomedical research and that the potential benefits to be gained in terms of diagnosis and treatment of disease outweigh the potential social costs of linking race or ethnicity with genetics (Burchard et al., 2003). Others, however, advocate that race should be abandoned as a tool for assessing the prevalence of disease genotypes and that race is not an acceptable surrogate for genetics in assessing the risk of disease and efficacy of treatment (Cooper et al., 2003). Arguments for the inclusion of race in biomedical research often focus on its use to identify single gene disorders and their medical outcome. The genetic basis of complex phenotypes such as athletic performance is poorly understood and more difficult to study. It is estimated that most human genetic variation is shared by all humans and that a marginal proportion (normally less than 10%) is specific to major continental groups (Cavalli-Sforza et al., 2003). Estimates from the human genome project and analysis of haplotype frequencies show that most haplotypes (i.e., linked segments of genetic variation, rarely subject to reassortment by recombination) are shared between two of the three major geographic populations: Europe, Asia, and Africa (International HapMap Consortium, 2005). It is estimated that the level of genetic diversity between human populations is not large enough to justify the use of the term race (Jobling et al., 2004). Consequently, any differences in physiology, biochemistry, and/or anatomy between groups defined solely by skin color are not directly applicable to their source populations, even if the differences found are indeed genetically determined. This point is well illustrated using an example from the recent literature. Bejan et al.

(2010) made reference to anthropometric literature that showed that the center of mass in "blacks" was 3% higher above the ground than in "whites" and extrapolated this to mean that "blacks" hold a 1.5% speed advantage in running and whites hold a 1.5% speed advantage in swimming. The authors also stated that "among athletes of the same height Asians are even more favored than whites in swimming but they are not setting records because they are not as tall."

The approach of comparing physiological characteristics between groups defined solely by skin color does not offer much insight into why some groups are more successful than others in certain sports. Even within groups of individuals with similar skin color, many ethnic and tribal groups exist. Over 70 languages are in everyday use in Ethiopia, while in Kenya, over 50 distinct ethnic communities speak close to 80 different dialects. Similarly, the Jamaican population, currently estimated to be about 2.8 million, trace their origins primarily to west Africa as a result of the transatlantic slave trade that began during the early 16th century and persisted until the passing of the Abolition of the Slave Trade Act in 1807. During this period, slaves were captured from a large number of geographical areas. In particular, slaves were captured from Senegal, Gambia, Guinea, Sierra Leone, Ivory Coast, Ghana, Benin (Dahomey), Nigeria, Cameroon, and Angola. The ethnic make up of the slaves transported to the Caribbean was therefore extremely diverse, comprising many main ethnic groups (e.g., Mandinka, Fulani, Wolof, Dyula [Jola], Mande, Dan, Kru, Asante [Cromanti], Fante, Ewe, Ga [dialect of Ewe], Yoruba, Igbo, Nziani, Agni, Fula, and Bantu). Furthermore, it was often the case during the slave trade period that the primarily white European slave owners would father children of their slaves, thus further influencing the resulting gene pool. This admixture is fittingly reflected in the motto of Jamaica "Out of many, one people." Consequently, the inadequate classification in the scientific literature of subjects into groups based on characteristics such as skin color will undoubtedly lead to equivocal results and serve only to augment existing stereotypes of genetically advantaged black athletes. Studies comparing black and white athletes offer some insight into the physiological determinants of elite performance but provide little, if any, insight into genetic influences on the disproportionate success of various ethnic groups in sport.

The capacity of *Homo sapiens* for endurance running is unique among the primates. This has led to the belief that endurance capabilities have been central to the recent evolutionary history of modern humans (Bramble and Lieberman, 2004). The unique endurance capacity of humans relative to the other primates is purported to be due to a number of traceable adaptations beneficial to endurance running in the *Homo* genus (Bramble and Lieberman, 2004). Although modern humans are young in evolutionary terms (~150,000 years) relative to the age of the *Homo* genus (~2 million years), human populations began to diverge into new environments outside Africa some 70,000 years ago (Macaulay et al., 2005). It is possible, therefore, that divergent human populations (in different continents for a significant portion of the age of the species) have accrued varying degrees of these adaptations due to different selection pressures, further specializing them for an endurance phenotype or otherwise. The varying adaptation to hypobaric hypoxia by geographically isolated populations represents a well-studied example of this, which may relate to endurance performance. Andean highlanders display higher levels of hemoglobin and saturation than Tibetans at similar altitude, while Ethiopian highlanders maintain oxygen delivery despite having hemoglobin levels and saturation typical of sea level ranges (Beall et al., 2002). There has been much conjecture surrounding such adaptations to altitude and their role in the evolution of human physiological responses (Hochachka et al., 1998). Given the origin of modern humans in east Africa (a higher, drier, and cooler environment than more distant ancestors may have inhabited), the ancestral form of modern humans would have been well adapted for altitude tolerance and, presumably, endurance performance, since the altitude tolerance phenotype is similar to the endurance performance phenotype (Hochachka et al., 1998). If the ancestral form developed in the highlands of east Africa was indeed an altitude/endurance phenotype,

east Africans may have further developed an up-regulated, high-capacity version (Hochachka et al., 1998), thus favoring them further for endurance performance. Other populations may have "diluted" the ancestral phenotype, or developed down-regulated, low-capacity versions through selection for other phenotypes during migration to different environments (e.g., Andean and Tibetan populations) (Hochachka et al., 1998). The genetic ancestry of elite east African runners was first tested using uniparentally inherited genetic markers such as mitochondrial DNA (mtDNA). Polymorphisms in mtDNA have been suggested to influence the variation in human physical performance as mtDNA encodes various subunits of enzyme complexes of oxidative phosphorylation, as well as components of the mitochondrial protein synthesis system. Mitochondrial DNA is inherited in its entirety from mother to child, only changing as new mutations arise, resulting in the accumulation of linked complexes of mutations down different branches of descent from a single ancestor (e.g., "Mitochondrial Eve"). This linear pattern of inheritance can also be used to trace the ancestral origins of individuals or populations and to construct phylogenetic trees back to "Mitochondrial Eve" (each branch of the tree being known as a haplogroup). The frequency of these haplogroups can be used to trace population movements and expansions and as such, different haplogroups are at widely varying frequencies in different regions of the world and even different regions of the same continent (Salas et al., 2002). A simplified version of an mtDNA phylogeny is shown in Figure 11.1. Initially, mtDNA analysis was applied to the cohort of Ethiopian distance runners (Scott et al., 2005b). Rather than the Ethiopian runners being restricted to one area of the tree, results revealed a wide distribution, similar to the general Ethiopian population (Figure 11.1) and in contrast to the concept that these runners are a genetically distinct group as defined by mtDNA. Furthermore, the mtDNA haplogroup diversity found in Ethiopian runners does not support a role for mtDNA polymorphisms in their success. It can be seen from Figure 11.1 that some of the Ethiopian athletes share a more recent common mtDNA ancestor with many Europeans. This finding does not support the hypothesis that the Ethiopian population, from which the athletes are drawn, has remained genetically isolated in east Africa but shows that it has undergone migration events and subsequent admixture during the development of the species. This opposes the possibility that Ethiopian athletes have maintained and further developed the ancestral endurance phenotype through having remained isolated in the east African highlands. It is likely that population movements within Africa as recently as a few thousand years ago have contributed to the peopling of east Africa, through the eastern path of the Bantu migrations. However, linguistic data show that Bantu languages are absent in Ethiopia (Scott et al., 2003) but frequent in Kenya (Onywera et al., 2006), indicating that the neighboring regions

may have been subject to widely different patterns of migration. Recent data on the mtDNA haplogroup frequencies of the Kenyan population and Kenyan runners (Scott et al., 2009) reveal that the mtDNA haplogroups found in Kenya are very different to those found in Ethiopia and show a lower frequency of Eurasian haplogroups M and R; these haplogroups are present at a frequency of 10% in Kenya compared to 45% in Ethiopia (Figure 11.1). It is interesting to note that these two regions which share success in distance running have such different ancestral contributions to their gene pool. In contrast to the Ethiopian cohort in which no differences in mtDNA haplogroup distribution were found between athletes and controls, international athletes from Kenya displayed an excess of

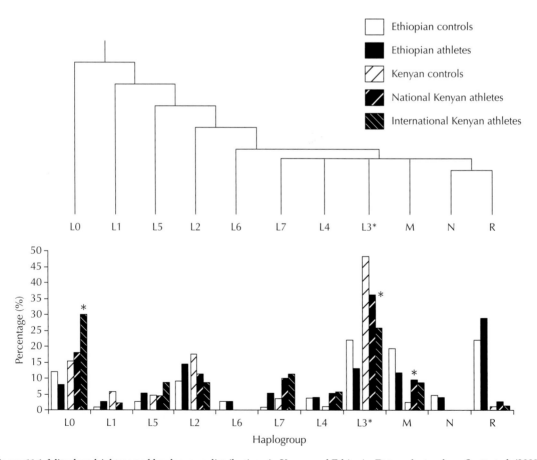

Figure 11.1 Mitochondrial tree and haplogroup distributions in Kenya and Ethiopia. Data redrawn from Scott et al. (2009).

L0 haplogroups, and a dearth of L3* haplogroups (Figure 11.1). National athletes from Kenya also showed differences from controls when each haplogroup was compared to the sum of all others, exhibiting an excess of M haplogroups (Figure 11.1). The association of mtDNA haplogroups L0 and M with Kenyan athlete status may suggest that these haplogroups contain polymorphisms which influence some aspect of endurance performance or its trainability. Higher resolution analysis is now necessary to establish which polymorphisms may be influential in the present association.

Y Chromosome Haplogroups

The idea that the elite east African runners studied to date do not arise from a limited genetic isolate is further supported by the analysis of the Y chromosome haplogroup distribution of Ethiopian runners (Moran et al., 2004). The Y chromosome can be considered as the male equivalent of mtDNA. The distribution of Y chromosome haplogroups of the Ethiopian runners relative to the general population is shown in Figure 11.2. Ethiopian runners differed significantly in their Y chromosome

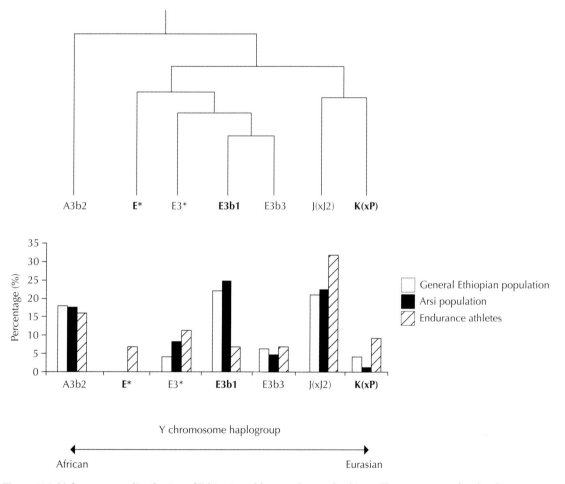

Figure 11.2 Y chromosome distribution of Ethiopian athletes and control subjects. The percentage of each subject group belonging to each haplogroup is shown. Haplogroups which differed in frequency between athletes and controls are shown in bold. Data redrawn from Moran et al. (2004).

distribution from both the general population and that of the Arsi region (Moran et al., 2004), which produces an excess of elite runners (Scott et al., 2003). The finding that Y chromosome haplogroups were associated with athlete status in Ethiopians suggests that either an element of Y chromosome genetics is influencing athletic performance, or that the Y chromosome haplogroup distributions were affected by population stratification (i.e., the population from which the athletes originate has a distinct Y chromosome distribution). However, the haplogroup distribution of the Arsi region did not differ from the rest of Ethiopia, suggesting that the observed associations were less likely to be a result of simple population stratification. Currently, these haplogroup frequencies are being assessed in the larger Kenyan cohort (Onywera et al., 2006). If the same haplogroups are found to be under- or over-represented, this would provide some evidence for a biological effect of the Y chromosome on running performance. However, despite the finding of a potential effect of the Y chromosome on endurance performance, the Y chromosome results show similar levels of diversity to those found using mtDNA. In addition, it can be seen from Figure 11.2, that a significant number of the athletes trace part of their male ancestry to outside Africa at some time during the age of our species. Collectively, the findings from Y chromosome and mtDNA studies imply that the phenomenon of east African running should best be considered as two distinct phenomena, that of Ethiopian and Kenyan success, at least from a genetic viewpoint.

The *ACE* and *ACTN3* Genes

Despite the idea being perpetuated that certain ethnic/racial groups are genetically adapted for athletic performance, only two candidate genes for human performance have been investigated in east African athletes (Scott et al., 2005a; Yang et al., 2007) and in sprint athletes from Nigeria (Yang et al., 2007), Jamaica, and USA (Scott et al., 2010). The first gene studied in elite east African runners was the angiotensin-converting enzyme (*ACE*) gene: the most studied of the candidate genes for human performance, where an insertion

polymorphism (I) is associated with lower levels of circulating and tissue ACE than the deletion polymorphism (D) (Danser et al., 1995). The *ACE* gene has been associated with a number of aspects of human performance (Jones et al., 2002). In general, the I allele has been associated with endurance performance and the D allele with power performance. The *ACE* I allele has also been associated with altitude tolerance (Montgomery et al., 1998) making it an ideal candidate gene to investigate in east African runners given the suggestion that the altitude at which these athletes live and train may largely account for their success. As such, *ACE* I/D genotype frequencies were tested in Kenyan runners relative to the general population (Scott et al., 2005a). Based on previous findings (Jones et al., 2002), it may have been expected that the elite runners would show an excess of the I allele. However, no significant differences were found in I/D genotype frequencies between runners and the general population (Scott et al., 2005a). Different levels of linkage disequilibrium in Africans and Caucasians (Zhu et al., 2000) meant that an additional, potentially causal variant (A22982G) was tested. However, no significant differences in A22982G genotype frequencies were found between runners and the general population (Scott et al., 2005a). Indeed 29% of Kenyan controls and only 17% of international Kenyan athletes had the putatively advantageous AA genotype (always found in concert with II in Caucasians) for endurance performance. Although controversy over the influence of *ACE* genotype on endurance performance continues, this study did not support a role for *ACE* gene variation in explaining the east African distance running phenomenon.

The only two genes to have been investigated in elite sprinters are the *ACE* gene and the alpha-actinin-3 (*ACTN3*) gene (Scott et al., 2010; Yang et al., 2007). The *ACTN3* gene has been associated with elite physical performance (Yang et al., 2003) and found at widely differing frequencies in different populations (Mills et al., 2001). In particular, a strong association has been found between the R577X polymorphism and elite athlete status in Australian Caucasian populations, with the *ACTN3*-deficient XX genotype being present at a lower frequency in sprint/power athletes, and at slightly higher frequency in elite female endurance athletes, relative to controls (Yang et al., 2003). This negative association of the XX genotype with sprint performance has subsequently been replicated in a cohort of elite Finnish track-and-field athletes (Niemi & Majamaa, 2005). Such data have helped establish the link between R577X and muscle strength and sprint performance. It was therefore of interest that the *ACTN3*-deficient XX genotype was found to be almost absent in controls and sprinters from Nigeria (Yang et al., 2007) and sprinters from the USA and Jamaica (Scott et al., 2010) (Table 11.2),

while there was no evidence for an association between the R577X polymorphism and endurance performance in east African runners (Yang et al., 2007); suggesting that *ACTN3* deficiency is not a major determinant of east African running success. As for the *ACE* gene, neither Jamaican sprinters nor sprinters from the USA differed from controls (Table 11.3). Sprinters from Jamaica did not differ from controls for A22982G genotype, although US sprinters did, displaying an excess of heterozygotes relative to controls, but no excess of GG homozygotes (Table 11.4). However, the finding of no excess in *ACE* DD or GG genotypes in these elite

Table 11.2 Genotype frequencies of *ACTN3* R577X polymorphism

Subjects	Genotype frequency No. (%)				Allele frequency No. (%)	
	RR	RX	XX	*n*	R	X
Jamaican						
Controls	232 (75)	73 (23)	6 (2)	311	537 (86)	85 (14)
Athletes	86 (75)	25 (22)	3 (3)	114	197 (86)	31 (14)
US African American						
Controls	126 (66)	57 (30)	7 (4)	190	309 (81)	71 (19)
Southwest	26 (59)	15 (34)	3 (7)	44	67 (76)	21 (24)
Northeast	50 (69)	20 (28)	2 (3)	72	120 (83)	24 (17)
Southeast	50 (68)	22 (30)	2 (2)	74	122 (82)	26 (18)
Athletes	79 (70)	32 (28)	2 (2)	113	190 (84)	36 (16)

Data from Scott et al. (2010).

Table 11.3 Genotype frequencies of *ACE* ID polymorphism

Subjects	Genotype frequency No. (%)				Allele frequency No. (%)	
	II	ID	DD	*n*	I	D
Jamaican						
Controls	53 (17)	143 (47)	108 (36)	304	249 (44)	359 (51)
Athletes	21 (19)	56 (52)	31 (29)	108	98 (45)	118 (55)
US African American						
Controls	32 (17)	99 (52)	59 (31)	190	163 (43)	217 (57)
Southwest	2 (5)	25 (57)	17 (39)	44	29 (33)	59 (67)
Northeast	13 (19)	38 (53)	21 (29)	72	64 (44)	80 (56)
Southeast	17 (23)	36 (49)	21 (28)	74	70 (47)	78 (53)
Athletes	12 (11)	62 (57)	36 (32)	110	86 (39)	134 (61)

Data from Scott et al. (2010).

Table 11.4 Genotype frequencies of *ACE* A22982G polymorphism

Subjects	Genotype frequency No. (%)				Allele frequency No. (%)	
	AA	AG	GG	*n*	A	G
Jamaican						
Controls	112 (40)	124 (44)	47 (17)	283	348 (62)	218 (38)
Athletes	37 (34)	58 (53)	15 (14)	110	132 (60)	88 (40)
US African American						
Controls	66 (37)	75 (42)	39 (22)	180	207 (58)	153 (43)
Southwest	11 (26)	17 (41)	14 (33)	42	39 (46)	45 (54)
Northeast	25 (37)	30 (44)	13 (19)	68	80 (59)	56 (41)
Southeast	30 (43)	28 (40)	12 (17)	70	88 (63)	52 (37)
Athletes	21 (24)	52 (58)	16 (18)	89	94 (53)	84 (47)

The US athletes differed significantly from controls ($P = 0.018$), showing an excess of heterozygotes. Data from Scott et al. (2010).

sprint athletes relative to controls suggests that *ACE* genotype is not a determinant of elite sprint athlete status. In summary, genotyping two of the key candidate genes for human performance in a cohort of the world's most successful endurance athletes and sprinters finds these genes not to be significant determinants of their success. Whether other nuclear DNA variants can help explain such phenomena remains to be determined. However, the extraordinary achievements of certain populations in sporting success must rely upon the successful integration of a number of physiological, biochemical, and biomechanical systems, which are themselves the product of a multitude of contributors. The success of these athletes is unlikely, therefore, to be the result of a single gene polymorphism; rather it is likely that elite athletes rely on the presence of a combination of advantageous genotypes at a multitude of loci. Currently, studies are underway to investigate the frequency of other candidate genes in these unique cohorts.

Non-genetic Explanations for Ethnic Differences in Sports Performance

Non-genetic explanations for the success of east African distance running include the suggestion that the distances east Africans run to school as children

serve them well for subsequent athletic success. A study by Scott et al. (2003) found that Ethiopian distance runners had traveled further to school as children, and more had done so by running. Many of the distances traveled were phenomenal with some children traveling upward of 20 km each day by running to and from school. A previous study by Saltin et al. (1995) has shown that east African children who had used running as a means of transport had a $\dot{V}O_2$max some 30% higher than those who did not; therefore, implicating distance traveled to school as a determinant of east African running success. Other studies have shown regional disparities in the production of Ethiopian and Kenyan distance runners (Onywera et al., 2006; Scott et al., 2003). In a study of the demographic characteristics of Ethiopian runners, 38% of marathon athletes were from the region of Arsi, which accounts for less than 5% of the Ethiopian population (Scott et al., 2003). These findings were mirrored in Kenya, where 81% of international Kenyan athletes originated from the Rift-Valley province, which accounts for less than 25% of the Kenyan population (Onywera et al., 2006). Although some believe that this geographical disparity is mediated by an underlying genetic phenomenon (Manners 1997), it is worth considering that both of these regions are altitudinous (Onywera et al., 2006; Scott et al., 2003) and that athletes have long used altitude training to induce further adaptations.

The phenomenal success of Jamaicans in sprinting has also been attributed to non-genetic factors. In Jamaica, there exists an excellent and unique model that focuses on identifying and nurturing athletic talent from junior to senior level (Robinson, 2007). It is argued that "...the real explanation of the outstanding achievements of the system is that all of its actors are moved by a spirit that unifies them to work to ensure that Jamaican athletics lives up to its rich history and tradition of excellence" (Robinson, 2007). This theme echoes what is found in Ethiopia and Kenya (Onywera et al., 2006; Scott et al., 2003).

Others have suggested that African or black athletes enjoy a psychological advantage, mediated through stereotype threat (Baker & Horton, 2003). A consequence of strengthening the stereotypical view of the superior black or African athlete is the development of a self-fulfilling prophecy of white athletes avoiding sporting events typically considered as favoring African or black athletes. This self-selection has resulted in a vicious cycle where the avoidance of these athletic events by whites has further strengthened the aforementioned stereotype to the extent that the unsubstantiated idea of the biological superiority of the African or black athlete becomes dogma. The reverse of this is also observed as black or African athletes tend to avoid sports such as swimming and skiing. This form of self-selection for sport is surprising given the lack of scientific evidence to support this stereotype.

Conclusions

The Beijing results further augment the idea that certain ethnic groups possess some inherent genetic advantage predisposing them to superior athletic performance. Despite this speculation, there is no genetic evidence to date to suggest that this is the case, although research is ongoing. The only available genetic studies of elite Ethiopian and Kenyan distance runners and sprinters from Jamaica, Nigeria, and the USA do not find these athletes possess unique genetic profiles; rather they serve to highlight a high degree of genetic diversity among these athletes. Although genetic contributions to the phenomenal success of black sprinters and east African distance runners cannot be excluded, results to date predominantly implicate environmental factors.

Acknowledgment

I acknowledge the important contribution of all my co investigators in the studies on which I have based this review (especially my doctoral student Dr Robert Scott) and Professor Craig Sharp for proof reading this chapter. I also acknowledge the invaluable assistance of the Ethiopian Olympic Committee, the Ethiopian Athletics Federation, Athletics Kenya, the Jamaica Amateur Athletic Association (JAAA), the Jamaica Olympic Association (JOA), the Olympians Association of Jamaica, the United States Olympic Committee (USOC), and United States of America Track and Field (USATF). The cooperation of all the athletics coaches, agents, and managers is also greatly appreciated. Most of all, however, I would like to thank the national and elite (i.e., the "living legends") athletes and controls who volunteered to take part in the genetic studies reviewed here for very little, if anything, in return other than sheer intrinsic interest in their sport.

References

Ama, P.F., Simoneau, J.A., Boulay, M.R., Serresse, O., Thériault, G. & Bouchard, C. (1986). Skeletal muscle characteristics in sedentary black and Caucasian males. *Journal of Applied Physiology* 61, 1758–1761.

Baker, J. & Horton, S. (2003). East African running dominance revisited: A role for stereotype threat? *British Journal of Sports Medicine* 37(6), 553–555.

Beall, C.M., Decker, M.J., Brittenham, G.M., Kushner, I., Gebremedhin, A. & Strohl, K.P. (2002). An Ethiopian pattern of human adaptation to high-altitude hypoxia. *The Proceedings of the National Academy of Sciences of the United States of America* 99(26), 17215–17218.

Bejan, A., Jones, E.C., & Charles, J.D. (2010). The evolution of speed in athletics: Why the fastest runners are black and swimmers white. *International Journal of Design & Nature and Ecodynamics* 5(3), 1–13.

Bosch, A.N., Goslin, B.R., Noakes, T.D. & Dennis, S.C. (1990). Physiological differences between black and white runners during a treadmill marathon. *European Journal of Applied Physiology* 61(1–2), 68–72.

Boulay, M.R., Ama, P.F. & Bouchard, C. (1988). Racial variation in work capacities and powers. *Canadian Journal of Sport Sciences* 13(2), 127–135.

Bramble, D.M. & Lieberman, D.E. (2004). Endurance running and the evolution of *Homo. Nature* 432(7015), 345–352.

Burchard, E.G., Ziv, E., Coyle, N., et al. (2003). The importance of race and ethnic background in biomedical research and clinical practice. *New England Journal of Medicine* 348(12), 1170–1175.

Cavalli-Sforza, L.L. & Feldman, M.W. (2003). The application of molecular genetic approaches to the study of human evolution. *Nature Genetics* Suppl. 33, 266–275.

Coetzer, P., Noakes, T.D., Sanders, B., et al. (1993). Superior fatigue resistance of elite black South African distance runners. *Journal of Applied Physiology* 75(4), 1822–1827.

Cooper, R.S., Kaufman, J.S. & Ward, R. (2003). Race and genomics. *New England Journal of Medicine* 348(12), 1166–1170.

Danser, A.H., Schalekamp, M.A., Bax, W.A., et al. (1995). Angiotensin-converting enzyme in the human heart. Effect of the deletion/insertion polymorphism. *Circulation* 92(6), 1387–1388.

Hochachka, P.W., Gunga, H.C. & Kirsch, K. (1998). Our ancestral physiological phenotype: An adaptation for hypoxia tolerance and for endurance performance? *The Proceedings of the National Academy of Sciences of the United States of America* 95(4), 1915–1920.

International HapMap Consortium (2005). A haplotype map of the human genome. *Nature* 437(7063), 1299–1320.

Jobling, M.A, Hurles, M.E. & Tyler-Smith, C. (2004). *Human Evolutionary Genetics.* Garland Publishing, London/New York.

Jones, A., Montgomery, H.E. & Woods, D.R. (2002). Human performance: A role for the ACE genotype? *Exercise and Sport Sciences Reviews* 30(4), 184–190.

Larsen, H.B. (2003). Kenyan dominance in distance running. *Comparative Biochemistry and Physiology* 136, 161–170.

Macaulay, V., Hill, C., Achilli, A., et al. (2005). Single, rapid coastal settlement of Asia revealed by analysis of complete mitochondrial genomes. *Science* 308(5724), 1034–1036.

Manners, J. (1997). Kenya's running tribe. *The Sports Historian* 17(2), 14–27.

Mills, M., Yang, N., Weinberger, R., et al. (2001). Differential expression of the actin-binding proteins, alpha-actinin-2 and -3, in different species: Implications for the evolution of functional redundancy. *Human Molecular Genetics* 10(13), 1335–1346.

Montgomery, H.E., Marshall, R., Hemingway, H., et al. (1998). Human gene for physical performance. *Nature* 393(6682), 221–222.

Moran, C.N., Scott, R.A., Adams, S.M., et al. (2004). Y chromosome haplogroups of elite Ethiopian endurance runners. *Human Genetics* 115(6), 492–497.

Morrison, E.Y. & Cooper, P.D. (2006). Some bio-medical mechanisms in athletic prowess. *West Indian Medical Journal* 55(3), 205–209.

Niemi, A-K. & Majamaa, K. (2005). Mitochondrial DNA and ACTN3 genotypes in Finnish elite endurance and sprint athletes. *European Journal of Human Genetics* 13, 965–969.

Onywera, V.O., Scott, R.A., Boit, M.K. & Pitsiladis, Y.P. (2006). Demographic characteristics of elite Kenyan endurance runners. *Journal of Sports Sciences* 24(4), 415–422.

Robinson, P. (2007). *Jamaican Athletics: A Model for the World.* Kingston, Marco Printers Ltd, Jamaica.

Salas, A., Richards, M., De la Fe, T., et al. (2002). The making of the African MtDNA landscape. *American Journal of Human Genetics* 71(5), 1082–1111.

Saltin, B., Larsen, H., Terrados, N., et al. (1995). Aerobic exercise capacity at sea level and at altitude in Kenyan boys, junior and senior runners compared with Scandinavian runners. *Scandinavian Journal of Medicine and Science in Sports* 5(4), 209–221.

Scott, R.A. & Pitsiladis, Y.P. (2007). Genotypes and distance running: Clues from Africa. *Sports Medicine* 37(4–5), 1–4.

Scott, R.A., Georgiades, E., Wilson, R.H., Goodwin, W.H., Wolde, B. & Pitsiladis, Y.P. (2003). Demographic characteristics of elite Ethiopian endurance runners. *Medicine and Science in Sports and Exercise* 35(10), 1727–1732.

Scott, R.A., Moran, C., Wilson, R.H., et al. (2005a). No association between angiotensin converting enzyme (ACE) gene variation and endurance athlete status in Kenyans. *Comparative Biochemistry and Physiology* 141(2), 169–175.

Scott, R.A., Wilson, R.H., Goodwin, W.H., M. et al. (2005b). Mitochondrial DNA lineages of elite Ethiopian athletes. *Comparative Biochemistry and Physiology* 140(3), 497–503.

Scott, R.A., Fuku, N., Onywera, V.O., et al. (2009). Mitochondrial haplogroups associated with elite Kenyan athlete status. *Medicine & Science in Sports & Exercise* 41(1), 123–128.

Scott, R.A., Irving, R., Irwin, L., et al. (2010). ACTN3 and ACE genotypes in elite Jamaican and US sprinters. *Medicine and Science in Sports and Exercise* 42, 107–112.

Weston, A.R., Karamizrak, O., Smith, A., Noakes, T.D. & Myburgh, K.H. (1999). African runners exhibit greater fatigue resistance, lower lactate accumulation, and higher oxidative enzyme activity. *Journal of Applied Physiology* 86(3), 915–923.

Weston, A.R., Mbambo, Z. & Myburgh, K.H. (2000). Running economy of African and Caucasian distance runners. *Medicine and Science in Sports and Exercise* 32(6), 1130–1134.

Yang, N., MacArthur, D.G., Gulbin, J.P., et al. (2003). ACTN3 genotype is associated with human elite athletic performance. *American Journal of Human Genetics* 73(3), 627–631.

Yang, N., MacArthur, D.G., Wolde, B., et al. (2007). The ACTN3 R577X polymorphism in East and West African athletes. *Medicine and Science in Sports and Exercise* 39(11), 1985–1988.

Yu, N., Chen, F.C., Ota, S., et al. (2002). Larger genetic differences within Africans than between Africans and Eurasians. *Genetics* 161(1), 269–274.

Zhu, X., McKenzie, C.A., Forrester, T., et al. (2000). Localization of a small genomic region associated with elevated ACE. *American Journal of Human Genetics* 67(5), 1144–1153.

Chapter 12

Selection Experiments in Rodents to Define the Complexity and Diversity of Endurance Capacity

STEVEN L. BRITTON AND LAUREN GERARD KOCH

Department of Anesthesiology, University of Michigan Medical School, Ann Arbor, MI, USA

Capacity to transfer energy for motion of mass at the atomic, molecular, cellular, tissue, and organismal levels of organization is the definition of "animal." For energy transfer, life evolved two types of respiration systems that utilize a terminal electron receptor. Anaerobic respiration developed first and uses terminal electron receptors other than oxygen. Subsequent evolution led to aerobic respiration that utilizes oxygen as an electron acceptor, produces a higher yield of ATP from food substrate, and was permissive for the development of complexity beyond single-cell organisms (Payne et al., 2009; Raymond & Segre, 2006). Aerobic capacity, as estimated from maximal oxygen consumption ($\dot{V}O_2max$) or an endurance run to exhaustion, is the summation of intrinsic factors (inborn, untrained) and that accrued in response to training. The statistical linkage of high aerobic capacity with both endurance performance and health drives large interest in resolving this phenotype mechanistically. To achieve this goal we were motivated to develop low and high forms of animal models for both the intrinsic and response to training phenotypes. In 1995, we first started working on the intrinsic models because they represent the more simplistic models for both development and evaluation. Here we describe progress across 20 generations of two-way (divergent) artificial selection and the phenotypes

and genotypes that have emerged in the low- and high-intrinsic models (LCR—low-capacity runners; HCR—high-capacity runners).

LCR and HCR of rats and mice have now been extensively studied with regard to behavioral, metabolic, histological, and organ system differences. These studies have found that the selections lead to the convergence of multiple peripheral and central phenotypic differences between the strains, including metabolic syndrome. We discuss the differences in physiology, behavior, oxygen transport, central fatigue, and skeletal muscle metabolism. LCR and HCR rodents show differences in many clinically relevant phenotypes, such as metabolic syndrome, fatty liver disease, kidney oxidative stress, oxidative damage, and pain thresholds. Translation of the phenotypic observations to molecular mechanisms has been done with mRNA profiles and candidate gene and pathway studies. Identification of cause and effect relationships between specific genetic polymorphisms and specific phenotype is a future challenge.

Artificial Selection for Endurance Exercise Capacity

Artificial selection is ideally started from a large founder population that has wide genetic heterogeneity for the trait of interest. For bidirectional divergent selection, low and high extremes for the trait of interest are identified from the founder population to serve as parents that initiate the contrasting lines. The response to selection (R) is estimated

Genetic and Molecular Aspects of Sport Performance, 1st edition.
Edited by Claude Bouchard and Eric P. Hoffman.
Published 2011 by Blackwell Publishing Ltd.

from the product of the selection differential (S) and the narrow-sense heritability (h^2):

$$R = S\,(h^2) \tag{12.1}$$

S is the mean difference between the phenotypic value of the selected parents and the phenotypic mean value of all the individuals in that generation. h^2 is the ratio of the additive genetic variance (V_A) and the total phenotypic variance as given by the sum of V_A and environmental variance (V_E):

$$h^2 = V_A/(V_A + V_E) \tag{12.2}$$

V_A is that portion of genetic variance that causes the offspring to resemble their parents and is considered the variance associated with the average effects of substituting one allele for another. Thus, success in genetic selection is directly dependent upon h^2.

Heritability

Because of the large-scale nature of artificial selection, we first estimated narrow-sense heritability (h^2) in a more tractable population of inbred strains of rats. This estimate for a heritable measure of a trait from a panel of inbred strains is based upon two assumptions related to the process of inbreeding. First, 20 generations of strict brother–sister mating produce populations of individuals that are similar genetically (i.e., almost "identical twins"). Trait variation demonstrated within an inbred strain is thus attributed to the environment such that an estimate of V_E can be obtained. Second, each inbred strain represents almost an exclusively homozygous genotype such that variation between inbred strains derives from allelic differences and can be used to estimate V_A.

An estimate of h^2 can be derived by partitioning V_A and V_E from Equation 12.2 into the between-group variance component (ΣB^2) and the within-group variance component (ΣW^2) from analysis of variance using these expressions:

$$\Sigma B^2 = \frac{(MSB - MSW)}{n} \tag{12.3}$$

$$\Sigma W^2 = MSW \tag{12.4}$$

where MSB is the mean square between strains, MSW the mean square within strains, and n is the number of animals in each strain. ΣB^2 equals $2V_A$, and V_E is approximated by ΣW^2. Combining Equations 12.2, 12.3, and 12.4 yields an estimate of h^2 (Hegmann & Possidente, 1981):

$$h^2 = \frac{\dfrac{1}{2}\left[\dfrac{MSB - MSW}{n}\right]}{\dfrac{1}{2}\left[\dfrac{MSB - MSW}{n}\right] + MSW} \tag{12.5}$$

Rats from 11 different inbred strains were tested for both maximal running capacity on a treadmill and isolated cardiac performance (Barbato et al., 1998). Running performance by both male and female rats was estimated from distance run to exhaustion using a standard treadmill running protocol. In addition, cardiac output, during constant preload (15 mmHg) and afterload (70 mmHg), was taken as a measure of cardiac performance from an isolated working heart preparation. We found rats of the Copenhagen strain (COP) were the lowest performers (distance = 298 m) and the Dark Agouti (DA) strain of rats ranked as the highest performers (distance = 840 m). Across the 11 strains, distance run correlated positively with isolated cardiac performance ($r = 0.87$) (Figure 12.1). Demonstration of a relatively high narrow-sense heritability ($h^2 = 0.42$) and segregation of a cardiac phenotype as partly determinant of the differences in running capacity among this panel consisting of 11 different genotypes encouraged us to start a large-scale artificial selection program.

Selective Breeding

Artificial selection for intrinsic aerobic endurance running capacity was started using the N:NIH outbred stock of rats as a founder population ($n = 168$). The availability of N:NIH rats from the Animal Resource Center of the National Institutes of Health for use as the base population was of major benefit. This stock of rats originated from the intentional crossbreeding of eight inbred strains representing the broadest phylogenetic spectrum of laboratory rats available: Maudsely Reactor (MR/N), Wistar (WN/N), Wistar Kyoto (WKY/N), Marshall 520 (M520), Fischer (F344), Ax C 9935

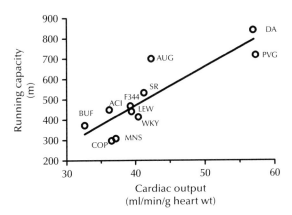

Figure 12.1 Correlation between isolated cardiac performance and distance run across 11 strains of inbred rats for males and females combined. Each 1 ml/min/g increase in cardiac output was associated with an increase of 19.03 m in distance run. (Reprinted from Barbato et al. (1998), with permission from *Journal of Applied Physiology*.)

Irish (ACI), Brown Norway (BN/SsN), and Buffalo (BUF/N) by Hansen and Spuhler in 1979 (Hansen & Spuhler, 1984). The crossbreeding produced a stock that has relatively wide genetic heterogeneity and is thus somewhat ideal as a starting population for artificial selection. Each rat in the founder population was of different parentage, so selection was not among brothers and sisters, which broadens the genetic variance.

The goal at each generation was to maintain trait and breeding value for running capacity within each line while keeping inbreeding at a minimum such that the greatest amount of original genetic heterozygosity was retained. This was accomplished by taking two offsprings (one female and one male) from each mating and using them as parents in the next generation (within-family selection) and a rotational breeding structure between 13 families in each line (Kimura & Crow, 1963). With equal representation of one male and one female offspring from each family, the rate of the inbreeding coefficient (F) per generation can be estimated using the exact relationship defined as $\Delta F = 1/(2N_e) = 1/(4N - 2)$. When male and female breeders (N) are chosen equally from all families, the effective number (N_e) of breeding

individuals is equivalent to $(2N - 1)$. Thus, for any generation (g) the cumulative ΔF is calculated as: $F_g = 1 - (1 - \Delta F)^g$. Thus with 26 parents in each selected line, the expected inbreeding for 20 generations of selective breeding is given by: $F_{20} = 1 - (1 - \Delta F)^{20} = 1 - (1 - 1/102)^{20} = (1 - 0.8212) = 0.1788$ (Falconer, 1976) or approximately 0.98% each generation. See Figure 12.2; for reference, the rate of inbreeding (ΔF) for strict brother–sister mating is also shown. In this case, the heterozygosity (H or $1 - \Delta F$) decreases with each subsequent generation in a sequence of 1, 0.75, 0.625, 0.5, 0.406, 0.328, etc., as given by these ratios 2/2, 3/4, 5/8, 8/16, 13/32, 21/64, etc., approaching zero as a limit. Interestingly, these ratios represent a Fibonacci sequence.

Selection for low and high capacity was based upon distance run to exhaustion on a motorized treadmill using a velocity-ramped running protocol devised by us such that an average laboratory rat would exhaust between 20 min and 30 min (Koch & Britton, 2001). The starting velocity was 10 m/min and was increased by 1 m/min every

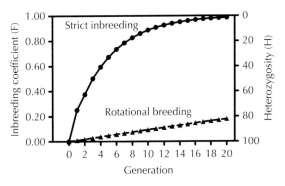

Figure 12.2 Theoretical changes heterozygosity across 20 generations of artificial selective breeding. A major goal was to retain as much heterozygosity at each generation to serve as selection substrate for the next generation. To accomplish this, we used a within-family rotational breeding paradigm with 13 breeding pairs in each line (26 parents). The triangles show a calculated (Kimura & Crow, 1963) decrease in heterozygosity across 20 generations of about 20% for the rotational breeding pattern. This contrast with strict sister–brother mating (circles) that drives homozygosity close to zero by 20 generations of selection.

2 min (slope was constant at 15°). Exhaustion was operationally defined as the third time a rat could no longer keep pace with the speed of the treadmill and remained on the shock grid for 2 s rather than run. Each rat was evaluated for maximal endurance running capacity on five consecutive days when at 11 weeks of age.

The single best trial of five was used because it was deemed the best indicator of capacity determined by the intrinsic genetic composition. This idea of estimating the genetic component from the one best day of performance, rather than the average for all trials, for example, has two origins. First, the environment can have an infinite negative influence upon capacity to reduce the value for a distance run to zero. Factors such as subtle differences in housing or daily handling could easily cause a genetically superior rat to perform below its norm on a given day. Second, the environment can have only a finite positive influence upon any test of capacity for a given genotype. While selecting on the best of five trials may reduce environmental error, it simultaneously created the possibility that a component of selection may be related to short-term adaptation. Indeed, for the five consecutive days of testing, day of run is not a predictor of best day for the LCR. For the HCR, however, the best run associates strongly with trial 5 (Koch & Britton, 2001). Although this may represent a minor component of aerobic biochemical adaptation for the high line, other factors such as differences in behavioral responses to running may also be involved.

Response to Selection

A total of 20 generations of selection (approximately 10 years) produced LCR and HCR that differed in running capacity by 532% (Figure 12.3). At generation 20, the rats from the LCR line averaged 302 ± 82 m while rats from the HCR line ran for 1617 ± 243 m for a treadmill test to exhaustion which is equivalent to a mean maximal running time of 20.4 ± 4 min for LCR compared to 63.4 ± 6 min for HCR. Selective breeding for running capacity produced a correlated change in body weight such that for both sexes, the low line became heavier and the high line became lighter.

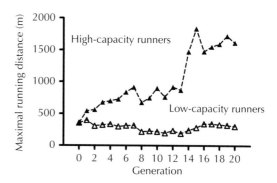

Figure 12.3 Response to selection for low- and high-running capacity for 20 generations. More response for high line is typical in two-way selection because most trait values cannot have a score below zero. At generation 20, the low- and high-selected lines differed 6.3-fold in endurance running capacity.

As early as generation 6, the body weights of the low and high females differed by 34 g (20%, P, 0.001) and the body weights of the low and high males differed by 40 g (16%, P, 0.001) (Koch & Britton, 2001). At generation 20, the average difference between an HCR female and LCR female was 33% (156 g versus 208 g, respectively). For males, the divide was greater (about 40%); the average body weight for HCR males was 234 g versus 327g for LCR males.

Physiological Differences Between LCR and HCR Rats

Because we had selected upon a surrogate for perhaps the most elemental feature of life, the expectation was that the LCR and HCR would display numerous differences for energy transfer at all levels of biological organization. Here we summarize differences obtained from studies performed in rats from generations 10 through 20 that at least partially explain the differences in running capacity.

Behavior

We reasoned that artificial selection for HCR would produce a less complicated phenotype relative to selection for LCR because of obligatory physiological and behavioral interactions. That is, a rat that demonstrates high capacity is both physiologically capable and behaviorally willing to run long distances. In contrast, LCR rats could have the physiological capacity, but not use this capacity because of complex behavioral features. In about 1995, Swallow et al. (1998) started artificial selection of outbred mice for high voluntary wheel-running capacity. Using mice from the 36th generation of selection, Rezende et al. (2006) provided strong evidence for a positive genetic relationship between aerobic capacity and voluntary exercise. They found that mice selected for high voluntary wheel running also had a consistently higher $\dot{V}O_2$max on an enforced treadmill run compared to control line mice.

The opposite side of this capacity/behavior relationship was explored by recording voluntary wheel running in females from generations 9 and 10 of the LCR and HCR models (Waters et al., 2008). During an 8-week trial, HCR animals exhibited 33% greater total wheel-running distance per day compared to LCR rats (16,838 m versus 12,666 m), which was due to the HCR rats exhibiting increases in both running speed and duration over LCR rats. Following the last wheel-running trial, values were lower in HCR for plasma corticosterone concentration ($P = 0.016$) and for striatal dopaminergic activity ($P = 0.031$), a divergence of physiological systems that could potentially influence locomotor behaviors in these lines.

Oxygen Transport

Peter Wagner and Norberto Gonzalez initiated a systematic study of the LCR/HCR rat model across generations to evaluate features of oxygen transport that could explain the difference for intrinsic running capacity. A four-component, series-coupled model as depicted graphically in Figure 12.4 served as theoretical framework (Wagner, 1993) to test the idea that divergence in maximum oxygen consumption between LCR and HCR rats was determined by differences in the quantitative interaction between diffusive and convective steps involved in the transport of oxygen. Maximal oxygen uptake was

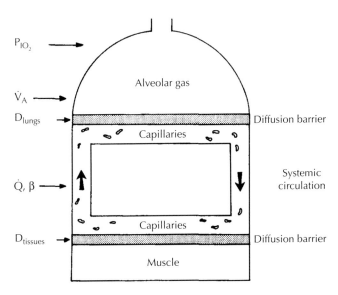

Figure 12.4 Four-component series-connected model of oxygen delivery from the atmosphere to cells. The two convective (movement of molecules within fluids) parts are alveolar ventilation (V_A) and the systemic circulation that delivers oxygenated blood to the capillaries (QO_2) as the product of cardiac output (Q) and arterial blood O_2 content (β). The two diffusive (movement of molecules from a region of higher concentration to one of lower concentration by random molecular motion) elements are at the capillary–interstitial interfaces for the lung and periphery. (Reprinted from Wagner (1993), with permission from *Respiration Physiology*.)

estimated using a progressive treadmill exercise running protocol during inspired oxygen partial pressure PO_2 of $\approx$145 Torr (normoxia) and at $\approx$70 Torr (hypoxia). Maximal oxygen uptake (in milliliter per minute per kilogram) was 57.6 in LCR and 64.4 in HCR during normoxia ($P < 0.05$) and 35.3 in LCR and 42.7 in HCR during hypoxia ($P < 0.05$). Somewhat surprising to us, no differences were found between the LCR and HCR for alveolar ventilation, alveolar-to-arterial PO_2 difference, lung O_2 diffusing capacity, or maximal rate of blood O_2 convection (cardiac output times arterial blood O_2 content, QO_2). Thus, neither ventilation nor efficacy of gas exchange nor peripheral oxygen delivery appeared to contribute to the difference in $\dot{V}O_2max$ between the lines in response to seven generations of divergent selection (Henderson et al., 2002).

Major differences, however, were observed for features that influence the capillary-to-tissue transfer of O_2. The O_2 extraction ratio was 0.74 in LCR and 0.81 in HCR ($P < 0.001$) and the tissue diffusion capacity (in units of milliliter per kilogram per Torr) was 0.92 in LCR and 1.18 in HCR ($P < 0.01$). Total capillary and fiber numbers in the medial gastrocnemius muscle were similar between HCR and LCR rats, but because fiber area was 37% lower in HCR, the number of capillaries per unit area (or mass) of muscle was higher in HCR by 32% ($P < 0.001$). A positive correlation ($r = 0.92$) was found between capillary density and muscle O_2 conductance. The skeletal muscle enzymes, citrate synthase and beta-hydroxyacyl-CoA dehydrogenase, were both approximately 40% higher in HCR compared to LCR, whereas phosphofructokinase was significantly lower in HCR. Resting muscle ATP, phosphocreatine, and glycogen contents were not different between the groups (Howlett et al., 2003). These data demonstrate that features associated with a higher transfer of O_2 at the tissue level (D_{tissue}, see Figure 12.4) were initially more responsive to selection for endurance exercise running capacity.

At generation 15 of selection, the low- and high-selected lines differed by 552% in running capacity. Wagner and Gonzalez (Gonzalez et al., 2006a) hypothesized that this marked difference would now utilize augmentation of oxygen delivery by

the cardiopulmonary system as well as continued improvement in skeletal muscle O_2 transfer. From generation 7 to 15, the $\dot{V}O_2max$ difference between lines increased from 13% to 50% because of both improvement in HCR and deterioration in LCR. The greater $\dot{V}O_2max$ in HCR was accompanied by a 41% increase in oxygen delivery by the cardiopulmonary system (QO_2max). The greater QO_2max of the HCR after 15 generations of selection was mostly accounted for by a 48% greater stroke volume compared to the LCR (Gonzalez et al., 2006b).

Evaluation of possible lung changes across the generations of selection is of large interest because, in general, lung function is not altered by exercise training. Despite the 25% difference in body size (HCR smaller), both lung volume and exercise diffusing capacity were similar in HCR and LCR (Kirkton et al., 2009). Lung volumes of LCR rat were consistent with published data on a predicted scaling relationship for mammalian lung volume as a function of body mass (Gehr et al., 1981) based on the regression from 141 animals (32 species) whereas that of the HCR rat was greater than that predicted for body mass. Indeed, on a per kilogram of body weight basis, alveolar ventilation at $\dot{V}O_2max$ was 78% higher, and total pulmonary vascular resistance was 30% lower in HCR relative to LCR. These data indicate that selection-sensitive variance for pulmonary function is present at the genomic level and underscore the importance of lung structure and function for attaining high levels of exercise capacity. Though most of the changes from generation 7 to 15 were related to O_2 delivery, HCR rats at generation 15 demonstrated an additional 36% more total muscle fiber and capillary number in the medial gastrocnemius relative to the LCR (Howlett et al., 2009).

Central Fatigue

While changes at the skeletal muscle level are known to influence one's ability to sustain endurance capacity, neural circuits regulating fatigue also contribute. Previous work suggests that increased brain serotonin (5-hydroxytryptamine, 5-HT) release can hasten the onset of exercise-induced fatigue and that increased brain dopamine (DA) release can delay the onset of fatigue and contribute to the "motivational" component of performing physical activity (Foley & Fleshner, 2008). These findings provided the general background to test for the proposed "central nervous system fatigue hypothesis" that perhaps, differences in brain 5-HT and DA pathways can partly explain the large differences in duration of exhaustive running between the LCR and HCR.

We evaluated for differences in key targets in central fatigue pathways between these two contrasting rat groups (Foley et al., 2006). Quantitative *in situ* hybridization was used to measure messenger RNA (mRNA) levels of tryptophan hydroxylase (TPH), 5-HT transporter (5-HTT), 5-HT1A and 5-HT1B autoreceptors, dopamine receptor-D2 (DR-D2) autoreceptors and postsynaptic receptors, and dopamine receptor-D1 (DR-D1) postsynaptic receptors, in 10 brain regions of LCR and HCR. Consistent with the hypothesis, HCR expressed higher levels of 5-HT1B autoreceptor mRNA in the raphe nuclei relative to LCR. Interestingly, HCR expressed higher levels of DR-D2 autoreceptor mRNA in the midbrain, while simultaneously expressing greater DR-D2 postsynaptic mRNA in the striatum compared to LCR. These data suggest that central serotonergic and dopaminergic systems may be involved in the mechanisms by which HCR have delayed onset of exercise-induced fatigue, but the complexity for this intermediate trait has yet to be resolved.

Skeletal Muscle Metabolism

Increased oxidative capacity, such as that obtained from endurance exercise training, is thought to protect against the development of obesity and type II diabetes. The LCR and HCR provide a model with inherent differences in aerobic capacity that allows for the testing of this supposition without the confounding environmental effects of a training stimulus. Noland et al. (2007) found that LCR rats on normal rat chow were heavier, hypertriglyceridemic, and less insulin-sensitive, and had lower skeletal muscle oxidative capacity compared with HCR rats. Upon exposure to a high-fat diet (HFD), LCR rats gained more weight and visceral adiposity, and their insulin resistant condition was exacerbated, despite consuming similar amounts of metabolizable energy as chow-fed controls. Of large interest, these metabolic variables remained unaltered in HCR rats. The HFD increased skeletal muscle oxidative capacity was similar in both strains, whereas hepatic oxidative capacity was diminished only in LCR rats. These results suggest that rats born with low-intrinsic capacity for exercise are predisposed to developing obesity and that expansion of skeletal muscle oxidative capacity does not prevent excess weight gain or the exacerbation of insulin resistance on an HFD. For rats born with high-intrinsic endurance capacity, an elevated basal skeletal muscle oxidative capacity and the ability to preserve liver oxidative capacity may protect them from HFD-induced obesity and insulin resistance.

Metabolic changes in skeletal muscle are considered highly important contributors to the improved insulin sensitivity following exercise training. However, the precise mechanisms in skeletal muscle by which exercise training enhances whole-body metabolic health are still not well understood. The simultaneous occurrence of several molecular adaptations in skeletal muscle following exercise training makes it difficult to determine which of these is essential for its insulin-sensitizing effects. Lessard and colleagues (2009) investigated skeletal muscle properties, *in vitro* and *in vivo*, that could underlie the intrinsic differences in the metabolic phenotypes observed in the sedentary condition in LCR and HCR rats derived from generation 16 of selection. Skeletal muscle ceramide and diacylglycerol content, basal $5'$ adenosine monophosphate-activated protein kinase (AMPK) activity, and basal lipolysis were similar between the low- and high-intrinsic capacity groups. However, stimulation of lipolysis in response to 10 μM isoproterenol was 70% higher in HCR. The impaired isoproterenol-sensitivity in LCR was associated with lower values for basal triacylglycerol lipase activity, Ser660 phosphorylation of hormone sensitive lipase, and β2-adrenergic receptor protein content in skeletal muscle. Expression of the orphan nuclear receptor Nur77, which is induced by β-adrenergic signaling and is associated with insulin sensitivity, was lower in LCR. Muscle protein content of Nur77 target genes, including uncoupling protein 3, fatty acid translocase/CD36, and the AMPK γ3 subunit were also lower in LCR. These results associate whole-body insulin resistance with impaired β-adrenergic response, and with reduced expression of genes that are critical regulators of glucose and lipid metabolism in skeletal muscle. We identify altered β-adrenergic signal transduction as a potential mechanism for impaired metabolic health in the LCR rats.

Novak et al. (2009) has put forth the general view that the "lean phenotype" is characterized by high endurance capacity, high activity levels, and high resting energy expenditure (Figure 12.5) that originates from altered skeletal muscle energetics. This is supported by their observations that: (1) leanness of the HCR relative to the

LCR cannot be accounted for increased body weight–corrected food consumption, (2) HCR are higher for V̇O$_2$max, daily cage activity, and resting energy expenditure relative to the LCR, (3) elevated skeletal muscle enzyme phosphoenolpyruvate carboxykinase (PEPCK) in HCR rats is in common with another lean selective rat model that is diet resistance and highly active, and (4) a similar three-way interaction between exercise capacity, activity, and resting metabolic rate is found between lean and overweight humans. Interestingly, this phenotypic triad parallels the evolutionary path by which the condition of being warm-blooded (endothermic) arose. The somewhat widely accepted "aerobic capacity"

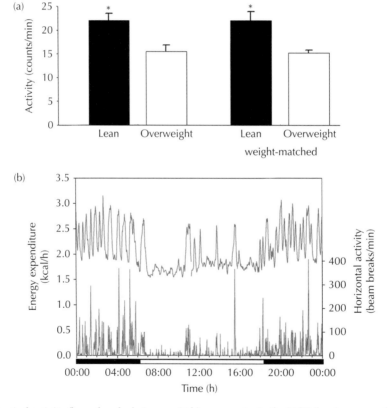

Figure 12.5 (a) Physical activity (beam breaks/min over 24 h) was greater in the HCR (lean) compared to the LCR (overweight) rats. The same difference was observed even for groups of LCR and HCR rats that did not differ in body weight. (b) Twenty-four hour sample trace from an HCR rat for energy expenditure and horizontal beam break activity. Activity and energy expenditure were highly correlated. (Reprinted from Novak et al. (2009), reprinted with permission from *PLoS ONE*.)

model of endothermy (Hayes & Garland, 1995) proposes that the relatively high basal metabolic rates that accompany endothermy evolved as a correlated response to the selection pressure on $\dot{V}O_2$max that is associated with locomotor endurance activity.

Clinically Relevant Differences

Metabolic Syndrome

The first test of the broader hypothesis that artificial selection of rats based on low- and high-intrinsic aerobic treadmill running exercise capacity would also yield models that contrast for disease risk was conducted in rats from generations 10 and 11 of selection (Wisloff et al., 2005). The HCR measured significantly higher than LCR for health factors such as $\dot{V}O_2$max (58%), heart function (contractility), endothelial nitric oxide formation, economy of oxygen use while running (17%), and levels of several transcription factors and enzymes associated with oxidative metabolism. The LCR scored higher on cardiovascular risks and features of the metabolic syndrome including random glucose, fasting glucose, insulin, visceral adiposity, triglycerides, and free fatty acids when still young adults (Table 12.1). These measures formed the initial basis for further exploration of clinically relevant phenotypes.

We also found that, relative to the HCRs, the LCR rats had higher mean blood pressures during the day (18%), at night (8%), and for the combined 24-h period (102 ± 6 mmHg versus 90 ± 7 mmHg) (Wisloff et al., 2005). Extrapolating from human data (Lewington et al., 2002), this 13% higher 24-h blood pressure suggests that the rats born with low-intrinsic capacity are twice as likely to develop cardiovascular disease.

For explanation of these blood pressure differences, we tested whether LCR had any evidence for dysregulation in GABAergic neurotransmission in the medullary cardiovascular-regulatory areas known to participate in the central control of blood pressure (Buck et al., 2007). We measured GAD65 and GAD67 mRNA expression via radioactive *in situ* hybridization in LCR and HCR brains. GAD65 and GAD67 mRNAs were widely expressed throughout the brainstem; quantification revealed increased GAD65 mRNA expression in LCR animals in the caudal nucleus tractus solitarius and rostral ventrolateral medulla as compared to HCR rats. These data are consistent with the notion that altered GABAergic neurotransmission in these medullary nuclei in LCR is at least partially responsible for the higher blood pressure.

Fatty Liver Disease

An increased deposition of fat in the liver is statistically associated with low aerobic fitness and high body mass index in humans; however, there is no mechanistic information detailing these connections. We evaluated sedentary 25-week-old rats from generation 17 to test whether LCR had reduced hepatic mitochondrial oxidative capacity and increased susceptibility to hepatic steatosis (Thyfault et al., 2009). The LCR livers had significantly reduced mitochondrial content, reduced oxidative capacity (Figure 12.6), and increased peroxisomal activity. In addition, LCR livers displayed a lipogenic phenotype with higher protein content of both sterol regulatory element binding protein-1c and acetyl CoA carboxylase. These differences were associated with hepatic steatosis in the LCR including (1) higher liver triglycerides, (2) increased percentage of hepatocytes associated with lipid droplets, and (3) increased hepatic lipid peroxidation compared to the HCR. Additionally, when rats aged to natural death, LCR livers had more hepatic injury (fibrosis and apoptosis). Thus, it appears that the LCR can serve as a model for increased susceptibility to both hepatic steatosis and liver injury.

Kidney Oxidative Stress

An association between decreased aerobic capacity and diminished insulin sensitivity has been demonstrated in individuals with early chronic kidney disease. Increased reactive oxygen species (ROS) formation has been implicated in many renal disorders and is thought to be a major contributor to diabetic kidney disease. Current evidence suggests increased oxidative stress leads to alterations in insulin signaling

Table 12.1 Metabolic and heart measures in LCR and HCR rats (*$P < 0.05$ for LCR vs. HCR)

	Metabolic		
	LCR	HCR	% difference LCR vs. HCR
Fasting glucose (mg/dl)	110 ± 9	92 ± 5	19%*
Random glucose (mg/dl)	86 ± 6	75 ± 12	15%*
Insulin (pM)	684 ± 195	296 ± 172	131%*
C-peptide (pM)	1590 ± 338	1077 ± 565	48% ns
C-peptide/Insulin ratio	2.4 ± 0.4	3.8 ± 1.2	−37%*
Visceral adiposity/Bwt (%)	1.5 ± 0.4	1.0 ± 0.3	63%*
Triglycerides (mg/dl)	67 ± 24	25 ± 4	167%*
Free fatty acids (mEq/l)	0.6 ± 0.2	0.3 ± 0.04	95%*

	Heart		
	LCR	HCR	% difference LCR vs. HCR
Cell shortening (%)	14.0 ± 1.2	17.1 ± 1.1	−18%*
Relative time to peak shortening (ms × PS^{-1})	2.7 ± 0.2	2.3 ± 0.2	22%*
Systolic [Ca^{2+}] (μM)	1.61 ± 0.03	1.73 ± 0.04	−7%*
Amplitude of [Ca^{2+}] transient (μM)	1.20 ± 0.03	1.38 ± 0.05	−13%*
Time to 50% re-lengthening (ms)	39.9 ± 1.2	35.2 ± 1.3	13%*
Diastolic [Ca^{2+}] (μM)	0.41 ± 0.02	0.35 ± 0.02	17%*
Time to 50% decay of [Ca^{2+}] transient (ms)	55.3 ± 1.4	45.91.3	20%*

Source: Reprinted from Wisloff et al. (2005), with permission from *Science*.

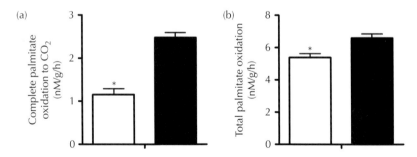

Figure 12.6 Hepatic mitochondrial fat oxidation. Both (a) complete palmitate oxidation to CO_2 and (b) total palmitate oxidation (CO_2 + acid soluble metabolites) were significantly lower in isolated mitochondria from the LCR rats than the HCR at 25 weeks of age. (Reprinted from Thyfault et al. (2009), with permission from *Journal of Physiology*.)

through serine phosphorylation of insulin receptor substrate-1 (IRS-1) in various insulin-sensitive tissues (Morino et al., 2006). Thus, we hypothesized that the kidneys of HCR rats would be protected

and LCR rats more susceptible to HFD-induced ROS and IRS-1 degradation (Morris et al. 2009). Eighteen-week-old LCR and HCR rats from generation 20 of selection were placed on an HFD or normal chow diet for 7 weeks. Intraperitoneal glucose tolerance, ROS, IR-beta, total IRS-1, and ubiquitination were measured.

We found the HCR displayed greater insulin sensitivity and were resistant to HFD-induced insulin resistance. In LCR kidneys, HFD increased ROS potential and reduced total IR-beta and IRS-1 without altering triacylglycerol content. The IRS-1 ubiquitination was higher in the LCR relative to HCR kidney. These data support that HFD-mediated kidney ROS is associated with reductions in IRS-1 and systemic insulin resistance. Furthermore, high-intrinsic aerobic capacity protected against IRS-1 degradation in the kidney following exposure to HFD.

Challenges to Oxygenation

Human epidemiological studies and experimental work with animals indicate that regular physical activity is associated with reductions in arrhythmias that underlie sudden cardiac death. Because HCR demonstrate reduced cardiovascular risk factors, Lujan et al. (Lujan et al. 2006) wanted to test specifically whether HCR rats would be less susceptible to ischemia-reperfusion-mediated ventricular tachyarrhythmias. Susceptibility was produced by 3 min of occlusion and reperfusion of the left main coronary artery in conscious LCR and HCR rats. Results document a significantly lower incidence of ventricular tachyarrhythmias in HCR (3 of 11, 27.3%) relative to LCR (6 of 7, 85.6%) rats. The decreased susceptibility to tachyarrhythmias in HCR rats was associated with a reduced cardiac metabolic demand during ischemia (lower rate–pressure product and ST segment elevation) as well as a wider range for the autonomic control of heart rate. The HCR and LCR models represent a unique substrate for evaluation of the intrinsic mechanisms underlying ischemia-mediated cardiac arrhythmogenesis.

Based on the early evidence of diverging susceptibilities for cardiovascular risk factors between LCR and HCR rats, Palpant et al. (2009) hypothesized that a low-intrinsic aerobic capacity portends an increased susceptibility to cardiac dysfunction during cardiac stresses such as oxygen deprivation. Using rats derived from 19 generations of selection that represented a 436% difference for intrinsic aerobic

running capacity, they found that when challenged by exposure to acute systemic hypoxia (7% oxygen), cardiac pump failure occurred significantly earlier in LCR rats compared to HCR animals. Acute cardiac decompensation in LCR rats was exclusively due to the development of intractable irregular ventricular contractions. Analysis of isolated cardiac ventricular myocytes showed significantly slower sarcomeric relaxation and delayed kinetics of calcium cycling in LCR-derived cells compared to HCR myocytes.

Pain

While it has been shown that aerobic exercise can provide temporary relief from a variety of painful stimuli known as exercise-induced analgesia, little is known about how one's level of fitness influences the perception and experience of pain. Pain perception was estimated in LCR and HCR rats before and after exercise to exhaustion by measuring paw withdrawal latency times to thermal pain (Hargreaves method). HCR animals had significantly higher pain thresholds compared to LCR animals before (27%) and after exercise (46%) (Geisser et al., 2008). Both LCR and HCR displayed evidence of hyperalgesia following exercise compared to baseline. Yet, the pain thresholds of HCR animals returned to baseline levels faster than LCR animals following exercise. These findings support the hypothesis that within the collection of phenotypes that determine one's intrinsic level of endurance running capacity, there are traits that play a role in the perception of pain. Thus, a higher level of fitness may serve as buffer to reduce or prevent the experience of clinical or chronic pain.

Retrieval of the LCR Phenotype

One use of the LCR is to test the efficacy of interventions that can ameliorate the disease risks. We compared the effectiveness for reducing cardiovascular risk factors by exercise training programs of different exercise intensities (Haram et al., 2009). Male LCR rats from generation 15 were subjected to either continuous moderate-intensity exercise (CME) or high-intensity aerobic interval training (AIT). AIT was more effective than CME at reducing cardiovascular risk factors linked to the metabolic syndrome (Figure 12.7). That is, AIT produced a larger stimulus than CME for increasing $\dot{V}O_2$max (45% vs. 10%), reducing systolic blood pressure (−20 mmHg vs. −6 mmHg), increasing levels of plasma HDL cholesterol (25% vs. 0%), and beneficially altering metabolism in fat, liver, and skeletal muscle tissues. Moreover, AIT had a positive effect over CME on sensitivity of aorta ring segments to acetylcholine (2.7- vs. 2.0-fold), partly because of intensity-dependent effects on expression levels of nitric oxide synthase and the density of caveolae. AIT also had a greater effect than CME on skeletal muscle Ca^{2+} handling (50% vs. 0%).

However, the two exercise training programs, AIT and CME, were equally effective at reducing body weight (5% vs. 7%) and retroperitoneal fat (54% and 55%). In comparison to the sedentary condition, these data from LCR rats strongly support the conclusion that high-intensity interval exercise training is more favorable than moderate-intensity continuous exercise training for reducing health risks consistent with the metabolic syndrome.

The neuronal isoform of nitric oxide synthase (NOS-1) may be an important regulator of cardiac contractility by modifying calcium release and uptake from sarcoplasmic reticulum. One working hypothesis from the high-intensity AIT study was

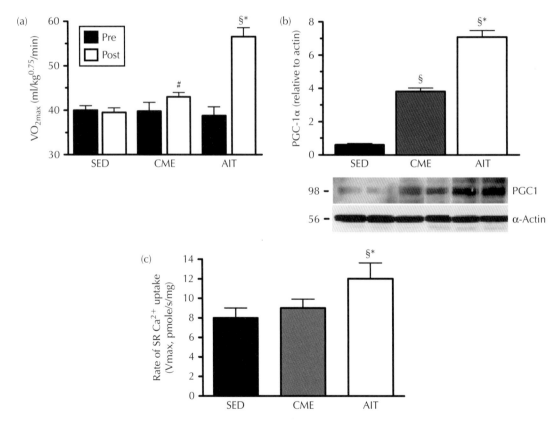

Figure 12.7 Aerobic interval training (AIT), relative to continuous moderate-intensity exercise (CME) and the sedentary (SED) condition, produced larger adaptational changes post-training. (a) Maximal oxygen uptake ($\dot{V}O_2$max). (b) Expression of peroxisome proliferator-activated receptor-γ co-activator-1α (PGC-1α) in samples from m. soleus. (c) Maximal rate of re-uptake of calcium into sarcoplasmic reticulum in m. soleus. (Reprinted from Haram et al. (2009), with permission from *Cardiovascular Research*.)

that NOS-1 modulates cardiomyocyte contractility more markedly in LCR versus HCR rats (Hoydal et al., 2007). Rats performed high-intensity interval treadmill running 5 days per week over 8 weeks and age-matched sedentary rats served as controls. At sedentary baseline before the training program, aerobic fitness measured as $\dot{V}O_2$max was 30% higher, and left ventricular cardiomyocyte contractility measured as fractional shortening, 42% higher in HCR than in LCR rats. Exercise training markedly increased aerobic fitness as well as cardiomyocyte contractility, relaxation, and corresponding changes in calcium transients in both lines. Selective inhibition of NOS-1 increased cardiomyocyte contractility (12–43%) and calcium transient amplitude (10–28%), prolonged time to 50% re-lengthening (13–52%), and time to 50% calcium decay (17–35%) in all groups (sedentary and trained). Interestingly, NOS-1-inhibition abolished the difference in systolic events between LCR and HCR rats whereas no such findings occurred in diastolic parameters. From these data, we conclude that NOS-1-derived nitric oxide production is a modulator of cardiomyocyte contractile performance and that calcium handling accounts for some of the inherent difference between LCR and HCR, whereas it contributes little during adaptation to exercise training.

Differences in Gene Expression

Alterations in gene expression were evaluated for cardiac and skeletal muscle tissue in generation 16 LCR and HCR rats in the sedentary condition and after 8 weeks on a high-intensity aerobic interval exercise program. Left ventricle gene expression was categorized by ontology analysis (Bye et al., 2008b). Out of 28,000 screened genes, 1540 were differentially expressed between sedentary HCR and LCR. Sedentary HCR expressed higher amounts of genes involved in lipid metabolism, whereas sedentary LCR expressed higher amounts of the genes involved in glucose metabolism. This suggests a switch in cardiac energy substrate utilization from normal mitochondrial fatty acid beta-oxidation in HCR to carbohydrate metabolism in LCR, an event that often occurs in diseased hearts. LCR were also associated with pathological growth signaling and cellular stress. Hypoxic

conditions seemed to be a common source for several of these observations, triggering hypoxia-induced alterations of transcription. Only one gene was found differentially expressed by exercise training, but this gene had unknown name and function.

In contrast to many genes being differentially expressed between the LCR and HCR rats for heart tissue, few differences were found for soleus (red fiber type) skeletal muscle (Bye et al., 2008a). Although sedentary HCR were able to maintain a 120% higher running speed at $\dot{V}O_2$max than sedentary LCR, only three transcripts were differentially expressed (false discovery rate, or = 0.05) between the two groups. Sedentary LCR expressed high levels of a transcript with strong homology to human leucyl-transfer RNA synthetase, of whose overexpression has been associated with a mutation linked to mitochondrial dysfunction. In terms of gene expression, the response to exercise training component in skeletal muscle was more pronounced in HCR than LCR. HCR rats upregulated several genes associated with lipid metabolism and fatty acid elongation, whereas LCR upregulated only one transcript after exercise training. The results indicate only minor differences in soleus muscle gene expression between sedentary HCR and LCR. However, the intrinsic level of fitness seems to influence the transcriptional adaptation to exercise, as more genes were upregulated after exercise training in HCR than LCR.

Conclusion

With about three billion nucleotides in the human genome, the number of nucleotide combinations that can influence the activity of genes is essentially infinite. The LCR and HCR models represent heterogenetic substrate for the divide between low- and high-intrinsic endurance capacity. As such, rats in each strain retain the complex networks of function that underlie endurance capacity. This contrasts with knockout and transgene approaches in inbred mice that are artificial, remove the modular nature of function, and do not emulate any feature of the integrated character of endurance capacity. The divide for disease and health is an attractive feature of these rat models and demonstrates the embedded condition of complex traits.

References

Barbato, J.C., Koch, L.G., Darvish, A., Cicila, G.T., Metting, P.J. & Britton, S.L. (1998). Spectrum of aerobic endurance running performance in eleven inbred strains of rats. *Journal of Applied Physiology*, 85, 530–536.

Buck, B.J., Kerman, I.A., Burghardt, P.R., et al. (2007). Upregulation of GAD65 mRNA in the medulla of the rat model of metabolic syndrome. *Neuroscience Letters*, 419, 178–183.

Bye, A., Hoydal, M.A., Catalucci, D., et al. (2008a). Gene expression profiling of skeletal muscle in exercise-trained and sedentary rats with inborn high and low $\dot{V}O_2$max. *Physiological Genomics*, 35(3), 213–221.

Bye, A., Langaas, M., Hoydal, M.A., et al. (2008b). Aerobic capacity-dependent differences in cardiac gene expression. *Physiological Genomics*, 33, 100–109.

Falconer, D.S. (1976). Genetic aspects of breeding methods. In: *The UFAW Handbook on the Care and Management of Laboratory Animals*, 5th edn. Churchill Livingstone, Edinburgh.

Foley, T.E. & Fleshner, M. (2008). Neuroplasticity of dopamine circuits after exercise: Implications for central fatigue. *Neuromolecular Medicine*, 10, 67–80.

Foley, T.E., Greenwood, B.N., Day, H.E., Koch, L.G., Britton, S.L. & Fleshner, M. (2006). Elevated central monoamine receptor mRNA in rats bred for high endurance capacity: Implications for central fatigue. *Behavioural Brain Research*, 174, 132–142.

Gehr, P., Mwangi, D.K., Ammann, A., Maloiy, G.M.O., Taylor, C.R. & Weibel, E.R. (1981). Design of the mammalian respiratory system. V. Scaling morphometric pulmonary diffusing capacity to body mass: Wild and domestic mammals. *Respiration Physiology*, 44, 61–86.

Geisser, M.E., Wang, W., Smuck, M., Koch, L.G., Britton, S.L. & Lydic, R. (2008). Nociception before and after exercise in rats bred for high and low aerobic capacity. *Neuroscience Letters*, 443, 37–40.

Gonzalez, N.C., Howlett, R.A., Henderson, K.K., et al. (2006a). Systemic oxygen transport in rats artificially selected for running endurance. *Respiratory Physiology & Neurobiology*, 151, 141–150.

Gonzalez, N.C., Kirkton, S.D., Howlett, R.A., et al. (2006b). Continued divergence in $\dot{V}O_2$max of rats artificially selected for running endurance is mediated by greater convective blood O_2 delivery. *Journal of Applied Physiology*, 101, 1288–1296.

Hansen, C. & Spuhler, K. (1984). Development of the National Institutes of Health genetically heterogeneous rat stock. *Alcoholism: Clinical and Experimental Research*, 8, 477–479.

Haram, P.M., Kemi, O.J., Lee, S.J., et al. (2009). Aerobic interval training versus continuous moderate exercise in the metabolic syndrome of rats artificially selected for low aerobic capacity. *Cardiovascular Research*, 81, 723–732.

Hayes, J.P. & Garland, T. (1995). The evolution of endothermy—testing the aerobic capacity model. *Evolution*, 49, 836–847.

Hegmann, J.P. & Possidente, B. (1981). Estimating genetic correlations from inbred strains. *Behavior Genetics*, 11, 103–114.

Henderson, K.K., Wagner, H., Favret, F., et al. (2002). Determinants of maximal O_2 uptake in rats selectively bred for endurance running capacity. *Journal of Applied Physiology*, 93, 1265–1274.

Howlett, R.A., Gonzalez, N.C., Wagner, H.E., et al. (2003). Skeletal muscle capillarity and enzyme activity in rats selectively bred for running endurance. *Journal of Applied Physiology*, 94, 1682–1688.

Howlett, R.A., Kirkton, S.D., Gonzalez, N.C., et al. (2009). Peripheral oxygen transport and utilization in rats following continued selective breeding for endurance running capacity. *Journal of Applied Physiology*, 106, 1819–1825.

Hoydal, M.A., Wisloff, U., Kemi, O.J., et al. (2007). Nitric oxide synthase type-1 modulates cardiomyocyte contractility and calcium handling: Association with low intrinsic aerobic capacity. *European Journal of Cardiovascular Prevention and Rehabilitation*, 14, 319–325.

Kimura, M. & Crow, J.F. (1963). On the maximum avoidance of inbreeding. *Genetical Research*, 4, 399–415.

Kirkton, S.D., Howlett, R.A., Gonzalez, N.C., et al. (2009). Continued artificial selection for running endurance in rats is associated with improved lung function. *Journal of Applied Physiology*, 106, 1810–1818.

Koch, L.G. & Britton, S.L. (2001). Artificial selection for intrinsic aerobic endurance running capacity in rats. *Physiological Genomics*, 5, 45–52.

Lessard, S.J., Rivas, D.A., Chen, Z.P., et al. (2009). Impaired skeletal muscle beta-adrenergic activation and lipolysis are associated with whole-body insulin resistance in rats bred for low intrinsic exercise capacity. *Endocrinology* 150, 4883–4891.

Lewington, S., Clarke, R., Qizilbash, N., Peto, R. & Collins, R. (2002). Age-specific relevance of usual blood pressure to vascular mortality: A meta-analysis of individual data for one million adults in 61 prospective studies. *The Lancet*, 360, 1903–1913.

Lujan, H.L., Britton, S.L., Koch, L.G. & Dicarlo, S.E. (2006). Reduced susceptibility to ventricular tachyarrhythmias in rats selectively bred for high aerobic capacity. *American Journal of Physiology—Heart and Circulatory Physiology*, 291, H2933–H2941.

Morino, K., Petersen, K.F. & Shulman, G.I. (2006). Molecular mechanisms of insulin resistance in humans and their potential links with mitochondrial dysfunction. *Diabetes*, 55, S9–S15.

Morris, E.M., Whaley-Connell, A.T., Thyfault, J.P., et al. (2009). Low aerobic capacity and high-fat diet contribute to oxidative stress and irs-1 degradation in the kidney. *American Journal of Nephrology* 30, 112–119.

Noland, R.C., Thyfault, J.P., Henes, S.T., et al. (2007). Artificial selection for high-capacity endurance running is protective against high-fat diet-induced insulin resistance. *American Journal of Physiology—Endocrinology and Metabolism*, 293, E31–E41.

Novak, C.M., Escande, C., Gerber, S.M., et al. (2009). Endurance capacity, not body size, determines physical activity levels: Role of skeletal muscle PEPCK. *PLoS ONE* 4, e5869.

Palpant, N.J., Szatkowski, M.L., Wang, W., et al. (2009). Artificial selection for whole animal low intrinsic aerobic capacity co-segregates with hypoxia-induced cardiac pump failure. *PLoS ONE* 4, e6117.

Payne, J.L., Boyer, A.G., Brown, J.H., et al. (2009). Two-phase increase in the maximum size of life over 3.5 billion years reflects biological innovation and environmental opportunity. *Proceedings of the National Academy of Sciences of the United States of America*, 106, 24–27.

Raymond, J. & Segre, D. (2006). The effect of oxygen on biochemical networks and the evolution of complex life. *Science* 311, 1764–1767.

Rezende, E.L., Garland, T., Jr., Chappell, M.A., Malisch, J.L. & Gomes, F.R. (2006). Maximum aerobic performance in lines of Mus selected for high wheel-running activity: Effects of selection, oxygen availability and the mini-muscle phenotype. *Journal of Experimental Biology* 209, 115–127.

Swallow, J.G., Carter, P.A. & Garland, T., Jr. (1998). Artificial selection for increased wheel-running behavior in house mice. *Behaviour Genetics* 28, 227–237.

Thyfault, J.P., Rector, R.S., Uptergrove, G.M., et al. (2009). Rats selectively bred for low aerobic capacity have reduced hepatic mitochondrial oxidative capacity and susceptibility to hepatic steatosis and injury, 587(Pt 8), 1805–1816.

Wagner, P.D. (1993). Algebraic analysis of the determinants of $\dot{V}O_2max$. *Respiratory Physiology* 93, 221–237.

Waters, R.P., Renner, K.J., Pringle, R.B., et al. (2008). Selection for aerobic capacity affects corticosterone, monoamines and wheel-running activity. *Physiology & Behavior* 93, 1044–1054.

Wisloff, U., Najjar, S.M., Ellingsen, O., et al. (2005). Cardiovascular risk factors emerge after artificial selection for low aerobic capacity. *Science*, 307, 418–420.

PART 3

CONTRIBUTIONS OF SPECIFIC GENES AND MARKERS

Chapter 13

Genes and Endurance Performance

BERND WOLFARTH

Department of Preventive and Rehabilitative Sports Medicine, Technical University Munich, Munich, Germany

A body of literature indicates that genes play an important role in the determination of endurance performance. Here, we summarize the important papers and findings, focusing on the genes and markers that were found to be significantly associated with endurance performance and were either replicated in independent studies or were generated from large, robust cohorts.

The current review is organized by study type: (i) case–control studies; (ii) cross-sectional association studies; (iii) association studies with training response phenotypes; and (iv) linkage studies. Taking into account the progression we have had in the last 20 years, we can foresee a much more definite picture of the genetic elements influencing aerobic performance phenotypes and trainability in the next decade. However, based on existing data, we have to conclude that the overall picture of genes involved in endurance performance as a target phenotype is not very sharp. In contrast to past thinking that perhaps only a dozen genes or fewer were involved, the current understanding is that a high number of genes are involved.

Background on the Genetics of Endurance

In early publications, family studies showed a significant familial aggregation for such endurance measures as maximal oxygen uptake ($\dot{V}O_2max$),

Genetic and Molecular Aspects of Sport Performance, 1st edition.
Edited by Claude Bouchard and Eric P. Hoffman.
Published 2011 by Blackwell Publishing Ltd.

with heritability levels up to 50% (Bouchard et al., 1986, 1998). In addition, twin studies suggested that the heritability could even be higher (Fagard et al., 1991; Maes et al., 1996). For trainability levels of different endurance performance phenotypes, a wide range of increase from about 5% to 60% of the initial level is described (Bouchard et al., 1998; Lortie et al., 1984). There is no doubt that genes are involved and account for differences in baseline as well as trainability levels of endurance traits. Progress in molecular genetic techniques gave us molecular markers to seek specific genes and polymorphic variations and to determine their implications for endurance and other performance-related phenotypes.

It has been suggested that attempts to identify genetic markers are more likely to succeed when using the magnitude of the response to training as a phenotype of interest, rather than the individual differences seen in a sedentary population. In the last two decades, scientists have been investigating these questions with batteries of genetic markers, using several different strategies: (i) to discriminate between endurance-trained elite athletes and matched sedentary controls; (ii) to account for the individual differences in aerobic performance among sedentary subjects; and (iii) to predict the pattern of response to aerobic training. Most results of these different approaches were generated from a handful of studies (Table 13.1).

To detect genetic differences, a few markers and techniques were used. Restriction fragment length polymorphisms (RFLP) for a variety of candidate

Table 13.1 Examples of major studies in the field of genes and endurance performance

Research group	Study	Design	Subjects	Intervention
Bouchard et al.	HERITAGE	Training study, family-based	100 white families ($n = 500$) and 100 black families ($n = 250$)	20-weeks controlled endurance training
Bouchard et al.	GENATHLETE	Case–control, endurance athletes vs. sedentary individuals	300 endurance athletes and 300 sedentary individuals	
Montgomery et al.		Training study	140 Caucasian males	10-weeks physical fitness program
Rogozkin et al.		Case–control, various athletes vs. sedentary individuals	750 mixed athletes and 1200 sedentary individuals	
He et al.		Training study	102 Chinese male	18-weeks endurance training

genes encoded in the nuclear DNA and mtDNA were initially investigated. More recently, single-nucleotide polymorphisms (SNPs) for the genes of interest have become prominent. With the availability of large cohorts and better statistical power, markers are combined in haplotype analysis and statistic-based genetic models. Microsatellite markers were also used widely in the recent past for linkage analysis or, if they were known to be in the gene of interest, for candidate gene approach investigations. Thus far, these approaches have not resulted in major advances. We are looking forward to studies based on large-scale screening technologies.

Several studies based on specific markers have been published (Bray et al., 2009). Most of the studies focused on candidate genes selected on the basis of our current understanding of the exercise physiology and exercise biochemistry of endurance performance traits. The list of potential candidate genes is large and increasing almost daily. So far, the candidate genes that have been considered belong more or less to one of the following groups: regulating hormones (*EPO, EPOR, ADR*), muscle metabolism (*CKMM, mtDNA*), lipid metabolism (*LPL, CPT, LDLR*), growth factors (*GH, GHR, IGF*), and others (e.g., *ACE, HSP70, TNFA*) (Table 13.2).

Table 13.2 Physiological targets of genetic variations

Target area	Genes of interest
Hormones	*ADR, ANG, EPO, EPOR*
Muscle	*CKM, CPT2, FABP3, mtDNA, Myostatin, Titin*
Lipoprotein metabolism	*Adiponectin, FABP2, LDLR, LEP, LEPR, LIPE, LPL*
Growth factors	*GH, IGF, IGFBP, TGF, VEGF*
Cellular mediators	*c-FOS, HIF1A, IL, c-JUN, PPAR*
Others	*ACE, BDKRB2, HBB, HSP70, NOS3, TNFalpha*

ADR, adrenergic receptors; ANG, angiogenin; EPO, erythropoietin; EPOR, erythropoietin receptor; CKM, creatine kinase muscle type; CPT2, carnitine palmitoyltransferse II; FABP3, fatty acid–binding protein 3; mtDNA, mitochondrial DNA, FABP2, fatty acid–binding protein 2; LDLR, low-density lipoprotein receptor; LEP, leptin; LEPR, leptin receptor; LIPE, hormone-sensitive lipase; LPL, lipoprotein lipase; GH, growth hormone; IGF, insulin-like growth factor; IGFBP, insulin-like growth factor–binding protein; TGF, transforming growth factor; VEGF, vascular endothelial growth factor; HIF1A, hypoxia-inducible factor 1; IL, interleukins; PPAR, peroxisome proliferative activated receptor family; ACE, angiotensin-converting enzyme; BDKRB2, bradykinin receptor B2; HBB, hemoglobin beta; HSP70, heat shock protein 70; NOS3, nitric oxide synthase 3; TNFalpha, tumor necrosis factor alpha.

It should be obvious by now that it is unlikely that a single gene locus (or a few loci) or a given DNA variant (or a few polymorphic sequences for that matter) will be sufficient to define the responder status or an endurance phenotype (Ruiz et al., 2009).

The heterogeneity of performance skills is demonstrated in decathletes. Data from world-class decathlon athletes show a negative correlation between performance in short-term disciplines (100 m sprint, shot put, long jump, and 110 m hurdles), intermediate duration disciplines (400 m), and more endurance-related disciplines such as the 1500 m race (Van Damme et al., 2002). This suggests that we have to distinguish between performances in some events (e.g., sprint) versus those in others (e.g., endurance). However, if we consider world-class athletes, even in the same discipline, we see different and sometimes opposing types of extraordinary abilities resulting in the same performance levels (Lippi et al., 2010). If we look, for example, at 1500 m runners, we see athletes from predominantly endurance events and others more from high-intensity activities, both being able to run the same distance in the same time, even close to world-record performances (Lippi et al., 2008). This obviously complicates the identification of relevant genes for human endurance performance, especially since each marker or gene is expected to make only a small contribution to the overall heritability of the performance phenotypes of interest (Lippi et al., 2010; Ruiz et al., 2009).

In addition, from a technical point of view, one must also recognize that for genetic (e.g., linkage) or physiological reasons, the genetic marker may be only a surrogate, and the true genetic determinants may lie in another gene. All these issues have to be considered when we discuss the genes and markers reported thus far.

The history of this field began about 30 years ago, when Claude Bouchard and his co-workers investigated athletes involved in Olympic events. In one of the first papers based on markers available at that time, it was shown that red blood cell antigen and enzyme polymorphisms were not associated with the status of elite athletes in endurance events (Chagnon et al., 1984). Moreover, 11 enzymes of the glycolytic pathway and 9 enzymes of the tricarboxylic acid cycle of the skeletal muscle did not exhibit any charge or mass variants in a large sample of subjects, thus supporting the notion that variations in the coding sequences of these genes are unlikely to account for the individuality of aerobic performance or trainability (Bouchard et al., 1988). Subsequently, suggestive evidence to the effect that sequence variations in mitochondrial DNA may contribute to individual differences in $\dot{V}O_2$max and its response to training was reported (Dionne et al., 1991). However, in a case–control study using the GENATHLETE cohort, the authors were not able to show an association between several mtDNA polymorphisms and the elite endurance athlete status (Rivera et al., 1998).

More than a decade after the first studies on markers and endurance performance, a paper was published in *Nature* with the highly promising title: "Human gene for physical performance." In this publication, Montgomery and his colleagues proposed that the angiotensin-converting enzyme (*ACE*) gene strongly influenced human physical performance (Montgomery et al., 1998). In two different experiments, they found associations between an insertion allele of the *ACE* insertion/deletion (I/D) polymorphism and two distinct performance-related phenotypes. The first experiment was conducted with mountaineers who had a history of ascents above 7000 m altitudes without supplementary oxygen. They compared the genotype distribution and allelic frequency of the *ACE* I/D polymorphism in 25 of these mountaineers to a control group of 1906 British males. For the allele frequency, they found a significant difference ($P = 0.003$) between climbers and controls with a relative excess of insertion allele carriers and deficiency of deletion allele carriers in the climber group. The second experiment came from a prospective 10-week training study with British army male recruits. Out of 123 males 87 completed a 10-week physical training program.

Along with other tests, they performed a muscular endurance test. Before and after training, the maximum duration for which the subjects were able to perform repetitive elbow-flexion while holding a 15-kg barbell was assessed. No difference was found at baseline between the different *ACE* I/D genotypes. But after training they found an 11-fold improvement in duration for the II compared to the DD genotype carriers ($P = 0.001$). A more detailed presentation on *ACE* is provided in Chapter 18, which is entirely devoted to this gene.

Genetic Markers for Endurance Performance

The main aim of this chapter is to summarize the current status of peer-reviewed papers dealing with different types of genetic markers in conjunction with endurance performance phenotypes. A fundamental resource for this task is the series "The Human Gene Map for Performance and Health-Related Fitness Phenotypes." It began in 2001 (Rankinen et al., 2001), followed by six updates until the last installment in 2009 (Bray et al., 2009). The final update contains tables of all publications with positive genetic associations to performance and fitness phenotypes. It included 214 autosomal gene entries and quantitative trait loci (QTLs) plus 7 others on the X chromosome. In addition, 18 mitochondrial genes were cited that influenced performance and fitness phenotypes (Bray et al., 2009).

In all issues of the fitness gene map, endurance performance was a major focus. In the evolution, in the status of the endurance-related phenotypes, the number of articles began increasing from the year 2003. From then on, about 10 articles per year were added and currently there are nearly 80 articles including 42 loci dealing with endurance performance phenotypes (Figure 13.1).

Case–Control Studies

The use of a case–control design, combined with the so-called candidate gene approach, is one of the most common scientific techniques to study the basis of genetics in the field of endurance performance.

The most extensively investigated candidate gene thus far is the *ACE* gene. The latest manuscripts reported significantly different genotype and allele distributions between elite marathon runners and other endurance athletes compared with sedentary controls (Hruskovicova et al., 2006). In endurance athletes from Israel, a higher number of D-allele carriers and DD genotypes were seen compared to controls, with an even higher level of significance comparing these athletes with elite sprinters (Amir et al., 2007). In the adenosine monophosphate deaminase 1 (*AMPD1*) gene, the C34T mutation showed a lower T allelic frequency in a cohort of 104 Spanish cyclists and runners compared to a control cohort of 100 Spanish non-athletes (Rubio et al., 2005).

Peroxisome proliferative activated receptor, gamma, coactivator 1, alpha (*PPARGC1*) Gly482Ser genotype distributions, in the same Spanish athlete cohort, were compared to a group of 100 unfit UK controls, with frequency analysis in 381 UK and 164 Spanish controls (37.5% and 36.9% respectively, $P = 0.83$). The authors found a significant difference between the two groups. The frequency of the minor Ser482 allele was significantly lower in the athletes than in the

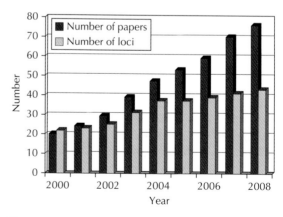

Figure 13.1 Cumulative publications and number of loci for endurance-related genetic papers in the last decade (publications prior to 2000 were summarized) (Rankinen et al., 2001).

unfit controls (29.1% vs. 40.0%, $P = 0.01$) (Lucia et al., 2005). The beta 2 adrenergic receptor (*ADRB2*) gene Arg16Gly polymorphism was investigated in a cohort of 300 elite endurance athletes compared to the same number of sedentary controls. The cohort showed an excess of Gly carriers in the sedentary group indicating a negative association of this allele with the endurance performance status (Wolfarth et al., 2007). In the same case–control cohort, an association was found between a specific allele of one microsatellite marker in the nitric oxide synthase (*NOS3*) gene and elite endurance status (Wolfarth et al., 2008). The *NOS3* locus showed significant association or linkage with endurance phenotypes in two additional studies including competition performance of triathletes and $\dot{V}O_2$max measures in a training study (Rico-Sanz et al., 2004; Saunders et al., 2006).

If we take into account the strengths and weaknesses of the case–control designs, there are two major issues to deal with in future studies: (i) sample size must be large enough so that statistical power does not become an issue and (ii) phenotypes must be reliably measured in both cases and controls.

Cross-sectional Association Studies

In 400 white subjects from the HERITAGE cohort, an association was seen between the rating of perceived exertion (RPE) in a submaximal-to-maximal exercise test and a C→T transition in the gene encoding for the AMP deaminase (*AMPD1*) (Rico-Sanz et al. 2003). The adrenergic receptor beta 2 (*ADRB2*) gene showed association with $\dot{V}O_2$max phenotypes in three different studies, including patients with heart failure (Wagoner et al., 2000) and postmenopausal women (McCole et al., 2004; Moore et al., 2001). These findings were replicated in part by a large case–control study of elite endurance athletes (Wolfarth et al. 2007) in which variations in this gene were investigated for their potential role in endurance traits in humans. Moreover, Saunders and his co-workers showed an association between a *NOS3* polymorphism and the performance in triathlon athletes (Saunders et al., 2006). Trends in results from case–control and linkage studies (Rico-Sanz et al., 2004; Wolfarth et al., 2008), although not from same polymorphisms but from markers in the same region, suggest that this chromosomal area is of high interest for further investigations. In an attempt to identify the optimum endurance polygenic profile, Ruiz and colleagues typed polymorphisms in six genes believed to be endurance related in a small cohort (46 cases, 123 controls). Using a so-called "total genotype score" they tried to capture an optimum endurance polygenetic profile. Although none of the athletes had the optimal profile, this may be a future research direction to explain polygenic endurance performance phenotypes (Ruiz et al., 2009).

Association Studies with Training Response Phenotypes

Several training intervention studies showed significant association to specific genes. In the HERITAGE Family Study, a creatine kinase (*CKMM*) polymorphism was associated with the training response of $\dot{V}O_2$max in white parents and offspring (Rivera et al. 1997). Similarly, significant associations between an ATPase, Na^+/K^+ transporting, alpha 2(+) polypeptide (*ATP1A2*) marker and both $\dot{V}O_2$max and maximal power output (W_{max}) were observed (Rankinen et al., 2000). Also in HERITAGE, the *AMPD1* C34T genotype showed associations for the changes measured after a 20-week controlled endurance training program with more than 400 white subjects (Rico-Sanz et al., 2003). From the same study, Hautala and colleagues (2007) found an association for changes in $\dot{V}O_2$max and W_{max} for a peroxisome proliferated activated receptor, delta (*PPARD*) polymorphism in African-American subjects. Two polymorphisms, located in the hemoglobin beta (*HBB*) gene on chromosome 11p15, were associated with running economy training response (comparing $\dot{V}O_2$max at constant speed versus incremental) in 102 Chinese army recruits following an 18-week training study (He et al., 2006).

In recent years, more and more studies on training response have been performed. Most of these studies lack adequate numbers of subjects. Again the HERITAGE Family Study with almost 750 subjects exposed to a sophisticated, rigorous, tightly monitored exercise regimen, including well-developed internal quality control, has been very helpful and should be emulated.

Linkage Studies

Thus far, all linkage analyses related to endurance performance phenotypes have come from the HERITAGE Family Study. The first linkage paper was published in 1997 and investigated a set of anonymous markers on chromosome 22 with no significance revealed for $\dot{V}O_2$max at baseline or its training response (Gagnon et al., 1997). The first, real genome-wide scan for $\dot{V}O_2$max measures from the HERITAGE cohort was published by Bouchard et al. (2000). They showed several suggestive linkages with markers on chromosomes 4, 8, 11, and 14 for baseline values and 1, 2, 4, 6, and 11 for the training response values. From these loci, the 11p15 signal could potentially overlap with the *HBB* polymorphism associated with running economy in a previously mentioned training response study (He et al., 2006). A second potential overlap involves the 14q21 locus in the region of the hypoxia-inducible factor 1 alpha (*HIF1A*) gene, located at 14q21-q24, which showed association with both baseline and training response $\dot{V}O_2$max values in a cohort of 125 older Caucasians and with lower $\dot{V}O_2$max in 91 African-American subjects before a training intervention (Prior et al., 2003). Subsequently, using the same genome-wide approach with more markers, several QTLs for maximal exercise capacity phenotypes and their respective response to the 20-week endurance training intervention were found in HERITAGE. Three markers located on the chromosome 7q32-26 QTL included the gene loci for leptin (*LEP*) and *NOS3* genes (Rico-Sanz et al., 2004). The latter was found to be associated with elite endurance athlete status in a case–control study (Wolfarth et al., 2008), as well as in an association study investigating competition performance in 443 triathlon athletes from South Africa (Saunders et al., 2006).

Genome-wide scans have now shifted to very large panels of SNPs without the need for family or pedigree data. The major challenge remains that of being able to assemble large cohorts of adequately phenotyped, informative individuals. In addition, there are many potential confounding variables, both known (such as sex, age) and unknown or unrecognized, that can complicate interpretations of genome-wide SNP panels (genome-wide association studies; GWAS). Hidden stratification, where unrecognized confounders can result in misleading data interpretations, is a particular challenge, such as complex ethnicities in cohorts of increasingly mixed heritage of study populations.

Future Research and Practical Implications

Investigation of genetic markers is an essential approach to exploring the basis of individual differences in human endurance performance and trainability. With the availability of new technologies and analytical methods as well as existing resources (Table 13.1) plus others to come, useful contributions on human endurance performance are likely to appear. Only through a combination of methods will we be able to identify a genetic profile associated with high- and low-endurance performance (Gomez-Gallego et al., 2009).

A practical implication of these findings is that talent detection based on genetic markers remains highly speculative. Moreover, several factors independent of genetic inheritance are involved in a superior endurance phenotype. However, the impact of genetic research in this field is likely to be significant as it should provide new insights into the physiology and pathways responsible for extraordinary endurance performance achievements. It can make major contributions to understanding the physiological basis for performance and training response patterns in humans. In addition, such findings will have practical implications in several associated fields including promising directions, such as the medical benefits of exercise in the context of preventive and rehabilitative medicine.

References

Amir, O., Amir, R., Yamin, C., et al. (2007). The ACE deletion allele is associated with Israeli elite endurance athletes. *Experimental Physiology* 92, 881–886.

Bouchard, C., Chagnon, M., Thibault, M.C., Boulay, M.R., Marcotte, M. & Simoneau, J.A. (1988). Absence of charge variants in human skeletal muscle enzymes of the glycolytic pathway. *Human Genetics* 78, 100.

Bouchard, C., Daw, E.W., Rice, T., et al. (1998). Familial resemblance for $\dot{V}O_2$max in the sedentary state: The HERITAGE Family Study. *Medicine and Science in Sports and Exercise* 30, 252–258.

Bouchard, C., Lesage, R., Lortie, G. et al. (1986). Aerobic performance in brothers, dizygotic and monozygotic twins. *Medicine and Science in Sports and Exercise* 18, 639–646.

Bouchard, C., Rankinen, T., Chagnon, Y.C., et al. (2000). Genomic scan for maximal oxygen uptake and its response to training in the HERITAGE Family Study. *Journal of Applied Physiology* 88, 551–559.

Bray, M.S., Hagberg, J.M., Perusse, L., et al. (2009). The human gene map for performance and health-related fitness phenotypes: The 2006–2007 update. *Medicine and Science in Sports and Exercise* 41, 35–73.

Chagnon, Y.C., Allard, C. & Bouchard, C. (1984). Red blood cell genetic variation in Olympic endurance athletes. *Journal of Sports Sciences* 2, 121–129.

Dionne, F.T., Turcotte, L., Thibault, M.C., Boulay, M.R., Skinner, J.S. & Bouchard, C. (1991). Mitochondrial DNA sequence polymorphism, $\dot{V}O_2$max, and response to endurance training. *Medicine and Science in Sports and Exercise* 23, 177–185.

Fagard, R., Bielen, E. & Amery, A. (1991). Heritability of aerobic power and anaerobic energy generation during exercise. *Journal of Applied Physiology* 70, 357–362.

Gagnon, J., Ho-Kim, M.A., Chagnon, Y.C., et al. (1997). Absence of linkage between $\dot{V}O_2$max and its response to training with markers spanning chromosome 22. *Medicine and Science in Sports and Exercise* 29, 1448–1453.

Gomez-Gallego, F., Santiago, C., Gonzalez-Freire, M., et al. (2009). Endurance performance: Genes or gene combinations? *International Journal of Sports Medicine* 30, 66–72.

Hautala, A.J., Leon, A.S., Skinner, J.S., Rao, D.C., Bouchard, C. & Rankinen, T. (2007). Peroxisome proliferator-activated receptor-delta polymorphisms are associated with physical performance and plasma lipids: The HERITAGE Family Study. *American Journal of Physiology—Heart and Circulatory Physiology* 292, H2498–H2505.

He, Z., Hu, Y., Feng, L., et al. (2006). Polymorphisms in the HBB gene relate to individual cardiorespiratory adaptation in response to endurance training. *British Journal of Sports Medicine* 40, 998–1002.

Hruskovicova, H., Dzurenkova, D., Selingerova, M., Bohus, B., Timkanicova, B., & Kovacs, L. (2006). The angiotensin converting enzyme I/D polymorphism in long distance runners. *Journal of Sports Medicine and Physical Fitness* 46, 509–513.

Lippi, G., Banfi, G., Favaloro, E.J., Rittweger, J. & Maffulli, N. (2008). Updates on improvement of human athletic performance: Focus on world records in athletics. *British Medical Bulletin* 87, 7–15.

Lippi, G., Longo, U.G. & Maffulli, N. (2010). Genetics and sports. *British Medical Bulletin* 93, 27–47.

Lortie, G., Simoneau, J.A., Hamel, P., Boulay, M.R. Landry, F. & Bouchard, C. (1984). Responses of maximal aerobic power and capacity to aerobic training. *International Journal of Sports Medicine* 5, 232–236.

Lucia, A., Gomez-Gallego, F., Barroso, I., et al. (2005). *PPARGC1A* genotype (Gly482Ser) predicts exceptional endurance capacity in European men. *Journal of Applied Physiology* 99, 344–348.

Maes, H.H., Beunen, G.P., Vlietinck, R.F. et al. (1996). Inheritance of physical fitness in 10-yr-old twins and their parents. *Medicine and Science in Sports and Exercise* 28, 1479–1491.

McCole, S.D., Shuldiner, A.R., Brown, M.D., et al. (2004). Beta2- and beta3-adrenergic receptor polymorphisms and exercise hemodynamics in postmenopausal women. *Journal of Applied Physiology* 96, 526–530.

Montgomery, H.E., Marshall, R., Hemingway, H., et al. (1998). Human gene for physical performance. *Nature* 393, 221–222.

Moore, G.E., Shuldiner, A.R., Zmuda, J.M., Ferrell, R.E., McCole, S.D. & Hagberg, J.M. (2001). Obesity gene variant and elite endurance performance. *Metabolism: Clinical and Experimental* 50, 1391–1392.

Prior, S.J., Hagberg, J.M., Phares, D.A., et al. (2003). Sequence variation in hypoxia-inducible factor 1alpha (HIF1A): Association with maximal oxygen consumption. *Physiological Genomics* 15, 20–26.

Rankinen, T., Perusse, L., Borecki, I., et al. (2000). The Na(+)-K(+)-ATPase alpha2 gene and trainability of cardiorespiratory endurance: The HERITAGE Family Study. *Journal of Applied Physiology* 88, 346–351.

Rankinen, T., Perusse, L., Rauramaa, R., Rivera, M.A., Wolfarth, B. & Bouchard, C. (2001). The human gene map for performance and health-related fitness phenotypes. *Medicine and Science in Sports and Exercise* 33, 855–867.

Rico-Sanz, J., Rankinen, T., Joanisse, D.R., et al. (2003). Associations between cardiorespiratory responses to exercise and the C34T *AMPD1* gene polymorphism in the HERITAGE Family Study. *Physiological Genomics* 14, 161–166.

Rico-Sanz, J., Rankinen, T., Rice, T., et al. (2004). Quantitative trait loci for maximal exercise capacity phenotypes and their responses to training in the HERITAGE Family Study. *Physiological Genomics* 16, 256–260.

Rivera, M.A., Dionne, F.T., Simoneau, J.A., et al. (1997). Muscle-specific creatine kinase gene polymorphism and $\dot{V}O_2$max in the HERITAGE Family Study. *Medicine and Science in Sports and Exercise* 29, 1311–1317.

Rivera, M.A., Wolfarth, B., Dionne, F.T., et al. (1998). Three mitochondrial DNA restriction polymorphisms in elite endurance athletes and sedentary controls. *Medicine and Science in Sports and Exercise* 30, 687–690.

Rubio, J.C., Martin, M.A., Rabadan, M., et al. (2005). Frequency of the C34T mutation of the *AMPD1* gene in world-class endurance athletes: Does this mutation impair performance? *Journal of Applied Physiology* 98, 2108–2112.

Ruiz, J.R., Gomez-Gallego, F., Santiago, C., et al. (2009). Is there an optimum endurance polygenic profile? *Journal of Physiology* 587, 1527–1534.

Saunders, C.J., Xenophontos, S.L., Cariolou, M.A., Anastassiades, L.C., Noakes, T.D. & Collins, M. (2006). The bradykinin beta 2 receptor (*BDKRB2*) and endothelial nitric oxide synthase 3 (*NOS3*) genes and endurance performance during Ironman Triathlons. *Human Molecular Genetics* 15, 979–987.

Van Damme, R., Wilson, R.S., Vanhooydonck, B. & Aerts, P. (2002). Performance constraints in decathletes. *Nature* 415, 755–756.

Wagoner, L.E., Craft, L.L., Singh, B., et al. (2000). Polymorphisms of the beta(2)-adrenergic receptor determine exercise capacity in patients with heart failure. *Circulation Research* 86, 834–840.

Wolfarth, B., Rankinen, T., Muhlbauer, S., et al. (2008). Endothelial nitric oxide synthase gene polymorphism and elite endurance athlete status: The Genathlete study. *Scandinavian Journal of Medicine and Science in Sports* 18, 485–490.

Wolfarth, B., Rankinen, T., Muhlbauer, S., et al. (2007). Association between a beta2-adrenergic receptor polymorphism and elite endurance performance. *Metabolism: Clinical and Experimental* 56, 1649–1651.

Chapter 14

Genes and Strength and Power Phenotypes

MARTINE A.I. THOMIS

Faculty of Kinesiology and Rehabilitation Sciences, Katholieke Universiteit Leuven, and Department of Biomedical Kinesiology, Research Centre for Exercise and Health, Leuven, Belgium

Gene variants (polymorphisms) causing interindividual variation in maximal isometric, dynamic strength, and power have been studied through family studies (linkage analyses), candidate-gene studies, and, most recently, genome-wide single-nucleotide polymorphism (SNP) association studies (GWAS). To date, there has been relatively little concordance between these three types of studies, and lack of replication may be due to the relatively small effect sizes of individual gene polymorphisms, differences in the study populations (age, sex, and ethnicity), and/or differences in phenotyping (baseline vs. response to intervention, or method to measure strength). To define the genetics of strength and response to strength training, longitudinal intervention studies are needed on large cohorts (>300 individuals). Genotyping of 1 million or more SNPs in such cohorts will likely define key gene polymorphisms contributing to baseline strength and response to training. Until such large studies are accomplished, the development and use of any potential gene profile for elite strength performance is highly speculative, if not misleading. Individual gene variants contribute only about 1–2% to the overall variation in strength phenotypes, with the possible exception of a variant in *IGF2* explaining as much as 10% of the observed variance. Only a limited number of studies have investigated gene–gene interaction effects.

Low-allele frequencies often prevent the study of these interactions in smaller study samples.

Great efforts have been made to review genetic linkage and association reports related to strength and power in a series of MSSE special reports "The Human Gene Map for Performance and Health-Related Fitness Phenotypes" (Bray et al., 2009; Perusse et al., 2003; Rankinen et al., 2001, 2002, 2004, 2006; Wolfarth et al., 2005). In this chapter, hypothesis-free linkage studies will be discussed separately from candidate-gene, hypothesis-driven studies. In this last approach, candidate genes will be grouped based on their hypothetical function in the muscle or broader system. Owing to space limitation, references to original studies will be limited to recent publications, and numbered references are cross-references to the reference number in Bray et al., 2009.

Linkage Studies

Only two research groups have published results on gene-pathway or whole genome linkage studies related to muscular strength phenotypes. The earliest reports (112, 113) were based on a subset of the Leuven Genes for Muscular Strength study, a sibling study in young adult males (17–36 years), and focused on microsatellite markers selected in the vicinity of 10 candidate genes within the myostatin pathway. Single-point linkage analysis by Huygens et al. (112, 113) was followed up by multipoint linkage analysis (114), and several genomic regions were identified with suggestive or significant linkage with knee extension and flexion torques (Table 14.1).

Genetic and Molecular Aspects of Sport Performance, 1st edition. Edited by Claude Bouchard and Eric P. Hoffman. Published 2011 by Blackwell Publishing Ltd.

Table 14.1 Summary[a] of linkage findings (LOD > 2.0) in strength and power characteristics

Chromosome	Region (cM)	Marker	Study	LOD-score (P-value)	Phenotype(s)
1	1q21.3(145.69)	rs13320	LGfMS-TG	2.326 (<0.0001)	Knee torque-length extension
2	2p24.2(41.12)	rs1445128	LGfMS-TG	2.57 (0.0097)	Knee torque 0.524 rad flexion
	2p23.3(50.2)	rs714513	LGfMS-TG	2.69 (<0.0001)	Knee torque-length extension
	2q14.3(139.88)	rs477449	LGfMS-TG	2.25 (<0.0001)	Knee slope extension
	2q32.2	D2S118 (GDF8)	LGfMS-1	2.63 (0.0002)	Knee extension (1.047 rad/s)
4	4p14(58.04)	rs1039559	LGfMS-TG	2.23 (0.0001)	Knee ratio extension
6	6p25.2(7.27)	rs9328112	LGfMS-TG	2.079 (<0.001)	Knee torque-length ext
7	7p12.3(70.13)	rs921630	LGfMS-TG	2.03 (0.0006)	Knee ratio flexion
8	(128–129)		FITSA	2.09	Lower leg MCSA
	(195)		FITSA	2.84	Leg extension power
9	9q21.32(81.01)	rs7845911	LGfMS-TG	2.00 (0.0003)	Knee torque-length flexion
	(173–174)		FITSA	2.65	Lower leg MCSA
10	10q26.13(153.12)	rs4962424	LGfMS-TG	2.524	Knee torque-length flexion
12	12q12–14	D12S1042/ D12S85	LGfMS-M	2.7–3.6 (0.00002)	Knee torque/work flexion (2.094 rad/s); work extension (1.047 rad/s)
	12q22–23	D12S78	LGfMS-M	2.6	Knee isometric flexion
13	13q14.2(45.55)	D13S153	LGfMS-1	2.78 (0.0002)	Trunk flexion (1.047 rad/s)
		D13S153– D13S1303	LGfMS-M	2.00–2.74 (0.0002)	Knee torque/work extension (1.047 rad/s); knee torque flexion (2.094 rad/s)
	(70)		FITSA	2.41	Maximal walking speed
14	14q24.3(76.68)	rs760267	LGfMS-TG	4.088 (<0.00001)	Knee torque-length flexion
	14q32.2(109.13)	rs926949	LGfMS-TG	2.998 (<0.005)	Mid-thigh MBA
15	(34)		FITSA	2.14	Isometric knee extension
	15q23(76.79)	rs1348318	LGfMS-TG	2.91 (<0.00001)	Knee slope extension
	15q32.2(99.82)	rs7175643	LGfMS-TG	2.05 (<0.01)	Mid-thigh MBA
18	18p11.31(21.53)	rs1941487	LGfMS-TG	2.321	Knee extension (1.047 rad/s)
	18q11.2(43.66)	rs1010800	LGfMS-TG	2.20 (<0.0001)	Knee torque-length extension
	18q23(118.79)	rs1866338	LGfMS-TG	2.08 (<0.001)	Knee ratio extension
20	(91)		FITSA	2.93	Lower leg MCSA

[a] QTLs selected based on LOD-value > 2.0 (only peak of a QTL region is reported). LGfMS-1: single-point microsatellite linkage analysis in the Leuven Genes for Muscular Strength Study (112; 113); LGfMS-M: multi-point microsatellite linkage analysis in LGfMS (114); LGfMS-TG: total genome SNP-panel linkage analysis in LGfMS (De Mars et al. 2008a, b); FITSA: total genome microsatellite linkage analysis in Finnish Twin Study on Aging (Tiainen et al., 2008). MBA: muscle+bone area, MCSA: muscle cross-sectional area.

Recently, the same research group published the results of a whole genome scan, based on a 6000 SNP-linkage panel with isometric and torque-length characteristics (De Mars et al., 2008b) and quantification of the torque-velocity relationship (De Mars et al., 2008a) in knee extension/flexion. Tiainen et al. (2008) reported on a satellite-marker-based total genome scan of walking speed, isometric knee strength, and leg power in older dizygotic twin pairs. An overview of the genomic regions identified as suggestive QTLs with LOD scores > 2.0 is given in Table 14.1.

To date, there is only limited overlap in chromosomal locations of different strength phenotypes as QTLs are located over the whole genome. Several candidate genes are encoded within these chromosomal regions. However, fine-mapping of QTL regions for muscular strength has been reported only for the 12q12-14 and 12q22-23 regions (Windelinckx et al., 2007).

Candidate Genes

Candidate genes for endurance performance can also be at times candidates for strength or power-related phenotypes, particularly under the assumption that endurance-decreasing alleles could be related to the strength phenotypes. Case–control studies have compared elite power athletes to control individuals or elite endurance athletes. Athlete status and type of discipline varies widely. For instance, power is related to very short explosive performance (e.g., power lifting) up to anaerobic energy demanding sprints up to 1000 m sprint skating. Population-based association studies also vary widely in studied populations: age, gender, or sample size; with a limited number of studies also investigating the association of specific alleles with response to strength training. Table 14.2 presents a summary table for case–control and population–based association studies, with candidate genes categorized by structural or physiological function.

Contractile and Structural Components of Sarcomere/Muscle

Within contractile and structural components of the sarcomere or skeletal muscle as a whole, the alpha-actinin 3 (*ACTN3*) R577X stop codon polymorphism has been most extensively studied since the first report by Yang et al. (350). The XX homozygotes lack the ACTN3 protein, a major actin-anchoring component of fast muscle Z-line. *ACTN3* 577 XX homozygotes were absent in a group of elite Australian sprinters. Other studies have attempted to replicate this observation with mixed results (198; 351; Druzhevskaya et al., 2008; Irving et al. 2009; Roth et al., 2008).

While ACTN3 deficiency seems to be detrimental for elite sprint/strength performance, other studies have not found associations or have only weak correlations with strength (184; 322; Norman et al., 2009; Walsh et al., 2008), contractile properties (McCauley et al., 2009; Norman et al., 2009), or anaerobic performance test (Norman et al., 2009). Two muscle biopsy studies in *ACTN3* 577XX versus RR homozygotes could not replicate each other's findings on fiber-type distribution (322) (Norman et al., 2009).

The *ACTN3* R577X genotype-training effect interaction was studied in response to a 12-week strength-training program in the FAMuSS study (37). Arm flexion one-repetition maximum gains were greatest in women homozygous for the X-allele compared with the R-allele homozygotes, with no associations in men. This might be partially related to a lower baseline isometric strength in these XX women. In an older sample exposed to a 10-week program of unilateral extensor knee training, opposite effects were found (56): female XX homozygotes exhibited greater baseline relative peak power (at 70% of one-repetition maximum [1RM]) than both RX and RR genotypes, with a lower change in relative peak power in response to the training. In men, no genotype differences were observed at baseline, but the change in absolute peak power in response to training tended to be higher in RR compared with XX genotypes. No evidence for a role of ACTN3 in the prevention of markers of muscle damage by eccentric exercise was found by Clarkson et al. (38). However, two polymorphisms in the myosin light chain kinase *MYLK* gene, which predominantly acts in type 2 fibers by phosphorylating myosin's regulatory light chain with functional effects on force development,

Table 14.2 Summary of case–control (CC), cross-sectional (pop), and genotype–training interaction studies (RT) in candidate genes for muscular strength and power phenotypes

Gene & variant	Sample type	Gene variant/ phenotype/result	P	References[a]
Sarcomeric structure-whole muscle				
Alpha-actinin-3 (*ACTN3*)				
R577X	107 sprinters; 436 controls (Australia)	R-allele frequency in sprinters 0.72, controls 0.23	$P < 0.001$	350
	23 top sprinters, 68 sprinters; 120 controls (Finland)	No XX homozygotes in top sprinters, inverse corr. between X-allele frequency and sprint performance	$P = 0.03$	198
	62 power athletes; 62 controls (Nigeria)	No XX genotypes in cases or controls	ns	351
	486 power athletes; 1197 controls (Russian)	X-allele underrepresented in power athletes (6.4% against 14.2%. Higher elite status, lower X-allele frequency	$P < 0.0001$	Druzhevskaya et al. (2008)
	75 power athletes (52 White, 23 Black); 876 controls (668 Whites, 208 Blacks) (US)	X-allele underrepresented in power athletes (all: 6.7% against 16.3%). No XX genotypes in black power athletes	$P = 0.005$	Roth et al. (2008)
	114 elite sprinter; 311 controls (Jamaica), 114 elite sprinters; 196 controls (African-American, US)	No difference in X-allele frequency in elites versus controls (X = 3% vs. 2% (JAM), 2% vs. 4% (AA, US)	ns	Irving et al. (2009)
	992 Greek adolescents	40-m sprint in males: RR homozygotes run 0.2 s faster (2.3% explained variance), other performance Phenotypes or endurance	$P = 0.003$ ns	184
	90 young males (43 with biopsies)	RR > RX > XX relative knee extensor torque XX have lower IIx fiber proportion and area	$P = 0.05$ $P = 0.04$–.03	322
	79 young males	Dynamic knee extension torque up to 4.19 rad/s Contractile properties (time to peak tension, half-relaxation time, peak rate of force development)	ns ns	McCauley et al. (2009)

Table 14.2 (Continued)

Gene & variant	Sample type	Gene variant/phenotype/result	P	References[a]
	848 men and women (22–90 years) BLSA	Women: XX < RX/RR: conc. knee extension 0.524 rad/s, 3.14 rad/s ecc.	$P = 0.011; 0.019$	Walsh et al. (2008)
		knee extension −0.524 rad/s, −3.14 rad/s	$P = 0.022; 0.019$	
		Lower total body fat-free mass	$P = 0.04$	
		lower limb fat-free mass	$P = 0.044$	
		Men:	ns	
	120 young adults; 21 men with biopsies	30-s Wingate power/fatigue index	ns	Norman et al. (2009)
		torque-velocity relationship	ns	
		Fiber type	ns	
		mRNA ACTN2 > ACTN3; ACTN2 mRNA XX > RR (%IIb fiber corrected)	$P = 0.037$	
	602 men and women FAMuSS; (18–40 years)(RT)	Women: baseline 1RM XX > RR	$P < 0.01$	37
		delta 1RM XX > RR	$P = 0.05–0.01$	
		Men:	ns	
	157 young adult men and women (acute eccentric contractions)	Baseline: XR with highest CK levels (3.5% var.)	$P = 0.035$	38
		delta phenotypes:	ns	
	157 older men and women (RT)	Women: baseline XX greater abs. & rel. peak power delta rel.	$P = 0.002–0.02$	56
		peak power XX < RR	$P = 0.02$	
		Men: baseline	ns	
		delta RR >XX	$P = 0.07$	
	1367 70–79 year (long.) (Health, Aging and Body Composition Study Cohort) and	Men: greater increase in 400 m walk time for XX over 5-year follow-up	$P = 0.03$	Delmonico et al. (2008)
		Women: XX 35% increased risk of incident persistent lower extremity limitation (hazard ratio = 0.65)	$P < 0.05$	
	1152 > 65 year (Osteoporotic Fractures in Men Study)	Baseline and 4-year follow-up	ns	
Myosin light chain kinase (*MLCK*)				
C49T	157 young adult men and women (acute eccentric contractions)	delta CK and Mb: TT > CT	$P < 0.01$	38
C3788A > CC	157 young adult men and women (acute eccentric contractions)	delta CK, delta strength loss: CA > CC	$P < 0.05$	38

(Continued)

Table 14.2 (Continued)

Gene & variant	Sample type	Gene variant/ phenotype/result	P	References[a]
Collagen, type I, alpha I (*COL1A1*)				
SpI(s<S)	273 Community-dwelling elderly men (>70 year)	Hand grip (21% difference ss < SS)	$P = 0.03$	318
		biceps strength (30% difference ss < SS)	$P = 0.04$	
Differentiation/growth factors; anabolic/catabolic factors				
Bone morphogenetic protein 2 (*BMP2*)				
rs15705 A/C	517 FAMuSS (RT)	Men: delta whole muscle volume: CC > A-carriers, 3.9% expl. var.	$P = 0.006$	Devaney et al. (2009)
Myostatin (*GDF8*)				
K153R	286 elderly women of which 55 African-American(70–79 years) WHAS II	Combined isometric hip, thigh and grip scores KK > RK > RR	$P = 0.04$	275
	450 older men and women	Non-age adjusted hand grip: R-allele carriers ($n = 6$) < KK ($n = 444$)		Corsi et al. (2002)
K153R 163 G > A	96 Caucasian, 93 African-American (RT)	delta muscle mass after 6 months RT	ns	Ferrell et al. (1999)
K153R	23 younger men and women 24 older men and women (RT)	delta strength after 9 weeks of RT	ns	117
2379 A > G 163 G > A	645 men and women from FAMuSS (18–40 years) (RT)	In African-Americans: Baseline MVC: 2379 AG/GG > AA.	$P < 0.05$	Kostek et al. (2009)
		delta 1RM, MVC or CSA	ns	
Follistatin (*FST*)				
-5003 A > T -833 G > T	645 men and women from FAMuSS (18–40 years) (RT)	In African-Americans: Baseline 1RM and CSA: TA/TT > AA	$P < 0.05$	Kostek et al. (2009)
		delta 1RM, MVC or CSA	ns	
Haplotype groups	315 men, women from BLSA (19–90 years)	Men: haplotype associated with leg fat-free mass, not with strength	$P = 0.012$	332
		Women:	ns	
Activin-type II receptor B (*ACVR2B*)				
Haplotype groups	315 men, women from BLSA (19–90 years)	Women: Hap Group 1 carriers less concentric quadriceps muscle strength than homozygous for Hap Group 2	$P = 0.036$	332
		Men:	ns	

Table 14.2 (Continued)

Gene & variant	Sample type	Gene variant/ phenotype/result	P	References[a]
Insulin-like growth factor 1 (*IGF1*)				
Promotor CA repeat (-192bp)	32 men, 35 women Caucasian (52–83 years)(RT)	Delta knee extensor torque: -192bp carriers > non-carriers	$P < 0.02$	140
		delta muscle volume: -192bp carriers > non-carriers	$P = 0.08$	
rs19779/rs40395/rs82681	32 men, 35 women Caucasian (52–83 years) (RT)		ns	140
-192bp	128 White and Black men and women (50–85 years)(RT)	Delta strength, Muscle Quality: -192bp main effect	$P < 0.01$	95
Calcineurin B (*PPP3R1*)				
Promoter I/D	128 White and Black men and women (50–85 years)(RT)	Delta strength/muscle quality:IGF1 -192bp x PPP3R1 (I/D) interaction	$P < 0.05$	95
Insulin-like growth factor binding protein 3 (*IGFBP3*)				
A -202C	128 White and Black men and women (50–85 years)(RT)	Baseline & delta strength or muscle quality	ns	95
A8618T-1592 A > C	151 subjects (78 women and 73 men) (acute eccentric muscle damage)	Men: baseline CK: −1592 AA > CC (14.3% expl. var.)	$P = 0.003$	59
		delta muscle damage	ns	
		Women:	ns	
Deiodinase, iodothyronine, type I (*DIO1*)				
D1a-T/C	350 predominantly Caucasian men (>70 year)	Isometric grip strength: D1a-T carr. >D1a-C hom.	$P = 0.047$	213
		maximum leg-extensor strength;	$P = 0.07$	
		lean body mass:	$P = 0.03$	
Insulin-like growth factor 2 (*IGF2*)				
Apal (G17200A)	397 men 296 women	Handgrip: GG < GA < AA (1% expl. var.)	$P = 0.05$	268
	2 cohorts BLSA; 94 men, 246 men and 239 women with follow-up	Women: total body FFM, isokinetic arm and leg strength (first visit and age 35): AA < GG	$P < 0.05$ $P < 0.05$	272
		Differences maintained at 65 year and across adult age span	$P < 0.05$	
		Rates of loss of grip strength over time	ns	
		Men: first visit isokinetic arm strength: AA < GG	$P < 0.05$	

(*Continued*)

Table 14.2 (Continued)

Gene & variant	Sample type	Gene variant/ phenotype/result	P	References[a]
	151 subjects (78 women and 73 men) (acute eccentric muscle damage)	Men: immediate strength loss: AA > GG (9.9% expl. var.)	$P = 0.019$	59
		soreness (4 days postexercise): (8.6%)	$P = 0.035$	
		CK (7 days postexercise): (15.5%)	$P = 0.002$	
		Women:	ns	
G12655A	151 subjects (78 women and 73 men) (acute eccentric muscle damage)	Men and women: delta muscle damage	ns	59
C16646T	151 subjects (78 women and 73 men) (acute eccentric muscle damage)	In men: soreness (4 days postexercise): CC > CT (5.5% expl. var.)	$P = 0.038$	59
C13790G	151 subjects (78 women and 73 men) (acute eccentric muscle damage)	Men: immediate strength loss: GG > CC (10.1% expl. var.)	$P = 0.017$	59
		soreness (3 days postexercise): (9.2%)	$P = 0.03$	
		CK (7 days postexercise): (9.3%)	$P = 0.025$	
		Women:	ns	
T13705C	151 subjects (78 women and 73 men) (acute eccentric muscle damage)	Women: baseline strength: TC($n = 1$) > TT($n = 77$)(6.2%)	$P = 0.01$	59
		Mb (4 days postexercise): (5.3%)	$P = 0.038$ $P = 0.018$	
		CK (4 days postexercise): (6.7%)		
		Men:	ns	
Insulin-like growth factor 2 (anti-sense) (*IGF2AS*)				
A1364C	151 subjects (78 women and 73 men) (acute eccentric muscle damage)	Women and men: immediate strength loss: AA > CC	$P = 0.014–0.025$	59
		Women: soreness (4 days postexercise): (16.4% expl. var.)	$P = 0.001$	
G11711T	151 subjects (78 women and 73 men) (acute eccentric muscle damage)	Men: soreness (3 days, 4 days postexercise): 11711 TT > TG (12.6–16.6% exp. var.)	$P = 0.007–0.001$	59
		strength loss (immediate, 4 days, 7 days postexercise): (9.5–16.8%)	$P = 0.02– <0.001$	
		CK (4 days, 7 days postexercise), Mb (4 days postexercise): (9.6–10.5%)	$P = 0.024–0.018$	

Table 14.2 (Continued)

Gene & variant	Sample type	Gene variant/ phenotype/result	P	References[a]
Insulin (*INS*)	151 subjects (78 women and 73 men) (acute eccentric muscle damage)			
C1045G		Women: CK (10 days postexercise):GG > CC (9.6% expl. var.)	$P = 0.017$	59
		Men:	ns	
Cytokines/inflammatory-catabolic action				
Tumor necrosis factor alpha (*TNFα*)				
-308G/A	214 > 60 years with knee osteoarthritis (18m walking, RT)	baseline walking distance: GG > A-carriers stair-climb time delta 6m stair-climb time: A-carriers > GG	$P = 0.02$ $P = 0.003$ $P = 0.007$	197
	509 women, 541 men from BLSA	Arm/appendicular skel. mass, leg muscle mass, ASM index	ns	Liu et al. (2008)
-1031T/C	509 women, 541 men from BLSA	arm/appendicular skel. mass, leg muscle mass, ASM index: C-carriers < TT	$P = 0.07–0.01$	Liu et al. (2008)
-863C/A	509 women, 541 men from BLSA	arm muscle mass: A carriers < CC	$P = 0.04$	Liu et al. (2008)
Haplotype	509 women, 541 men from BLSA	arm and trunk, appendicular muscle mass, ASM index	$P = 0.08–0.01$	Liu et al. (2008)
Tumor necrosis factor alpha receptor 2 (*TNFR2*)				
T676G	214 > 60 years with knee osteoarthritis (18m walking, RT)	baseline walking distance, stair-climb time: TT > G-carriers	$P = 0.02–0.06$	197
Interleukin-6 (*IL-6*)				
-174G > C	214 > 60 years with knee osteoarthritis (18m walking, RT)	baseline walking distance: GG > C-carrier	$P < 0.05$	197
	222 males, 304 females InChianti (65 ± 15 years)	Hand grip; knee extension power, hip strength (+ 11 other SNPs)	ns	Barbieri et al. (2003)
-174G > C +11 SNPs (haplotypes)	436 WHAS I&II; 266 InChianti women (70–79 years)	Grip, knee strength	ns	Walston et al. (2005)

(Continued)

Table 14.2 (Continued)

Gene & variant	Sample type	Gene variant/phenotype/result	P	References[a]
Interleukin-15 (*IL-15*)				
C7336T	748 (448 females) of FAMuSS (RT)	Baseline & response to RT	ns	Pistilli et al. (2008)
A96679T	748 (448 females) of FAMuSS (RT)	Males: delta 1 RM T-carriers > AA	$P = 0.02$	Pistilli et al. (2008)
Interleukin-15 receptor alpha (*IL15-Rα*)				
C25449A	748 (448 females) of FAMuSS (RT)	Females: Delta isometric strength: CC > A-carriers	$P = 0.014$	Pistilli et al. (2008)
	76 men and 77 women (whole body RT)	Delta lean mass, arm & leg circumference: AA > AC > CC	$P < 0.05$	249
		delta arm & leg muscle quality: CC > AC > AA additional effect of rs3136617 on hypertrophy (3.5%)	$P < 0.05$	
A1755C	748 (448 females) of FAMuSS (RT)	Males: Baseline muscle volume:AA > C-carriers muscle quality: AA < C-carriers	$P = 0.04$	Pistilli et al. (2008)
G18447A	748 (448 females) of FAMuSS (RT)	Baseline & response to RT	ns	Pistilli et al. (2008)
	76 men and 77 women (whole body RT)	delta lean mass, arm & leg circumference delta arm & leg muscle quality	ns	249
Neurotrophic factors				
Ciliary neurotrophic factor (*CNTF*)				
G1357A	363 women WHAS I&II (70–79 years)	Hand grip: GG/GA > AA (1.5% expl. var.) – haplotype analysis with 7 other SNPs point to 1357 G > A (rs1800169) as causal SNP	$P = 0.006$	13
	494 healthy men and women (20–90 years) BLSA	Conc. knee strength, muscle quality: GG < GA Ecc. knee strength: GG = GA > AA (age, gender, FFM corrected)	$P < 0.05$	257
	138 females (38–44 years), 102 (60–80 years) 154 males (45–49 years), 99 (60–78 years)	Females: middle-aged KF 3.14 rad/s: GG > AA older female group Isom KF: GG = AA > GA Men:	$P = 0.04$ $P = 0.02$ ns	49
	754 ($n = 452$ females) from FAMuSS (RT)	delta MVC in women: GG > A-carriers delta 1 RM in women: GG > A-carriers in untrained arm (delta) CSA	$P < 0.05$ $P < 0.05$ ns	Walsh et al. (2009)

Table 14.2 (Continued)

Gene & variant	Sample type	Gene variant/ phenotype/result	P	References[a]
Ciliary neurotrophic factor receptor (*CNTFR*)				
C174T	465 men and women (20–90 years) from BLSA	lower limb FFM; KE concentric and eccentric peak torque at 0.524 rad/ sand 3.14 rad/s: CC < T-carriers (association disappears when corrected for lower limb FFM)	$P = 0.011$ $P = 0.002$	256
	138 females (38–44 years), 102 (60–80 years) 154 males (45–49 years), 99 (60–78 years)	isometric & dynamic knee strength measures (conc & eccentric)(1–4% expl. var.):CC > T-carriers	$P = 0.013$– 0.0065	49
C-1703T	465 men and women (20–90 years) from BLSA	isom & dynamic strength measures, FFM	ns	256
	138 females (38–44 years), 102 (60–80 years) 154 males (45–49 years), 99 (60–78 years)	male (older) isom KE/ KF: T-carriers > CC	$P = 0.03$–0.003	49
T1069A	465 men and women (20–90 years) from BLSA	isom & dynamic strength measures, FFM	ns	256
	138 females (38–44 years), 102 (60–80 years) 154 males (45–49 years), 99 (60–78 years)	Female (middle) conc KF at 3.14 rad/s/isom: TT > A-carriers	$P = 0.04$–0.01	49
Vitamin-related				
Vitamin D receptor (*VDR*)				
BsmI(B/b)	501 women > 70 years	Quadriceps strength: BB < bb grip strength	$P = 0.04$ ns	82
	313 postmeno-pausal women > 70 years	Quadriceps strength: BB < bb grip strength	$P = 0.01$ ns	Vandevyver et al. (1999)
	297 Caucasian men (58–93 years)	Quadriceps strength	ns	258
	175 healthy Swedish women (20–39 years)	Hamstrings strength: BB > bb Grip strength, quadriceps strength	$P = 0.03$ ns	86
	109 young Han Chinese women	Knee Flexion at 3.14 rad/s: BB + Bb > bb	$P = 0.03$	334

(Continued)

Table 14.2 (Continued)

Gene & variant	Sample type	Gene variant/ phenotype/result	P	References[a]
	107 stable COPD patients 104 healthy age-matched controls	Patients: quadriceps maximal voluntary contraction force: bb >bB + BB controls:	$P = 0.0005$ ns	Hopkinson et al. (2008)
	Osteoporosis and Ultrasound cohort: max 2393 women (67 ± 7 years)	Chair rise test, maximal leg power, maximal leg force, grip strength: bb > Bb + BB	$P = 0.098$– $P < 0.03$	Barr et al. (2009)
ApaI(A/a)	109 young Han Chinese women	2.094 rad/s KE ecc: AA < Aa and aa 0.524, 2.094 rad/s EF conc: AA < aa	$P < 0.004$ $P = 0.03–0.04$	334
BsmI/TaqI haplotype	253 men (54.9 ± 10.2 years) 240 women (41.5 ± 13.2 years)	Men:isometric KE, concentric KE: Bt/Bt > bT/bT + Bt/bT Women	$P = 0.02–0.05$ ns	342
ApaI/TaqI haplotype	271 men > 70 years 137 control men (20–50 years)	Grip strength, isometric biceps strength, lower limb strength	ns	Van Pottelbergh et al. (2002)
FokI (F/f)	271 men > 70 years 137 control men (20–50 years)	Grip strength, isometric biceps strength, lower limb strength control: Lower limb strength: FF > Ff + ff	ns $P = 0.01$	Van Pottelbergh et al. (2002)
	302 Caucasian men (58–93 years)	Quadriceps strength: FF < ff; ns after adjustment for appendicular muscle mass Grip strength	$P = 0.03$ ns	258
	253 men (55 ± 10 years)	Men: Handgrip, KE/KF isometric, eccentric and concentric	ns	342
	240 women (41.5 ± 13 years)	Women: isom (at 1.571 rad) and concentric quadriceps strength ff > Ff + FF	$P < 0.05$	
	107 stable COPD patients 104 controls	total sample: Knee Extension: CC < CT + TT control sample: Quadr. volunt. max. contr.: CC < CT + TT patients:	$P = 0.01$ $P = 0.052$ $P = 0.02$	Hopkinson et al. (2008)
	Osteoporosis and Ultrasound cohort: max 2393 women (67 ± 7 years)	Chair rise test, maximal leg power, maximal leg force, grip strength	ns	Barr et al. (2009)
Hormone(receptors) and metabolites				
Estrogen receptor 1 (*ESR1*)				
TA repeat e:<16;E:17>	175 women (20–39 years)	Body composition or quadriceps, hamstrings or grip strength	ns	Grundberg et al. (2005)

Table 14.2 (Continued)

Gene & variant	Sample type	Gene variant/ phenotype/result	P	References[a]
PvuII	313 postmeno-pausal women (76 ± 5 years)	Grip, quadriceps strength	ns	Vandevyver et al. (1999)
	434 female twins from FITSA (63–76 years)	mCSA, knee extension, leg power, grip strength strength	ns	Ronkainen et al. (2008)
Catechol-O-methyltransferase (*COMT*)				
Val158Met	434 female twins from FITSA (63–76 years)	Lower leg muscle cross-sectional area: LL > HL strength	$P < 0.004$ ns	Ronkainen et al. (2008)
Glucocorticoid receptor (*NR3C1*)				
ER22/23EK > non-carriers	286–337 men and women from the Amsterdam Growth and Health Longi-tudinal Study (13 years up to 36 years)	Men: total lean mass, thigh circumference, arm pull, vertical jump: ER22/23EK > non-carriers women:	$P < 0.04$ ns	319
Energy metabolism (anaerobic)				
Adenosine monophosphate deaminase 1 (*AMPD1*)				
c.34C > T (TT = deficient < TC;CC)	$n = 7$ AMPD1 deficient, $n = 7$ controls (fatiguing protocol)	adductor pollicis muscle function: delta strength of fatigued muscle	ns	De Ruiter et al. (2000)
	$n = 8$ AMPD1 deficient, $n = 8$ controls (fatiguing protocol)	delta strength of fatigued quadriceps muscle; 5-min half-relaxation time: TT > Controls	$P = 0.002$	De Ruiter et al. (2002)
	$n = 7$ deficient, 7 CC, 3 CT	Muscle biopsy after 30-s Wingate: TT do not deplete ATP, no accumulation of IMP, Power output	ns	Norman et al. (2001)
	$n = 3$ deficient, $n = 4$ heterozygote $n = 12$ control	Muscle biopsy after graded cycle test: deficiency is not associated with impairment of TCAi anaplerosis, exercise capacity, phosphocreatine hydrolysis or cellular energy charge	ns	Tarnopolsky et al. (2001)
	139 healthy young adult	30-s Wingate test: TT (deficient): lower mean power , higher 15-s, 30-s FI: CC = CT > TT	$P = 0.0035$– 0.0006	69
Muscle creatine kinase (*CKM*); Adenylate kinase (AK1M)				
Isoelectric focusing of CKM (CKM1M2) and AK1M (AK1M*1/*2)	295 healthy young adults	10-s, 90-s cycle test 60-s adductor pollicis fatiguing test: CKM variant had trend to lower strength decrease	ns	Bouchard et al. (1989)

[a] References: Number of the reference as can be found in Bray et al. (2009).

were associated with baseline muscle strength and with creatine kinase and myoglobin responses in strength loss after an eccentric exercise bout. The role of the *ACTN3* polymorphism is discussed in greater detail in Chapter 18 of this book.

Collagen alpha 1 chain type I forms the fibrils of tendon, ligaments, and bones. In elderly men, the collagen alpha 1 chain type I (*COL1A1*) SpI polymorphism was related to upper-limb muscle strength. The presence of the s-allele was associated with lower grip and biceps strength (318).

Differentiation/Growth Factors and Anabolic and Catabolic Factors

A large set of anabolic and catabolic factors both during muscle development and in regeneration or repair of adult muscle interact to maintain, increase, or decrease muscle mass.

Very early on in the development of the embryo, pluripotent mesenchymal cells differentiate toward bone, fat, or muscle depending on the concentration of bone morphogenetic protein 2. This concentration-dependency might be regulated by an A to C transversion at the highly conserved regulatory element of the *BMP2* gene. A recent study (Devaney et al., 2009) showed an association of increased muscle volume after 12 weeks of RT in the male FAMuSS sample (not in females); however, no baseline differences were observed.

The "mighty mouse" or double muscle phenotype in several animals directed the search for muscle mass determining gene variants in the myostatin (*MSTN, GDF8*) gene and genes coding for interacting factors like follistatin (*FST*), or its receptor activin-type II receptor B (*ACVR2B*).

Several polymorphic variations in *MSTN* were tested for association with strength phenotypes in 286 women. The R153 allele (K153R polymorphism) was more common in African-American women (17% vs. 1.3% in Caucasians), and was associated with lower strength (275). In an Italian older population, the R allele was also associated with lower strength. However, the association was eliminated when correcting for age (Corsi et al., 2002). Two studies found no association of the K153R allele with muscle hypertrophy or strength increases

after long duration (Ferrell et al., 1999) or short duration (117) RT. Similarly, *MSTN* 163 G > A was not associated with changes induced by RT (Ferrell et al., 1999). A recent study within the FAMuSS cohort (Kostek et al., 2009) has reported baseline maximum voluntary contractions to be associated with the *MSTN* R153 allele in African-Americans. Genetic variation in the follistatin (*FST*) gene, which expresses a secreted glycoprotein that inhibits MSTN activity, was studied by Walsh et al. (332) using haplotypes and by Kostek and colleagues (2009) (-5003 A > T and -833 G > T). In the FAMuSS study, FST^{T-5003} allele carriers had greater baseline 1RM and muscle cross-sectional area (CSA) than AA homozygotes (only in African-Americans). Although an *FST* haplotype was associated with leg fat-free mass in male Baltimore Longitudinal Study of Aging subjects, no association was observed with muscle strength (332). Walsh et al. also studied haplotypes within the myostatin receptor, the activin-type II receptor B (*ACVR2B*) gene, and observed only in females significant associations with knee extensor concentric peak torque (332).

Insulin-like growth factors (IGF) 1 and 2 play an important role in muscle mass homeostasis, muscle repair, and hypertrophic responses to exercise. Insulin-like growth factor 1 produced in skeletal muscle can stimulate muscle hypertrophy through activation of satellite cells and increased protein synthesis rates. Given the role of IGF2 in satellite cell proliferation and the age-associated decrease in insulin-like growth factor 2, enhanced gene expression in response to muscle damage provides a protective role in age-associated losses in muscle mass and strength in humans.

Kostek et al. (140) studied 67 older Caucasian men and women before and after a 10-week unilateral strength-training program for associations between a CA-repeat microsatellite polymorphism in the *IGF1* promoter. Carriers of the 192 allele showed greater quadriceps-muscle strength gains compared with non-carriers, with no differences observed in muscle quality response to training. Other polymorphisms in the *IGF1* gene were not associated with any muscle phenotypes. The same research group has examined the interaction between promoter region polymorphisms in *IGF1* (CA 192 allele); its binding protein, insulin-like growth factor binding protein 3 (*IGFBP3*; A-202C); and calcineurin B (*PPP3R1*; 5-bp insertion-deletion polymorphism) in relation to the response of muscle strength and muscle volume to strength training in 128 older men and women (95). The changes in strength and muscle quality defined as torque/mass) were associated with a main effect of the *IGF1* 192 allele plus an *IGF1* 192 allele × *PPP3R1* I/D interaction effect (4.5% and 5.5% of variance for 1RM and MQ change, respectively). The -202 A > C polymorphism in *IGFBP3* was not associated with any phenotype.

There is a complex relationship between the thyroid, growth hormone, and IGF1 axis. Peeters and colleagues (213) reported higher isometric grip strength and leg-extensor strength in 350 predominantly Caucasian men (>70 years) who carried the D1a-T allele of the deiodinase, iodothyronine, type I (*DIO1*) gene compared with D1a-C allele homozygotes.

A G/A polymorphism in the *IGF2* gene (ApaI) was associated with grip strength in 397 older British men (268) (1% explained variance), but not in 296 women of the same age. Schrager and colleagues (272) reported significantly lower arm and leg isokinetic strength measures in women homozygous for the A allele compared with GG homozygotes in the BLSA sample. Devaney et al. (59) studied the association of several polymorphisms in the *IGF2* gene in relation to exertional muscle damage of the elbow flexors in 151 young men and women. Multiple polymorphisms in the *IGF2*, *IGFBP-3*, and *IGF2AS* gene regions each explained about 9–16% of variation in response phenotypes (strength loss, soreness, postexercise creatine kinase [CK], and myoglobin [Mb] levels).

Catabolic Actions of Cytokines

Inflammatory cytokines (interleukin-6 [IL-6], interleukin15 [IL-15], and tumor necrosis factor alpha [TNF-α]) and their receptors (IL-15Ra, TNFR1, and TNFR2) have been studied for their role in muscle mass atrophy and age-related sarcopenia.

Nicklas et al. (197) examined associations between several cytokine gene markers and physical function before and after exercise training in older men and women (>60 years). Stair-climbing performance improved in response to training more in A-allele carriers of the A-308G polymorphism in *TNF* gene compared with GG homozygotes. This polymorphism was not associated with muscle mass, grip or knee strength in a large sample of the BLSA. However, male A carriers of the -863C > A and C carriers of the -1013T > C genotype (or the combination of both alleles) had low muscle mass indicators (Liu et al., 2008), an observation which is consistent with the catabolic effect of TNFα.

Twelve SNPs (and related haplotypes) in the *IL-6* gene were not associated with strength measures or frailty in older women (Walston et al., 2005). Similarly, the *IL-6* C-174G polymorphism was not associated with grip strength or leg extension strength in elderly Italian subjects (Barbieri et al., 2003). Riechman and coworkers (249) reported no association between the interleukin-15 receptor gene (*IL15RA*) and the gain in arm or leg 1RM strength in response to strength training in 153 young men and women, although significant associations with changes in arm or leg muscle mass and muscle quality were reported. A recent study (Pistilli et al., 2008) reports variants in *IL15* and its receptor to be associated with changes in 1RM strength, baseline whole muscle volume, and muscle quality in the FAMuSS study.

Neurotrophic Factors

The neuroregulatory cytokine ciliary neurotrophic factor (CNTF) has neurotrophic and myotrophic characteristics and supports survival and differentiation in a variety of neuronal cell types including motor neurons. Null allele carriers (rs1800169) have been shown to present lower isometric strength in elderly and middle-aged populations (13; 49) and gender-specificity (GA > GG or AA) has been reported (257). Associations with changes in strength in the untrained arm (transfer to co-lateral limb) in the FAMuSS study might be explained by the neurotrophic action of CNTF (Walsh et al., 2009). Three polymorphisms have been studied in the CNTF receptor (*CNTFR*) gene. The 174T carriers in the BLSA cohort had increased eccentric slow and fast velocity knee extensor peak torque; however, an underlying association with fat-free mass of the lower limb indicates that the association probably works through muscle mass determination. On the other hand, opposite allelic effects (T alleles with lower knee extensor strength) and lack of association with muscle mass was also reported. Similarly inconclusive findings for two other SNPs (C-1703T and T1069A) in *CNTFR* were found (49; 256).

The Renin–Angiotensin System

A separate chapter in this book deals with the study of the angiotensin-converting enzyme (*ACE*) insertion/deletion polymorphism in human performance. D-allele carriers have higher levels of ACE activity, which has been related to overload-induced cardiac hypertrophy and smooth-muscle hypertrophy. Possibly the effect of the *ACE* I/D on regulation of hypertrophic responses to overload (in rats) or increased type IIb fiber distribution (Zhang et al., 2003) works through the angiotensin II receptor AT1.

Vitamin-Related

Vitamin D status is negatively associated with muscle myopathy and positively with muscle strength and physical performance. Vitamin D acts through a receptor present in the nucleus of muscle cells (which influences gene transcription and protein synthesis) as well as a cell surface receptor. In women, the BsmI B allele has been associated with reduced quadriceps strength (82; Vandevyver et al., 1999) and overall lower limb strength (Barr et al., 2009) but also with increased hamstring strength (86; 334). ApaI AA homozygotes showed lower quadriceps and concentric elbow flexor strength than a-carriers (334). The TaqI or BsmI/TaqI haplotype did not influence muscle strength in some groups of women (334; 342). In men, homozygous bb chronic obstructive pulmonary disease (COPD) patients had higher quadriceps strength than bB/BB patients (Hopkinson et al., 2008); however, Roth et al. (258) did not

report significant associations. No association between the ApaI/TaqI haplotype and different muscle strength characteristics was shown (Van Pottelbergh et al., 2002), whereas for the Bt/Bt homozygotes, higher quadriceps strength was observed compared to bT carriers (342).

In the *VDR* gene, the FokI T/C transition in exon 2 creates a shorter but transcriptionally more active VDR protein, suggesting that individuals with the C allele ("F") would have improved muscle strength. Against this hypothesis, the C allele has been associated with reduced lower limb or quadriceps strength in healthy adult and elderly men (Van Pottelbergh et al., 2002; 258), women (342), and in older COPD patients (Hopkinson et al., 2008). Lack of association with falls, balance, and muscle power was found in two independent samples of postmenopausal women (Barr et al., 2009).

Hormones and Their Receptors

Negative reports at this moment exclude an important role for polymorphisms in the estrogen receptor 1 (*ESR1*) gene to be associated with strength phenotypes (Grundberg et al., 2005; Ronkainen et al., 2008; Vandevyver et al., 1999). In the gene for an estrogen metabolizing enzyme catechol-O-methyltransferase (*COMT*), a polymorphism that increased estrogen availability was associated with muscle mass; however, no effect on muscle strength was reported (Ronkainen et al., 2008). The glucocorticoid receptor (*NR3C1*) ER22/23EK polymorphism was related to decreased glucocorticoid sensitivity and higher muscle mass and arm pull strength in a Dutch male sample (319).

Energy Metabolism

Although individual variation in anaerobic energy metabolism is not a limiting factor in maximal strength performances, it is of high importance in anaerobic power performances (e.g., 10- to 90-s Wingate test). The adenosine monophosphate deaminase 1 (*AMPD1*) gene catalyzes hydrolysis of AMP to inosine monophosphate (IMP) and ammonia. The AMPD reaction further displaces the adenylate kinase reaction in the direction of ATP formation during exercise, preventing a large increase in ADP. This also serves to maintain a high ATP-to-ADP ratio that is advantageous for sustained muscle work. The *AMPD1* mutant T allele is associated with low AMP deaminase activity and lack of change in skeletal muscle ATP and IMP concentration after a 30-s high-intensity exercise bout (Norman et al., 2001). The latter response was also found by after maximal exercise (Tarnopolsky et al., 2001). There is no effect of the polymorphism on isometric, peak power (De Ruiter et al., 2000; Norman et al., 2001; Tarnopolsky et al., 2001) or 30-s Wingate test, and other parameters of muscle metabolism (Tarnopolsky et al., 2001). In contrast, AMPD1-deficiency was associated with a more rapid decline in power output and lower mean power in a 30-s Wingate cycle test (69), or after a 20-min isometric quadriceps fatiguing test (De Ruiter et al., 2002).

In one of the first studies exploring genetic variants in anaerobic capacity, measured by 10-s and 90-s cycling power output, no difference in performance between carriers and non-carriers of skeletal muscle creatine kinase (*CKM*) and adenylate kinase (*AK1*) variants were noted. However, decline in force production of the adductor pollicis muscle during a 60-s maximum voluntary contraction test tended to be less in the carriers of the CKM variant (Bouchard et al., 1989).

In relation to glucose and lipid metabolism, variants within the peroxisome proliferator-activated receptor alpha (*PPARA*) and gamma (*PPARG*) genes and resistin have been studied. *HIF1A1* has also gained some attention. However, since these metabolic parameters are primarily relevant for long-duration contraction, they will not be discussed in this chapter.

Conclusions

A clear understanding of the genetic underpinnings for interindividual differences in baseline strength or power, or response to training, awaits GWAS studies in larger cohorts of normal volunteers. To date, most gene variants have only hypothetical effects on gene products or are in linkage disequilibrium with a true variant, few polymorphisms are true null alleles. An approach using multiple genes studied within a targeted pathway is only starting to emerge, and the field has not systematically screened genes which are known to play a role in, for example, muscle repair (Bamman et al., 2007). Some gene variants have been studied by different independent research groups (e.g., *ACTN3* R577X) and inconclusive results or contradictory (lack of or opposite associations) have been reported. Although there is considerable public interest in gene test panels able to identify athletic potential, the available knowledge is too limited to justify development and implementation of such a panel at this point in time.

References

Bamman, M.M., Petrella, J.K., Kim, J.S., Mayhew, D.L. & Cross, J.M. (2007). Cluster analysis tests the importance of myogenic gene expression during myofiber hypertrophy in humans. *Journal of Applied Physiology* 102, 2232–2239.

Barbieri, M., Ferrucci, L., Ragno, E., et al. (2003). Chronic inflammation and the effect of IGF-I on muscle strength and power in older persons. *American Journal of Physiology Endocrinology and Metabolism* 284, E481–E487.

Barr, R., Macdonald, H., Stewart, A., et al. (2009). Association between vitamin D receptor gene polymorphisms, falls, balance and muscle power: Results from two independent studies (APOSS and OPUS). *Osteoporosis International* 21(3), 457–466.

Bouchard, C., Chagnon, M., Thibault, M.C., et al. (1989). Muscle genetic variants and relationship with performance and trainability. *Medicine and Science in Sports and Exercise* 21, 71–77.

Bray, M.S., Hagberg, J.M., Perusse, L., et al. (2009). The human gene map for performance and health-related fitness phenotypes: The 2006–2007 update. *Medicine and Science in Sports and Exercise* 41, 35–73.

Corsi, A.M., Ferrucci, L., Gozzini, A., Tanini, A. & Brandi, M.L. (2002). Myostatin polymorphisms and age-related sarcopenia in the Italian population. *Journal of the American Geriatric Society* 50, 1463.

De Mars, G., Windelinckx, A., Huygens, W., et al. (2008a). Genome-wide linkage scan for contraction velocity characteristics of knee musculature in the Leuven Genes for Muscular Strength Study. *Physiological Genomics* 35, 36–44.

De Mars, G., Windelinckx, A., Huygens, W., et al. (2008b). Genome-wide linkage scan for maximum and length-dependent knee muscle strength in young men: Significant evidence for linkage at chromosome 14q24.3. *Journal of Medical Genetics* 45, 275–283.

De Ruiter, C.J., May, A.M., van Engelen, B.G., Wevers, R.A., Steenbergen-Spanjers, G.C. & de Haan, A. (2002). Muscle function during repetitive moderate-intensity muscle contractions in myoadenylate deaminase-deficient Dutch subjects. *Clinical Science (London)* 102, 531–539.

De Ruiter, C.J., Van Engelen, B.G.M., Wevers, R.A. & de Haan, A. (2000). Muscle function during fatigue in myoadenylate deaminase-deficient Dutch subjects. *Clinical Science (London)* 98, 579–585.

Delmonico, M.J., Zmuda, J.M., Taylor, B.C. et al. (2008). Association of the ACTN3 genotype and physical functioning with age in older adults. *The Journals of Gerontology Series A, Biological Sciences and Medical Sciences* 63, 1227–1234.

Devaney, J.M., Tosi, L.L., Fritz, D.T., et al. (2009). Differences in fat and muscle mass associated with a functional human polymorphism in a post-transcriptional BMP2 gene regulatory element. *Journal of Cellular Biochemistry* 107, 1073–1082.

Druzhevskaya, A.M., Ahmetov, I.I., Astratenkova, I.V. & Rogozkin, V.A. (2008). Association of the ACTN3 R577X polymorphism with power athlete status in Russians. *European Journal of Applied Physiology* 103, 631–634.

Ferrell, R.E., Conte, V., Lawrence, E.C., Roth, S.M., Hagberg, J.M. & Hurley, B.F. (1999). Frequent sequence variation in the human myostatin (GDF8) gene as a marker for analysis of muscle-related phenotypes. *Genomics* 62, 203–207.

Grundberg, E., Ribom, E.L., Brandstrom, H., Ljunggren, O., Mallmin, H. & Kindmark, A. (2005). A TA-repeat polymorphism in the gene for the estrogen receptor alpha does not correlate with muscle strength or body composition in young adult Swedish women. *Maturitas* 50, 153–160.

Hopkinson, N.S., Li, K.W., Kehoe, A., et al. (2008). Vitamin D receptor genotypes influence quadriceps strength in chronic obstructive pulmonary disease. *American Journal of Clinical Nutrition* 87, 385–390.

Irving, R., Scott, R., Irwin, L., et al. (2009). The Actn3 R577X polymorphism in elite Jamaican and USA sprinters. *Medicine and Science in Sports and Exercise* 41, 165.

Kostek, M.A., Angelopoulos, T.J., Clarkson, P.M., et al. (2009). Myostatin and follistatin polymorphisms interact with muscle phenotypes and ethnicity. *Medicine and Science in Sports and Exercise* 41, 1063–1071.

Liu, D., Metter, E.J., Ferrucci, L. & Roth, S.M. (2008). TNF promoter polymorphisms associated with muscle phenotypes in humans. *Journal of Applied Physiology* 105, 859–867.

McCauley, T., Mastana, S.S., Hossack, J., Macdonald, M. & Folland, J.P. (2009). Human angiotensin-converting enzyme I/D and alpha-actinin 3 R577X genotypes and muscle functional and contractile properties. *Experimental Physiology* 94, 81–89.

Norman, B., Esbjornsson, M., Rundqvist, H., Osterlund, T., von Walden, F. & Tesch, P.A. (2009). Strength, power, fiber types, and mRNA expression in trained men and women with different ACTN3 R577X genotypes. *Journal of Applied Physiology* 106, 959–965.

Norman, B., Sabina, R.L. & Jansson, E. (2001). Regulation of skeletal muscle ATP catabolism by AMPD1 genotype during sprint exercise in asymptomatic subjects. *Journal of Applied Physiology* 91, 258–264.

Perusse, L., Rankinen, T., Rauramaa, R., Rivera, M.A., Wolfarth, B. & Bouchard, C. (2003). The human gene map for performance and health-related fitness phenotypes: The 2002 update. *Medicine and Science in Sports and Exercise* 35, 1248–1264.

Pistilli, E.E., Devaney, J.M., Gordish-Dressman, H., et al. (2008). Interleukin-15 and interleukin-15R alpha SNPs and associations with muscle, bone, and predictors of the metabolic syndrome. *Cytokine* 43, 45–53.

Rankinen, T., Bray, M.S., Hagberg, J.M., et al. (2006). The human gene map for performance and health-related fitness phenotypes: The 2005 update. *Medicine and Science in Sports and Exercise* 38, 1863–1888.

Rankinen, T., Perusse, L., Rauramaa, R., Rivera, M.A., Wolfarth, B. & Bouchard, C. (2001). The human gene map for performance and health-related fitness phenotypes. *Medicine and Science in Sports and Exercise* 33, 855–867.

Rankinen, T., Perusse, L., Rauramaa, R., Rivera, M.A., Wolfarth, B. & Bouchard, C. (2002). The human gene map for performance and health-related fitness phenotypes: The 2001 update. *Medicine and Science in Sports and Exercise* 34, 1219–1233.

Rankinen, T., Perusse, L., Rauramaa, R., Rivera, M.A., Wolfarth, B. & Bouchard, C. (2004). The human gene map for performance and health-related fitness phenotypes: The 2003 update. *Medicine and Science in Sports and Exercise* 36, 1451–1469.

Ronkainen, P.H., Pollanen, E., Tormakangas, T., et al. (2008). Catechol-o-methyltransferase gene polymorphism is associated with skeletal muscle properties in older women alone and together with physical activity. *PLoS One* 3, e1819.

Roth, S.M., Walsh, S., Liu, D., Metter, E.J., Ferrucci, L. & Hurley, B.F. (2008). The ACTN3 R577X nonsense allele is under-represented in elite-level strength athletes. *European Journal of Human Genetics* 16, 391–394.

Tarnopolsky, M.A., Parise, G., Gibala, M.J., Graham, T.E. & Rush, J.W. (2001). Myoadenylate deaminase deficiency does not affect muscle anaplerosis during exhaustive exercise in humans. *Journal of Physiology* 533, 881–889.

Tiainen, K.M., Perola, M., Kovanen, V.M., et al. (2008). Genetics of maximal walking speed and skeletal muscle characteristics in older women. *Twin Research and Human Genetics* 11, 321–334.

Van Pottelbergh, I., Goemaere, S., De Bacquer, D., De Paepe, A. & Kaufman, M. (2002). Vitamin D receptor gene allelic variants, bone density, and bone turnover in community-dwelling men. *Bone* 31, 631–637.

Vandevyver, C., Vanhoof, J., Declerck, K., et al. (1999). Lack of association between estrogen receptor genotypes and bone mineral density, fracture history, or muscle strength in elderly women. *Journal of Bone and Mineral Research* 14, 1576–1582.

Walsh, S., Kelsey, B.K., Angelopoulos, T.J., et al. (2009). CNTF 1357 G --> A polymorphism and the muscle strength response to resistance training. *Journal of Applied Physiology* 107, 1235–1240.

Walsh, S., Liu, D., Metter, E.J., Ferrucci, L. & Roth, S.M. (2008). ACTN3 genotype is associated with muscle phenotypes in women across the adult age span. *Journal of Applied Physiology* 105, 1486–1491.

Walston, J., Arking, D.E., Fallin, D., et al. (2005). IL-6 gene variation is not associated with increased serum levels of IL-6, muscle, weakness, or frailty in older women. *Experimental Gerontology* 40, 344–352.

Windelinckx, A., Vlietinck, R., Aerssens, J., Beunen, G. & Thomis, M.A. (2007). Selection of genes and single nucleotide polymorphisms for fine mapping starting from a broad linkage region. *Twin Research and Human Genetics* 10, 871–885.

Wolfarth, B., Bray, M.S., Hagberg, J.M., et al. (2005). The human gene map for performance and health-related fitness phenotypes: The 2004 update. *Medicine and Science in Sports and Exercise* 37, 881–903.

Zhang, B., Tanaka, H., Shono, N., et al. (2003). The I allele of the angiotensin-converting enzyme gene is associated with an increased percentage of slow-twitch type I fibers in human skeletal muscle. *Clinical Genetics* 63, 139–144.

Chapter 15

Genes and Response to Training

TUOMO RANKINEN, MARK A. SARZYNSKI AND CLAUDE BOUCHARD

Human Genomics Laboratory, Pennington Biomedical Research Center, Baton Rouge, LA, USA

As evidenced by several consensus meetings and expert panel reports [no authors listed] 1996; Bouchard, 2001; Bouchard & Blair, 1999; Leon, 1997; US Department of Health and Human Services, 1996), the body of scientific evidence regarding the effects of regular physical activity and sedentary behavior on risk factors for common diseases, health outcomes, and mortality rates is already impressive and growing. However, the effects of regular exercise and habitual physical activity have been almost always tested and reported in terms of main effects and group differences. Consequently, the interpretations and conclusions have been based on the average effects observed in groups of subjects. There are considerable individual differences in risk-factor responses to regular physical activity, even when all subjects are exposed to the same volume of exercise, adjusted for their own tolerance level (Bouchard & Rankinen, 2001).

Although the research on molecular genetics of physical activity, health-related fitness, and health-related outcomes is still in its infancy and no genes for exercise-related phenotypes have been confirmed at the level recently recommended for genetic association studies (Chanock et al., 2007), we need to recognize early that some alleles at key genes are likely to play an important role in the ability to benefit from regular exercise. The sooner we incorporate this advance in our thinking and move in the direction of fully integrated molecular

epidemiology research, the sooner we will be able to understand the true relation between a sedentary lifestyle or poor fitness and the risk of disease. Moving along this path will provide some of the building blocks that are necessary to bring us eventually in the era of individualized, and hopefully more efficacious, public health recommendations and preventive medicine measures.

Human Gene Map for Performance Traits

The concept of heterogeneity in responsiveness to standardized exercise programs was first introduced in the early 1980s (Bouchard, 1983). In a series of carefully controlled and standardized exercise training studies conducted with young and healthy adult volunteers, it was shown that the individual differences in training-induced changes in several physical performance and health-related fitness phenotypes were large, with the range between low and high responders reaching several folds (Bouchard, 1983, 1995; Bouchard et al., 1992; Lortie et al., 1984; Simoneau et al., 1986). The most comprehensive data on the individual differences in trainability come from the HERITAGE Family Study, where 742 healthy but sedentary subjects followed a highly standardized, well-controlled, laboratory-based endurance-training program for 20 weeks. The training program induced several beneficial changes in cardiorespiratory fitness and other cardiovascular and type 2 diabetes mellitus risk-factor phenotypes. However, these changes were characterized by marked interindividual

Genetic and Molecular Aspects of Sport Performance, 1st edition.
Edited by Claude Bouchard and Eric P. Hoffman.
Published 2011 by Blackwell Publishing Ltd.

differences. For example, the average increase in maximal oxygen consumption ($\dot{V}O_2$max) was 384 ml O_2 with a standard deviation of 202. The training responses varied from no change to increases of more than 1000 ml O_2 per minute (Bouchard & Rankinen, 2001; Bouchard et al., 1999; Skinner et al., 2000). The same pattern of variation was evident for several other training response traits (Bouchard & Rankinen, 2001; Wilmore et al., 2001). Similar heterogeneity in responsiveness to exercise training has been reported also in other populations (Hautala et al., 2003; Kohrt et al., 1991).

As reviewed in more detail elsewhere in this book (Chapter 11), the evidence from genetic epidemiology studies suggests that there is a genetically determined component affecting exercise training response phenotypes. In fact, familial aggregation has been shown to be the strongest predictor of interindividual variation in training responsiveness in the HERITAGE Family Study. However, since these traits are complex and multifactorial in nature, the search for genes and mutations responsible for the genetic regulation must target not only several families of phenotypes, but also consider the phenotypes in response to exercise training. It is also obvious that the research on molecular genetics of exercise-related phenotypes is still in its infancy.

The latest update of the Human Gene Map for Performance and Health-Related Fitness Phenotypes included 214 autosomal and 7 X chromosome gene entries and quantitative trait loci (QTLs) (Bray et al., 2009). Moreover, there were 18 mitochondrial genes in which sequence variants have been shown to influence relevant fitness and performance phenotypes. These findings were reported in 361 peer-reviewed research articles (Table 15.1). A total of 49 unique genes from 81 studies have been investigated in relation to exercise training-induced changes in hemodynamic (17 genes, 23 reports), body composition (19 genes, 22 reports), insulin and glucose metabolism (15 genes, 16 reports), and plasma lipid, lipoprotein, and hemostatic (15 genes, 20 reports) phenotypes. In addition, 15 autosomal genes and 1 gene encoded by mitochondrial DNA were reported in at least one study to be associated with physical performance-related phenotypes: 12 genes were associated with endurance phenotypes,

Table 15.1 Number of research articles and genetic loci summarized in the 2007 update of the Human Gene Map for Performance and Health-Related Fitness Phenotypes[a]

Phenotypes	No. of papers	No. of loci
Endurance	72	47
Strength and anaerobic	40	26
Hemodynamics	67	80
Body composition	44	39
Insulin and glucose	21	30
Lipids, inflammation, and hemostatic	38	25
Chronic diseases	7	7
Exercise intolerance	66	37
Physical activity	9	18

[a] See reference Bray et al. (2009) for details.

whereas 5 genes were associated with speed and muscle strength-related traits (Bray et al., 2009). As a point of reference, the latest update of a gene map for obesity-related phenotypes, which does not contain the latest genome-wide association studies (GWAS), included more than 600 loci (Rankinen et al., 2006).

Most of the genes summarized in the human fitness gene map are based on only one study with positive findings. For example, the genes associated with body composition, plasma lipid, and hemostatic phenotype training responses were all based on a single study. Similarly, most of the genes associated with physical performance training responses were reported only in one study with the angiotensin-converting enzyme (*ACE*; see Chapter 17 of this book for details) and apolipoprotein E (*APOE*) genes being the only exceptions. However, the associations between $\dot{V}O_2$max training response and *APOE* genotype are difficult to interpret, because one study found that the E2 and E4 genotypes are associated with the lowest and highest, respectively, training responses (Hagberg et al., 1999), while the other study reported that carriers of the epsilon 2 allele showed significantly greater training-induced improvement than the E3/3 homozygotes (Thompson et al., 2004). However, with hemodynamic phenotypes some candidate gene findings have been replicated in at least two studies. These include angiotensinogen (*AGT*) and *ACE*, central components of the renin–angiotensin system. An association between blood pressure training response and the *AGT* M235T polymorphism has been reported both in the HERITAGE Family Study and the DNASCO study (Rankinen et al., 2000; Rauramaa et al., 2002). In white HERITAGE males, the *AGT* M235M homozygotes showed the greatest reduction in submaximal exercise diastolic blood pressure following a 20-week endurance-training program (Rankinen et al., 2000), whereas in middle-aged Eastern Finnish men, M235M homozygotes had the most favorable changes in resting systolic and diastolic blood pressure during a 6-year exercise intervention trial (Rauramaa et al., 2002).

Similarly, an association between the *ACE* I/D polymorphism and exercise training-induced left ventricular (LV) growth has been reported in two studies (Figure 15.1) (Montgomery et al., 1997; Myerson et al., 2001). Montgomery and coworkers reported in 1997 that the *ACE* D-allele is associated with greater increases in LV mass and septal and posterior wall thickness after 10 weeks of physical training in British Army recruits (Montgomery et al., 1997). In 2001, the same group reported that the training-induced increase in LV mass in another cohort of Army recruits was 2.7 times greater in the D/D genotype as compared to the I/I homozygotes.

It was also reported that the association between the *ACE* genotype and LV mass training response was not affected by angiotensin II type 1 receptor inhibitor treatment (Myerson et al., 2001).

A gene encoding alpha-actinin 3 (*ACTN3*) has become a popular candidate gene for exercise performance traits. A C/T transition in codon 577 of the *ACTN3* gene replaces an arginine residue (R577) with a premature stop codon (X577) resulting in a nonfunctional gene product. The stop codon variant is quite common in humans with allele frequencies ranging from 10% in African populations to 50% in Caucasians and Asians. Although cross-sectional studies have provided some evidence of associations between the *ACTN3* R577X variant and physical performance traits (see Chapter 18 for details), results from exercise training studies are mixed. In 352 young adult Caucasian and Asian women, the X577X homozygotes showed lower baseline values but greater increases in dynamic muscle strength after 12 weeks of strength training, while no differences were found in training responses between the genotypes among 247 men (Clarkson et al., 2005). However, a strength-training study in elderly men

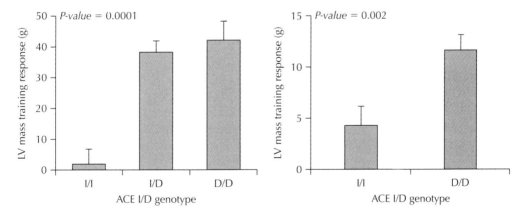

Figure 15.1 Angiotensin-converting enzyme (*ACE*) I/D polymorphism is associated with left ventricular mass training response. Left panel summarizes data from 140 healthy army recruits who participated in a 10-week basic training program (Montgomery et al., 1997). Right panel shows data from a replication study using a similar training program with 141 healthy recruits (Myerson et al., 2001). Reprinted, with permission, from T. Rankinen and C. Bouchard, 2007, Genetic differences in the relationships among physical activity, fitness, and health. In *Physical activity and health*, edited by C. Bouchard, S.N. Blair, and W.L. Haskell (Champaign, IL: Human Kinetics), 351.

and women found exactly the opposite. In women (n = 86), the X577X homozygotes showed significantly higher baseline knee extensor concentric peak power than the heterozygotes and R577R homozygotes, whereas the improvements brought about by resistance training tended to be greater in the R577R homozygotes than in the stop codon homozygotes (Delmonico et al., 2007). Thus, more studies are needed to clarify whether the *ACTN3* locus modifies the effects of resistance training on muscle strength.

Hypothesis-Free Gene Finding Approaches and Exercise Training Phenotypes

The majority of the exercise-related genetic association studies have relied on the physiological candidate gene approach. The main weakness of this strategy is that the gene selection is limited to our current understanding of the physiology regulating the trait of interest. It is safe to say that such an approach is less than optimal to identify all relevant candidate genes. The problem is further compounded by the fact that even if the protein encoded by the selected candidate gene is central for the phenotype regulation, there are no guarantees that DNA sequence variation in the gene locus contributes to genetic variation of the trait. In fact, quite often mutations/sequence variations that alter function of the key proteins induce markedly compromised phenotype or even reduced survival.

An alternative approach for gene identification is a systematic screening of the entire genome for genes and genomic regions that are linked to or associated with the trait of interest. Genomic screening may utilize either DNA sequence variants (single-nucleotide polymorphisms [SNPs] and/or microsatellites/copy number variants) or gene expression signatures. The advantage of the genome-wide approaches is that they do not require a specific hypothesis regarding the underlying genes and therefore are not limited by the gaps in our understanding of the biology/ physiology of the trait.

Genome-wide linkage analysis has been a powerful method to identify genes that cause Mendelian disorders and monogenic diseases. Success with multifactorial and oligogenic/polygenic traits has been less spectacular, although the genomic region containing a transcription factor 7-like 2 (*TCF7L2*) gene, a major disposing gene for type 2 diabetes, was originally identified through linkage analysis (Duggirala et al., 1999; Grant et al., 2006; Reynisdottir et al., 2003).

The HERITAGE Family Study has used genome-wide linkage analysis to find genes for exercise training response phenotypes. QTLs for training-induced changes in submaximal exercise (50 watts) stroke volume (ΔSV50) and heart rate (ΔHR50) were found on chromosomes 10p11 and 2q33.3-q34, respectively (Rankinen et al., 2002; Spielmann et al., 2007). The ΔSV50 QTL on 10p11 was narrowed down to a 7-Mb region using dense microsatellite mapping. Genes within the region were tested for associations by genotyping a dense panel of SNPs within the gene loci. Among the linkage-positive families (family-specific LOD score >0.025), the strongest associations were found with SNPs in the kinesin heavy chain (*KIF5B*) gene locus (Argyropoulos et al., 2009). Resequencing of the *KIF5B* revealed several DNA sequence variants especially in the putative promoter region. The SNP that showed the strongest association with ΔSV50 was found to modify the *KIF5B* promoter activity. Furthermore, analogous inhibition and overexpression studies showed that changes in KIF5B expression level alter mitochondrial localization and biogenesis: KIF5B inhibition led to diminished biogenetic and perinuclear accumulation of mitochondria, while overexpression enhanced mitochondrial biogenesis (Argyropoulos et al., 2009).

The QTL for ΔHR50 on chromosome 2q33.3-q34 was localized within a 10-Mb region. A dense panel of almost 1500 SNPs was genotyped to achieve a uniform SNP coverage of the entire region. The strongest evidence of association was detected with two SNPs located in the 5'-region of the *CREB1* gene locus (peak P-value = 1.6×10^{-5}), and the associations remained significant even when multiple testing was taken into account (Rankinen et al., 2010). The most significant SNP explained almost 5% of the variance in ΔHR50, and the common allele homozygotes and heterozygotes had about a 57% and 20%, respectively, greater decrease in HR50 than the minor allele homozygotes. Furthermore, the same SNP, which is located about 2.6 kb upstream of the first exon of *CREB1*, was shown to modify promoter activity *in vitro*. Given the role of CREB1 in cardiac memory formation (Rosen & Cohen, 2006), it serves as an excellent positional and functional candidate gene for ΔHR50.

Global gene expression profiling was used in the HERITAGE Family Study to identify genes associated with insulin sensitivity training response (ΔS_I) (Teran-Garcia et al., 2005). Total RNA was extracted from vastus lateralis muscle biopsies from 16 subjects: eight of them were ΔS_I high responders, while the remaining eight were age, sex, and body mass index (BMI)-matched ΔS_I nonresponders. RNA samples were pooled within each responder group, labeled with fluorescent dyes, and hybridized onto *in situ*-generated microarrays containing 18,861 genes. A total of 47 and 361 transcripts were differentially expressed (at least 1.4-/0.7-fold difference) before and after the training program, respectively. Five genes (*CTBP1, FHL1, PDK4, SKI,* and *TTN*) that exhibited at least 50% difference in expression between high and nonresponders either at baseline or post-training were selected for validation experiments. Quantitative real-time PCR confirmed the microarray-based expression patterns for four of the five genes: *PDK4* expression was 1.8-fold greater in high responders than in nonresponders at baseline, while high responders showed significantly greater *FHL1, TTN,* and *SKI* expression levels after the training program (Teran-Garcia et al., 2005). Association of the *FHL1* gene with exercise training-induced changes in insulin metabolism phenotypes was further investigated by genotyping three SNPs in the *FHL1* locus (Xq26) (Teran-Garcia et al., 2007). In white women, SNP rs9018 was associated with disposition index ($P = 0.016$) and glucose disappearance index ($P = 0.008$) changes, while in the white males, the same SNP showed suggestive association with fasting insulin training response ($P = 0.04$). Another SNP (rs2180062) was associated with fasting insulin ($P = 0.012$), insulin sensitivity ($P = 0.046$), disposition index ($P = 0.006$), and glucose disappearance index ($P = 0.03$) training responses in white males (Teran-Garcia et al., 2007).

Experimental Versus Observational Studies

An obvious problem with exercise training response-related genetic studies is a paucity of studies with sufficiently large sample size and a carefully monitored training program combined with a high level of subject compliance, mainly due to the relatively high cost of such studies. Therefore, genotype-by-physical activity interaction studies in existing large epidemiological cohorts have been proposed as a less-expensive alternative to intervention trials. While the interaction studies clearly have the potential to provide valuable insights as to how regular physical activity may affect genetic predisposition to various health outcomes, especially if physical activity levels of the subjects have been documented reliably, it is obvious that they add little to our understanding on the genetics of responsiveness to exercise training. A good example comes from recent studies dealing with physical activity—*FTO* (fat mass and obesity-associated) gene interactions on body weight and body composition. *FTO* was the first gene that was identified using high-density GWAS as a strong candidate for obesity-related phenotypes in Caucasians (Frayling et al., 2007; Scott et al., 2007; Zeggini et al., 2007). Minor alleles of obesity-risk SNPs located in the first intron of the gene were associated with about a 65% greater risk of obesity in the homozygote state: the homozygotes of the minor allele were 3–4 kg heavier and heterozygotes had 1–2 kg greater body weight than the common allele homozygotes. The population-attributable risk of *FTO* for obesity has been estimated to be as high as 20%. The initial finding has been replicated in several large cohorts and the association holds both in children and in adults (Loos & Bouchard, 2008).

Consequently, two studies have reported that the *FTO* gene-related risk of obesity was modified by physical activity level of the subjects (Andreasen et al., 2008; Rampersaud et al., 2008). In a large population-based cohort, Andreasen and coworkers reported that the greater BMI level of the *FTO* risk-allele homozygotes was particularly evident in those who were sedentary, while the difference in BMI between the risk-allele and common allele homozygotes was not significant among physically active individuals (Andreasen et al., 2008). A similar interaction was reported in the Old Order Amish (Rampersaud et al., 2008). Although the *FTO* SNP that showed the strongest association with body weight and fatness was different in the Amish than in the Danish cohort (rs1861868 vs. rs9939609; $r^2 = 0.11$ in the HapMap CEU data), the pattern of interaction was identical: the inverse relationship between daily physical activity level and BMI was significantly steeper in the homozygotes of the high-fatness allele than in the other genotypes (Rampersaud et al., 2008). Given the cross-sectional nature of both studies, these findings could be interpreted in at least two ways: either physical activity helps to prevent weight gain or it accelerates weight loss in the risk-allele homozygotes. The latter interpretation was fairly frequently cited especially in the popular media.

The weight loss hypothesis was formally tested in the HERITAGE Family Study, the largest exercise intervention trial conducted so far and the only trial designed *a priori* to investigate genetics of responsiveness to training (Rankinen et al., 2009). The HERITAGE results did not support the weight loss hypothesis. In fact, the data suggested that the risk-allele homozygotes are resistant to training-induced changes in adiposity (Figure 15.2). After 20 weeks of

carefully supervised endurance training with 100% compliance, the *FTO* risk-allele homozygotes did not lose fat mass, while the homozygotes for the nonrisk allele showed a significant reduction in total adiposity. Thus, the experimental data would suggest that regular physical activity does not help the *FTO* risk-allele carriers to lose weight that has already been gained. However, the weight gain prevention hypothesis is a very interesting one, but remains to be tested in controlled clinical trials. The *FTO* example described above also emphasizes the importance of testing (genetic) hypotheses related to responsiveness to exercise training in appropriately designed studies.

Future Directions

We believe that there are compelling reasons for exercise scientists to take into account individual differences in responsiveness to regular physical activity in their models and incorporate genetic information in studies designed to understand the relationships between physical activity, health, and disease. However, it is vital that the genetic hypotheses are taken into account already when designing future studies. The importance of adequate sample size is becoming even more evident now that

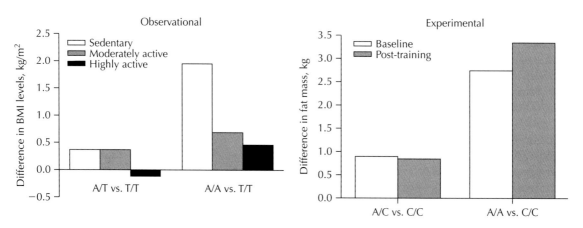

Figure 15.2 Different outcomes of *FTO* genotype-by-physical activity interactions on adiposity in observational and experimental studies. In a cross-sectional population study (left panel; see reference Andreasen et al., 2008 for details), the obesity-risk allele of the *FTO* gene (rs9939609 A-allele) was associated with high BMI in sedentary but not in physically active individuals. In a 20-week exercise training study (right panel; see reference Rankinen et al., 2009 for details), the difference in total adiposity observed in the sedentary state (baseline) between the *FTO* risk-allele (rs8050136 A-allele) and common allele (C) homozygotes became even greater after the training program due to smaller training-induced fat mass loss among the A/A homozygotes (−0.2 kg vs. −0.8 kg).

technologies for detailed candidate gene studies and GWAS with a large number of DNA markers are available. Use of hundreds of thousands of SNPs in a typical GWAS leads to inflated multiple testing problems. Consequently, the sample sizes required to maintain adequate statistical power while keeping the experiment-wise false-positive rate in check will be thousands of individuals. The GWAS provides a very powerful tool for geneticists and exercise scientists to study the genetic basis of the interindividual variation in responsiveness to exercise training. However, we must be sure that the studies are adequately designed to take full advantage of this power.

References

(No authors listed) (1996). Physical activity and cardiovascular health. NIH Consensus Development Panel on Physical Activity and Cardiovascular Health. *JAMA* 276, 241–246.

Andreasen, C.H., Stender-Petersen, K.L., Mogensen, M.S., et al. (2008). Low physical activity accentuates the effect of the *FTO* rs9939609 polymorphism on body fat accumulation. *Diabetes* 57, 95–101.

Argyropoulos, G., Stutz, A.M., Ilnytska, O., et al. (2009). *KIF5B* gene sequence variation and response of cardiac stroke volume to regular exercise. *Physiological Genomics* 36, 79–88.

Bouchard, C. (1983). Human adaptability may have a genetic basis. In: F. Landry (ed.) *Risk Reduction and Health Promotion. Proceedings of the 18th Annual Meeting of the Society of Prospective Medicine*, pp. 463–476. Canadian Public Health Association, Ottawa.

Bouchard, C. (1995). Individual differences in the response to regular exercise. *International Journal of Obesity and Related Metabolic Disorders* 19, S5–S8.

Bouchard, C. (2001). Physical activity and health: Introduction to the dose-response symposium. *Medicine and Science in Sports and Exercise* 33, S347–S350.

Bouchard, C. & Blair, S.N. (1999). Introductory comments for the consensus on physical activity and obesity. *Medicine and Science in Sports and Exercise* 31, S498–S501.

Bouchard, C. & Rankinen, T. (2001). Individual differences in response to regular physical activity. *Medicine and Science in Sports and Exercise* 33, S446–S451.

Bouchard, C., Dionne, F.T., Simoneau, J.A. & Boulay, M.R. (1992). Genetics of aerobic and anaerobic performances. *Exercise and Sport Sciences Reviews* 20, 27–58.

Bouchard, C., An, P., Rice, T., et al. (1999). Familial aggregation of $\dot{V}O_2$max response to exercise training: Results from the HERITAGE Family Study. *Journal of Applied Physiology* 87, 1003–1008.

Bray, M.S., Hagberg, J.M., Perusse, L., et al. (2009). The human gene map for performance and health-related fitness phenotypes: The 2006–2007 update. *Medicine and Science in Sports and Exercise* 41, 34–72.

Chanock, S.J., Manolio, T., Boehnke, M., et al. (2007). Replicating genotype–phenotype associations. *Nature* 447, 655–760.

Clarkson, P.M., Devaney, J.M., Gordish-Dressman, H., et al. (2005). ACTN3 genotype is associated with increases in muscle strength in response to resistance training in women. *Journal of Applied Physiology* 99, 154–163.

Delmonico, M.J., Kostek, M.C., Doldo, N.A., et al. (2007). Alpha-actinin-3 (*ACTN3*) R577X polymorphism influences knee extensor peak power response to strength training in older men and women. *Journals of Gerontology Series A: Biological Sciences and Medical Sciences* 62, 206–212.

Duggirala, R., Blangero, J., Almasy, L., et al. (1999). Linkage of type 2 diabetes mellitus and of age at onset to a genetic location on chromosome 10q in Mexican Americans. *American Journal of Human Genetics* 64, 1127–1140.

Frayling, T.M., Timpson, N.J., Weedon, M.N., et al. (2007). A common variant in the FTO gene is associated with body mass index and predisposes to childhood and adult obesity. *Science* 316, 889–894.

Grant, S.F., Thorleifsson, G., Reynisdottir, I., et al. (2006). Variant of transcription factor 7-like 2 (TCF7L2) gene confers risk of type 2 diabetes. *Nature Genetics* 38, 320–323.

Hagberg, J.M., Ferrell, R.E., Katzel, L.I., Dengel, D.R., Sorkin, J.D. & Goldberg, A.P. (1999). Apolipoprotein E genotype and exercise training-induced increases in plasma high-density lipoprotein (HDL)- and HDL2-cholesterol levels in overweight men. *Metabolism* 48, 943–945.

Hautala, A.J., Makikallio, T.H., Kiviniemi, A., et al. (2003). Cardiovascular autonomic function correlates with the response to aerobic training in healthy sedentary subjects. *American Journal of Physiology—Heart and Circulatory Physiology* 285, H1747–H1752.

Kohrt, W.M., Malley, M.T., Coggan, A.R., et al. (1991). Effects of gender, age, and fitness level on response of $\dot{V}O_2$max to training in 60–71 yr olds. *Journal of Applied Physiology* 71, 2004–2011.

Leon, A. (1997). *Physical Activity and Cardiovascular Health: A National Consensus*. Human Kinetics Publishers, Champaign, IL.

Loos, R.J. & Bouchard, C. (2008). *FTO*: The first gene contributing to common forms of human obesity. *Obesity Reviews* 9, 246–250.

Lortie, G., Simoneau, J.A., Hamel, P., Boulay, M.R., Landry, F. & Bouchard, C. (1984). Responses of maximal aerobic power and capacity to aerobic training. *International Journal of Sports Medicine* 5, 232–236.

Montgomery, H.E., Clarkson, P., Dollery, C.M., et al. (1997). Association of angiotensin-converting enzyme gene I/D polymorphism with change in left ventricular mass in response to physical training. *Circulation* 96, 741–747.

Myerson, S.G., Montgomery, H.E., Whittingham, M., et al. (2001). Left ventricular hypertrophy with exercise and *ACE* gene insertion/deletion polymorphism: A randomized controlled trial with losartan. *Circulation* 103, 226–230.

Rampersaud, E., Mitchell, B.D., Pollin, T.I., et al. (2008). Physical activity and the association of common FTO gene variants with body mass index and obesity. *Archives of Internal Medicine* 168, 1791–1797.

Rankinen, T., An, P., Perusse, L., et al. (2002). Genome-wide linkage scan for exercise stroke volume and cardiac output in the HERITAGE Family Study. *Physiological Genomics* 10, 57–62.

Rankinen, T., Gagnon, J., Perusse, L., et al. (2000). AGT M235T and ACE ID polymorphisms and exercise blood pressure in the HERITAGE Family Study. *American Journal of Physiology— Heart and Circulatory Physiology* 279, H368–H374.

Rankinen, T., Rice, T., Teran-Garcia, M., Rao, D.C. & Bouchard, C. (2009). *FTO* genotype is associated with exercise training-induced changes in body composition. *Obesity (Silver Spring)* 18, 322–326.

Rankinen, T., Zuberi, A., Chagnon, Y.C., et al. (2006). The human obesity gene map: The 2005 update. *Obesity* 14, 529–644.

Rankinen, T., Argyropoulos, G., Rice, T., Rao, D.C. & Bouchard, C. (2010). CREB1 is a strong genetic predictor of the variation in exercise heart rate response to regular exercise: The HERITAGE Family Study. *Circulation. Cardiovascular Genetics* 3, 294–299.

Rauramaa, R., Kuhanen, R., Lakka, T.A., et al. (2002). Physical exercise and blood pressure with reference to the angiotensinogen M235T polymorphism. *Physiological Genomics* 10, 71–77.

Reynisdottir, I., Thorleifsson, G., Benediktsson, R., et al. (2003). Localization of a susceptibility gene for type 2 diabetes to chromosome 5q34-q35.2. *American Journal of Human Genetics* 73, 323–335.

Rosen, M.R. & Cohen, I.S. (2006). Cardiac memory—new insights into molecular mechanisms. *Journal of Physiology* 570, 209–218.

Scott, L.J., Mohlke, K.L., Bonnycastle, L.L., et al. (2007). A genome-wide association study of type 2 diabetes in Finns detects multiple susceptibility variants. *Science* 316, 1341–1345.

Simoneau, J.A., Lortie, G., Boulay, M.R., Marcotte, M., Thibault, M.C. & Bouchard, C. (1986). Inheritance of human skeletal muscle and anaerobic capacity adaptation to high-intensity intermittent training. *International Journal of Sports Medicine* 7, 167–171.

Skinner, J.S., Wilmore, K.M., Krasnoff, J.B., et al. (2000). Adaptation to a standardized training program and changes in fitness in a large, heterogeneous population: The HERITAGE Family Study. *Medicine and Science in Sports and Exercise* 32, 157–161.

Spielmann, N., Leon, A.S., Rao, D.C., et al. (2007). Genome-wide linkage scan for submaximal exercise heart rate in the HERITAGE family study. *American Journal of Physiology—Heart and Circulatory Physiology* 293, H3366–H3371.

Teran-Garcia, M., Rankinen, T., Koza, R.A., Rao, D.C. & Bouchard, C. (2005). Endurance training-induced changes in insulin sensitivity and gene expression. *American Journal of Physiology— Endocrinology and Metabolism* 288, E1168–E1178.

Teran-Garcia, M., Rankinen, T., Rice, T., et al. (2007). Variations in the four and a half LIM domains 1 gene (*FHL1*) are associated with fasting insulin and insulin sensitivity responses to regular exercise. *Diabetologia* 50, 1858–1866.

Thompson, P.D., Tsongalis, G.J., Seip, R.L., et al. (2004). Apolipoprotein E genotype and changes in serum lipids and maximal oxygen uptake with exercise training. *Metabolism* 53, 193–202.

US Department of Health and Human Services. (1996). *Physical Activity and Health: A Report of the Surgeon General.* US Dept of Health and Human Services, Centers for Disease Control and Prevention, National Center for Chronic Disease Prevention and Health Promotion, Atlanta, GA.

Wilmore, J.H., Stanforth, P.R., Gagnon, J., et al. (2001). Heart rate and blood pressure changes with endurance training: The HERITAGE Family Study. *Medicine and Science in Sports and Exercise* 33, 107–116.

Zeggini, E., Weedon, M.N., Lindgren, C.M., et al. (2007). Replication of genome-wide association signals in UK samples reveals risk loci for type 2 diabetes. *Science* 316, 1336–1341.

Chapter 16

Genetic Determinants of Exercise Performance: Evidence from Transgenic and Null Mouse Models

EUNHEE CHUNG AND LESLIE A. LEINWAND

Department of Molecular, Cellular, and Developmental Biology, University of Colorado, Boulder, CO, USA

Endurance exercise capacity is determined by multiple factors including cardiovascular, pulmonary, musculoskeletal, metabolic, psychological, and genetic. Human familial and twin studies have suggested a strong genetic component to exercise performance and to the adaptation to exercise training. For example, sedentary state $\dot{V}O_2max$ is highly correlated among siblings (Bouchard et al., 1998). Twin studies showed that after 15 weeks of endurance training, there were 5 to 10 times more variation between twin pairs than within-pairs in terms of the oxidative enzyme activity of the muscle and $\dot{V}O_2max$ response (Hamel et al., 1986). Supporting the genetic determinants of performance in twin and familial studies, mice selectively bred for high voluntary wheel running (VW) over generations resulted in higher $\dot{V}O_2max$ and elevated treadmill running capacity (Meek et al., 2009). Consistent with this finding, rats selected for high treadmill capacity have higher $\dot{V}O_2max$ than low capacity rats, and their treadmill performance is strongly correlated with their VW capacity (Waters et al., 2008). On the other hand, our laboratory showed that six different inbred mouse strains varied considerably in their ability to exercise but, perhaps more interestingly, there was no correlation in their ability to run on a voluntary wheel with forced treadmill running exercise (Lerman et al.,

2002). Taken together, most studies suggest that variability in response to exercise is highly associated with genetic background in both humans and animal models. However, it is still unclear which genes are responsible for the "endurance phenotype."

Classically, exercise physiology research has been carried out at the integrative organ levels, and the investigation of the underlying mechanisms has been carried out by mechanical, histological, biochemical, and cellular approaches. The introduction of molecular biology into the field of exercise science has provided data along the lines of candidate gene approaches and more global gene expression array analysis. Moreover, the use of genetic manipulation by overexpression or genetic inactivation of certain genes has great potential for testing whether a gene has a role in exercise endurance and/or adaptation. Functional evaluation performed at different levels, from isolated muscle contractile measurements to animal treadmill exercise performance and VW capacity, can reveal phenotypes in animals that are not observed at rest. The purpose of this chapter is to summarize the current research utilizing transgenic and null mice to elucidate the important genes that contribute to exercise performance.

Methods for Generating Mouse Models

Animals

Genetically modified animals are created by transgenic, inactivation or knock-in techniques,

Genetic and Molecular Aspects of Sport Performance, 1st edition.
Edited by Claude Bouchard and Eric P. Hoffman.
Published 2011 by Blackwell Publishing Ltd.

and the majority of studies have involved laboratory mice. Transgenic animals can be produced such that a transgene is inserted into the genome to determine the role of overexpression of a normal gene product. If one wants to assess the effect of a mutant protein on exercise performance, a transgenic approach can also be used. However, the mutant needs to function as a dominant mutation since in both cases (overexpression or mutant) the endogenous wild-type gene is still expressed. If a mutant is recessive, then homologous recombination can be used to create a "knock-in" mouse in which the mutant replaces both copies of the wild-type gene. Null mouse models have been used to address the contribution of a particular gene to exercise using homologous recombination methods. The conventional null techniques in which the gene is inactivated in all tissues at all developmental stages cannot be used if embryonic lethality occurs. In this latter case, the study may require conditional and/or organ-targeted strategies. The Cre/loxP recombination system can be a tool to eliminate an endogenous gene in a tissue-specific and/or time-dependent manner. The bacteriophage loxP (loxP) site is a 34-bp sequence that is inserted both upstream and downstream of a key exon of the gene of interest. The loxP sequence has no function in the mammalian genome, so the gene of interest functions normally despite the presence of loxP sites. These animals would then be crossed with a line of mice expressing the cre-recombinase (cre) gene in a tissue- and/or developmental-specific manner. For example, a cardiac and skeletal muscle-specific null mouse can be produced by a line of mice in which cre expression is driven by a promoter active only in heart and in muscle, such as the muscle creatine kinase (MCK) promoter. In skeletal muscle, MCK levels are first detectable in the fetus around day 16 after the heart has formed and rapidly increase more than five-fold after 28 days post-birth (Trask & Billadello, 1990). Thus, gene deletion by the MCK promoter–crerecombinase would delete the gene of interest in late gestation and would be present throughout life. The myosin light chain 1 promoter has been used to confer cre expression predominantly in fast-twitch skeletal muscle fibers (Rosenthal

et al., 1989), and the myosin light chain 2v promoter has been used to delete genes in a cardiac-specific manner (Shen et al., 1991).

Measurements of Exercise Capacity

Exercise capacity is the maximal amount of work achieved during exercise. Classically, maximal oxygen consumption ($\dot{V}O_2$max) or peak $\dot{V}O_2$ has been used to assess exercise performance. However, the validity of $\dot{V}O_2$max or peak $\dot{V}O_2$ is controversial due to the difficulty of determining the true metabolic limit or the time when the subject can no longer exercise. Currently, graded treadmill exercise (GTE) test or VW is best used to evaluate exercise capacity in laboratory animals. Exercise capacity is often quantified by parameters such as total duration, distance and speed in both GTE and VW, or the number of beam brakes at the back of the treadmill in the case of GTE. To test fatigability, grip strength or *in vitro* contractile measurements are often used. However, these variables are highly influenced by the animal's motivation, and electric shock is often used to motivate animals to run in the GTE. Thus, caution is required in comparing exercise capacity data derived from VW and treadmill exercise testing since they may involve different sets of genes. This is exemplified by data from gene expression arrays. Our laboratory, in collaboration with the National Heart, Lung and Blood Institute (NHLBI) NHLBI-sponsored Programs for Genomic Applications (PGA) at Harvard Medical School, has expression array data on hearts from FVB one of inbred strain mice female mice that underwent VW (21 days) and forced treadmill running (14 days).These time points were used because they resulted in similar degree of hypertrophy. The VW resulted in 102 up-regulated and 45 down-regulated genes, but treadmill exercise resulted in the least number of changes with 20 up-regulated and 33 down-regulated genes (Luckey et al., unpublished data). More surprisingly, there were no genes that similarly changed between the two running protocols. Consistent with our findings, hearts from rats trained for 11 weeks by treadmill exercise had 27 genes differentially expressed compared to sedentary controls (Diffee et al., 2003).

Cardiovascular Involvement in Exercise Performance

Endurance exercise training has been shown to elicit positive adaptations in the cardiovascular system. Among these adaptations, increased submaximal and maximal stroke volume (SV) is a hallmark of adaptation of the heart to endurance exercise training. Previous studies have suggested that the increased SV resulting from training can be achieved by increases in ventricular filling (end-diastolic volume: EDV) partially due to cardiac hypertrophy, myocardial contractile function, or a decrease in arterial resistance (afterload). In addition, increases in capillary density (the number of capillaries/mm^2) and myoglobin content following exercise training are well documented (Moore & Korzick, 1995).

Cardiac Hypertrophy

Cardiac hypertrophy associated with normal or enhanced cardiac function is commonly observed in well-trained humans or animal models. In animal models, heart mass or the heart weight to body weight ratio have been widely used as evidence of hypertrophy in response to training. Increased maximal SV is highly correlated with an increase in ventricular weight in response to treadmill exercise. However, not all studies showed an increase in ventricular mass with an increase in SV (Scheuer & Tipton, 1977). Thus, the role of cardiac hypertrophy as a factor for determining exercise capacity has been an important area of research to investigate. Many genetic models have been developed that show decreased or increased heart size, and some of these mouse models have been tested for their exercise capacity and hypertrophic response to exercise.

Previous studies suggest that exercise-induced cardiac hypertrophy is mediated by signaling through the phosphatidylinositol-3-kinase (PI3K) cascade. Mice expressing a cardiac-specific dominant-negative p110 alpha mutant do not show exercise-induced hypertrophy but have preserved cardiac function in response to 4 weeks of swim training (McMullen et al., 2003). They have

a training effect similar to that of wild-type mice, indicated by elevated citrate synthase activity following swim training and similar exercise performance in GTE. Similarly, protein kinase B1 (*Akt1*) null mice do not undergo swim training-induced cardiac hypertrophy (DeBosch et al., 2006). On the other hand, mice with cardiac-specific overexpression of myristoylated Akt (myr-Akt) have significant cardiac hypertrophy and exercise to the same extent in VW as non-transgenic controls (NTC). They undergo significant amounts of exercise-induced hypertrophy despite having extremely hypertrophic hearts at baseline (Luckey et al., in preparation). Although the PI3K cascade has been implicated as an important mediator of exercise-induced cardiac hypertrophy, not all downstream molecules of PI3K activation appear to be involved. Mice null for downstream effectors of PI3K signaling, S6 kinases 1 and 2, have similar swimming training capacity with normal cardiac hypertrophy (McMullen et al., 2004). Mice expressing a constitutively active GSK3 beta (caGSK3β), which is known to be an inhibitor of pathologic cardiac hypertrophy (Antos et al., 2002), have smaller hearts, but show cardiac hypertrophy following 21 days of VW and have similar capacity to the NTC (Luckey et al., in preparation). The results from others and our laboratories suggest that cardiac hypertrophy may not be a prerequisite for enhancing exercise performance.

Adrenergic Receptors

Studies with cultured myocytes and transgenic mouse models suggest that the alpha1-adrenergic receptor (AR) plays a critical role in cardiac hypertrophy in response to various stimuli, but in a sex-dependent manner (O'Connell et al., 2003). Only male mice with a double null of alpha$_{1A/C}$ and alpha$_{1B}$ AR (*ABKO*) have smaller hearts and significantly decreased exercise capacity both measured by VW and treadmill exercise. Impaired exercise capacity is likely due to a decrease in cardiac output through the decrease in heart rate and heart size. Female *ABKO* mice have normal heart size, but the consequence of female *ABKO* on exercise capacity was not determined.

On the other hand, beta-ARs are important regulators of myocardial contractile properties and energy metabolism. Beta$_1$-ARs, expressed at higher levels in the mouse heart than beta$_2$-ARs, are responsible for inotropic and chronotropic responses. Beta$_1$-AR null mice have normal heart rate and blood pressure at rest, but do not increase heart rate in response to exercise (Rohrer et al., 1998). However, their exercise capacity is not impaired in that they run the same distance at a constant speed on the treadmill and respond in a similar way to a GTE test compared to NTC. These studies suggest that compensatory mechanisms are likely to occur in beta$_1$-AR null mice. The beta$_2$-AR plays a major role in vascular relaxation. Beta$_2$-AR null mice are normotensive and have normal resting heart rate. Their exercise capacity is enhanced such that they run longer than wild-type littermates with a lower respiratory exchange ratio, suggesting that they utilize fat as major energy source (Chruscinski et al., 1999). An increased exercise capacity seen in beta$_2$-AR null mice suggests that fat utilization may cause a carbohydrate-sparing effect during exercise, resulting in their capacity to run longer distances than NTC. Double null mice, similar to beta$_1$-AR null mice, are able to run and have similar maximal exercise capacity to the NTC. However, they show decreased $\dot{V}O_2$ and VCO_2 at all workloads, indicating a blunted cardiovascular and respiratory response to a GTE (Rohrer et al., 1999). Thus, even though double null mice have a similar capacity to exercise, impaired cardiovascular and respiratory response may hinder prolonged exercise.

Another effect of adrenergic stimulation is the alteration of kinetics in the sarcoplasmic reticulum (SR). Phospholamban (PLB) is a membrane protein that regulates the SR calcium ATPase in cardiac muscle. Phosphorylation of PLB removes inhibition of the SR calcium ATPase, enhancing calcium reuptake into the SR. PLB null mice show increases in myocardial contractility and relaxation at rest, but attenuate the stimulatory effects of beta-AR agonists. However, PLB null mice are able to run same duration and have normal $\dot{V}O_2$ response compared to wild type (Desai et al., 1999).

Although PLB null mice are unable to increase ventricular contractility in response to exercise, increased $\dot{V}O_2$ during exercise could occur by compensatory mechanisms, such as decreased afterload or enhanced ventricular filling which has been shown in beta$_1$-AR null mice.

Capillary Density

Vascular endothelial growth factor (VEGF) is important for vasculogenesis and angiogenesis during embryonic and early postnatal life. Increased capillary density has been reported in response to exercise training, and VEGF mRNA levels are elevated after a single bout of exercise (Prior et al., 2004). Thus, it could be speculated that VEGF plays an important role in exercise performance. Since global or even one allele null of VEGF results in embryonic lethality, muscle-specific VEGF null mice were used to test its role in exercise performance. VEGF levels are reduced by 80% and >90% and result in the reduction of capillary density (the number of capillary per mm^2) by 61% and nearly 50% in cardiac and skeletal muscles, respectively (Olfert et al., 2009). Although compensatory mechanisms undoubtedly occur in skeletal muscle of conditional VEGF null mice, including increases in oxidative and glycolytic regulatory enzymes, reduction of muscle VEGF limits exercise capacity, as evidenced by much lower maximal running speed and submaximal endurance treadmill running compared to NTC. In addition, these mice exhibit cardiac dysfunction such as 15% lower fractional shortening. Thus, a minimum number of muscle capillaries are required to perform exercise. The contribution of genes in the cardiovascular system to exercise capacity is summarized in Table 16.1.

Skeletal Muscle and Exercise Capacity

Skeletal muscle is composed of heterogeneous myofibers that adapt to various demands that enable them to perform a variety of functions. Myofibers are classified depending on their metabolic and contractile properties, and fatigue susceptibility.

Table 16.1 Genetic models to examine cardiovascular components of exercise capacity

Mouse models	Pathways function	Tissue specificity	Phenotype	Types of exercise	Results
dnp110α (McMullen et al., 2003)	Cardiac hypertrophy	Cardiac muscle	↓ HW	GTE	≈Total distance run
ABKO (O'Connell et al., 2003)	Cardiac hypertrophy; smooth muscle contraction	Cardiac muscle	Normal function at rest ↓ HW (only male)	VW GTE	↓ Total distance, duration, and speed (V) ↑ Beam brakes (GTE)
β1-AR−/− (Rohrer et al., 1998)	Cardiac contractility	Cardiac muscle	Normal at rest ↓ HR during exercise	GTE	≈Total distance run ≈$\dot{V}O_2$ response during exercise
β2-AR−/− (Chruscinski et al., 1999)	Vascular relaxation; metabolism	Cardiac/smooth muscle	Normal at rest ↑ BP during exercise	GTE	↑ Total distance run ↓ Beam brakes
β1/β2-AR−/− (Rohrer et al., 1999)	Cardiac contractile function; vascular relaxation; metabolism	Cardiac/smooth muscle	Normal at rest ↓ HR during exercise	GTE	≈Total distance run ↓$\dot{V}O_2$ response during exercise
PLB−/− (Desai et al., 1999)	Cardiac contraction and relaxation	Cardiac muscle	↑ Basal cardiac function; ↓ inotropic and lusitropic responses to β-adrenergic stimulation	GTE	≈Time to exhaustion ≈$\dot{V}O_2$ response
mVEGF−/− (Olfert et al., 2009)	Regulate capillary density	Cardiac/skeletal muscle	↓ BW, HW/BW, muscle mass; ↓ FS by 15% ↓ Capillary density	GTE	↓ Maximal running speed ↓ Time to exhaustion

dnp110α, cardiac-specific dominant-negative p110α mutant mice; ABKO, double null of $\alpha_{1A/C}$ & α_{1B} adrenergic receptor mice; AR, adrenergic receptor; PLB, phospholamban; VEGF, vascular endothelial growth factor; GTE, graded treadmill exercise test; VW, voluntary wheel running; BW, body weight; HW, heart weight; BP, blood pressure; HR, heart rate; FS, % fractional shortening; ↑, increase; ↓, decrease.

Type I fibers preferentially metabolize fatty acids, contain slow isoforms of contractile proteins, and are fatigue resistant. Type II fibers preferentially use glycolytic metabolism, are involved in rapid bursts of contraction, express fast isoforms of contractile proteins, and are susceptible to fatigue. Type II fibers are further classified as IIA, IID/X, and IIB. Type IIA fibers have fast contractile properties but use both oxidative and glycolytic metabolism. In contrast, type IIB fibers are glycolytic with rapid fatigability, and IID/X lie between type IIA and IIB. Exercise training results in shifts in myosin heavy chain (MyHC) expression in skeletal muscle from fast to slower isoforms (IIb→IId/x→IIa→I) (Baldwin & Haddad, 2001).

Genes Involved in Skeletal Muscle Remodeling and Mitochondrial Biogenesis

Calcium-dependent signaling pathways that involve calcineurin, various kinases, transcription factors, and transcriptional regulators are involved in myofiber remodeling that promotes the transformation of fast, glycolytic fibers into slow, oxidative fibers (Bassel-Duby & Olson, 2006). Upon activation of calcineurin, nuclear factor of activated T-cells (NFAT) is activated and results in translocation of NFAT from the cytoplasm to the nucleus. Thus, the modulation of calcineurin or its downstream effectors play important roles in determining muscle fiber types. For example, when the inhibitory effect of calcineurin binding protein is inactivated by myozinin 1 (*Myoz1*), the amount of nuclear NFAT is significantly increased, resulting in a fast to slow fiber-type transformation compared to wild-type littermates. In addition, they have elevated oxidative enzymes and enhanced exercise performance such that they run faster and longer in both voluntary and treadmill exercise (Frey et al., 2008). Myocyte enhancer factor 2 (Mef2) also plays an important role in skeletal remodeling and fiber-type differentiation. Skeletal muscle-specific deletion of Mef2c or Mef2d results in a reduction in slow fibers within the soleus (Potthoff et al., 2007). In adult skeletal muscle, Mef2 is tightly regulated by class II histone deacetylases (HDACs). Phosphorylation of class II HDACs from various kinases, such as protein kinase D1 (PKD1), results in their export from the nucleus and activates MEF2-dependent genes. Overexpression of Mef2c in mice (Potthoff et al., 2007) or constitutively active PKD1 (Kim et al., 2008) promotes the formation of slow and oxidative myofibers resulting in improved exercise capacity and fatigue resistance. These results support that the calcineurin axis is an important mediator for skeletal muscle remodeling and altered fiber type with increased oxidative enzyme and enhanced exercise performance.

In conjunction with muscle fiber-type shifts, the oxidative capacity of skeletal muscle plays a major role in determining exercise capacity. Recent evidence implicates the transcriptional coactivator, peroxisome proliferator-activated receptor (PPAR)-gamma coactivator 1 (PGC-1alpha), in the maintenance of muscle fiber-type integrity, energy metabolism, and thus exercise capacity. For example, muscle-specific overexpression of PGC-1alpha (MCK-PGC-1alpha) results in a muscle fiber shift from glycolytic to oxidative with up-regulation of genes encoding enzymes in oxidative and fat metabolism (Calvo et al., 2008). Consequently, MCK-PGC-1alpha mice ran longer distances in both voluntary and forced treadmill exercise (Calvo et al., 2008), and showed decreased fatigability during *ex vivo* electrical stimulation (Lin et al., 2002). In addition, they have significantly higher peak $\dot{V}O_2$, and lower respiratory exchange ratio during peak $\dot{V}O_2$ challenge, indicating that they are utilizing fatty acid as fuel during exercise. Supporting the

role of PCG-1alpha in enhancing exercise capacity, muscle-specific PGC-1alpha null mice display the converse shift in muscle fiber types from oxidative type I and IIA toward glycolytic type IID/X and IIB fibers (Handschin et al., 2007). They perform less treadmill exercise compared to NTCs such that they run less distance, time and perform less work. Most of the results from transgenic and null models of PCG-1alpha are consistent with the idea that PCG-1alpha is a performance enhancer. However, one study demonstrated that overexpression of PCG-1alpha by a tetracycline response element promoter (TRE-PCG-1alpha) impairs exercise capacity (Wende et al., 2007). Although TRE-PCG-1alpha mice have similar characteristics to the MCK-PGC-1alpha mice, they show decreased exercise capacity during a high-intensity treadmill test due to an inability to utilize muscle glycogen.

Transgenic and null studies demonstrate that PPARdelta and AMP-activated protein kinases (AMPK) are important regulators of type I fiber determination and skeletal muscle metabolism. Overexpression of a constitutively active PPARdelta (VP-PPARdelta) in skeletal muscle recapitulates the response of skeletal muscle to exercise adaptation, switching muscle fiber types to type I, increasing oxidative enzymes, and promoting mitochondrial biogenesis. Consequently, VP-PPARdelta mice run longer distances and time whereas PPARdelta null mice run significantly less distance and time in treadmill exercise (Wang et al., 2004). Surprisingly, the muscles of VP-PPARdelta mice showed significantly increased AMPK activity without exercise suggesting cross talk between AMPK and PPARdelta (Narkar et al., 2008). Indeed, transgenic mice overexpressing an inactivated kinase AMPK alpha2 in cardiac and skeletal muscle (AMPKalpha2-KD) exhibit reduced capacity for voluntary running (Mu et al., 2001). Genes related to mitochondrial biogenesis and fiber-type phenotypes and their contribution to exercise are summarized in Table 16.2.

Myosin Heavy Chain

Given that exercise training shifts muscle fiber types, we can speculate that MyHCs play a major role in determining exercise capacity. Harrison et al. (2002)

Table 16.2 Genes related to mitochondrial biogenesis and fiber-type phenotypes and their contribution to exercise

Mouse models	Tissue specificity	Phenotype/mechanisms	Types of exercise	Results
Myoz1−/− (Frey et al., 2008)	Fast-twitch muscle	↓ BW ↓ Fast-twitch muscle mass ↑ Oxidative enzymes	VW GTE	↑ Running distance and speeds (VW) ↑ Running distance and time (GTE)
MEF2C-VP16 (Potthoff et al., 2007)	Skeletal muscle	↑ Slow fibers	GTE	↑ Running time and distance
MCK-caPDK1(Kim et al., 2008)	Fast-twitch muscle	↓ BW ↑ MyHC I & IIa; ↓ IIb ↑ Oxidative gene expression	*In vitro* contractile measurement	Fatigue resistance
MCK-PGC-1α (Calvo et al., 2008)	Skeletal/cardiac muscle (expressed preferentially in type I fibers)	↑ Type I fibers in type II fibers (i.e., white vastus and plantaris) ↑ Genes encoding fatty acid oxidation, oxidative phosphorylation ↑ Muscle glycogen	VW GTE	↑ Running distance
TRE-PGC-1α (Wende et al., 2007)	Skeletal/cardiac muscle (expressed preferentially in type I fibers)	↓ Ability to utilize glycogen during exercise	GTE	↓ Running distance in high-intensity exercise
PGC-1α−/− (Handschin et al., 2007)	Skeletal/cardiac muscle (expressed preferentially in type I fibers)	↓ # of mitochondrial genes ↑ Type IIX/D and IIB fibers ↓ Type I and IIA fibers	GTE Grip strength	↓ Running time, distance, and work (GTE) ↓ Grip strength by 60%
VP16-PPARδ (Wang et al., 2004)	Skeletal/cardiac muscle (10-fold less in heart)	↑ # of type I fibers ↑ Oxidative enzymes, mito-chondrial biogenesis	GTE	↑ Running time and distance

Myoz1−/−, myozinin 1 null mice; MEF2C-VP16, overexpression of Mef2c mice; MCK-caPDK1, muscle-specific overexpression of constitutively active protein kinase D1; MCK-PGC-1α, muscle-specific overexpression of PGC-1α; TRE-PGC-1α, overexpression of PCG-1α by a tetracycline response element promoter ; PGC-1α−/−, null mice; VP16-PPARδ, overexpression of a constitutively active PPARδ.

tested the role of fast skeletal MHC isoform genes by using either IId/x or IIb null mice. Both IId/x and IIb null mice engage in VW exercise, but average speed and distance are lowest in IId/x null mice and moderately impaired in IIb null mice compared to wild-type mice. This may be due to the compensatory mechanisms in IIb null mice such that increased citrate synthase activity may improve exercise more than in IId/x mice. In addition, muscle oxidative capacity and muscle fiber cross-sectional area are elevated after 4 weeks VW training in both IId/x and IIb null mice similar to the wild-type control mice despite lower exercise capacity in both null mice. Even with much lower

exercise performance in IId/x null mice, MyHC isoform shifts occurred (increased I and IIa and decreased IIb). In contrast, no MyHC isoform shifts occurred in response to exercise in the IIb null mice. This study suggests that individual MyHC isoforms, especially IId/x, is required for optimal exercise performance (see Table 16.3).

Structural Proteins

It has been suggested that the extracellular matrix and intracellular structural proteins are necessary for normal exercise capacity in mice. Desmin is a major intermediate filament protein expressed in

Table 16.3 Contractile proteins and exercise capacity

Mouse models	Pathway function	Tissue specificity	Phenotype	Types of exercise	Results
MyHC IId/x−/− (Harrison et al., 2002)	Skeletal muscle contraction	Skeletal muscle	Not significantly different	VW	↓↓Lowest in mean distance, duration, and speed compared to NTC
MyHC IIb−/− (Harrison et al., 2002)	Skeletal muscle contraction	Skeletal muscle	↑ Citrate activity ↓ Total fiber number in tibialis anteria	VW	↓Lower in mean distance, duration, and speed
Desmin−/− (Haubold et al., 2003)	Intermediate filament protein/protect muscles from mechanical stress	Skeletal, cardiac, and smooth muscle	Skeletal and cardiac myopathy	VW and GTE	↓ Running time, speed, and distance No training effect after 3 weeks of VW ↓ Treadmill stress and endurance tests
Syncoilin−/− (McCullagh et al., 2008)	Intermediate filament protein	Skeletal/cardiac muscle	No overt phenotype	VW and *in vitro* contractile measurement	No effects on voluntary exercise capacity ↓ By 28% in maximal force from EDL, but not soleus Predispose muscle damage

MyHC, myosin heavy chain.

cardiac, skeletal, and smooth muscle in vertebrates. It is located on the periphery of Z disks of the sarcomere which connects Z disks to one another. It has been suggested that desmin plays key roles in protecting muscles from mechanical stress, and involved in sarcomeric integrity and organization. Desmin null mice run less time, distance, and speed in VW and are unable to tolerate the treadmill stress and endurance test compared to wild-type littermates (Haubold et al., 2003). Syncoilin, another intermediate filament protein that has been shown to interact with both desmin and α-dystrobrevin, is also highly expressed in skeletal and cardiac muscle. Although desmin null mice lose syncoilin from the Z disk, a mouse null for syncoilin does not affect the expression of desmin or α-dystrobrevin. Unlike the desmin null mice, syncoilin null mice do not show altered voluntary exercise capacity (McCullagh et al., 2008). However, *in vitro* contractile measurements showed that maximal force was reduced nearly 30% in extensor digitorum longus (EDL) of syncoilin null mice compared to wild-type

littermates, but not soleus. This result suggests that syncoilin may play a role in force generation in fast-type fibers, but overall the loss of syncoilin may be compensated for by other intermediate filament proteins, such as desmin. The effects of MyHC isoforms and structural proteins on exercise are summarized in Table 16.3.

Summary

With the use of null and transgenic technology, there are increasing number of models to study the importance of genes involved in exercise performance. These studies suggest that certain genes are required to achieve optimal exercise performance, but others are physiologically redundant and thus compensatory mechanisms exist. However, caution is required to interpret the data from the genetically manipulated mouse models. First, MCK-driven conditional null mice or transgenic mice (depending on the promoter used to drive transgene expression) can affect both cardiac and skeletal muscle, but the majority of studies have investigated only one or the other but not usually both. Thus, the resulting phenotypes could result not only from skeletal muscle, but are also affected by modification of cardiac function. Second, strain differences can be significant confounding factors in exercise capacity, and genetically modified mouse models may involve multiple strains. Third, sex differences in exercise performance and adaptation have been well described in wild-type mice (Konhilas et al., 2004) as well as selectively bred mice (Meek et al., 2009). However, many exercise studies present results from mixtures of sexes or only one sex. In the future, studies need to include both sexes when identifying genes that are important in exercise performance and response to exercise.

References

Antos, C.L., McKinsey, T.A., Frey, N., et al. (2002). Activated glycogen synthase-3beta suppresses cardiac hypertrophy in vivo. *Proceedings of the National Academy of Sciences of the United States of America* 99, 907–912.

Baldwin, K.M. & Haddad, F. (2001). Plasticity in skeletal, cardiac, and smooth muscle: Invited review: Effects of different activity and inactivity paradigms on myosin heavy chain gene expression in striated muscle. *Journal of Applied Physiology* 90, 345–357.

Bassel-Duby, R. & Olson, E.N. (2006). Signaling pathways in skeletal muscle remodeling. *Annual Review of Biochemistry* 75, 19–37.

Bouchard, C., Daw, E.W., Rice, T., et al. (1998). Familial resemblance for $\dot{V}O_2$max in the sedentary state: The HERITAGE Family Study. *Medicine and Science in Sports and Exercise* 30, 252–258.

Calvo, J.A., Daniels, T.G., Wang, X., et al. (2008). Muscle-specific expression of PPAR{gamma} coactivator-1{alpha} improves exercise performance and increases peak oxygen uptake. *Journal of Applied Physiology* 104, 1304–1312.

Chruscinski, A.J., Rohrer, D.K., Schauble, E., Desai, K.H., Bernstein, D. & Kobilka, B.K. (1999). Targeted disruption of the beta 2 adrenergic receptor gene. *The Journal of Biological Chemistry* 274, 16694–16700.

DeBosch, B., Treskov, I., Lupu, T.S., et al. (2006). Akt1 is required for physiological cardiac growth. *Circulation* 113, 2097–2104.

Desai, K.H., Schauble, E., Luo, W., Kranias, E. & Bernstein, D. (1999). Phospholamban deficiency does not compromise exercise capacity. *American Journal of Physiology—Heart and Circulatory Physiology* 276, H1172–H1177.

Diffee, G.M., Seversen, E.A., Stein, T.D. & Johnson, J.A. (2003). Microarray expression analysis of effects of exercise training: Increase in atrial MLC-1 in rat ventricles. *American Journal of Physiology—Heart and Circulatory Physiology* 284, H830–H837.

Frey, N., Frank, D., Lippl, S., et al. (2008). Calsarcin-2 deficiency increases exercise capacity in mice through calcineurin/NFAT activation. *The Journal of Clinical Investigation* 118, 3598–3608.

Hamel, P., Simoneau, J.A., Lortie, G., Boulay, M.R. & Bouchard, C. (1986). Heredity and muscle adaptation to endurance training. *Medicine and Science in Sports and Exercise* 18, 690–696.

Handschin, C., Chin, S., Li, P., et al. (2007). Skeletal muscle fiber-type switching, exercise intolerance, and myopathy in PGC-1{alpha} muscle-specific knockout animals. *The Journal of Biological Chemistry* 282, 30014–30021.

Harrison, B.C., Bell, M.L., Allen, D.L., Byrnes, W.C. & Leinwand, L.A. (2002). Skeletal muscle adaptations in response to voluntary wheel running in myosin heavy chain null mice. *Journal of Applied Physiology* 92, 313–322.

Haubold, K.W., Allen, D.L., Capetanaki, Y. & Leinwand, L.A. (2003). Loss of desmin leads to impaired voluntary wheel running and treadmill exercise performance. *Journal of Applied Physiology* 95, 1617–1622.

Kim, M.-S., Fielitz, J., McAnally, J., et al. (2008). Protein kinase D1 stimulates MEF2 activity in skeletal muscle and enhances muscle performance. *Molecular and Cellular Biology* 28, 3600–3609.

Konhilas, J.P., Maass, A.H., Luckey, S.W., Stauffer, B.L., Olson, E.N. & Leinwand, L.A. (2004). Sex modifies exercise and cardiac adaptation in mice. *American Journal of Physiology—Heart and Circulatory Physiology* 287, H2768–H2776.

Lerman, I., Harrison, B.C., Freeman, K., et al. (2002). Genetic variability in forced and voluntary endurance exercise performance in seven inbred mouse strains. *Journal of Applied Physiology* 92, 2245–2255.

Lin, J., Wu, H., Tarr, P.T., et al. (2002). Transcriptional co-activator PGC-1α drives the formation of slow-twitch muscle fibres. *Nature* 418, 797–801.

McCullagh, K.J., Edwards, B., Kemp, M.W., Giles, L.C., Burgess, M. & Davies, K.E. (2008). Analysis of skeletal muscle function in the C57BL6/SV129 syncoilin knockout mouse. *Mammalian Genome* 19, 339–351.

McMullen, J.R., Shioi, T., Zhang, L., et al. (2003). Phosphoinositide 3-kinase (p110{alpha}) plays a critical role for the induction of physiological, but not pathological, cardiac hypertrophy. *Proceedings of the National Academy of Sciences of the United States of America* 100, 12355–12360.

McMullen, J.R., Shioi, T., Zhang, L., et al. (2004). Deletion of ribosomal S6 kinases does not attenuate pathological, physiological, or insulin-like growth factor 1 receptor-phosphoinositide 3-kinase-induced cardiac hypertrophy. *Molecular and Cellular Biology* 24, 6231–6240.

Meek, T.H., Lonquich, B.P., Hannon, R.M. & Garland, T., Jr. (2009). Endurance capacity of mice selectively bred for high voluntary wheel running. *The Journal of Experimental Biology* 212, 2908–2917.

Moore, R.L. & Korzick, D.H. (1995). Cellular adaptations of the myocardium to chronic exercise. *Progress in Cardiovascular Diseases* 37, 371–396.

Mu, J., Brozinick, J.T., Jr., Valladares, O., Bucan, M. & Birnbaum, M.J. (2001). A role for AMP-activated protein kinase in contraction- and hypoxia-regulated glucose transport in skeletal muscle. *Molecular Cell* 7(5), 1085–1094.

Narkar, V.A., Downes, M., Yu, R.T., et al. (2008). AMPK and PPAR[delta] agonists are exercise mimetics. *Cell* 134(3), 405–415.

O'Connell, T.D., Ishizaka, S., Nakamura, A., et al. (2003). The {alpha}1A/C- and {alpha}1B-adrenergic receptors are required for physiological cardiac hypertrophy in the double-knockout mouse. *The Journal of Clinical Investigation* 111, 1783–1791.

Olfert, I.M., Howlett, R.A., Tang, K., et al. (2009). Muscle-specific VEGF deficiency greatly reduces exercise endurance in mice. *The Journal of Physiology* 587, 1755–1767.

Potthoff, M.J., Wu, H., Arnold, M.A., et al. (2007). Histone deacetylase degradation and MEF2 activation promote the formation of slow-twitch myofibers. *The Journal of Clinical Investigation* 117, 2459–2467.

Prior, B.M., Yang, H.T. & Terjung, R.L. (2004). What makes vessels grow with exercise training? *Journal of Applied Physiology* 97, 1119–1128.

Rohrer, D.K., Schauble, E.H., Desai, K.H., Kobilka, B.K. & Bernstein, D. (1998). Alterations in dynamic heart rate control in the beta 1-adrenergic receptor knockout mouse. *American Journal of Physiology—Heart and Circulatory Physiology* 274, H1184–H1193.

Rohrer, D.K., Chruscinski, A., Schauble, E.H., Bernstein, D. & Kobilka, B.K.

(1999). Cardiovascular and metabolic alterations in mice lacking both beta 1- and beta 2-adrenergic receptors. *The Journal of Biological Chemistry* 274, 16701–16708.

Rosenthal, N., Kornhauser, J.M., Donoghue, M., Rosen, K.M. & Merlie, J.P. (1989). Myosin light chain enhancer activates muscle-specific, developmentally regulated gene expression in transgenic mice. *Proceedings of the National Academy of Sciences of the United States of America* 86, 7780–7784.

Scheuer, J. & Tipton, C.M. (1977). Cardiovascular adaptations to physical training. *Annual Review of Physiology* 39, 221–251.

Shen, R.A., Goswami, S.K., Mascareno, E., Kumar, A. & Siddiqui, M.A. (1991). Tissue-specific transcription of the cardiac myosin light-chain 2 gene is regulated by an upstream repressor element. *Molecular*

and Cellular Biology 11, 1676–1685.

Trask, R.V. & Billadello, J.J. (1990). Tissue-specific distribution and developmental regulation of M and B creatine kinase mRNAs. *Biochimica et Biophysica Acta* 1049, 182–188.

Wang, Y.X., Zhang, C.L., Yu, R.T., et al. (2004). Regulation of muscle fiber type and running endurance by PPARdelta. *PLoS Biology* 2, 1532–1539.

Waters, R.P., Renner, K.J., Pringle, R.B., et al. (2008). Selection for aerobic capacity affects corticosterone, monoamines and wheel-running activity. *Physiology & Behavior* 93, 1044–1054.

Wende, A.R., Schaeffer, P.J., Parker, G.J., et al. (2007). A role for the transcriptional coactivator PGC-1{alpha} in muscle refueling. *The Journal of Biological Chemistry* 282, 36642–36651.

Chapter 17

The *ACE* Gene and Performance

JAMES R.A. SKIPWORTH, ZUDIN A. PUTHUCHEARY,
JAI RAWAL AND HUGH E. MONTGOMERY

UCL Institute for Human Health and Performance, London, UK

Peptidyl dipeptidase angiotensin-converting enzyme (ACE) is a key enzymatic component of the renin-angiotensin system, having critical endocrine and paracrine roles in the regulation of blood pressure and salt/water balance. The *ACE* gene shows a polymorphic insertion of an alu repetitive element in intron 16, and this 287-bp insertion/deletion (I/D) polymorphism has been directly linked to inter-individual variation in plasma and tissue ACE activity. The clear functional associations of the I/D polymorphism has led to many studies of association with cardiovascular and exercise-related phenotypes in human populations. In this chapter, we review this extensive literature. The I/D polymorphism is associated with differences in response to training, including muscle endurance, and anabolic strength improvements. The I/D polymorphism has also been associated with proportion of type 1 fibers in muscle tissue, and cardiac muscle growth. The associations of *ACE* polymorphisms with fitness ($\dot{V}O_2$max), and elite sports performance remain more controversial.

Human Endocrine and Paracrine Renin–Angiotensin Systems

The renally derived protease renin cleaves angiotensinogen to yield decapeptide angiotensin I, acted upon by the peptidyl dipeptidase angiotensin-converting enzyme (ACE) to generate octapeptide angiotensin II (ang II). The vascular effects of ang II seem to be mediated through two specific human receptors (AT_1R and AT_2R) (Timmermans & Smith, 1994), although other receptors (at which ang II degradation products may be especially active) also exist. Stimulation of the AT_1R by ang II increases blood pressure through vasoconstriction and through salt and water retention secondary to adrenal aldosterone release. Levels of (vasodilator) bradykinin are also inversely related to ACE activity (Brown et al., 1998; Murphey et al., 2000). Increasing ACE activity therefore drives hypertensive responses (increased AT_1R receptor activation) and diminishes hypotensive responses (reduced BK_2 receptor activation), thereby playing a crucial role in the regulation of human blood pressure and salt and water balance (Kem & Brown, 1990). In addition to this endocrine renin–angiotensin system (RAS), however, local RAS also exist in disparate cells and organs and tissues (Dzau, 1989; Nazarov et al., 2001; Paul et al., 2006) where they serve a variety of functions, many of which are related to the regulation of tissue growth and injury responses.

Functional polymorphic variants have been identified in the genes of RAS components, including renin, angiotensinogen, ang II, and bradykinin receptors (Table 17.1). The most well-studied of them, however, is a polymorphism in the human angiotensin-1 converting enzyme gene. Plasma ACE levels are very stable within individuals, but marked inter-individual variations exist

Genetic and Molecular Aspects of Sport Performance, 1st edition.
Edited by Claude Bouchard and Eric P. Hoffman.
Published 2011 by Blackwell Publishing Ltd.

Table 17.1 Polymorphisms of other RAS genes are similarly associated with differences in RAS activity

Gene	Polymorphism/Alleles	Gene location	Functional effect
ACE (17q22-24) (Costerousse et al., 1993; Danser et al., 1995; Rigat et al., 1990; Villard et al., 1996)	287-bp insertion (I)/ deletion (D) C>T (position 4656)	Intron 16 3′ UTR	Protein levels Protein levels
Angiotensinogen (1q42-q43) (Inoue et al., 1997; Ishigami et al., 1997; Jeunemaitre et al., 1992a–c; Paillard et al., 1999; Sato et al., 1997)	M>T (position 235) A>C (position 20) A>G (position 6) C>T (position 532)	Exon2 (+704) 5′ UTR promoter 5′ UTR promoter 5′ UTR	Plasma protein levels Plasma protein levels Protein levels Protein levels
Renin (1q32-q32) (Mansego et al., 2008)	rs5707 (T>G)	Intron 4	Protein levels
Angiotensin II type 1 receptor (3q21-q25) (21-23 Bonnardeaux et al., 1994; Miller et al., 1999; Poirier et al., 1998)	A>C (position 1166) T>A (position 810)	3′UTR Promoter	Receptor sensitivity Unknown
Angiotensin II type 2 receptor (Xq22-q23) (Martin & Elton, 1995)	G>A (position 1675)	Intron 1	Protein levels

(Alhenc-Gelas et al., 1991). The absence (Deletion "D") rather than the presence (Insertion "I") of a 287 bp alu repeat sequence within intron 16 of the *ACE* gene is associated with elevated plasma (Rigat et al., 1990) and tissue (Costerousse et al., 1993; Danser et al., 1995) ACE activity: those homozygous for the deletion allele demonstrate an elevation of ACE activity of almost 75% in the heart and white blood cells when compared to those of II/ID genotype (Costerousse et al., 1993; Danser et al., 1995).

The *ACE* I/D Polymorphism and Human Muscle

The *ACE* I/D polymorphism was the first specific gene variant to be associated with human physical performance (Montgomery et al., 1998). Since that time, the *ACE* gene has become by far the best-studied locus in this regard. The maximum duration of standardized repetitive elbow flexion with a 15-kg barbell was studied in young Caucasian military recruits. At baseline, performance prior to training was found to be independent of *ACE* genotype. However, after 10 weeks of military training, the duration of exercise was

significantly prolonged in the 66 individuals of either II or ID genotype (79.4 ± 25.2 and 24.7 ± 8.8 s: $P = 0.005$ and 0.007 respectively) but not for the 12 DD homozygotes (7.1 ± 14.9 s: $P = 0.642$). Thus, II homozygotes showed an 11-fold greater improvement when compared to those with the DD genotype ($P = 0.001$) (Montgomery et al., 1998).

Such effects may be mediated through increased ang II synthesis or kinin degradation, and through a variety of downstream mechanisms. In this regard, allele-specific effects on "delta efficiency" (DE)—the ratio of external mechanical work to internal work performed by skeletal muscle—may contribute to functional performance changes. In a subsequent study of young Caucasian military recruits, DE increased significantly over a period of training in those of II, but not DD, genotype (Williams et al., 2000). In keeping with an effect on "metabolic efficiency," the I-allele seems associated with a relative anabolic response among military recruits under conditions of intensive training and high calorie expenditure during training (Montgomery et al., 1999). Similarly, in postmenopausal women, the I-allele is associated with greater increases in adductor pollicis muscle strength gain in response to hormone replacement therapy (HRT) (Woods et al., 2001b). Such effects may, in turn, be partly dependent on genotype-differences in skeletal muscle fiber-type: the I-allele is associated with a higher proportion of type I (slow-twitch, fatigue-resistant) fibers, as compared to the D allele (associated with a lower proportion of type IIb) (Scott et al., 2001; Zhang et al., 2003).

However, the *ACE* I/D polymorphism may exert influence on muscle function in other ways. In this regard, differences in muscle fiber *size and growth* may also be important: ang II is not only a trophic agent for vascular smooth muscle (Berk et al., 1989) but also for cardiac muscle (Liu et al., 1998; Wollert & Drexler, 1999). Alterations in cardiac size has been associated with both increased (in race horses) and decreased performance/excess mortality (in human disease states) (Buhl et al., 2005; Koren et al., 1991; Levy et al., 1990; Vakili et al., 2001; Young et al., 2005). A local myocardial RAS exists (Danser et al., 1999), which may mediate both physiological and pathophysiological cardiac growth. Thus, expression of myocardial RAS components increases during ventricular hypertrophy (Schunkert et al., 1990), while ACE inhibition attenuates such growth (Lievre et al., 1995) in a variety of animal models. It may also lead to a greater regression in left ventricular (LV) mass than other similarly hypotensive agents (Cruickshank et al., 1992; Schmieder et al., 1996, 1998). In keeping for a role for ACE in mediating human LV growth, the deletion (D) rather than the insertion (I) polymorphic variant of the *ACE* gene is associated with greater LV ACE activity (Danser et al., 1995) and a greater LV growth response (Myerson et al., 2001). In further support, the *ACE* genotype appears to be associated with pathological LV hypertrophy, in conditions as diverse as diabetes (Estacio et al., 1999), hypertension (Kuznetsova et al., 2000), hypertrophic cardiomyopathy (Yoneya et al., 1995), and aortic stenosis (Dellgren, 1999). However, the *ACE* D-allele also seems associated with an exaggerated LV growth response under training conditions. In 140 Caucasian male military recruits exposed to 10 weeks of standardized physical training, echocardiographic LV mass increased by 18%, and the magnitude of this rise was significantly associated with *ACE* I/D genotype (mean LV mass changing by +2.0, +38.5, and +42.3 g in the II, ID, and DD groups respectively ($P < 0.0001$)). Further, the prevalence of LVH, as defined by echocardiogram, also increased significantly among those of *DD* genotype only ($P < 0.01$) (Montgomery et al., 1997). Similarly, Di Mauro et al. (2010) investigated 74 white healthy male endurance athletes to show that a DD group of athletes had significantly higher LVMI, as well as a higher prevalence of subjects with LVH (LVMI >131 g/m^2). However, 15 athletes with *ACE* DD and *AGTR1* AC/CC genotypes demonstrated the highest LVMI (150 ± 23 g/m^2); whereas the lowest LVMI was in a patient with *ACE* ID and *AGTR1* AA (127 ± 18 g/m^2) genotype.

Further gene–environment interaction studies have since confirmed these results in studies of LV mass in alternate endurance athlete groups (Hernandez et al., 2003), in wrestlers (Kasikcioglu et al., 2004), and in football players (Fatini et al., 2000). Those studies that have failed to identify such associations have generally studied diverse populations or those exposed to a variety of hypertrophic stimuli over variable periods of time (Karjalainen et al., 1999; Lindpaintner et al., 1996; Staessen et al., 1997; Yildiz et al., 2000).

These hypertrophic effects may partly be mediated through increased synthesis of Ang II (Beinlich et al., 1991; Liu et al., 1998), which activates the AT$_1$ receptor, resulting in hypertrophy of cardiomyocytes (Liu et al., 1998; Wollert & Drexler, 1999). However, the administration of non-hypotensive doses of antagonists to the AT$_1$ receptor seems to have little impact on physiological (exercise-driven) LV growth (Myerson et al., 2001). Conversely, increased ACE activity may promote cardiac growth through the increased degradation of growth inhibitory kinins (Brull et al., 2001; Murphey et al., 2000; Linz & Scholkens, 1992a, b). A polymorphic variant of the bradykinin 2 receptor (Brull et al., 2001) gene exists, whereby the absence (−) rather than the presence (+) of a 9-bp deletion in the gene is associated with greater gene transcription (Braun et al., 1996; Lung et al., 1997) and receptor response (Houle et al., 2000). In a group of 109 military recruits undergoing

10 weeks of basic physical training, both the *ACE* and *B2BKR* genotypes interacted biologically in an additive way, with those of a genotype likely to be associated with lowest kinin activity (*ACE* DD, *B2BKR* +9/+9) exhibiting the greatest LV growth. In these, mean LV growth was 15.7 g, compared to −1.37 g in those homozygous for *ACE* I and *B2BKR* −9 genotypes ($P = 0.003$ for trend across genotypes) (Brull et al., 2001).

The role of such genotype-dependent alterations in human cardiac growth on human performance is unclear. However, *ACE* genotypes may also influence human *skeletal* muscle growth, in which angiotensin II may play a key role in transducing mechanical load to yield growth responses (Gordon et al., 2001). In support, quadriceps muscle strength has been associated with the D-allele in patients with chronic obstructive pulmonary disease (Hopkinson et al., 2004), and increases in quadriceps muscle strength in response to training appear to be D-allele related in young adult men (Folland et al., 2000).

However, ang II may also have contrary actions, regulating the synthesis of cytokines which may mediate muscle wasting—including interlukin-6 (IL6) (Pueyo et al., 2000; Hernandez-Presa et al., 1997)—or inhibit production of other cellular growth factors. Thus, rodents exposed to ang II infusion suffer marked muscle wasting to which inhibition of the insulin-like growth factor system may contribute (Ailhaud, 1999; Brink et al., 1996, 2001).

Further, ang II may influence muscle metabolism, modulating the action of insulin at the muscle cell level (Di Bari et al., 2004), while ACE may also influence substrate delivery and the efficiency of skeletal muscle and subsequent conservation of energy stores (Moran et al., 2006). Meanwhile, RAS expression also plays an important role in the regulation of fat storage in adipocytes (Ailhaud, 1999; Shenoy & Cassis, 1997). The *ACE* I-allele is thus associated with the ability to recognize and avoid severe hypoglycemic events in diabetics (Pedersen-Bjergaard et al., 2001), and with high altitude (hypoxic) physical performance (Montgomery et al., 1998; Thompson et al., 2007; Tsianos et al., 2005). However, RAS effects on muscle may also be mediated through bradykinin's action at its type 1 and 2 receptors (Carter et al.,

2005). Bradykinin has metabolic effects, influencing glycogen levels, lactate concentration (Linz et al., 1996), the availability of glucose/free fatty acid substrates (Wicklmayr, 1980), and the expression of the GLUT4 glucose transporter (Taguchi et al., 2000). Among patients suffering chronic obstructive pulmonary disease, those homozygous for the +9 bradykinin receptor polymorphism (see earlier) exhibit reduced fat-free mass and quadriceps strength (Hopkinson et al., 2006).

Given these effects of *ACE* (and *RAS*) genotype on human cardiac and skeletal muscle, what associations exist with more global measures of physical performance?

ACE and $\dot{V}O_2$max

A study of 47 postmenopausal women demonstrated that *ACE* II genotype carriers had a 6.3 ml/kg/min higher $\dot{V}O_2$max ($P < 0.05$) than the *ACE* DD group and a 3.3 ml/kg/min higher $\dot{V}O_2$max ($P < 0.05$) than the *ACE* ID genotype group (Hagberg et al., 1998). Similar findings have been reported in 57 smoking patients with ischemic heart disease and dilated cardiomyopathy (Abraham et al., 2002). However, the validity of any association between *ACE* genotype and $\dot{V}O_2$max remains unproven. Day et al. (2007) studied $\dot{V}O_2$max in 62 sedentary females undergoing submaximal cycle ergometry, and found no association with *ACE* genotype. Similarly, Roltsch et al. (2005) found no relationship between *ACE* genotype and $\dot{V}O_2$max in 77 sedentary and exercise-trained young females undergoing maximal treadmill exercise tests. Further, no significant association was found between *ACE* genotype, peak oxygen uptake, or other measures of performance in 147 US army recruits undergoing 8 weeks of basic training (Sonna et al., 2001), although heterogeneity in the study population makes reliable interpretation difficult. These results were corroborated by the HERITAGE Family Study which utilized a genomic scan incorporating all 22 pairs of autosomes (Bouchard et al., 1995, 2000) but observed no evidence of linkage between the *ACE* locus (17q23) and baseline $\dot{V}O_2$max or its change after 20 weeks of standardized endurance training. Woods et al. (2002b) similarly identified no significant association when assessing the response to training in 58 army recruits.

ACE and Elite Performance

Thus, although the *ACE* genotype is not convincingly associated with alterations in $\dot{V}O_2$max, associations with other sporting intermediate phenotypes do exist (see earlier). Other data suggest that these translate into differences in more global elite performance. In general, the I-allele has been associated with elite athletic performance in endurance-orientated events (Myerson et al., 1999; Nazarov et al., 2001), and the D allele with strength/power-orientated performance (Myerson et al., 1999; Nazarov et al., 2001; Woods et al., 2001a). Thus, an excess I-allele frequency was found in Australian national rowers (Gayagay et al., 1998), Russian athletes (Nazarov et al., 2001), and elite long-distance cyclists (Alvarez et al., 2000). Similarly, among British Olympic standard runners, I-allele frequency increased with the distance run: for the three distance groups 200, 400–3000, and 5000 m, the I-allele frequencies were 0.35, 0.53, and 0.62 (*P* = 0.009 for linear trend), respectively (Myerson et al., 1999).

A number of other studies have suggested an association of the *ACE* genotype with various elite performance groups. The D allele is thus associated with elite (especially shorter-distance) swimming (Tsianos et al., 2004), and the I-allele with performance in the South African Ironman Triathlon (Collins et al., 2004). Conflicting data do exist, but generally seem to be explained by the study of heterogeneous (often mixed race and sex, and sporting discipline) subject groups (Karjalainen et al., 1999; Montgomery & Dhamrait, 2002; Sonna et al., 2001; Rankinen et al., 2000; Taylor et al., 1999; Woods et al., 2000).

Some of these effects may, in part, be mediated through alterations in bradykinin metabolism. In a study of 115 healthy men and women and of running distance in 81 Olympic standard track athletes, Williams et al. (2004) showed delta-efficiency (DE) to be strongly associated with the *B2BKR* genotype (23.84 ± 2.41 vs. 24.25 ± 2.81 vs. $26.05 \pm 2.26\%$ for those of +9/+9 vs. +9/−9 vs. −9/−9 genotype;

P = 0.0008). There was also evidence for an interaction with the *ACE* I/D genotype, with those who were of *ACE* II and *B2BKR* −9/−9 genotype having the highest baseline DE and be significantly associated with endurance (predominantly aerobic) among the elite athletes, suggesting that at least part of the association of *ACE* and fitness phenotypes is via higher kinin activity (Williams et al., 2004).

ACE Genotype and the Response to Hypoxia

The I-allele is also associated with elite mountaineering status (Montgomery et al., 1998), success in rapid ascent to high altitude (Tsianos et al., 2005) and success in ascending beyond 8000 m (Tsianos et al., 2005). Such effects may be dependent upon I-allele associated gains in $\dot{V}O_2$max, or in metabolic efficiency (Bischoff et al., 2003). However, RAS exist in the carotid body and ventral medulla, and the *ACE* I-allele may also be associated with an enhanced exertional ventilatory response to acute hypoxia (Patel et al., 2003), and thus with preserved arterial oxygenation at high altitude (Woods et al., 2002a).

Thus, genotype can influence sporting intermediate phenotypes as well as more global measures of sporting performance.

Conclusion

Differences in human phenotypes result from the interaction of genetic variation with environmental stimuli. This holds true of human sporting phenotypes, whether intermediate (e.g., muscle strength) or final (sporting "prowess"), as has been demonstrated for the *ACE* I/D polymorphism. The investigation of genetic influences is of more than prurient interest. It allows the exploration of human physiology and pathophysiology related to sport, perhaps opening the way for the development of new therapeutic strategies. Further, such knowledge may be both translational and transferable—opening the way to the better understanding of disease processes (e.g., those causing progressive muscle wasting, bone weakness and fracture, or excessive cardiac growth).

References

Abraham, M.R., Olson, L.J., Joyner, M.J., Turner, S.T., Beck, K.C. & Johnson, B.D. (2002). Angiotensin-converting enzyme genotype modulates pulmonary function and exercise capacity in treated patients with congestive stable heart failure. *Circulation* 106(14), 1794–1799.

Ailhaud, G. (1999). Cell surface receptors, nuclear receptors and ligands that regulate adipose tissue development. *Clinica Chimica Acta* 286(1–2), 181–190.

Alhenc-Gelas, F., Richard, J., Courbon, D., Warnet, J.M. & Corvol, P. (1991). Distribution of plasma angiotensin I-converting enzyme levels in healthy men: Relationship to environmental and hormonal parameters. *Journal of Laboratory and Clinical Medicine* 117(1), 33–39.

Alvarez, R., Terrados, N., Ortolano, R., et al. (2000). Genetic variation in the renin-angiotensin system and athletic performance. *European Journal of Applied Physiology* 82(1–2), 117–120.

Beinlich, C.J., White, G.J., Baker, K.M. & Morgan, H.E. (1991). Angiotensin II and left ventricular growth in newborn pig heart. *Journal of Molecular and Cellular Cardiology* 23(9), 1031–1038.

Berk, B.C., Vekshtein, V., Gordon, H.M. & Tsuda, T. (1989). Angiotensin II-stimulated protein synthesis in cultured vascular smooth muscle cells. *Hypertension* 13(4), 305–314.

Bischoff, H.A., Stahelin, H.B., Dick, W., et al. (2003). Effects of vitamin D and calcium supplementation on falls: A randomized controlled trial. *Journal of Bone Mineral Research* 18(2), 343–351.

Bonnardeaux, A., Davies, E., Jeunemaitre, X., et al. (1994). Angiotensin II type 1 receptor gene polymorphisms in human essential hypertension. *Hypertension* 24(1), 63–69.

Bouchard, C., Leon, A.S., Rao, D.C., Skinner, J.S., Wilmore, J.H. & Gagnon, J. (1995). The HERITAGE family study. Aims, design, and measurement protocol. *Medicine and Science in Sports and Exercise* 27(5), 721–729.

Bouchard, C., Rankinen, T., Chagnon, Y.C., et al. (2000). Genomic scan for maximal oxygen uptake and its response to training in the HERITAGE Family Study. *Journal of Applied Physiology* 88(2), 551–559.

Braun, A., Kammerer, S., Maier, E., Bohme, E. & Roscher, A.A. (1996). Polymorphisms in the gene for the human B2-bradykinin receptor.

New tools in assessing a genetic risk for bradykinin-associated diseases. *Immunopharmacology* 33(1–3), 32–35.

Brink, M., Wellen, J. & Delafontaine, P. (1996). Angiotensin II causes weight loss and decreases circulating insulin-like growth factor I in rats through a pressor-independent mechanism. *Journal of Clinical Investigation* 97(11), 2509–2516.

Brink, M., Price, S.R., Chrast, J., et al. (2001). Angiotensin II induces skeletal muscle wasting through enhanced protein degradation and down-regulates autocrine insulin-like growth factor I. *Endocrinology* 142(4), 1489–1496.

Brown, N.J., Blais, C., Jr., Gandhi, S.K. & Adam, A. (1998). ACE insertion/deletion genotype affects bradykinin metabolism. *Journal of Cardiovascular Pharmacology* 32(3), 373–377.

Brull, D., Dhamrait, S., Myerson, S., et al. (2001). Bradykinin *B2BKR* receptor polymorphism and left-ventricular growth response. *Lancet* 358(9288), 1155–1156.

Buhl, R., Ersbøll, A.K., Eriksen, L. & Koch, J. (2005). Changes over time in echocardiographic measurements in young Standardbred racehorses undergoing training and racing and association with racing performance. *Journal of the American Veterinary Medical Association* 226(11), 1881–1887.

Carter, C.S., Onder, G., Kritchevsky, S.B. & Pahor, M. (2005). Angiotensin-converting enzyme inhibition intervention in elderly persons: effects on body composition and physical performance. *Journals of Gerontology. Series A, Biological Sciences and Medical Sciences* 60(11), 1437–1446.

Collins, M., Xenophontos, S.L., Cariolou, M.A., et al. (2004). The *ACE* gene and endurance performance during the South African Ironman Triathlons. *Medicine and Science in Sports and Exercise* 36(8), 1314–1320.

Costerousse, O., Allegrini, J., Lopez, M. & Alhenc-Gelas, F. (1993). Angiotensin I-converting enzyme in human circulating mononuclear cells: Genetic polymorphism of expression in T-lymphocytes. *Biochemical Journal* 290, 33–40.

Cruickshank, J.M., Lewis, J., Moore, V. & Dodd, C. (1992). Reversibility of left ventricular hypertrophy by differing types of antihypertensive therapy. *Journal of Human Hypertension* 6(2), 85–90.

Danser, A.H., Schalekamp, M.A., Bax, W.A., et al. (1995). Angiotensin

converting enzyme in the human heart. Effect of the deletion/insertion polymorphism. *Circulation* 92(6), 1387–1388.

Danser, A.H., Saris, J.J., Schuijt, M.P. & van Kats, J.P. (1999). Is there a local renin-angiotensin system in the heart? *Cardiovascular Research* 44(2), 252–265.

Day, S.H., Gohlke, P., Dhamrait, S.S. & Williams, A.G. (2007). No correlation between circulating ACE activity and V̇O₂max or mechanical efficiency in women. *European Journal of Applied Physiology* 99(1), 11–18.

Dellgren, G., Eriksson, M.J., Blange, I., Brodin, L.A., Radegran, K. & Sylven, C. (1999). Angiotensin-converting enzyme gene polymorphism influences degree of left ventricular hypertrophy and its regression in patients undergoing operation for aortic stenosis. *American Journal of Cardiology* 84(8), 909–913.

Di Bari, M., van de Poll-Franse, L.V., Onder, G., et al. (2004). Antihypertensive medications and differences in muscle mass in older persons: The Health, Aging and Body Composition Study. *Journal of the American Geriatrics Society* 52(6), 961–966.

Di Mauro, M., Izzicupo, P., Santarelli, F., et al. (2010). *ACE* and *AGTR1* polymorphisms and left ventricular hypertrophy in endurance athletes. *Medicine and Science in Sports and Exercise* 42(5), 915–921.

Dzau, V.J. (1989). Multiple pathways of angiotensin production in the blood vessel wall: Evidence, possibilities and hypotheses. *Journal of Hypertension* 7(12), 933–936.

Estacio, R.O., Jeffers, B.W., Havranek, E.P., Krick, D., Raynolds, M. & Schrier, R.W. (1999). Deletion polymorphism of the angiotensin converting enzyme gene is associated with an increase in left ventricular mass in men with type 2 diabetes mellitus. *American Journal of Hypertension* 12(6), 637–642.

Fatini, C., Guazzelli, R., Manetti, P., et al. (2000). RAS genes influence exercise-induced left ventricular hypertrophy: An elite athletes study. *Medicine and Science in Sports and Exercise* 32(11), 1868–1872.

Folland, J., Leach, B., Little, T., et al. (2000). Angiotensin-converting enzyme genotype affects the response of human skeletal muscle to functional overload. *Experimental Physiology* 85(5), 575–579.

Gayagay, G., Yu, B., Hambly, B., et al. (1998). Elite endurance athletes and the

ACE I allele—the role of genes in athletic performance. *Human Genetics* 103(1), 48–50.

Gordon, S.E., Davis, B.S., Carlson, C.J. & Booth, F.W. (2001). ANG II is required for optimal overload-induced skeletal muscle hypertrophy. *American Journal of Physiology. Endocrinology and Metabolism* 280(1), E150–E159.

Hagberg, J.M., Ferrell, R.E., McCole, S.D., Wilund, K.R. & Moore GE. (1998). V̇O₂max is associated with *ACE* genotype in postmenopausal women. *Journal of Applied Physiology* 85(5), 1842–1846.

Hernandez, D., de la Rosa, A., Barragan, A., et al. (2003). The *ACE*/DD genotype is associated with the extent of exercise-induced left ventricular growth in endurance athletes. *Journal of the American College of Cardiology* 42(3), 527–532.

Hernandez-Presa, M., Bustos, C., Ortego, M., et al. (1997). Angiotensin-converting enzyme inhibition prevents arterial nuclear factor-kappa B activation, monocyte chemoattractant protein-1 expression, and macrophage infiltration in a rabbit model of early accelerated atherosclerosis. *Circulation* 95(6), 1532–1541.

Hopkinson, N.S., Nickol, A.H., Payne, J., et al. (2004). Angiotensin converting enzyme genotype and strength in chronic obstructive pulmonary disease. *American Journal of Respiratory and Critical Care Medicine* 170(4), 395–399.

Hopkinson, N.S., Eleftheriou, K.I., Payne, J., et al. (2006). +9/+9 Homozygosity of the bradykinin receptor gene polymorphism is associated with reduced fat-free mass in chronic obstructive pulmonary disease. *American Journal of Clinical Nutrition* 83(4), 912–917.

Houle, S., Landry, M., Audet, R., Bouthillier, J., Bachvarov, D.R. & Marceau, F. (2000). Effect of allelic polymorphism of the B (1) and B (2) receptor genes on the contractile responses of the human umbilical vein to kinins. *Journal of Pharmacology and Experimental Therapeutics* 294(1), 45–51.

Inoue, I., Nakajima, T., Williams, C.S., et al. (1997). A nucleotide substitution in the promoter of human angiotensinogen is associated with essential hypertension and affects basal transcription in vitro. *Journal of Clinical Investigation* 99(7), 1786–1797.

Ishigami, T., Umemura, S., Tamura, K., et al. (1997). Essential hypertension and 5′ upstream core promoter region of human angiotensinogen gene. *Hypertension* 30(6), 1325–1330.

Jeunemaitre, X., Lifton, R.P., Hunt, S.C., Williams, R.R. & Lalouel, J.M. (1992a). Absence of linkage between the angiotensin converting enzyme locus and human essential hypertension. *Nature Genetics* 1(1), 72–75.

Jeunemaitre, X., Rigat, B., Charru, A., Houot, A.M., Soubrier, F. & Corvol, P. (1992b). Sib pair linkage analysis of renin gene haplotypes in human essential hypertension. *Human Genetics* 88(3), 301–306.

Jeunemaitre, X., Soubrier, F., Kotelevtsev, Y.V., et al. (1992c). Molecular basis of human hypertension: role of angiotensinogen. *Cell* 71(1), 169–180.

Karjalainen, J., Kujala, U.M., Stolt, A., et al. (1999). Angiotensinogen gene M235T polymorphism predicts left ventricular hypertrophy in endurance athletes. *Journal of the American College of Cardiology* 34(2), 494–499.

Kasikcioglu, E., Kayserilioglu, A., Ciloglu, F., et al. (2004). Angiotensin-converting enzyme gene polymorphism, left ventricular remodeling, and exercise capacity in strength-trained athletes. *Heart Vessels* 19(6), 287–293.

Kem, D.C. & Brown, R.D. (1990). Renin—from beginning to end. *New England Journal of Medicine* 323(16), 1136–1137.

Koren, M.J., Devereux, R.B., Casale, P.N., Savage, D.D. & Laragh, J.H. (1991). Relation of left ventricular mass and geometry to morbidity and mortality in uncomplicated essential hypertension. *Annals of Internal Medicine* 114(5), 345–352.

Kuznetsova, T., Staessen, J.A., Wang, J.G., et al. (2000). Antihypertensive treatment modulates the association between the D/I ACE gene polymorphism and left ventricular hypertrophy: A meta-analysis. *Journal of Human Hypertension* 14(7), 447–454.

Levy, D., Garrison, R.J., Savage, D.D., Kannel, W.B. & Castelli, W.P. (1990). Prognostic implications of echocardiographically determined left ventricular mass in the Framingham Heart Study. *New England Journal of Medicine* 322(22), 1561–1566.

Lievre, M., Gueret, P., Gayet, C., et al. (1995). Ramipril-induced regression of left ventricular hypertrophy in treated hypertensive individuals. HYCAR Study Group. *Hypertension* 25(1), 92–97.

Lindpaintner, K., Lee, M., Larson, M.G., et al. (1996). Absence of association or genetic linkage between the angiotensin-converting-enzyme gene and left ventricular mass. *New England Journal of Medicine* 334(16), 1023–1028.

Linz, W. & Scholkens, B.A. (1992a). A specific B2-bradykinin receptor antagonist HOE 140 abolishes the antihypertrophic effect of ramipril. *British Journal of Pharmacology* 105(4), 771–772.

Linz, W. & Scholkens, B.A. (1992b). Role of bradykinin in the cardiac effects of angiotensin-converting enzyme inhibitors. *Journal of Cardiovascular Pharmacology* 20(Suppl. 9), S83–S90.

Linz, W., Wiemer, G. & Scholkens, B.A. (1996). Role of kinins in the pathophysiology of myocardial ischemia. In vitro and in vivo studies. *Diabetes* 45(Suppl. 1), S51–S58.

Liu, Y., Leri, A., Li, B., et al. (1998). Angiotensin II stimulation in vitro induces hypertrophy of normal and postinfarcted ventricular myocytes. *Circulation Research* 82(11), 1145–1159.

Lung, C.C., Chan, E.K. & Zuraw, B.L. (1997). Analysis of an exon 1 polymorphism of the B2 bradykinin receptor gene and its transcript in normal subjects and patients with C1 inhibitor deficiency. *Journal of Allergy and Clinical Immunology* 99(1 Pt 1), 134–146.

Mansego, M.L., Redon, J., Marin, R., et al. (2008). Renin polymorphisms and haplotypes are associated with blood pressure levels and hypertension risk in postmenopausal women. *Journal of Hypertension* 26(2), 230–237.

Martin, M.M. & Elton, T.S. (1995). The sequence and genomic organization of the human type 2 angiotensin II receptor. *Biochemical and Biophysical Research Communications* 209(2), 554–562.

Miller, J.A., Thai, K. & Scholey, J.W. (1999). Angiotensin II type 1 receptor gene polymorphism predicts response to losartan and angiotensin II. *Kidney International* 56(6), 2173–2180.

Montgomery, H. & Dhamrait, S. (2002). *ACE* genotype and performance. *Journal of Applied Physiology* 92(4), 1774–1775.

Montgomery, H.E., Clarkson, P., Dollery, C.M., et al. (1997). Association of angiotensin-converting enzyme gene I/D polymorphism with change in left ventricular mass in response to physical training. *Circulation* 96(3), 741–747.

Montgomery, H.E., Marshall, R., Hemingway, H., et al. (1998). Human gene for physical performance. *Nature* 393(6682), 221–222.

Montgomery, H., Clarkson, P., Barnard, M., et al. (1999). Angiotensin-converting-enzyme gene insertion/deletion polymorphism and response to physical training. *Lancet* 353(9152), 541–545.

Moran, C.N., Vassilopoulos, C., Tsiokanos, A., et al. (2006). The associations of *ACE* polymorphisms with physical, physiological and skill parameters in adolescents. *European Journal of Human Genetics* 14(3), 332–339.

Murphey, L.J., Gainer, J.V., Vaughan, D.E. & Brown, N.J. (2000). Angiotensin-converting enzyme insertion/deletion polymorphism modulates the human in vivo metabolism of bradykinin. *Circulation* 102(8), 829–832.

Myerson, S., Hemingway, H., Budget, R., Martin, J., Humphries, S. & Montgomery, H. (1999). Human angiotensin I-converting enzyme gene and endurance performance. *Journal of Applied Physiology* 87(4), 1313–1316.

Myerson, S.G., Montgomery, H.E., Whittingham, M., et al. (2001). Left ventricular hypertrophy with exercise and ACE gene insertion/deletion polymorphism: A randomized controlled trial with losartan. *Circulation* 103(2), 2226–2230.

Nazarov, I.B., Woods, D.R., Montgomery, H.E., et al. (2001). The angiotensin converting enzyme I/D polymorphism in Russian athletes. *European Journal of Human Genetics* 9(10), 797–801.

Paillard, F., Chansel, D., Brand, E., et al. (1999). Genotype-phenotype relationships for the renin-angiotensin-aldosterone system in a normal population. *Hypertension* 34(3), 423–429.

Patel, S., Woods, D.R., Macleod, N.J., et al. (2003). Angiotensin-converting enzyme genotype and the ventilatory response to exertional hypoxia. *European Respiratory Journal* 22(5), 755–760.

Paul, M., Poyan Mehr, A. & Kreutz, R. (2006). Physiology of local renin-angiotensin systems. *Physiological Reviews* 86(3), 747–803.

Pedersen-Bjergaard, U., Agerholm-Larsen, B., Pramming, S., Hougaard, P. & Thorsteinsson, B. (2001). Activity of angiotensin-converting enzyme and risk of severe hypoglycaemia in type 1 diabetes mellitus. *Lancet* 357(9264), 1248–1253.

Poirier, O., Georges, J.L., Ricard, S., et al. (1998). New polymorphisms of the angiotensin II type 1 receptor gene and their associations with myocardial infarction and blood pressure: The ECTIM Study. Etude Cas-Temoin de l'Infarctus du Myocarde. *Journal of Hypertension* 16(10), 1443–1447.

Pueyo, M.E., Gonzalez, W., Nicoletti, A., Savoie, F., Arnal, J.F. & Michel, J.B. (2000). Angiotensin II stimulates endothelial vascular cell adhesion molecule-1 via nuclear factor-kappaB activation induced by intracellular oxidative stress. *Arteriosclerosis, Thrombosis, and Vascular Biology* 20(3), 645–651.

Rankinen, T., Wolfarth, B., Simoneau, J.A., et al. (2000). No association between the angiotensin-converting enzyme ID polymorphism and elite endurance athlete status. *Journal of Applied Physiology* 88(5), 1571–1575.

Rigat, B., Hubert, C., Alhenc-Gelas, F., Cambien, F., Corvol, P. & Soubrier, F. (1990). An insertion/deletion polymorphism in the angiotensin I-converting enzyme gene accounting for half the variance of serum enzyme levels. *Journal of Clinical Investigation* 86(4), 1343–1346.

Roltsch, M.H., Brown, M.D., Hand, B.D., et al. (2005). No association between *ACE* I/D polymorphism and cardiovascular hemodynamics during exercise in young women. *International Journal of Sports Medicine* 26(8), 638–644.

Sato, N., Katsuya, T., Rakugi, H., et al. (1997). Association of variants in critical core promoter element of angiotensinogen gene with increased risk of essential hypertension in Japanese. *Hypertension* 30(3 Pt 1), 321–325.

Schmieder, R.E., Martus, P. & Klingbeil, A. (1996). Reversal of left ventricular hypertrophy in essential hypertension. A meta-analysis of randomized double-blind studies. *JAMA* 275(19), 1507–1513.

Schmieder, R.E., Schlaich, M.P., Klingbeil, A.U. & Martus, P. (1998). Update on reversal of left ventricular hypertrophy in essential hypertension (a meta-analysis of all randomized double-blind studies until December 1996). *Nephrology, Dialysis, Transplantation* 13(3), 564–569.

Schunkert, H., Dzau, V.J., Tang, S.S., Hirsch, A.T., Apstein, C.S. & Lorell, B.H. (1990). Increased rat cardiac angiotensin converting enzyme activity and mRNA expression in pressure overload left ventricular hypertrophy. Effects on coronary resistance, contractility, and relaxation. *Journal of Clinical Investigation* 86(6), 1913–1920.

Scott, W., Stevens, J. & Binder-Macleod, S.A. (2001). Human skeletal muscle fiber type classifications. *Physical Therapy* 81(11), 1810–1816.

Shenoy, U. & Cassis, L. (1997). Characterization of renin activity in brown adipose tissue. *American Journal of Physiology* 272(3 Pt 1), C989–C999.

Sonna, L.A., Sharp, M.A., Knapik, J.J., et al. (2001). Angiotensin-converting enzyme genotype and physical performance during US army basic training. *Journal of Applied Physiology* 91(3), 1355–1363.

Staessen, J.A., Wang, J.G., Ginocchio, G., et al. (1997). The deletion/insertion polymorphism of the angiotensin converting enzyme gene and cardiovascular-renal risk. *Journal of Hypertension* 15(12 Pt 2), 1579–1592.

Taguchi, T., Kishikawa, H., Motoshima, H., et al. (2000). Involvement of bradykinin in acute exercise-induced increase of glucose uptake and GLUT-4 translocation in skeletal muscle: Studies in normal and diabetic humans and rats. *Metabolism* 49(7), 920–930.

Taylor, R.R., Mamotte, C.D., Fallon, K. & van Bockxmeer, F.M. (1999). Elite athletes and the gene for angiotensin-converting enzyme. *Journal of Applied Physiology* 87(3), 1035–1037.

Thompson, J., Raitt, J., Hutchings, L., et al. (2007). Angiotensin-converting enzyme genotype and successful ascent to extreme high altitude. *High Altitude Medicine and Biology* 8(4), 278–285.

Timmermans, P.B. & Smith, R.D. (1994). Angiotensin II receptor subtypes: Selective antagonists and functional correlates. *European Heart Journal* 15(Suppl. D), 79–87.

Tsianos, G., Sanders, J., Dhamrait, S., Humphries, S., Grant, S. & Montgomery, H. (2004). The *ACE* gene insertion/deletion polymorphism and elite endurance swimming. *European Journal of Applied Physiology* 92(3), 360–362.

Tsianos, G., Eleftheriou, K.I., Hawe, E., et al. (2005). Performance at altitude and angiotensin I-converting enzyme genotype. *European Journal of Applied Physiology* 93(5–6), 630–633.

Vakili, B.A., Okin, P.M. & Devereux, R.B. (2001). Prognostic implications of left ventricular hypertrophy. *American Heart Journal* 141(3), 334–341.

Villard, E., Tiret, L., Visvikis, S., Rakotovao, R., Cambien, F. & Soubrier, F. (1996). Identification of new polymorphisms of the angiotensin I-converting enzyme (ACE) gene, and study of their relationship to plasma ACE levels by two-QTL segregation-linkage analysis. *American Journal of Human Genetics* 58(6), 1268–1278.

Wicklmayr, M., Dietze, G., Gunther, B., et al. (1980). The kallikrein-kinin system and muscle metabolism—clinical aspects. *Agents and Actions* 10(4), 339–343.

Williams, A.G., Rayson, M.P., Jubb, M., et al. (2000). The *ACE* gene and muscle performance. *Nature* 403(6770), 614.

Williams, A.G., Dhamrait, S.S., Wootton, P.T., et al. (2004). Bradykinin receptor gene variant and human physical performance. *Journal of Applied Physiology* 96(3), 938–942.

Wollert, K.C. & Drexler, H. (1999). The renin-angiotensin system and experimental heart failure. *Cardiovascular Research* 43(4), 838–849.

Woods, D.R., Humphries, S.E. & Montgomery, H.E. (2000). The ACE I/D polymorphism and human physical performance. *Trends in Endocrinology and Metabolism* 11(10), 416–420.

Woods, D., Hickman, M., Jamshidi ,Y., et al. (2001a). Elite swimmers and the D allele of the *ACE* I/D polymorphism. *Human Genetics* 108(3), 230–232.

Woods, D., Onambele, G., Woledge, R., et al. (2001b). Angiotensin-I converting enzyme genotype-dependent benefit from hormone replacement therapy in isometric muscle strength and bone mineral density. *Journal of Clinical Endocrinology and Metabolism* 86(5), 2200–2204.

Woods, D.R., Pollard, A.J., Collier, D.J., et al. (2002a). Insertion/deletion polymorphism of the angiotensin I-converting enzyme gene and arterial oxygen saturation at high altitude. *American Journal of Respiratory and Critical Care Medicine* 166(3), 362–366.

Woods, D.R., World, M., Rayson, M.P., et al. (2002b). Endurance enhancement related to the human angiotensin I-converting enzyme I-D polymorphism is not due to differences in the cardiorespiratory response to training. *European Journal of Applied Physiology* 86(3), 240–244.

Yildiz, A., Akkaya, V., Hatemi, A.C., et al. (2000). No association between deletion-type angiotensin-converting enzyme gene polymorphism and left-ventricular hypertrophy in hemodialysis patients. *Nephron* 84(2), 130–135.

Yoneya, K., Okamoto, H., Machida, M., et al. (1995). Angiotensin-converting enzyme gene polymorphism in Japanese patients with hypertrophic cardiomyopathy. *American Heart Journal* 130(5), 1089–1093.

Young, L.E., Rogers, K. & Wood, J.L. (2005). Left ventricular size and systolic function in thoroughbred racehorses and their relationships to race performance. *Journal of Applied Physiology* 99(4), 1278–1285.

Zhang, B., Tanaka, H., Shono, N., et al. (2003). The I allele of the angiotensin-converting enzyme gene is associated with an increased percentage of slow-twitch type I fibers in human skeletal muscle. *Clinical Genetics* 63(2), 139–144.

Chapter 18

The *ACTN3* Gene and Human Performance

DANIEL G. MACARTHUR[1,2] AND KATHRYN N. NORTH[1,3]

[1]*Discipline of Paediatrics and Child Health, Faculty of Medicine, University of Sydney, Sydney, NSW, Australia*
[2]*Human Evolution, Wellcome Trust Sanger Institute, Cambridge, UK*
[3]*Institute for Neuroscience and Muscle Research, Children's Hospital at Westmead, Sydney, NSW, Australia*

Our understanding of the genetic influences on human physical performance is evolving rapidly in the post-genomic era. This chapter presents a recent but growing body of evidence that one particular genetic polymorphism—the R577X variation in the *ACTN3* gene—represents a common genetic influence on athletic performance and skeletal muscle function. There is one clear message from the genetic association data presented: it is extremely likely that α-actinin-3 deficiency reduces the performance of fast skeletal muscle fibers in both elite athletes and in the nonathlete population. A negative association between the α-actinin-3-deficient XX genotype and elite sprint athlete status has now been observed in six separate studies (Druzhevskaya et al., 2008; Eynon et al., 2009; Niemi & Majamaa, 2005; Papadimitriou et al., 2008; Roth et al., 2008; Yang et al., 2003), while multiple large independent studies have found that the XX genotype is associated with lower baseline muscle strength (Clarkson et al., 2005a; Walsh et al., 2008) and with poorer sprint performance (Moran et al., 2007) in nonathlete cohorts. All of these studies support the notion that α-actinin-3 deficiency inhibits the performance of the fast glycolytic muscle fibers responsible for rapid, forceful contraction, making this one of the more consistently supported associations with a broad physical performance trait for any genetic variant.

We also discuss the limitations of studies published to date, and the remaining questions that require future research. How much predictive power does the R577X genotype provide in terms of identifying athletic potential? What role does the variation play in human health and fitness in nonathletes? Some studies have reported sex differences in the associations between specific muscle phenotypes and R577X, suggesting need for more studies in sex-stratified populations for a broader range of performance traits.

Structure and Function of α-Actinin-3

In skeletal muscle, the contractile apparatus is composed of repeating units, or sarcomeres, that contain highly organized arrays of actin-containing thin filaments and myosin-containing thick filaments. The Z lines are dense three-dimensional structures that run perpendicular to the myofibrils and anchor the thin filaments. The sarcomeric α-actinins are a major protein component of the Z line and cross-link actin thin filaments to maintain the ordered myofibrillar array. In mammals, the *ACTN2* and *ACTN3* genes encode the skeletal muscle α-actinins, α-actinin-2 and α-actinin-3, respectively, which differ in sequence adjacent to the actin-binding domain and have high sequence similarity elsewhere (81% identical and 91% similar) (Beggs et al., 1992). The sarcomeric α-actinins contain a number of functional domains: an N-terminal actin-binding domain, a central rod domain, and a C-terminal

Genetic and Molecular Aspects of Sport Performance, 1st edition. Edited by Claude Bouchard and Eric P. Hoffman. Published 2011 by Blackwell Publishing Ltd.

region that contains two potential calcium-binding EF hand motifs (Figure 18.1) (Blanchard et al., 1989; MacArthur & North, 2004). α-Actinin-2 is present in all human skeletal muscle fibers, whereas α-actinin-3 is present only in type 2 (fast, glycolytic) fibers (North & Beggs, 1996).

The sarcomeric α-actinins were initially identified as actin-binding proteins (hence the name), and their major role was considered to be primarily structural. *In vitro* studies support a broader role for the α-actinins in structural maintenance and coordination of myofiber contraction. The C-terminus of sarcomeric α-actinin plays a crucial role in the maintenance of Z-line integrity and myofibrillar organization (Lin et al., 1998); this region is the site of binding for a number of important interacting proteins, including the Z-line protein myotilin (Salmikangas et al., 2003). In addition, the α-actinins

link dystrophin (deficient in Duchenne muscular dystrophy) with the integrin adhesion system (Hance et al., 1999) and are thought to contribute to the maintenance of integrity of the myofiber. However, the role of the α-actinins in skeletal muscle appears to be more complex, since they also interact with a wide range of signaling and metabolic proteins associated with the Z-line (reviewed in MacArthur & North, 2004). Thus, differential expression of the sarcomeric α-actinins, and particularly the specialized expression of α-actinin-3 in fast muscle fibers, may also influence skeletal muscle metabolism and/or fiber-type specification.

Deficiency of α-Actinin-3 Is Common in Humans

Several years ago, our team identified a common genetic variation in the *ACTN3* gene that results

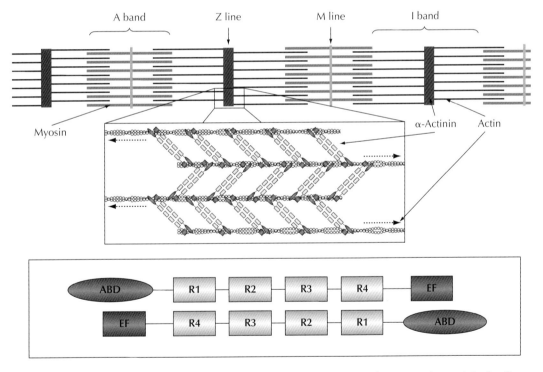

Figure 18.1 Localization and structure of the sarcomeric α-actinins. Upper image: the sarcomeric α-actinins localize to the sarcomeric Z-line, where they form head-to-tail dimers cross-linking actin filaments from adjacent sarcomeres. Lower image: each α-actinin molecule contains an N-terminal actin-binding domain (ABD), a C-terminal EF hand domain (EF), and four spectrin-like repeats (R1-4). In muscle, α-actinin molecules are always present in the form of head-to-tail dimers.

in the replacement of an arginine (R) with a stop codon (X) at amino acid 577 (R577X) (North et al., 1999). This variation creates two different versions of the *ACTN3* gene, both common in the general population: the 577R version (or *allele*) is the normal, functional version of the gene, while the 577X allele contains a sequence change abolishing the production of functional α-actinin-3 protein.

Since every human inherits two copies of the *ACTN3* gene—one maternal and one paternal copy—there are three possible combinations, or *genotypes*, of R577X alleles. In individuals of European descent, less than a third of the population have two copies of the functional R allele (the RR genotype), while just over half the population have one copy of each of the two alleles (the RX genotype). Remarkably, the remaining 18% of the healthy European population—and, we estimate, more than a billion people worldwide—have two copies of the nonfunctional 577X variant (the XX genotype), resulting in complete deficiency of α-actinin-3 protein in their skeletal muscle (Mills et al., 2001).

This widespread deficiency of a potentially important skeletal muscle protein was profoundly unexpected (North, 2008). The fact that more than a billion humans lack this protein without suffering from overt muscle disease suggests that α-actinin-3 is at least partially redundant—in other words, that many of its functions can be compensated for by other factors, most likely including the closely related protein α-actinin-2. However, the specialized expression pattern and strong sequence conservation of α-actinin-3 over 300 million years of evolutionary time (Mills et al., 2001) suggest that this protein does in fact possess some roles in fast fibers that cannot be completely taken over by α-actinin-2. In addition, the region around the 577X stop codon variant displays an unusual pattern of genetic variation suggestive of recent positive natural selection acting over the last 50,000 years of human evolution (MacArthur et al., 2007), indicating that this polymorphism has conferred some effect on human fitness during our recent evolutionary history.

Given these lines of evidence, it seems plausible that the absence of α-actinin-3 in the fast fibers of skeletal muscle may affect their functional properties, and thus contribute to variation in muscle function in the general population. We explore this hypothesis in the studies described in the remainder of this chapter.

Approaches to Exploring the Association Between *ACTN3* and Performance

We and other groups have performed genetic association studies testing for correlations between the different genotypes of the R577X polymorphism and a variety of performance-related traits. These studies utilize two major approaches: case–control analysis of elite athletes and cross-sectional analysis of nonathletes.

In case–control studies, R577X genotypes are collected from groups of elite athletes, and the frequency of each genotype in these groups are compared with the frequency in a large population of sedentary (nonelite) controls. If one genotype is more favorable to elite athletic performance than the other genotypes, it will be found at a higher frequency in elite athletes than in controls.

In cross-sectional association studies, large groups of individuals are tested for one or more traits (such as muscle strength or sprint speed), and then the group is broken down by R577X genotype. If one genotype is beneficial for the studied trait, then individuals with that genotype will tend to have higher values for that trait than individuals with other genotypes.

Case–Control Association Studies in Elite Athletes

Initial Studies: Support for an Effect on Sprint/Power Performance

The first association study to explore the influence of *ACTN3* genotype on athletic performance was published by our team (Yang et al., 2003). We studied DNA samples from a large group of elite athletes from the Australian Institute of Sport (AIS) competing in a wide variety of sports, along with a cohort of 436 nonathlete controls drawn from the general population. All athletes and controls in this study were of European ancestry, since we have also shown that the frequency of the R577X polymorphism varies in different ethnic groups (Mills et al., 2001). We hypothesized that α-actinin-3 deficiency would predominantly affect the function of fast muscle fibers, and thus have a greater influence on athletes competing in sprint or power events specifically requiring optimal fast fiber performance, rather than on athletes competing in endurance events who rely predominantly on slow muscle fibers. We thus asked the AIS to classify the athletes based on their specialized area of performance: the final cohort consisted of a group of 107 athletes competing in sprint or power events such as short-distance running, swimming, and cycling, and a group of 194 athletes competing in endurance events such as long-distance running, cycling, and cross-country skiing. We then determined the R577X genotype for each of the DNA samples to allow comparison of the frequency of α-actinin-3 deficiency between athlete groups and controls.

The results of this analysis (Yang et al., 2003) are shown in Figure 18.2. The frequency of the α-actinin-3-deficient XX genotype was almost 20% in our control cohort, similar to the levels previously determined in individuals of European ancestry (Mills et al., 2001). In power athletes, this frequency was dramatically reduced: in the power athlete cohort as a whole, the frequency of α-actinin-3 deficiency was only 6%, about a third of the levels in controls, while *none* of the Olympians or female power athletes had an XX (α-actinin-3 deficient) genotype. This reduction in frequency was specific to power athletes; if anything, endurance athletes showed a slight *increase* in the frequency of α-actinin-3 deficiency, although this was significant only in females.

These data suggested that the presence of α-actinin-3 is required for optimal fast fiber performance in power athletes, whereas the *absence* of α-actinin-3 may provide some sort of advantage for endurance athletes. The R577X polymorphism thus joined the growing list of genetic factors reported to influence athletic performance (Wolfarth et al., 2005).

The association between R577X and athletic performance certainly has biological plausibility: the R577X polymorphism has a clear biochemical effect, completely eliminating the production of functional α-actinin-3 protein; in addition, the localization of α-actinin-3 to fast muscle fibers is consistent with a negative effect of α-actinin-3 deficiency on sprint/power performance.

Nevertheless, an isolated genetic association study must be treated with caution, as there are numerous ways in which such studies can generate false-positive results (Romero et al., 2002). The association of R577X with athletic performance has now been supported by the independent replication of our findings in a large number of subsequent studies.

Firstly, a study of elite Finnish athletes (Niemi & Majamaa, 2005) compared the frequency of α-actinin-3 deficiency in a group of 68 sprint athletes, 40 endurance athletes, and 120 ethnically matched controls, and found a very similar pattern to our own study of Australian athletes: a marked

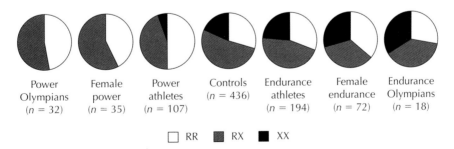

| Power Olympians (*n* = 32) | Female power (*n* = 35) | Power athletes (*n* = 107) | Controls (*n* = 436) | Endurance athletes (*n* = 194) | Female endurance (*n* = 72) | Endurance Olympians (*n* = 18) |

☐ RR ◪ RX ◼ XX

Figure 18.2 R577X genotype frequencies in controls and elite sprint and endurance athletes from Yang et al. (2003). The frequency of the 577XX (α-actinin-3 deficient) genotype is significantly lower in the total power athlete group (6%) than in controls (18%), and significantly higher in female endurance athletes (29%) than in female controls. The power Olympian and female power athlete groups both contain no 577XX individuals.

decrease in the frequency of α-actinin-3 deficiency in sprint athletes, and a slight (but not significant) increase in the frequency of the XX genotype in endurance athletes. In both cases, these trends were most apparent in athletes who had competed at an international level.

Since then, significantly decreased frequency of the XX genotype relative to controls has been observed in 34 sprinters and 73 power athletes from Greece (Papadimitriou et al., 2008), 486 Russian power athletes (Druzhevskaya et al., 2008), 52 white bodybuilders and power lifters (Roth et al., 2008), and 81 Israeli sprinters (Eynon et al., 2009). Taken together, these studies provide convincing support for the notion that the presence of α-actinin-3 is important for the optimal performance of fast fibers in sprint or power activities.

Figure 18.3 shows the results of a meta-analysis of results of all studies of sprint/power athletes performed in South-Western and North-Eastern European athlete cohorts. Separation of these two groups is necessary due to the consistent difference in 577X allele frequencies in controls from these two regions. Both groups show consistent and highly significant associations between 577X allele frequency and elite sprint or power athlete status.

Inconsistent Association with Endurance Athlete Status

The possible positive association of α-actinin-3 deficiency with endurance athletes seen in our study (Yang et al., 2003) is not statistically significant in the Finnish study (Niemi & Majamaa, 2005), although the Finnish endurance athletes do show a higher frequency of the XX genotype than controls. In addition, the recent analysis of Israeli athletes mentioned earlier (Eynon et al., 2009) found a pattern strikingly consistent with both the initial Australian study and the Finnish analysis: a reduced frequency of the XX genotype in sprinters and a higher frequency in endurance athletes (this time statistically significant).

However, other studies have failed to replicate the association with endurance athlete status. An analysis comparing 50 elite male endurance cyclists and 52 Olympic-level endurance runners with 123 sedentary male controls (Lucia et al., 2006) found no significant differences in genotype frequencies between controls and either of the two athlete groups (although the frequency of α-actinin-3 deficiency was slightly higher in the endurance cyclist cohort than in controls—26% vs. 17.9%). There was also no association between R577X genotype and a common measure of endurance performance—maximal oxygen uptake ($\dot{V}O_2max$)—in either of the athlete groups. Most puzzlingly, a recent analysis of 456 elite Russian endurance athletes found that the frequency of the XX genotype was actually significantly *lower* in the athletes relative to controls (Ahmetov et al., 2008). This observation is inconsistent with any previous study of *ACTN3* genotypes in elite endurance athletes, but is also theoretically the most well powered to detect an association.

Given these overall findings, the potential association between R577X genotype and elite endurance athlete status must be regarded as unproven, and—if it exists—is certainly much weaker than the strong and consistent negative association between

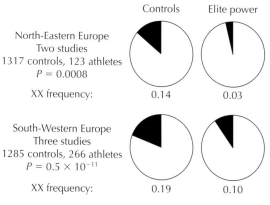

	Controls	Elite power
North-Eastern Europe Two studies 1317 controls, 123 athletes $P = 0.0008$		
XX frequency:	0.14	0.03
South-Western Europe Three studies 1285 controls, 266 athletes $P = 0.5 \times 10^{-11}$		
XX frequency:	0.19	0.10

Figure 18.3 Combined analysis provides strong evidence for an association between the R577X polymorphism and sprint/power athlete status. Studies of northern/eastern and southern/western European populations were analyzed separately due to differing allele frequencies in controls; both populations show highly significant combined evidence for association. Area in black indicates the frequency of the XX (α-actinin-3-deficient) genotype in each group. Studies included: NE Europe: (Druzhevskaya et al., 2008; Niemi & Majamaa, 2005); SW Europe: (Papadimitriou et al., 2008; Roth et al., 2008; Yang et al., 2003).

the 577X null allele and sprint/power athlete status. Further studies of homogeneous groups of elite athletes will be required to determine if there is a weak association restricted only to specific subgroups of endurance athletes, explaining the heterogeneity described earlier.

Cross-Sectional Association Studies in Nonathlete Cohorts

In addition to the studies in elite athletes, several cross-sectional association studies have examined the correlation of R577X genotype with a variety of measures of muscle function in nonathlete populations. These studies have provided evidence that the effect of α-actinin-3 deficiency on performance is not restricted to elite athletes, and that R577X is one of the many genetic factors that influence variation in muscle function in the general population.

The first study to assess the association of R577X with muscle function in nonathletes examined elbow flexor strength, both at baseline and following a 12-week resistance training protocol, in a large cohort of 247 males and 355 females aged 18–40 (Clarkson et al., 2005a). Isometric strength was estimated using a measure of maximal voluntary contraction (MVC), while dynamic strength was measured using a one repetition maximum (1-RM) protocol. The α-actinin-3-deficient (577XX) genotype was associated with significantly lower baseline MVC in females (but not males); unexpectedly, α-actinin-3-deficient females also showed a significantly *greater* response to training for the 1-RM measurement. These associations were significant in both the European and Asian subsets of the female cohort.

The authors argue that their results are consistent with a role for α-actinin-3 in the maintenance of sarcomeric integrity, which fits with the known interactions of the α-actinins with a variety of structural proteins (MacArthur & North, 2004). Greater exercise-induced muscle damage in the absence of α-actinin-3 may result in poorer overall power generation, but it may also stimulate adaptive remodeling of the sarcomere, resulting in a more rapid response to training. This explanation is speculative, and further studies will be needed both to confirm the positive effect of α-actinin-3 deficiency on the response to strength training and to dissect out the mechanisms underlying this effect.

Another study by the same group assessed the effect of the R577X polymorphism on an indirect measure of muscle damage (Clarkson et al., 2005b). In this study, 78 males and 79 females with a mean age of 24 performed 50 maximal eccentric contractions of the elbow flexor muscles, a protocol designed to produce subclinical muscle damage. Serum creatine kinase (CK), a well-characterized marker of muscle membrane disruption, was assayed both at baseline and at various time-points following the exercise protocol. No association was seen between R577X genotype and the change in serum CK levels following exercise-induced injury. Unexpectedly, there was a (weakly significant) lower level of *baseline* serum CK levels in α-actinin-3-deficient individuals compared to RR and RX subjects; the authors speculated that α-actinin-3 deficiency may be associated with lower baseline muscle mass or physical activity levels. Given the small sample size and the weakness of the association, this finding should be regarded as tentative, but certainly warrants further study.

The largest study to date on the association of *ACTN3* with muscle performance traits examined R577X genotype in 525 males and 467 females aged between 11 and 18 (Moran et al., 2007). These individuals had previously been examined for a variety of traits related to body composition, strength/power, and endurance performance. The authors showed a significant association between R577X genotype and performance in a 40-m sprint, with XX (α-actinin-3 deficient) males—but not females—taking significantly longer to run the distance compared to their RR or RX counterparts. Interestingly, this association was highly specific: no association was found between R577X genotype and other strength/power phenotypes (such as handgrip strength, basketball throw, and vertical jump height), or with a proxy measure for aerobic capacity, the shuttle run test. In addition, R577X genotype did not correlate significantly with measures of body composition such as body mass index and skinfold widths. The authors argue that the specific association of R577X genotype with sprint performance is consistent with a primary role for α-actinin-3 in the protection of the sarcomere from mechanical damage during repetitive force generation, as sprinting requires repeated cycles of muscle contraction, unlike the other strength/power phenotypes tested (such as vertical jump), which require only a single, forceful muscle contraction.

The male-specific association between R577X and sprint performance observed by Moran et al. (2007) contrasts with the stronger female associations seen in our athlete study (Yang et al., 2003) and in the nonathlete cohort studied by Clarkson et al. (2005a). We have previously suggested that the apparent differential effect of R577X genotype on performance in females and males may relate to differences in the relative influence of endogenous steroid hormones on muscle performance between the sexes (Yang et al., 2003). If so, we would expect this effect to be more pronounced in adults, and the contrasting effect seen by Moran et al. (2007) may be due to the fact that their cohort is largely peri-pubescent.

Several more recent studies have provided broadly consistent support for a reduction in muscle strength in XX individuals. In a study of fast velocity knee extension strength in a group of 90 young men, relative dynamic quadriceps torques at 300°/s was lower in XX individuals ($P < 0.05$) (Vincent et al., 2007), and in a large female cohort with a wide age range ($n = 394$, 22–90 years), both knee extensor concentric peak torque and eccentric peak torque were lower in women with the XX genotype (Walsh et al., 2008). In a study designed to assess the genotype effect on aging progression over a 5-year period, women aged between 70 and 79 with the XX genotype had about a 35% greater risk to develop difficulties in walking and climbing stairs compared with those with RR genotypes, and old men (70–79 years) with XX genotypes had greater increase in times to complete 400-m walking (Delmonico et al., 2008).

Finally, the *ACTN3* genotype has also been shown to influence exercise capacity in women with the metabolic myopathy known as McArdle disease; those carrying the X allele had a higher $\dot{V}O_2$ peak and a higher oxygen uptake ($\dot{V}O_2$) at the ventilatory threshold than RR homozygotes (Lucia et al., 2007).

Mechanistic Studies

In order to dissect the structural and biochemical changes underlying the effects of α-actinin-3 deficiency on muscle function, our group has generated an *Actn3* knockout (KO) mouse to serve as a model of human XX individuals (MacArthur et al., 2007). The KO mice are morphologically indistinguishable from their wild-type (WT) littermates, show grossly normal muscle ultrastructure under light and electron microscopy, and do not demonstrate substantial loss of fast (2B) fibers as defined by staining for myosin heavy chain isoforms. Unlike humans, α-actinin-2 is not normally expressed in all fast fibers in WT mice (Mills et al., 2001). In *Actn3* KO mice, however, α-actinin-2 is up-regulated and expressed in all fiber types, similar to the expression pattern seen in human muscle (MacArthur et al., 2007). Thus, KO mice display an expression pattern very similar to that seen in humans with the *ACTN3* XX genotype, in whom α-actinin-2 is the only sarcomeric α-actinin expressed in all fiber types.

In the following section, the results from the KO mice will be compared to those seen in studies of nonathlete humans. These data can be summarized briefly in the following way: in *Actn3* KO mice, the absence of α-actinin-3 results in a shift in the properties of fast muscle fibers toward those characteristic of slow muscle fibers, including changes in metabolism, fiber size, contractile properties, and force generation. These changes provide plausible mechanistic explanations for the effects of α-actinin-3 deficiency on human muscle function, but further studies will be required to directly confirm that similar changes occur in α-actinin-3-deficient human muscle.

Muscle Metabolism and Endurance Performance

Biochemical analyses suggest a shift in muscle energy metabolism in *Actn3* KO mice toward the more efficient aerobic/oxidative pathway. Muscles from the *Actn3* KO mouse display significantly decreased activity in the anaerobic glycolytic pathway (typical of fast muscle fibers) and increased activity in the aerobic oxidative pathway (typical of slow muscle fibers), but without a marked shift in fiber-type distribution as defined by myosin heavy chain isoform (MacArthur et al., 2007, 2008). Specifically, the activity of the anaerobic enzyme lactate dehydrogenase is reduced by 16% in KO muscle, while the activities of the mitochondrial enzymes citrate synthase, succinate dehydrogenase, and cytochrome *c* oxidase are increased by 22–39%. Two enzymes involved in fatty acid oxidation, β-hydroxyacyl-CoA dehydrogenase and medium-chain acyl-CoA dehydrogenase, also show 30–42% higher levels in *Actn3* KO muscles.

Physiological studies in isolated fast-twitch muscles indicate that KO muscles have longer twitch half-relaxation times and enhanced force recovery following fatigue compared to muscles from WT animals (Chan et al., 2008; MacArthur et al., 2008). As a consequence, *Actn3* KO mice are able to run for a distance over 33% longer than their WT littermates before reaching exhaustion (MacArthur et al., 2007), a finding compatible with the proposed (but inconsistent) association between the XX genotype and human elite endurance athlete status described earlier.

Muscle Strength and Power

A deficiency in α-actinin-3 results in a subtle but significant decrease in muscle strength in both mice and humans, although in both cases, muscle strength in α-actinin-3-deficient individuals lies within the normal range, consistent with *ACTN3* genotype contributing to *normal variation* rather than to a disease phenotype.

Grip strength in *Actn3* KO mice is 7.4% lower in males and 6.0% lower in females compared with the WT littermates (MacArthur et al., 2008). This finding is entirely consistent both with the negative association between the XX genotype and sprint/power athlete status, and with the reported associations between the XX genotype and human muscle strength described earlier.

The contractile profile for isolated KO muscles described in the preceding section is consistent with a shift in the contractile properties of fast muscle fibers (those typically responsible for generating rapid contractile force) toward those characteristic of slow fibers (which are weaker but metabolically more efficient). Since slow fibers are associated with lower force production, such a shift would also potentially explain the reduced muscle strength associated with α-actinin-3 deficiency in both mice and humans.

Muscle Mass, Fiber Diameter, and Fiber-Type Proportions

Detailed studies of *Actn3* KO mice suggest that the reduced muscle strength resulting from α-actinin-3 deficiency is at least partially the result of reduced muscle mass. KO mice display a reduction in lean muscle mass using dual-energy X-ray absorptiometry scans, and a lower weight for individual muscles: all muscles that contain fast-twitch fibers (and hence express α-actinin-3) have significant mass reduction in KO compared with WT (MacArthur et al., 2008).

Interestingly, one study using dual-energy X-ray absorptiometry in humans also found a significant reduction in lean mass as well as fat mass in women with the XX genotype compared with women with combined RR and RX genotypes (Walsh et al., 2008). In this cohort, the body mass index of women with the XX genotype was also significantly lower than in women with the RR genotype. In a study examining the cross-sectional area (CSA) of human thigh muscle, postmenopausal women with the XX genotype had significantly reduced CSA compared with women with the RR genotype (Delmonico et al., 2008). However, these findings conflict with an earlier study reporting no effect of R577X genotype on either body mass index or biceps CSA (Clarkson et al., 2005a).

In *Actn3* KO mice, the reduction in muscle size is primarily attributable to a specific decrease in the diameter of fast-twitch (2B) muscle fibers—again reflecting a shift toward the phenotype of slow-twitch (type 1) fibers (MacArthur et al., 2008). There is no difference in the diameter of type 1 or 2A fibers, in which α-actinin-3 is not usually expressed.

Thus far, only a single study has explored the hypothesis that the absence of α-actinin-3 may influence fiber diameter and fiber type in humans (Vincent et al., 2007). In a cohort of young men (18–29 years), muscle biopsies were compared between 22 XX and 22 RR individuals. The percentage surface area and number of type 2X fibers (equivalent to 2B fibers in the mouse) were significantly less in those with the XX genotype compared with RR, while there were no significant differences in the size or number of type 1 and 2A fibers. This provides preliminary evidence that α-actinin-3 deficiency may influence muscle fiber properties in humans.

Limitations and Future Studies

Much more work remains to be done in this area to answer a number of major questions. Firstly, what physiological aspects of human performance (in both athletes and nonathletes) are most strongly influenced by R577X? Secondly, exactly how much predictive power does the R577X genotype provide in terms of identifying athletic potential? Thirdly, precisely *how* does α-actinin-3 deficiency affect the function of skeletal muscle? And finally, from a biomedical perspective, what role does this variation play in influencing human health, fitness, and disease?

There are also more tentative associations between R577X and other muscle traits that warrant further detailed investigation. In particular, it appears that α-actinin-3 deficiency may have a positive influence on endurance performance, although this has so far been significant in only two studies (Eynon et al., 2009; Yang et al., 2003). Notably, the association of R577X with endurance performance seen by Yang et al. (2003) was strikingly gender-specific, being significant only in females. The gender breakdown of several endurance athlete cohorts (Eynon et al., 2009; Niemi & Majamaa, 2005) is unclear, and Lucia et al. (2006) studied only male athletes, so we would argue that further studies targeted toward female endurance athletes would be worthwhile. In addition, a single study has suggested that α-actinin-3 deficiency might boost the response of muscle to training (Clarkson et al., 2005a). Further large genetic association studies, involving both elite athletes from a wider variety of sports and nonathletes examined for a broader range of performance traits, will be required to investigate these areas further.

Once the influence of R577X on athletic performance and muscle function has been clearly elucidated,

this information may be useful for athlete talent identification programs. Indeed several companies, including Genetic Technologies, 23andMe, and CyGene, are already marketing *ACTN3* gene tests directly to consumers. Given the growing support described earlier for the notion that R577X influences muscle function, it would intuitively appear that testing R577X may be useful for coaches and sporting bodies, or for young hopefuls deciding whether they will excel in a career as an elite sprint athlete—but how predictive will R577X genotype information really be?

The answer to this question is still unclear, for a number of reasons. Firstly, many different genetic and environmental factors influence physical performance, with R577X genotype determining only a small proportion of overall variation. The cross-sectional association studies cited earlier estimate that R577X accounts for only 2.2% of the total variance in baseline muscle strength in adult women (Clarkson et al., 2005a) and 2.6% of the total variance in 40-m sprint speed in adolescent males (Moran et al., 2007). These values are fairly rough estimates due to the sample sizes in these studies, and these proportions are likely to be substantially higher in elite athletes, as a reduction of environmental variance in this group will increase the relative importance of genetic influences. Nonetheless, these figures emphasize that R577X is just one of a myriad of complex, interacting factors that influence muscle performance. Secondly, it is uncertain whether R577X genotype actually adds any further information to existing tests used in talent identification. Even though this genetic variant does appear to influence skeletal muscle function, it may well be that existing direct tests of muscle power—such as vertical jump, dynamometry, and sprint tests—*already* capture this information. Sporting bodies and young athletes should thus await the results of further research before using *any* genetic information to guide decisions about talent identification or sports selection.

Another area in which considerable uncertainty exists is the mechanistic basis for the effect of α-actinin-3 deficiency on muscle function. In this chapter, we have described extensive studies carried out on the effects of α-actinin-3 deficiency in a KO mouse model, which have suggested that the loss of this protein results in a shift in the properties of fast fibers toward those characteristic of slow fibers. If similar changes can be confirmed in human muscle, this transformation would provide a powerful mechanistic explanation for the negative association between α-actinin-3 deficiency and muscle strength and sprint performance, and also supports the notion that the loss of α-actinin-3

may benefit endurance performance (a hypothesis inconsistently supported in the human literature). Further detailed investigations of the structural and biochemical changes in human muscle as a result of α-actinin-3 deficiency are urgently required to explore the similarities and differences between XX humans and the *Actn3* KO mouse model, and to directly clarify the mechanistic effects of α-actinin-3 deficiency in human muscle.

References

Ahmetov, I.I., Druzhevskaya, A.M., Astratenkova, I.V., Popov, D.V., Vinogradova, O.L. & Rogozkin, V.A. (2008). The *ACTN3* R577X polymorphism in Russian endurance athletes. *British Journal of Sports Medicine* 44, 649–652.

Beggs, A.H., Byers, T.J., Knoll, J.H., Boyce, F.M., Bruns, G.A. & Kunkel, L.M. (1992). Cloning and characterization of two human skeletal muscle alpha-actinin genes located on chromosomes 1 and 11. *The Journal of Biological Chemistry* 267, 9281–9288.

Blanchard, A., Ohanian, V. & Critchley, D. (1989). The structure and function of alpha-actinin. *Journal of Muscle Research and Cell Motility* 10, 280–289.

Chan, S., Seto, J.T., Macarthur, D.G., Yang, N., North, K. & Head, S. (2008). A gene for speed: Contractile properties of isolated whole EDL muscle from an α-actinin-3 knockout mouse. *The American Journal of Physiology* 295, C897–C904.

Clarkson, P.M., Devaney, J.M., Gordish-Dressman, H., et al. (2005a). ACTN3 genotype is associated with increases in muscle strength in response to resistance training in women. *Journal of Applied Physiology* 99, 154–163.

Clarkson, P.M., Hoffman, E.P., Zambraski, E., et al. (2005b). ACTN3 and MLCK genotype associations with exertional muscle damage. *Journal of Applied Physiology* 99, 564–569.

Delmonico, M.J., Zmuda, J.M., Taylor, B.C., et al. (2008). Association of the *ACTN3* genotype and physical functioning with age in older adults. *Journal of Gerontology* 63A, 1227–1234.

Druzhevskaya, A.M., Ahmetov, I.I., Astratenkova, I.V. & Rogozkin, V.A. (2008). Association of the *ACTN3* R577X polymorphism with power athlete status in Russians. *European Journal of Applied Physiology* 103, 631–634.

Eynon, N., Duarte, J.A., Oliveira, J., et al. (2009). *ACTN3* R577X polymorphism and Israeli top-level athletes. *International Journal of Sports Medicine* 30, 695–698.

Hance, J.E., Fu, S.Y., Watkins, S.C., Beggs, A.H. & Michalak, M. (1999). alpha-Actinin-2 is a new component of the dystrophin-glycoprotein complex. *Archives of Biochemistry and Biophysics* 365, 216–222.

Lin, Z., Hijikata, T., Zhang, Z., et al. (1998). Dispensability of the actin-binding site and spectrin repeats for targeting sarcomeric alpha-actinin into maturing Z bands in vivo: Implications for in vitro binding studies. *Developmental Biology* 199, 291–308.

Lucia, A., Gomez-Gallego, F., Santiago, C., et al. (2006). ACTN3 genotype in professional endurance cyclists. *International Journal of Sports Medicine* 27, 880–884.

Lucia, A., Gómez-Gallego, F., Santiago, C., et al. (2007). The 577X allele of the *ACTN3* gene is associated with improved exercise capacity in women with McArdle's disease. *Neuromuscular Disorders* 17, 603–610.

MacArthur, D.G. & North, K.N. (2004). A gene for speed? The function and evolution of α-actinin-3. *Bioessays* 26, 786–795.

MacArthur, D.G., Seto, J.T., Raftery, J.M., et al. (2007). Loss of *ACTN3* gene function alters mouse muscle metabolism and shows evidence of positive selection in humans. *Nature Genetics* 39, 1261–1265.

MacArthur, D.G., Seto, J.T., Chan, S., et al. (2008). An *Actn3* knockout mouse provides mechanistic insights into the association between α-actinin-3 deficiency and human athletic performance. *Human Molecular Genetics* 17, 1076–1086.

Mills, M., Yang, N., Weinberger, R., et al. (2001). Differential expression of the

actin-binding proteins, α-actinin-2 and -3, in different species: Implications for the evolution of functional redundancy. *Human Molecular Genetics* 10, 1335–1346.

Moran, C.N., Yang, N., Bailey, M.E.S., et al. (2007). Association analysis of the *ACTN3* R577X polymorphism and complex quantitative body composition and performance phenotypes in adolescent Greeks. *European Journal of Human Genetics* 15, 88–93.

Niemi, A.K. & Majamaa, K. (2005). Mitochondrial DNA and ACTN3 genotypes in Finnish elite endurance and sprint athletes. *European Journal of Human Genetics* 13, 965–969.

North, K. (2008). Why is alpha-actinin-3 deficiency so common in the general population? The evolution of athletic performance. *Twin Research and Human Genetics* 11, 384–394.

North, K.N. & Beggs, A.H. (1996). Deficiency of a skeletal muscle isoform of alpha-actinin (alpha-actinin-3) in merosin-positive congenital muscular dystrophy. *Neuromuscular Disorders* 6, 229–235.

North, K.N., Yang, N., Wattanasirichaigoon, D., Mills, M., Easteal, S. & Beggs, A.H. (1999). A common nonsense mutation results in α-actinin-3 deficiency in the general population. *Nature Genetics* 21, 353–354.

Papadimitriou, I.D., Papadopoulos, C., Kouvatsi, A. & Triantaphyllidis, C. (2008). The *ACTN3* gene in elite Greek track and field athletes. *International Journal of Sports Medicine* 29, 352–355.

Romero, R., Kuivaniemi, H., Tromp, G. & Olson, J. (2002). The design, execution, and interpretation of genetic association studies to decipher complex diseases. *American Journal of Obstetrics and Gynecology* 187, 1299–1312.

Roth, S.M., Walsh, S., Liu, D., Metter, E.J., Ferrucci, L. & Hurley, B.F. (2008).

The *ACTN3* R577X nonsense allele is under-represented in elite-level strength athletes. *European Journal of Human Genetics* 16, 391–394.

Salmikangas, P., van der Ven, P.F., Lalowski, M., et al. (2003). Myotilin, the limb-girdle muscular dystrophy 1A (LGMD1A) protein, cross-links actin filaments and controls sarcomere assembly. *Human Molecular Genetics* 12, 189–203.

Vincent, B., De Bock, K., Ramaekers, M., et al. (2007). ACTN3 (R577X) genotype is associated with fiber type distribution. *Physiological Genomics* 32, 58–63.

Walsh, S., Liu, D., Metter, E.J., Ferrucci, L. & Roth, S.M. (2008). *ACTN3* genotype is associated with muscle phenotypes across the adult age span. *Journal of Applied Physiology* 105, 1486–1491.

Wolfarth, B., Bray, M.S., Hagberg, J.M., et al. (2005). The human gene map for performance and health-related fitness phenotypes: The 2004 update. *Medicine and Science in Sports and Exercise* 37, 881–903.

Yang, N., MacArthur, D.G., Gulbin, J.P., et al. (2003). *ACTN3* genotype is associated with human elite athletic performance. *American Journal of Human Genetics* 73, 627–631.

Chapter 19

Mitochondrial DNA Sequence Variation and Performance

SHINYA KUNO[1], HIROFUMI ZEMPO[1] AND HARUKA MURAKAMI[2]

[1]*Department of Sports Medicine, University of Tsukuba, Ibaraki, Japan*
[2]*National Institute of Health and Nutrition, Tokyo, Japan*

Mitochondria play an important role as energy-producing organelles, synthesizing a large amount of adenosine triphosphate (ATP) from glucose and fatty acids. Glucose is converted to pyruvic acid in cytoplasm and enters mitochondria, followed by conversion from acetyl CoA to a large amount of ATP through oxidative phosphorylation. Fatty acid is converted to acetyl CoA, and then it follows the same pathway as carbohydrates.

Only a few studies have reported on the relationships between mitochondrial DNA (mtDNA) polymorphisms and exercise performance. Many more studies involving various races will be necessary to generalize the relationships among mtDNA polymorphisms, the performance of athletes, and trainability. Generalization by haplogroup-based classification is also difficult because findings vary among countries, as shown by the comparisons between Finnish and Spanish and between Ethiopian and Kenyan athletes and sedentary controls. One cause of the different findings between countries may be the presence of mtDNA polymorphisms influencing mitochondrial function downstream from the major haplogroup. It should be a goal of future research to identify the sites of mtDNA polymorphisms influencing the mitochondrial function and their mechanisms.

Genetic and Molecular Aspects of Sport Performance, 1st edition.
Edited by Claude Bouchard and Eric P. Hoffman.
Published 2011 by Blackwell Publishing Ltd.

ATP Synthesis and Degradation

ATP synthesis in mitochondria mainly takes place in the matrices and cristae. The matrix is the inner space of mitochondria, where catabolism and anabolism of acetyl CoA centering on the tricarboxylic acid (TCA) cycle take place. The cristae are folding structures formed by invaginations of the inner membrane of the mitochondrial double membrane. The folding creates more surface area to increase the efficiency of ATP synthesis. ATP is synthesized from adenosine diphosphate (ADP) and phosphoric acid (Pi) when H^+ passes through ATP synthetase and enters the matrix, utilizing energy of the H^+ electrochemical potential gradient between the matrix and intermembrane space. The H^+ electrochemical gradient between the matrix and intermembrane space is formed by the electron transport chain, consisting of respiratory complexes I–IV, ubiquinone, and cytochrome *c* localized in the inner membrane.

When the energy requirement increases due to exercise, ATP degradation to sustain muscular contraction progresses, and the ADP level in the cytosol rises. The increase in the ADP level promotes mitochondrial respiration and subsequent ATP biosynthesis. This phenomenon shows that mitochondrial respiration is dependent on the levels of available ADP and Pi. The creatine phosphate (PCr)/creatine (Cr) ratio was also reported to be another factor influencing mitochondrial respiration. An increase in the cytoplasmic Cr level promotes mitochondrial respiration, and this process is controlled by

215

mitochondrial creatine kinase (CKmit). CKmit is present on the inner mitochondrial membrane, and the reaction of this enzyme elevates the local level of free ADP in the presence of Cr. Mitochondrial energy production during exercise is therefore systematically controlled through these mechanisms.

Mitochondrial DNA

Mitochondrial DNA (mtDNA), independent from nuclear DNA, is a double-stranded circular molecule of 16,569 bp and is present in the matrix. mtDNA encodes the following molecules necessary for the oxidative phosphorylation system: 13 essential peptide subunits of the oxidative phosphorylation system, 2 ribosomal RNAs (rRNAs), and 22 transfer RNAs (tRNAs). The mtDNA sequence starts at the center of the D-loop, which contains the promoter region with the mitochondrial transcription factor A (TFAM)-binding site. However, this region is known to have a higher frequency of mutations than other regions. The mtDNA encodes 13 proteins that are part of complexes I (NADH dehydrogenase, ND), III (cytochrome b, CYTB), and IV (cytochrome c oxidase, COX). Because the coding capacity of mtDNA is limited, nuclear genes encode most of the numerous gene products of mitochondria. The complex II (SDHA, SDHB, SDHC, and SDHD), a part of IV (CYCS), and other proteins are encoded by nuclear DNA. These are transcribed from nuclear DNA, translated, and transferred into mitochondria.

Coordinated Expression of Mitochondrial Genes

The coordinated expression and translation of proteins encoded by mtDNA and nuclear DNA, followed by transfer into mitochondria and assembly, are necessary for the functional action of mitochondria. COX is a multi-subunit enzyme encoded by both mtDNA and nuclear DNA, and the coordinated assembly of subunits encoded by mtDNA and nuclear DNA is necessary for its biological function. Hood (1990) investigated the level of mRNA at steady state of Cox subunit III (encoded by mtDNA) and VIc (encoded by nuclear DNA) in various rat tissues (heart, brain, liver, kidney, and skeletal muscle). They found that the two subunits were maintained at the same level in each tissue. Similar findings were

also observed in human tissues (skeletal muscle, liver, heart, kidney, brain, and white blood cells).

A common transcription factor may be regulating a coordinated expression of the distant genomes in the nucleus and mitochondria.

Many have attempted to identify a nuclear DNA-encoded transcription factor necessary for the coordinated expression, using the rat cytochrome c gene. Many cis-elements are present in the cytochrome c gene. One cis-element is the nuclear respiratory factor 1 (NRF-1)-recognition site. Previous studies identified NRF-1-binding sites in many genes involved in mitochondrial functions, including TFAM involved in mtDNA transcription and replication. The expression of nuclear DNA-encoded mitochondrial respiratory complex genes or TFAM occurs through NRF-1, and mtDNA expression subsequently occurs through the increased TFAM level, thus realizing coordinated expression. Table 19.1 lists some genes of mitochondrial respiration targeted by NRF-1. Expression of not only the mitochondrial respiratory complex but also of other gene classes is involved. The targets include mitochondria and cytoplasmic enzymes of the heme synthesis system and the mitochondrial protein import and assembly machinery, which illustrates the fact that NRF-1 plays an integrative role between the nucleus and mitochondria. Peroxisome proliferator-activated receptor γ (PPARγ) coactivator-1 (PGC-1) is a coactivator of PPARγ, which is a nuclear receptor, and PPARγ regulates the expression of genes involved in fatty acid oxidation. It was shown that PGC-1 is also involved in mitochondrial biogenesis. PGC-1 specifically interacts with NRFs, as many mitochondrial genes contain recognition sites and promote gene expression.

Performance and mtDNA

Aerobic Capacity and Mitochondria

Mitochondrial oxidative ability was evaluated in many studies using enzyme activity, oxygen consumption, and rate of the ATP synthesis. The relationship between mitochondrial functions and whole-body exercise performance has been investigated. Suter et al. (1995) investigated the relationship between mitochondrial volume density in skeletal muscle using electron microscope and maximum oxygen

Table 19.1 NRF-1 and NRF-2 recognition sites in nuclear genes required for respiratory chain expression and function (Scarpulla, 2002)

	NRF-1a	NRF-2a		NRF-1a	NRF-2a
Oxidative phosphorylation			*mtDNA transcription and replication*		
Rat cytochrome *c*	+		Human Tfam	+	+
Human cytochrome *c*	+		Mouse Tfam		+
			Rat Tfam		+
Complex I			Mouse MRP RNA	+	
Human NADH dehydrogenase subunit 8 (TYKY)	+		Human MRP RNA	+	
			Human mtTFB	+	+
Complex II					
Human succinate dehydrogenase subunit B	+	+	*HEME biosynthesis*		
Human succinate dehydrogenase subunit C	+	+	Rat 5-aminolivulinate synthase	+	
Human succinate dehydrogenase subunit D	+	+	Mouse uroporphyrinogen III synthase	+	+
Complex III			*Protein import and assembly*		
Human ubiquinone-binding protein	+		Human TOM 20	+	+
Human core protein I	+		Mouse chaperonin 10	+	
			Human SURF-1		+
Complex IV					
Rat cytochrome oxidase subunit IV		+	*Ion channels*		
Mouse cytochrome oxidase subunit IV		+	Human VDAC3	+	
Mouse cytochrome oxidase subunit Vb	+	+	Mouse VDAC3	+	
Rat cytochrome oxidase subunit Vb	+	+	Human VDAC1		+
Human/primate cytochrome oxidase subunit Vb	+	+			
Rat cytochrome oxidase subunit VIc	+		*Shuttles*		
Human cytochrome oxidase subunit VIaL	+	+	Human glycerol phosphate dehydrogenase		+
Bovine cytochrome oxidase subunit VIIaL	+	+			
Human cytochrome oxidase subunit VIIaL		+	*Translation*		
Bovine cytochrome oxidase subunit VIIc		+	Human mitochondrial ribosomal S12	+	+
Complex V					
Bovine ATP synthase γ subunit	+				
Human ATP synthase c subunit	+				
Human ATP synthase β subunit		+			

consumption ($\dot{V}O_2$max), and power output at a 4-mM lactic acid level. They found that the mitochondrial volume density and performance were significantly correlated. In addition, after 6 months of endurance training, the rate of increase in mitochondrial volume density was related to the gains in $\dot{V}O_2$max and power output. It has also been reported that the muscular mitochondria density in endurance athletes was higher than that in untrained subjects, and that the mitochondrial gene expression level was also higher (Puntschart et al., 1995). Thus, the mitochondrial volume density in human skeletal muscle is related to whole-body exercise performance. In addition, a strong correlation was noted between mitochondrial ATP production rate and $\dot{V}O_2$max ($r = 0.84$). Moreover, it has reported that ATP production rate was increased by endurance training (Wibom et al., 1992).

mtDNA Polymorphism in Athletes

Each cell has multiple copies of mtDNA, and all copies in all cells are inherited from the mother's oocyte (maternal inheritance or matrilinear). All mtDNA copies in an individual have the same sequence (homoplasmic), yet individuals differ in their mtDNA based upon polymorphisms associated with their matrilinear ancestry. The different versions of mtDNA within a population can be defined by distinct sets of polymorphisms, and these are called haplogroups. Haplogroups serve as markers of genetic as well as geographic clusters. Because mtDNA is inherited in a matrilineal fashion, mitochondrial haplogroup classification can be used to depict maternal pedigrees (Cann et al., 1987).

Haplogroups have also been used to define human migration pathway from Africa, Europe, and Asia (Figure 19.1). Based on mitochondrial pedigrees, the common human ancestor branched off haplogroup L0 and L3 in Africa. Haplogroup M and H, which are frequent types in Asians and Europeans respectively, are derived from L3. L is the most prevalent type in Africans (more than 70%). In people of European descent, H is the most prevalent haplogroup (40%), followed by J, K, I, and U. In individuals of Asian descent, the M type is the most common (more than 50%), followed by C, D, and A (MITOMAP, 2009). The revised Cambridge reference sequence (rCRS; Figure 19.1), close to the haplogroup H seen well in Europeans, has been referred as the world standard mtDNA sequence.

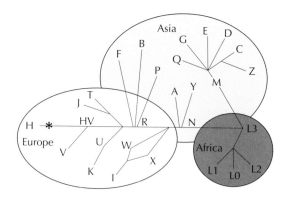

Figure 19.1 Simplified mtDNA lineages. Multiple mutations occurred in individual mtDNA and formed haplogroups and genetic populations. Haplogroups of modern humans expanded from the ancestor L0 to mainly L in Africa, M in Asian races, and H in European races. * denotes the position of the revised Cambridge reference sequence (rCRS) (MITOMAP, 2009).

Six studies compared the distributions of mtDNA polymorphisms between athletes and sedentary controls. Many researchers investigated the frequencies of mtDNA haplogroups, whereas Rivera et al. (1998) investigated the frequencies of four single nucleotide polymorphisms (SNPs) located in ND5 and D-loop of endurance athletes and sedentary controls. However, there was no significant difference in the frequencies of SNPs at 13,364, 13,470, 12,406, and 16,133 bp (Table 19.2).

The first study to compare the frequency of mtDNA haplogroups between athletes and sedentary controls was reported in 2005. The distribution of mtDNA haplogroups between endurance athletes and sprint athletes was compared in Finnish subjects (Table 19.3; Niemi & Majamaa, 2005). The frequencies of haplogroups J and K were lower in endurance than in sprint athletes.

Table 19.2 The frequency of mtDNA polymorphisms between endurance athletes and healthy controls

Reference	Subject	n	Gene	Location	P value
Rivera et al. (1998)	Endurance athletes Controls	125 65	ND5 ND5 ND5 D-Loop	12406 13364 13470 16133	ns ns ns ns
Castro et al. (2007)	Endurance athletes Controls	95 275	ND5	13368	0.012 (E vs. C: 1% vs. 8%)

ND, NADH dehydrogenase subunits; ns, not significant; E, endurance; C, control.

Table 19.3 The frequencies of mitochondrial haplogroups in athletes

Reference and nation	Haplogroup	Cohort (n)	Cohort (n)	P value
		Frequency (%)	Frequency (%)	
Europe				
Niemi and Majamaa (2005)		Endurance (n = 52)	Controls (n = 1060)	
Finland	H	52.0	48.0	ns
	V	5.8	4.8	
	U	21.0	24.0	
	K	0.0	4.5	
	T	5.8	3.6	
	J	1.9	4.8	
	W	5.8	4.4	
	I	7.7	2.8	
	X	0.0	1.1	
	Other	0.0	2.1	
		Sprint (n = 89)	Controls (n = 1060)	
	H	47.0	48.0	ns
	V	7.9	4.8	
	U	15.0	24.0	
	K	9.0	4.5	
	T	4.5	3.6	
	J	6.7	4.8	
	W	6.7	4.4	
	I	0.0	2.8	
	X	2.3	1.1	
	Other	1.1	2.1	
		Sprint (n = 89)	Endurance (n = 52)	
	H	47.0	52.0	0.039
	V	7.9	5.8	
	U	15.0	21.0	
	K	9.0	0.0	
	T	4.5	5.8	
	J	6.7	1.9	
	W	6.7	5.8	
	I	0.0	7.7	
	X	2.3	0.0	
	Other	1.1	0.0	
Castro et al. (2007)		Athletes (n = 95)	Controls (n = 275)	
Spain	H	43.0	42.0	ns
	I	1.0	1.0	
	J	15.0	15.0	
	K	12.0	9.0	
	T[a,b]	1.0	8.0	
	U	13.0	12.0	
	V	5.0	5.0	
	W	4.0	3.0	
	X	2.0	2.0	
	Other	5.0	3.0	

(Continued)

Table 19.3 (Continued)

Reference and nation	Haplogroup	Cohort (*n*)	Cohort (*n*)	*P* value
		Frequency (%)	Frequency (%)	
Africa				
Scott et al. (2005)		Endurance (*n* = 76)	Controls (*n* = 108)	
Ethiopia	L1	15.8	18.5	ns
	L2	14.5	7.4	
	L3A	25.0	26.9	
	M	14.5	19.4	
	E1	7.9	9.3	
	E2	22.4	18.5	
Scott et al. (2009)		Athletes (National + International) (*n* = 291)	Controls (*n* = 85)	
Kenya				
	L0	20.9	15.0	0.003
	L1	1.5	6.0	
	L2	10.5	18.0	
	L3	33.6	48.0	
	L4	5.2	1.0	
	L5	6.0	5.0	
	L7	10.2	4.0	
	M	9.8	2.0	
	R	2.5	1.0	
		Athletes (National only) (*n* = 221)	Controls (*n* = 85)	
	L0	18.0	15.0	0.023
	L1	2.0	6.0	
	L2	11.0	18.0	
	L3[c]	36.0	48.0	
	L4	5.0	1.0	
	L5	5.0	5.0	
	L7	10.0	4.0	
	M[d]	10.0	2.0	
	R	3.0	1.0	
		Athletes (International only) (*n* = 70)	Controls (*n* = 85)	
	L0[e]	30.0	15.0	< 0.001
	L1	0.0	6.0	
	L2	9.0	18.0	
	L3[f]	26.0	48.0	
	L4	6.0	1.0	
	L5	9.0	5.0	
	L7	11.0	4.0	
	M[g]	9.0	2.0	
	R	1.0	1.0	
		Athletes (International) (*n* = 70)	Athletes (National) (*n* = 221)	
	L0	30.0	18.0	ns
	L1	0.0	2.0	
	L2	9.0	11.0	
	L3	26.0	36.0	

Table 19.3 (Continued)

Reference and nation	Haplogroup	Cohort (*n*)	Cohort (*n*)	*P* value
		Frequency (%)	Frequency (%)	
	L4	6.0	5.0	
	L5	9.0	5.0	
	L7	11.0	10.0	
	M	9.0	10.0	
	R	1.0	3.0	

[a] T vs. all the other haplogroups; $P = 0.012$.
[b] T vs. H; $P = 0.02$.
[c] L3 vs. all the other haplogroups; $P = 0.068$.
[d] M vs. all the other haplogroups; $P = 0.05$.
[e] L0 vs. all the other haplogroups; $P = 0.032$.
[f] L3 vs. all the other haplogroups; $P = 0.005$.
[g] M vs. all the other haplogroups; $P = 0.141$.

Haplogroup I was not detected in the sprint athletes, whereas its frequency was high in the endurance athletes. Haplogroups J (De Benedictis et al., 1999) and K (Ross et al., 2001) have been reported to associate with longevity; however, they were also high in persons with Leber's hereditary optic neuroretinopathy (LHON). Marcuello et al. (2009) observed a lower efficiency of the electron transport chain accounting for lower $\dot{V}O_2$max in Spanish subjects with haplogroup J. A low efficiency of the electron transport chain leads to low-level energy production, suggesting that haplogroup J is not suitable for endurance performance. One Spanish study reported a lower frequency of haplogroup T in endurance athletes than in sedentary controls (Table 19.3; Castro et al., 2007). Interestingly, haplogroup T was significantly more frequent among Spanish patients with morbid left ventricular hypertrophy (Castro et al., 2006), which may explain in part the low frequency of haplogroup T in endurance athletes. Although the above studies were performed in Europeans, there were discrepancies in the association of haplogroups with endurance performance: no haplogroup K was detected in any endurance athlete in the Finnish study, whereas its frequency was slightly higher in endurance athletes than in control subjects in the Spanish study. In contrast, the frequency of haplogroup T was low in Spanish endurance athletes but high in Finnish endurance athletes. The reasons for these differences between the countries are not known. They may relate to

ethnic and/or geographical area–associated variations in the distributions of haplotypes or of polymorphisms downstream of the major haplogroups.

Some studies focusing on Africans were recently reported. Many top-level endurance athletes are from East Africa, particularly Kenyan and Ethiopian elite athletes, whereas Jamaican and many American top-level sprint athletes are originally from West Africa. mtDNA haplogroups were determined in 76 Ethiopian endurance athletes and 108 control subjects. There were no significant differences in the frequencies of mitochondrial haplogroups between the groups (Table 19.3; Scott et al., 2005). In contrast, a Kenyan study showed that the frequencies of haplogroups L0 and M were higher in internationally acclaimed endurance athletes than in control subjects, whereas the frequency of haplogroup L3 was low (Table 19.3; Scott et al., 2009). These Ethiopian and Kenyan data showed differences in the haplogroups that were advantageous for endurance athletes even between adjacent countries. However, Scott et al. observed that the distribution of haplogroup in Ethiopians was different from that in Kenyans, but the haplogroups more prevalent in Kenyan athletes were also present in Ethiopian non-athletes at reasonable frequencies.

mtDNA Polymorphism and Endurance Capacity in the Sedentary State

It has been reported that endurance capacity varies markedly even in healthy sedentary people, and genetic factors are partly responsible for these interindividual differences. In an early study, Dionne et al. (1991) investigated the associations of mtDNA polymorphisms with $\dot{V}O_2$max and its trainability in 46 sedentary young adult male (Table 19.4). The $\dot{V}O_2$max in the untrained state was lower in subjects with the non-rCRS sequence at 13,365 (*ND5*), 13,470 (*ND5*), and 15,925 (*tRNA*) than those with rCRS sequence. The response to training was lower in subjects with the non-rCRB sequence at 12,406 (*ND5*). We investigated the relationships among polymorphisms in the mtDNA D-loop region, $\dot{V}O_2$max, citrate synthase (CS) activity in skeletal muscle, and mtDNA content in 55 adult men (Murakami et al., 2002) (Table 19.4). Our result showed that the $\dot{V}O_2$max was higher in subjects with 16,298C, 16,325T, and 199C.

Since the D-loop contains regions involved in the replication and transcription of mtDNA, itis possible that the mtDNA polymorphisms identified in our study may influence mitochondrial oxidative capacity by affecting mtDNA replication and transcription. When we compared the data of polymorphisms reported by Dionne et al. with our findings, there was no subject carrying their polymorphisms associated with endurance performance in our subjects; in other words, as shown in the previous studies of mitochondrial haplogroups, there are large racial differences in mtDNA variation. Thus,

even if an association of a certain polymorphism with performance is observed in one group, it may be other polymorphisms in other groups.

One study reported that haplogroups were associated with endurance capacity in sedentary subjects. Marcuello et al. (2009) investigated the relationship between mtDNA haplogroups and $\dot{V}O_2$max in young Spanish men. $\dot{V}O_2$max was significantly lower in haplogroup J than in non-J; however, no significant difference was noted in maximal heart rate. They speculated that the lower endurance capacity in haplogroup J was related to a lower efficiency of the electron transport chain.

Studies involving racial and ethnic comparisons covering several geographical areas and based on much larger sample sizes are needed to clarify whether mtDNA polymorphisms are associated with endurance capacity, and the pathways leading to the variability in the response to exercise training.

Verification of the Mechanism Based on Cell Culture Model

Research focusing on the relationship between mtDNA polymorphisms and endurance capacity has not yet clarified the influence of the mtDNA polymorphisms on endurance capacity. $\dot{V}O_2$max is determined by several factors, such as skeletal

Table 19.4 mtDNA polymorphisms, aerobic capacity, and mitochondrial copy number and function in non-athletes

Reference	Subject	n	Gene	Location	Phenotype	P value
Dionne et al. (1991)	Male	46	*ND2*	4740	$\dot{V}O_2$max	< 0.07
			ND5	13365	$\dot{V}O_2$max	< 0.05
			ND5	13470	$\dot{V}O_2$max	< 0.05
			tRNA	15925	$\dot{V}O_2$max	< 0.05
			ND5	12406	$\Delta \dot{V}O_2$max	< 0.05
Murakami et al. (2002)	Male	55	D-Loop	16298	$\dot{V}O_2$max	0.008
			D-Loop	16325	$\dot{V}O_2$max	0.03
			D-Loop	199	$\dot{V}O_2$max	0.04
			D-Loop	194	Citrate synthase activity	0.032
			D-Loop	514	mtDNA number	0.032
			D-Loop	16223	$\%\Delta\dot{V}O_2$max	0.047
			D-Loop	16362	$\%\Delta\dot{V}O_2$max	0.025
			D-Loop	16519	$\%\Delta\dot{V}O_2$max	0.027
			D-Loop	16519	$\%\Delta$Citrate synthase activity	0.025
Marcuello et al. (2009)	Male	114		Haplotype J	$\dot{V}O_2$max	0.02
					HRmax	ns

ND, NADH dehydrogenase subunits; ns, not significant.

muscle mitochondrial oxidative capacity, cardio-respiratory function, muscle mass, and others. Accordingly, the mechanisms linking mtDNA poly-morphisms and $\dot{V}O_2$max have yet to be investigated more directly. We investigated whether mtDNA polymorphisms induce variations in mitochondrial function using a cell culture model (Soma et al., 2001; Murakami et al., 2001). To verify the effect of mtDNA polymorphisms on mitochondrial function, it is desirable that mtDNA be derived from various subjects but that the nuclear DNA is identical. ρ^0 HeLa cells contain only the nuclear DNA sequence and have no mtDNA. ρ^0 HeLa cells were fused with platelets containing only a subject's mtDNA and no nuclear DNA (Figure 19.2). These unique cells con-taining the ρ^0 HeLa nuclear DNA and a given sub-ject's mtDNA are called cybrids (Soma et al., 2001).

This model has been employed to investigate the relationships between mtDNA mutations and cancer (Hayashi et al., 1986), Parkinson disease (Aomi et al., 2001), and age-related disorder of mitochondrial function (Hayashi et al., 1994). We prepared cybrids with platelets derived from athletes with a high aerobic capacity and control subjects, and measured oxygen consumption and COX activity level. There was no significant difference between athlete- and control-derived cybrids (Soma et al., 2001).

In another study, sedentary male subjects were divided into two groups according to the level of $\dot{V}O_2$max; a high-performing (HP) group showing higher $\dot{V}O_2$max (1 SD above the average) and a low-performing (LP) group (1 SD below the average). Cybrids were prepared from platelets of subjects in both groups. There was no significant difference in the oxygen consumption or COX activity level in cybrids (Murakami et al., 2001). Although, no association was found between the properties of cybrids and $\dot{V}O_2$max in our previous cell culture model studies, there was variability in both oxygen consumption and level of COX activity. Moreover, O_2 consumption of cybrids represents state 4 in mitochondrial respiratory flow. We need to investi-gate state 3 mitochondrial respiration, which is more analogous to mitochondrial respiration in skeletal muscle during maximal exercise. Therefore, further investigations are required to clarify whether there is a true association between mtDNA polymorphisms and $\dot{V}O_2$max or submaximal endurance capacity.

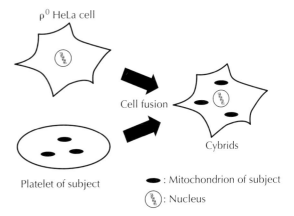

Figure 19.2 Schematic representation of the introduction of mtDNA into ρ^0 HeLa cells. Platelets of the subjects purified by centrifuging venous blood were fused with ρ^0 HeLa cells in the presence of polyethylene glycol. After the fusion, mtDNA-repopulated cybrids were isolated in selective medium. The mitochondrial enzymes of cybrids were encoded in both the nuclear DNA from HeLa cells and the mtDNA from the subject. Since cybrids share the same nuclear background as HeLa cells, they can be used for examining the influence of mtDNA polymorphisms on mitochondrial respiratory function. (Reproduced with permission from Soma et al., 2001).

Exercise Intolerance Associated with mtDNA Mutation

Mutations in mitochondrial genes have been shown to cause a wide range of disease conditions, including neonatal death (Leigh disease), progres-sive muscular dystrophy (mitochondrial myopa-thy), eye disease (LHON), and many others. The polymorphisms described in haplogroups above are considered benign; however, there are likely mtDNA changes that are intermediate between benign polymorphisms and disease-associated pathogenic mutations. These intermediate changes may not always cause disease, but rather result in a predisposition to a problem. For example, some mtDNA mutations were reported to cause exercise intolerance. Exercise intolerance may result from a number of conditions, including diminished aero-bic respiration due to mitochondrial dysfunction and/or increased anaerobic respiration with con-comitant rise in lactic acid metabolism, leading to discontinuation of exercise. Mutations of *CYTB* (mtDNA 14,747–15,887) are frequently reported as mtDNA mutations inducing exercise intolerance.

A mutation at mtDNA 14,849 changed a serine codon to a proline codon. As a result, the peptide is unable to be bound by an electron transporter, ubiquinone (CoQ) (Schuelke et al., 2002). A mutation at 15,059 produced a stop codon, which disrupts *CYTB* formation and results in respiratory enzyme complex III deficiency, causing muscular disorder accompanied by a chronically high serum lactic acid level (Andreu et al., 1999a). A mutation at 15,170 also specified a stop codon, which disrupts *CYTB* formation and reduces activities of respiratory enzyme complexes I and III, with the same exercise intolerance (Bruno et al., 2003). A mutation at 15,762 changes the glycine codon to a glutamic acid codon, which elevates serum lactic acid level only slightly in a resting state. However, immediately after starting exercise, the serum level of lactic acid sharply increases, making it difficult to continue exercise (Andreu et al., 1998). Mutations of the *tRNA* (Campos et al., 1995; Uusimaa et al., 2007), *ND4* (Andreu et al., 1999b), *COX I* (Karadimas et al., 2000), and *COX III* (Hanna et al. 1998) genes were also reported as exercise intolerance-inducing mtDNA mutations.

Changes in Aerobic Capacity Induced by Training and mtDNA

Mitochondrial Adaptation Induced by Endurance Training

Repeated muscle contraction increases the number of mitochondria in skeletal muscle and mitochondrial enzyme activity. As a result, those changes enhance the efficiency of ATP synthesis in mitochondria at the same absolute power output level, and make the continuation of exercise possible. An increase in the number of mitochondria plays an important role in endurance exercise performance. Endurance training causes an increase in mitochondrial density, enzyme activity (Hoppeler et al., 1985; Suter et al., 1995), and ATP synthesis rate (Wibom et al., 1992) in human skeletal muscle. Adaptations of the mitochondrial structure and capacity were reported to change in parallel with an increase in $\dot{V}O_2$max, an indicator of whole-body endurance (Suter et al., 1995). In a rat muscle contraction model, continuous electrical stimulation of the skeletal muscle induced parallel increases in the mRNA levels of *COX III* and *COX VIc* encoded by mtDNA and nuclear DNA, respectively (Hood et al., 1989). In addition, not only mRNA-level adaptations but also mtDNA-level adaptations were reported. Williams et al. (1986) performed electrical stimulation on rabbit skeletal muscle for 5 or 21 days and investigated changes in mtDNA level. No change was observed on day 5, but an about eight-fold increase was found on day 21. In a study in which rats actually performed endurance training for 3, 6, and 12 weeks, no change was observed in the mtDNA level at 3 or 6 weeks, but the level started to increase at 12 weeks. Therefore, training enhances mitochondrial biogenesis and oxidative ability through mtDNA transcription and replication.

Training-Induced Adaptation of Mitochondrial Gene Transcription Factors

As described above, PGC1, NRF-1, and TFAM were identified as factors facilitating the coordinated expression of mitochondrial genes. Training induces adaptation also in these transcription factors.

PGC-1 controls many genes involved in fatty acid and glucose metabolism as well as mitochondrial oxidative capacity. *PGC1* gene expression was enhanced in rats after a single bout of swimming for 6 h and with swimming training for 2 h over 3 and 7 days. The peak of *PGC1* gene expression after the single bout of swimming was detected after 18 h (Baar et al., 2002). Regarding NRF-1, 5-day treadmill running did not enhance *NRF1* gene expression in rats. However, when a single bout of exercise was performed in rats, peak *NRF1* expression was detected after 6 h, which returned to baseline level after 24 h (Murakami et al., 1998). *NRF1* expression is transiently enhanced by exercise and returned to a steady level thereafter. After electrical stimulation of a rat skeletal muscle for 3 days, *Tfam* mRNA, protein and binding to mtDNA levels were investigated after 1, 2, 3, 4, 5, 7, and 14 days (Gordon et al., 2001). The *Tfam* mRNA level increased at 4 days, and the protein and DNA-binding level increased at 7 days. The *Cox3* mRNA and Cox activity levels were also increased after 7 days. In a human study, 4-week one-leg cycling exercise training increased *TFAM* level (Bengtsson et al., 2001).

Training induces mtDNA gene expression and replication in a coordinated manner and may play an important role in endurance performance and trainability. From a genetic view, a relationship between a *PGC1A* gene polymorphism and endurance capacity was reported (Lucia et al., 2005; Stefan et al., 2007). Regulated pathways involved in mitochondrial proliferation are likely to be related to endurance performance and perhaps trainability. The D-loop contains transcription and replication regions and experiences a higher frequency of mutation than nuclear genes. Mutations in the gene promoter region influence gene expression level. Therefore, polymorphisms in the mtDNA D-loop may influence the efficiency of binding of Tfam, and subsequently, mtDNA gene expression and mitochondrial oxidative ability.

mtDNA Polymorphism and Trainability

It is known that the response to training is characterized by observed individual differences. Dionne et al. (1991) investigated the relationship between mtDNA polymorphisms and trainability in 46 sedentary adult male (Table 19.4). The subjects with the non-rCRB sequence at 12,406 (*ND5*) exhibited a lower trainability profile than rCRB subjects. We have reported the associations between mtDNA polymorphisms and $\dot{V}O_2$max gains with training, with the trainability being highest in subjects with 16,223T and 16,262C (Murakami et al., 2002). These positions are in the D-loop region, and it is possible that the mtDNA polymorphisms identified in our study influence mitochondrial aerobic capacity by affecting mtDNA replication and transcription.

We also verified the polymorphisms reported by Dionne et al. in our subjects. The associated D-loop site reported by Dionne et al. was 16,133 (response to training and unadjusted), but we could not detect any polymorphism at this site in our study of Japanese subjects (Murakami et al., 2002). However, as highlighted above for the haplogroups, there are large racial differences in mtDNA variation, and this lack of replication should not be surprising.

Summary

In summary, mitochondrial function is strongly associated with energy status in muscle cells and with muscle performance. Mitochondria respond to muscle activity and training by adjusting both the number of mitochondria in myofibers, and the gene expression of the mitochondria. Many of the proteins in mitochondria are encoded by mtDNA, and polymorphisms or mutations in mtDNA can disrupt mitochondrial function and muscle function. Alterations in some mtDNA mutations impair mitochondrial functions leading to a reduced exercise capacity exceeding the natural variation observed among healthy individuals, and reaching at times a morbid level. Other polymorphisms influence trainability. However, associations between elite athletes and specific mtDNA haplogroups have been inconsistent. Attempts to identify mtDNA haplogroups associated with athleticism have been complicated by the differing patterns in different ethnic subpopulations, where associations may be due more to ethnicity than true differences in mitochondrial function (hidden stratification).

References

Andreu, A.L., Bruno, C., Dunne, T.C., et al. (1999a). A nonsense mutation (G15059A) in the cytochrome b gene in a patient with exercise intolerance and myoglobinuria. *Annals of Neurology* 45, 127–130.

Andreu, A.L., Tanji, K., Bruno, C., et al. (1999b). Exercise intolerance due to a nonsense mutation in the mtDNA ND4 gene. *Annals of Neurology* 45, 820–823.

Andreu, A.L., Bruno, C., Shanske, S., et al. (1998). Missense mutation in the mtDNA cytochrome b gene in a patient with myopathy. *Neurology* 51, 1444–1447.

Aomi, Y., Chen, C.S., Nakada, K., et al. (2001). Cytoplasmic transfer of platelet mtDNA from elderly patients with Parkinson's disease to mtDNA-less HeLa cells restores complete mitochondrial respiratory function. *Biochemical and Biophysical Research Communications* 280, 265–273.

Baar, K., Wende, A.R., Jones, T.E., et al. (2002). Adaptations of skeletal muscle to exercise: Rapid increase in the transcriptional coactivator PGC-1. *FASEB Journal* 16, 1879–1886.

Bengtsson, J., Gustafsson, T., Widegren, U., Jansson, E. & Sundberg, C.J. (2001). Mitochondrial transcription factor A and respiratory complex IV increase in response to exercise training in humans. *Pflugers Archiv* 443, 61–66.

Bruno, C., Santorelli, F.M., Assereto, S., et al. (2003). Progressive exercise intolerance associated with a new muscle-restricted nonsense mutation (G142X) in the mitochondrial cytochrome b gene. *Muscle Nerve* 28, 508–511.

Campos, Y., Bautista, J., Gutiérrez-Rivas, E., et al. (1995). Clinical heterogeneity in two pedigrees with the 3243 bp tRNA(Leu(UUR)) mutation of mitochondrial DNA. *Acta Neurologica Scandinavica* 91, 62–65.

Cann, R.L., Stoneking, M. & Wilson, A.C. (1987). Mitochondrial DNA and human evolution. *Nature* 325, 31–36.

Castro, M.G., Huerta, C., Reguero, J.R., et al. (2006). Mitochondrial DNA haplogroups in Spanish patients with hypertrophic cardiomyopathy. *International Journal of Cardiology* 112, 202–206.

Castro, M.G., Terrados, N., Reguero, J.R., Alvarez, V. & Coto, E. (2007). Mitochondrial haplogroup T is negatively associated with the status of elite endurance athlete. *Mitochondrion* 7, 354–357.

De Benedictis, G., Rose, G., Carrieri, G., et al. (1999). Mitochondrial DNA inherited variants are associated with successful aging and longevity in humans. *FASEB Journal* 13, 1532–1536.

Dionne, F.T., Turcotte, L., Thibault, M.C., Boulay, M.R., Skinner, J.S. & Bouchard, C. (1991). Mitochondrial DNA sequence polymorphism, $\dot{V}O_2$max, and response to endurance training. *Medicine and Science in Sports and Exercise* 23, 177–185.

Gordon, J.W., Rungi, A.A., Inagaki, H. & Hood, D.A. (1991). Effects of contractile activity on mitochondrial transcription factor A expression in skeletal muscle. *Journal of Applied Physiology* 90, 389–396.

Hanna, M.G., Nelson, I.P., Rahman, S., et al. (1998). Cytochrome c oxidase deficiency associated with the first stop-codon point mutation in human mtDNA. *American Journal of Human Genetics* 63, 29–36.

Hayashi, J., Werbin, H. & Shay, J.W. (1986). Effects of normal human fibroblast mitochondrial DNA on segregation of HeLaTG mitochondrial DNA and on tumorigenicity of HeLaTG cells. *Cancer Research* 46, 4001–4006.

Hayashi, J., Ohta, S., Kagawa, Y., et al. (1994). Nuclear but not mitochondrial genome involvement in human age-related mitochondrial dysfunction. Functional integrity of mitochondrial DNA from aged subjects. *Journal of Biological Chemistry* 269, 6878–6883.

Hood, D.A. (1990). Co-ordinate expression of cytochrome c oxidase subunit III and VIc mRNAs in rat tissues. *Biochemical Journal* 269, 503–506.

Hood, D.A., Zak, R., Pette, D. (1989). Chronic stimulation of rat skeletal muscle induces coordinate increases in mitochondrial and nuclear mRNAs of cytochrome-c-oxidase subunits. *European Journal of Biochemistry* 179, 275–280.

Hoppeler, H., Howald, H., Conley, K., et al. (1985). Endurance training in humans: aerobic capacity and structure of skeletal muscle. *Journal of Applied Physiology* 59, 320–327.

Karadimas, C.L., Greenstein, P., Sue, C.M., et al. (2000). Recurrent myoglobinuria due to a nonsense mutation in the COX I gene of mitochondrial DNA. *Neurology* 55, 644–669.

Lucia, A., Gómez-Gallego, F., Barroso, I., et al. (2005). PPARGC1A genotype (Gly482Ser) predicts exceptional endurance capacity in European men. *Journal of Applied Physiology* 99, 344–348.

Marcuello, A., Martínez-Redondo, D., Dahmani, Y., et al. (2009). Human mitochondrial variants influence on oxygen consumption. *Mitochondrion* 9, 27–30.

MITOMAP (2009). A human mitochondrial genome database. Available at http://www.mitomap.org and http://www.mitomap.org/pub/MITOMAP/MITOMAPFigures/simple-tree-mitomap2008.pdf.

Murakami, H., Soma, R., Hayashi, J., et al. (2001). Relationship between mitochondrial DNA polymorphism and the individual differences in aerobic performance. *Japanese Journal of Physiology* 51, 563–568.

Murakami, H., Ota, A., Simojo, H., Okada, M., Ajisaka, R. & Kuno, S. (2002). Polymorphisms in control region of mtDNA relates to individual differences in endurance capacity or trainability. *Japanese Journal of Physiology* 52, 247–256.

Murakami, T., Shimomura, Y., Yoshimura, A., Sokabe, M. & Fujitsuka, N. (1998). Induction of nuclear respiratory factor-1 expression by an acute bout of exercise in rat muscle. *Biochimica et Biophysica Acta* 1381, 113–122.

Niemi, A.K. & Majamaa, K. (2005). Mitochondrial DNA and *ACTN3* genotypes in Finnish elite endurance and sprint athletes. *European Journal of Human Genetics* 13, 965–969.

Puntschart, A., Claassen, H., Jostarndt, K., Hoppeler, H. & Billeter, R. (1995). mRNAs of enzymes involved in energy metabolism and mtDNA are increased in endurance-trained athletes. *American Journal of Physiology* 269, C619–C625.

Rivera, M.A., Wolfarth, B., Dionne, F.T., et al. (1998). Three mitochondrial DNA restriction polymorphisms in elite endurance athletes and sedentary controls. *Medicine and Science in Sports and Exercise* 30, 687–690.

Ross, O.A., McCormack, R., Curran, M.D., et al. (2001). Mitochondrial DNA polymorphism: Its role in longevity of the Irish population. *Experimental Gerontology* 36, 1161–1178.

Scarpulla, R.C. (2002). Nuclear activators and coactivators in mammalian mitochondrial biogenesis. *Biochimica et Biophysica Acta* 1576, 1–14.

Schuelke, M., Krude, H., Finckh, B., et al. (2002). Septo-optic dysplasia associated with a new mitochondrial cytochrome b mutation. *Annals of Neurology* 51, 388–392.

Scott, R.A., Wilson, R.H., Goodwin, W.H., et al. (2005). Mitochondrial DNA lineages of elite Ethiopian athletes. *Comparative Biochemistry and Physiology—Part B: Biochemistry & Molecular Biology* 140, 497–503.

Scott, R.A., Fuku, N., Onywera, V.O., et al. (2009). Mitochondrial haplogroups associated with elite Kenyan athlete status. *Medicine and Science in Sports and Exercise* 41, 123–128.

Soma, R., Murakami, H., Hayashi, J., et al. (2001). The effects of cytoplasmic transfer of mtDNA in relation to whole-body endurance performance. *Japanese Journal of Physiology* 51, 475–480.

Stefan, N., Thamer, C., Staiger, H., et al. (2007). Genetic variations in PPARD and PPARGC1A determine mitochondrial function and change in aerobic physical fitness and insulin sensitivity during lifestyle intervention. *The Journal Clinical Endocrinology and Metabolism* 92, 1827–1833.

Suter, E., Hoppeler, H., Claassen, H., et al. (1995). Ultrastructural modification of human skeletal muscle tissue with 6-month moderate-intensity exercise training. *International Journal of Sports Medicine* 16, 160–166.

Uusimaa, J., Moilanen, J.S., Vainionpää, L., et al. (2007). Prevalence, segregation, and phenotype of the mitochondrial DNA 3243A>G mutation in children. *Annals of Neurology* 62, 278–287.

Wibom, R., Hultman, E., Johansson, M., et al. (1992). Adaptation of mitochondrial ATP production in human skeletal muscle to endurance training and detraining. *Journal of Applied Physiology* 73, 2001–2010.

Williams, R.S. (1986). Mitochondrial gene expression in mammalian striated muscle. Evidence that variation in gene dosage is the major regulatory event. *Journal of Biological Chemistry* 261, 12390–12394.

Chapter 20

Genes, Exercise, and Lipid Metabolism

YOHAN BOSSÉ

Department of Molecular Medicine, Laval University, and Institut Universitaire de Cardiologie et de Pneumologie de Québec, Québec, QC, Canada

Lipids are an important part of the fuel mix metabolized to make athletic records possible. An athlete's ability to perform in long-distance events depends in part on the capacity to store, mobilize, transport, and utilize lipids. Scientific evidence also supports the concept that athletic performance is attributable to a significant extent to the genetic background of an individual. During the last decades, key molecular pathways that govern lipid's mobilization, transport, and oxidation have been elucidated. Genetic variants that alter the level of expression or other functional properties of gene products involved in these pathways are likely to play a role in the innate ability to combust lipids during exercise. However, much remains to be discovered about genes and alleles contributing to lipid metabolism during exercise and in response to training. The first part of this chapter introduces the basic physiological principles of lipid metabolism at rest and during exercise. The second part reviews molecular pathways and gene products involved in lipid metabolism as well as some of the adaptations that occur with training. Finally, the last section summarizes our current understanding of the genetic factors that modulate lipid metabolism responsiveness to exercise training, which is mostly derived from studies on the effects of exercise on health-related outcomes such as obesity and dyslipidemia.

Genetic and Molecular Aspects of Sport Performance, 1st edition. Edited by Claude Bouchard and Eric P. Hoffman. Published 2011 by Blackwell Publishing Ltd.

Lipid Classification and Functions

Lipids constitute a broad group of water-insoluble molecules. They can simplistically be classified in three main groups: simple lipids, compound lipids, and derived lipids. *Simple lipids* consist mostly of triglycerides, which are the major form of fat storage in the body. Triglycerides are composed of a single molecule of glycerol and three fatty acid molecules. Although the glycerol molecule is fixed, the size and chemical structure of fatty acids vary and include saturated and unsaturated molecules. *Compound lipids* include phospholipids and lipoproteins. They are composed of triglycerides and other chemical moieties. Lipoproteins constitute the main form of lipid transport in the blood (see later). Finally, *derived lipids* consist of simple lipids chemically modified to generate other bioactive molecules. Cholesterol and other steroid compounds are the most widely known derived lipids.

Lipids serve many essential roles in the body. Lipids are essential constituents of the membranes of every cell in the body. Depots of lipid act as insulator against heat loss and as a cushion for the protection of vital organs, such as the heart, liver, and brain. Lipids also act as carrier of fat-soluble vitamins, A, D, E, and K. Of particular relevance for this chapter, lipids serve as an energy source and reserve. Lipids are energy-dense compared to other energy substrates (carbohydrates and proteins). One gram of lipid contains approximately 9 cal, which is more than twice the energy found in the same quantity of carbohydrates and proteins. The body's lipid content varies substantially among individuals, but represents about 15% and 25% of the total body mass of an average male and female, respectively. For a 70-kg male, the total energy stored as lipid is estimated at about 95,000 cal, or enough energy to run about 1350 km, assuming a rough energy expenditure of 1 cal/kg/km. This caloric reservoir is impressive compared to the 2000 or so calories stored as carbohydrate. Thus, lipids are by far the largest, readily available energy reserve of the human body.

A Quick Glance at Lipid Metabolism

The metabolism of lipids is a dynamic and highly regulated process influenced by the person's nutritional state and fitness level, as well as by the level of energy expenditure, including the intensity and duration of physical activity. The metabolic state of an individual determines the commitment of a particular fuel substrate toward production of energy or fat synthesis and storage. Figure 20.1 provides a schematic overview of lipid metabolism at rest

and during exercise. Dietary lipids are absorbed in the gastrointestinal tract. Lipids are then packed in lipoproteins called chylomicrons within the intestinal cells and released into the lymphatic system before entering the bloodstream. Triglycerides contained in chylomicrons are hydrolyzed by the lipoprotein lipase (LPL), releasing fatty acids for uptake by peripheral tissues including skeletal muscles and adipose tissues. Chylomicron remnants are taken up by the liver. At rest, particularly in a postprandial state, fatty acids in muscles and adipose tissues undergo esterification, where three fatty acids bind with glycerol to form triglycerides. A limited amount of triglycerides is stored in muscles. Intramuscular lipids are used as energy sources and replenished in the postexercise recovery phase. In contrast, adipose tissues store a practically unlimited quantity of triglycerides. During exercise, triglycerides are hydrolyzed in the cells' cytosol into their individual components by a process known as lipolysis. The process that mediates hydrolysis of triglycerides and release of glycerol and fatty acids

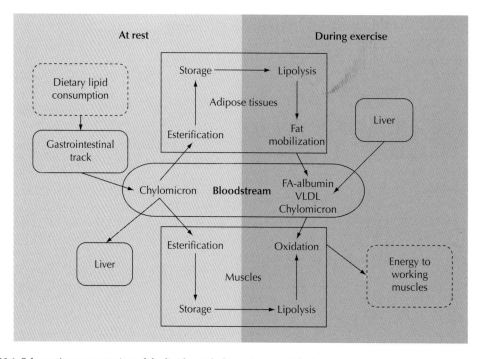

Figure 20.1 Schematic representation of the lipid metabolism at rest and during exercise.

from adipose tissues is called lipid mobilization. Fat cells or adipocytes generate large quantities of fatty acids that are delivered to the working muscles via the bloodstream. Chylomicrons and very low-density lipoproteins (VLDL) released by the liver are also rich sources of lipids during exercise. In muscles, fatty acid uptake from the circulation or derived from intramuscular lipolysis undergoes oxidation to produce energy-rich substrates for the working muscles.

Lipid and Exercise

Lipid Utilization During Exercise

The use of lipids as a fuel during exercise prevents depletion of the limited glycogen reserves of the body and spares the proteins. The contribution of lipids to the fuel mix is greatly influenced by the intensity and duration of exercise. At light-to-moderate intensity, lipids constitute the main source of energy. With increased intensity, there is a progressive shift toward a greater reliance on carbohydrates. This shift is most evident at intensities at which glucose oxidation provides the extra energy to sustain high-intensity power output. However, total fat oxidation is lowered at high-intensity (85% $\dot{V}O_2$max) compared to moderate-intensity exercise (65% $\dot{V}O_2$max), which may suggest an impaired capacity to release, transport, and oxidize fatty acids at high-intensity exercise (Romijn et al., 1993). On average, the maximal rate of fat oxidation occurs at an exercise intensity of 50–65% of the $\dot{V}O_2$max depending on the level of fitness (Nordby et al., 2006). The proportion of lipids used in the total fuel mix also depends on the duration of exercise. A seminal study conducted 75 years ago revealed that lipids constitute approximately 60–80% of the total energy expended during the last 2 h of a 6-h period of continuous exercise (Edwards et al., 1934). This contrasts with the 20–40% of total energy expenditure provided by lipids during the first 2 h of the same exercise session. Thus, there is a progressive increase in the combustion of lipids as a source of energy with a concomitant decline in carbohydrate utilization as exercise duration increases. The transient shift in energy substrates with increased exercise intensity and duration is associated with elevation and reduction of the respiratory quotient during the exercise period, respectively.

Exercise Training and Lipid Utilization

The level of fitness of an individual also has a profound effect on the metabolism of lipids during exercise. Aerobic exercise training results in an augmentation of fat utilization at rest and during exercise. The contribution of lipids as energy source is greater in trained individuals compared to untrained individuals (Saltin & Astrand, 1993). This shift is not mediated by a single adaptation, but from the integration of multiple physiological and molecular adjustments that occur with training. These include the proliferation of capillaries and increased density of mitochondria in trained muscles, which enhance fatty acids delivery and increase fat oxidation capacity, respectively. At the molecular level, key enzymes are up-regulated in trained muscles including carnitine palmitoyltransferases that facilitate fatty acid transport across the mitochondria and fatty acid–binding proteins (FABPs) that solubilize intracellular fatty acids and accelerate their movement in myocytes (see later). Overall, physical training induces changes that favor high lipid utilization by accelerating transport and uptake by the contracting muscles as well as by upgrading the tissue metabolic machinery for lipolysis and lipid oxidation.

Molecular Pathways that Govern Lipid Metabolism

Triglycerides in adipose tissue are the largest nutrient store of potential energy for biologic work. To use this rich source of energy during exercise, triglycerides must first be hydrolyzed in the adipose tissues, and the resultant fatty acids must be delivered to the working muscles. Throughout the last decades, we have gained substantial insights about the molecular processes that take place between the first step of fat mobilization in the adipose tissues and the last step of fat oxidation in the mitochondria of working muscles.

Lipid Storage and Mobilization in Adipose Tissues

Mobilization of lipids from adipose tissue in a state of high energy demand, such as aerobic exercise, requires the coordination of key molecular pathways that suppress adipose tissue triglycerides synthesis and increase triglycerides hydrolysis. For many years, adipocytes have been regarded as fat-storage organs devoid of other biological functions. However, research has recently revealed a multiplicity of endocrine, paracrine, and autocrine effects of the large number of intracellular and secreted proteins produced by the adipocytes. It is also now well established that adipocytes possess a large variety of receptors that interact with and react to several hormonal, neuronal, and nutritional factors. Figure 20.2 shows the main biochemical and molecular pathways that govern the biology of the adipocytes at rest and during exercise. The main neurohumoral mediators found in the bloodstream that are involved in controlling the balance between triglyceride breakdown and synthesis are illustrated. Studying these pathways is extremely important for the understanding of the biology of the adipocytes. Insights into these pathways are useful to elucidate the roles of adipose tissue in metabolic diseases, such as obesity and type II diabetes, and to learn about fuel expenditure and fuel economy.

At rest or during the postprandial period, antilipolytic pathways dominate and favor triglyceride synthesis and storage. Insulin is the main antilipolytic hormone. Insulin receptor (INSR)-mediated signaling inhibits lipolysis via a cAMP-dependent pathway (Figure 20.2). The cellular level of cAMP is a major integrator for the regulation of fat storage and mobilization. The insulin signaling cascade, through phosphatidylinositol-3-kinase (PI3-K) and protein kinase B (PKB) activation, leads to the phosphorylation of phosphodiesterase 3B (PDE3B) and the hydrolysis of cAMP. This results in a lack of

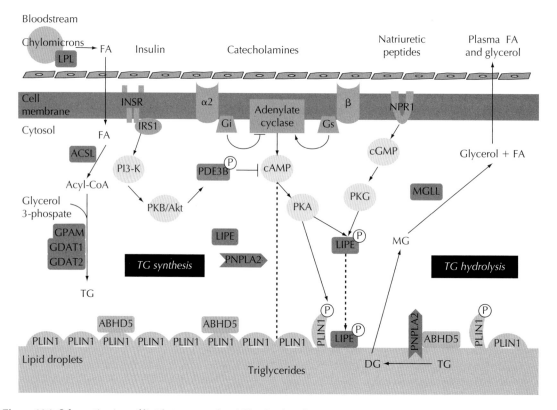

Figure 20.2 Schematic view of lipid storage and mobilization in adipocytes.

cAMP-mediated protein kinase A (PKA) activation and maintains the hormone-sensitive lipase (LIPE) in an inactive state in the cytosol. Catecholamines also have the capacity to inhibit lipolysis by activating the α_2-adrenergic receptors. The latter receptors are coupled with inhibitory G proteins (Gi) that restrain adenylate cyclase activity and prevent the conversion of ATP to cAMP. It should be noted that other antilipolytic pathways have been elucidated in adipocytes but it is beyond the scope of this chapter to describe them.

The resulting action of antilipolytic pathways is: (1) the uptake of fatty acids from the bloodstream; (2) triglyceride synthesis in the adipocytes; and (3) storing of triglycerides in lipid droplets. LPL is a key enzyme involved in the uptake of fatty acids. LPL is synthesized and secreted by the adipocytes. The enzyme subsequently translocates to the luminal side of small capillaries surrounding the adipocytes. LPL binds to heparin sulfate proteoglycans on the surface of endothelial cells and hydrolyze the triglycerides in lipoproteins (Figure 20.2). Fatty acids liberated by LPL from triglyceride-rich lipoproteins cross the endothelial lining of adipose tissue capillaries and are taken up by the adipocytes where they serve as substrates for triglyceride synthesis. In the adipocytes, long-chain fatty acids are converted into acyl-CoA by the action of the acyl-CoA synthetase (ACSL). On the other hand, glycerol is phosphorylated into glycerol-3-phosphate and reacts with acyl-CoA. Sequential enzymatic reactions lead to the production of triglycerides, which is carried out by different enzymes including glycerol-3-phosphate acyltransferase (GPAM) and diacylglycerol acyltransferase 1 and 2 (GDAT1 and GDAT2).

During exercise, the adipocytes shift from an antilipolytic to a lipolytic state. Exercise is associated with higher blood level of catecholamines (epinephrine and norepinephrine) and lower level of insulin with a net balance, which favor the activation of the lipolytic cascade. Binding of catecholamines to β-adrenoreceptors (ADRB1–3) on the plasma membrane of the adipocytes signals to a stimulatory G protein (G_s) and catalyzes the conversion of ATP to cAMP by adenylate cyclase (Figure 20.2). This second messenger

(cAMP) initiates the breakdown of triglycerides by activating PKA. It should be noted, however, that lipolysis can occur in adipocytes independent of cAMP production and PKA activity. Atrial natriuretic peptides produced by the heart during exercise can also stimulate lipolysis via the cGMP/PKG-signaling pathway (Moro et al., 2004) (Figure 20.2). The production of cAMP and cGMP phosphorylates LIPE and other accessory proteins that activate lipolysis. Three enzymes are sequentially involved in the liberation of fatty acids from triglycerides: (1) patatin-like phospholipase domain containing 2 (PNPLA2), also known as the adipose triglyceride lipase; (2) LIPE; and (3) monoglyceride lipase (MGLL). The specific role of these lipases in adipose tissues during exercise in humans remains to be determined. However, a recent study conducted in mice suggests that the disruption of the *LIPE* gene results in a reduced capacity to perform aerobic exercise (Fernandez et al., 2008). In fact, *LIPE*-null mice have a reduced endurance capacity during submaximal aerobic treadmill running. Although the relevance of results obtained from mice can be questioned, the study provides a proof of the concept that proper fat mobilization is essential to reach optimal performance during aerobic exercise. It also suggests that genetic variants that alter the expression or activity of LIPE are likely to influence endurance performance.

The activity of lipases in adipocytes is highly dependent on other proteins. Abhydrolase domain containing protein 5 (ABHD5), also known as CGI-58, interacts directly with PNPLA2 to enhance lipolysis (Lass et al., 2006). Mutations in the *ABHD 5* gene have been associated with Chanarin–Dorfman syndrome, which is characterized by the presence of intracellular lipid droplets in most tissues (Lefevre et al., 2001). Similarly, perilipins encoded by the *PLIN1* gene are essential for LIPE lipolytic action. Unphosphorylated PLIN1 creates a barrier that prevents cytosolic lipases from hydrolyzing triglycerides inside the lipid droplets. cAMP-dependent protein kinase activation by catecholamines leads to the phosphorylation of LIPE and PLIN1. Phosphorylation promotes the translocation of LIPE to the lipid droplets and the formation of a complex (LIPE and PLIN1) that facilitates the hydrolysis of triglycerides. The absence of PLIN1 in mice results in a reduced capacity to store fat due to aberrant adipose tissues lipolysis (Martinez-Botas et al., 2000). It is still unknown whether naturally occurring human genetic variants in genes encoding components of the fat-storage and mobilization processes in adipose tissue influence athletic performance.

Lipid Transport and Lipoprotein Metabolism

Figure 20.3 shows a schematic representation of our current understanding of the lipoprotein metabolism. Briefly, it is a highly regulated multi-organ system that is connected by the bloodstream and controlled by the liver, intestine, and peripheral tissues. The system ensures proper delivery of lipids throughout the body. Being insoluble, lipids are transported through the circulation in spheroidal macromolecules called lipoproteins. On the structural level, a lipoprotein is composed of a hydrophobic core containing mainly cholesterol esters and triglycerides along with a hydrophilic coat

that contains free cholesterol, phospholipids, and apolipoproteins. Lipoproteins are extremely heterogeneous in terms of size, density, chemical composition, electric charge, and functional properties. Accordingly, many classes of lipoproteins can be identified.

The main triglyceride-carrying lipoproteins are VLDL and chylomicrons. The latter are formed in intestinal cells by packaging triglycerides, cholesterol, and apolipoprotein B (APOB) isoform B48. In contrast, VLDL particles are formed in liver cells and contain triglycerides, cholesterol, and apolipoprotein B isoform B100. The main cholesterol-carrying lipoproteins are low-density lipoproteins (LDL) and high-density lipoproteins (HDL). LDL particles are remnants of VLDL. Triglycerides contained in VLDL are hydrolyzed by LPL in the circulation, releasing fatty acids for peripheral tissues and producing intermediate-density lipoproteins (IDL). The latter particles are further hydrolyzed by the hepatic lipase (LIPC) that leads to the production of LDL. The resulting smaller and denser lipoprotein

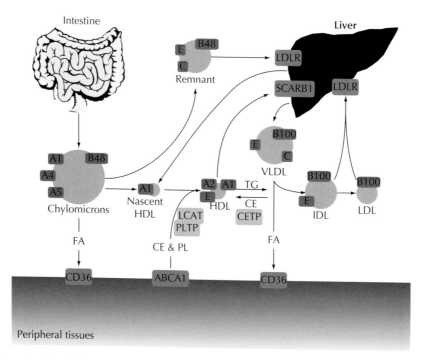

Figure 20.3 Schematic view of the lipoprotein metabolism. CE, cholesterol; PL, phospholipids. (Adapted from Lusis and Pajukanta, 2008.)

particles are subsequently taken up by the liver via LDL receptors (LDLR). Hence, the VLDL–LDL loop ensures the transport of lipids from the liver to the periphery. In contrast, HDL mediate reverse cholesterol transport, from the periphery to the liver. Nascent HDL particles are released from the liver and intestine. They are then built up from lipids from peripheral tissues and from the exchanges of lipids with triglyceride-rich lipoproteins, which are mediated by the action of different enzymes such as the cholesterol ester transfer protein (CETP), the lecithin-cholesterol acyltransferase (LCAT), and phospholipid transfer protein (PLTP). HDL particles are taken up by the liver via scavenger receptor class B member 1 (SCARB1). Different apolipoproteins on the surface of LDL and HDL act as ligands for lipoprotein receptors. Overall, lipid and lipoprotein levels in blood are orchestrated by many enzymes, receptors, and apolipoproteins. The genes encoding these modulators are good candidates for human variation in baseline and training-induced changes in lipid–lipoprotein levels.

Lipid Storage and Oxidation in Skeletal Muscles

Lipids that serve as fuel for working muscles originate from different sources, including the hydrolysis of triglyceride-rich lipoproteins and albumin-bound fatty acids in the plasma as well as the lipolysis of intramuscular triglycerides. To be used by the muscles, fatty acids derived from the circulation must reach the mitochondrial matrix. From circulation to the mitochondrial matrix, fatty acids must pass different membranes and compartments including the capillary endothelium, the interstitial space, the plasma membrane, the cytosol, and the outer and inner mitochondrial membranes (Figure 20.4). Throughout this path, fatty acids are chemically modified and chaperoned by different enzymes and transporters. Membrane-bound transporters are increasingly being recognized as facilitators of

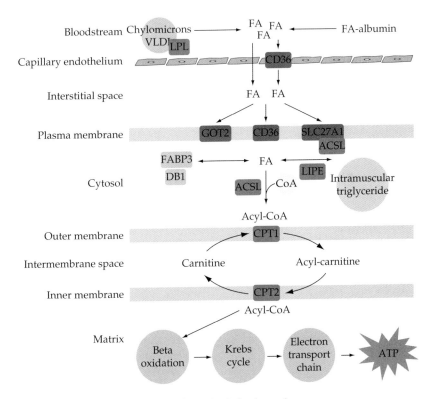

Figure 20.4 Schematic view of lipid uptake and oxidation in skeletal muscle.

the transport of fatty acids between compartments. Components of the chain of events that lead to the oxidation of fatty acids in skeletal muscles provide multiple sites of regulation to control locally the overall rate of lipid uptake and oxidation.

Approximately 55–65% of whole-body lipid oxidation during submaximal exercise comes from plasma fatty acids (Helge et al., 2001). This includes fatty acids bound to albumin and from triglyceride-rich lipoproteins degraded by LPL. Similar to adipose tissues, LPL is the rate-limiting enzyme for the hydrolysis of triglyceride-rich lipoproteins in skeletal muscles. LPL is first synthesized and secreted by the myocytes and then translocated to the lumen of capillaries. The fatty acids liberated by the action of LPL are available for uptake by working muscles. Exercise is known to increase the activity, transcription, and translation of this enzyme (Seip et al., 1997). The higher LPL activity level in trained individuals suggests that this enzyme is an important point of regulation for fatty acid oxidation during exercise.

Fatty acid transport from circulation to cytosol is assured by both passive diffusion and plasma membrane protein-mediated processes (Kiens, 2006). A number of fatty acid transporters have been identified, including fatty acid translocase CD36; glutamic-oxaloacetic transaminase 2 (GOT2), also known as plasma membrane-bound fatty acid–binding protein (FABPpm); and soluble carrier family 27 proteins (SLC27A1-6), also known as fatty acid transport proteins (FATP1-6). Little is known about SLC27A1-6 in skeletal muscles, but some members of the family, SLC27A1 and 4, possess ACSL activity, which allows fatty acids to be trapped inside the cells. Further support for a role of the SLC27A1-6 family in skeletal muscles comes from SLC27A1 knockout mice (Kim et al., 2004). These mice are characterized by a marked reduction in intramuscular triglyceride content following chronic high-fat feeding. Additional studies will be required to elucidate the role of each member of this family at rest and during exercise. In contrast, the role of CD36 is better understood (Holloway et al., 2008). CD36 is regulated by muscle contraction and plays a crucial role in the clearance of circulating fatty acids. The amount of CD36 is correlated with intramuscular triglyceride content. This transporter is found in capillary endothelium and plasma membrane (Figure 20.4). Finally, GOT2 is also induced by muscle contraction and appears to be expressed in proportion to the rate of fatty acid oxidation in muscles. Interestingly, GOT2 is up-regulated following an endurance-training program (Kiens et al., 1997), an adaptation that may partially explain the increased capacity to uptake lipids in trained individuals. Thus, skeletal muscles express a number of transmembrane proteins that facilitate transport of fatty acids into the myocytes. Whether they act in concert or individually at rest, during and after exercise remain to be determined.

Once in the cytosol, fatty acids are catalyzed by ACSL to form acyl-CoA. Very low levels of free fatty acids and acyl-CoA are found in the cytosol, and most are bound to cytosolic lipid–binding proteins such as FABP3 and diazepam-binding inhibitor (DB1), also known as acyl-Coenzyme A–binding protein. Mice lacking FABP3 are characterized by a 42–45% reduction in fatty acid uptake in skeletal muscles (Luiken et al., 2003), suggesting a permissive role of this intracellular receptor protein in fatty acid uptake into muscle tissues. However, neither FABP3 nor DB1 was changed at the mRNA or protein levels in humans by training (Kiens et al., 2004). Lipid-binding proteins ensure intracellular trafficking of lipids. In a state of low energy demand, fatty acids are directed toward resynthesis into intramuscular triglycerides. Intramuscular triglycerides are lipid droplets stored within muscle cells. This source of fatty acids can contribute as much as 10–50% of total fat oxidation during exercise depending on intensity (Horowitz, 2003). During high-intensity exercise, the muscles rely predominantly on intramuscular store as a source of lipids. The breakdown of triglycerides during exercise in skeletal muscles is poorly understood. Lipolysis of intramuscular triglycerides is believed to involve mechanisms akin to triglyceride degradation in adipocytes with LIPE as a rate-limiting enzyme. Phosphorylation and allosteric regulations of this enzyme determine the fate of fatty acids toward either oxidation by the mitochondria or resynthesis into triglycerides (Kiens, 2006).

The importance of LPL, fatty acid transporters, and binding proteins, and the breakdown of intracellular pool of triglycerides as regulatory sites for the utilization of fatty acids during exercise is still a matter of debate. In contrast, regulatory steps that control the passage of acyl-CoA into the mitochondria may be more deterministic of the overall fatty acid combustion during exercise. The acyl-CoA molecules are unable to pass the mitochondrial membranes directly. The transport is ensured by carnitine palmitoyltransferase I (CPT1) that is present on the outer mitochondrial membrane. This enzyme generates acyl-carnitine that subsequently crosses the inner mitochondrial membrane by the CPT2/translocase system (Figure 20.4). The passage of acyl-CoA across the mitochondrial membrane during exercise is controlled by many metabolites including, among other factors, malonyl CoA and carnitine concentrations (Roepstorff et al., 2004). These metabolites alter the activity of CPT1, which is considered the central gatekeeper for the entry of acyl-CoA into the mitochondria. Within the mitochondrial matrix, acyl-CoA is successively cleaved into two-carbon molecules by the process known as β-oxidation. This process generates acetyl-CoA that feeds the Krebs cycle to produces hydrogen atoms that are subsequently oxidized through the electron transport chain to generate ATP.

Genetic Variants Affecting Lipid Metabolism During Exercise and in Response to a Training Program

It is becoming apparent that a large number of genes are involved in pathways that regulate lipid metabolism during exercise and in response to a training regimen. Naturally occurring genetic variants within these genes are likely to influence the level or functional properties of encoding proteins. Genes described in the previous sections are prime candidates for the genetic component affecting an individual's ability to store, mobilize, transport, and oxidize lipids. Very few studies have investigated the role of genetic variants on lipid metabolism during exercise or in response to physical activity, especially in highly trained athletes.

Advances in this field during the last two decades were mostly derived from studies on health-related outcomes such as obesity and dyslipidemia. The next section summarizes the main findings on genetic variants influencing the responsiveness of blood lipid–lipoprotein phenotypes and adiposity indices to exercise. It is assumed that the information gained from these studies is also relevant for elite athletes as it reveals common genetic variants that may influence the capacity to store, mobilize, transport, uptake, and oxidize lipids.

Genetics of Blood Lipid–Lipoprotein Phenotypes and Exercise

A cardioprotective effect of regular exercise emerges from studies investigating the effects of exercise training on blood lipid levels (Leon & Sanchez, 2001). Most studies on the effects of exercise on blood lipids have focused on the mean changes in study participants. However, marked individual differences in lipid and lipoprotein responses to exercise training were observed, even when the dose of exercise was tightly controlled (Leon et al., 2002). Genetic epidemiology studies suggested that a large part of this variability is attributable to genetic factors (Bouchard et al., 1994; Rice et al., 2002). These observations motivated the search to identify the molecular candidates.

Most genetic studies were conducted using a candidate gene strategy testing the effects of specific polymorphisms (Bray et al., 2009). Figure 20.5 provides an overview of gene polymorphisms associated with blood lipid–lipoprotein responses to acute and chronic exercise or physical activity. Three main kinds of studies were performed. First, a few studies were conducted to verify the influence of gene polymorphisms on lipid responses to acute exercise. They were typically conducted by screening individuals for a particular genotype. Lipid responses were then compared across genotype groups after a single bout or up to a few days of submaximal or maximal exercise. Few polymorphisms located in the following genes were found to influence the lipid responses to acute exercise: neuropeptide Y (*NPY*), beta-2-adrenergic receptor

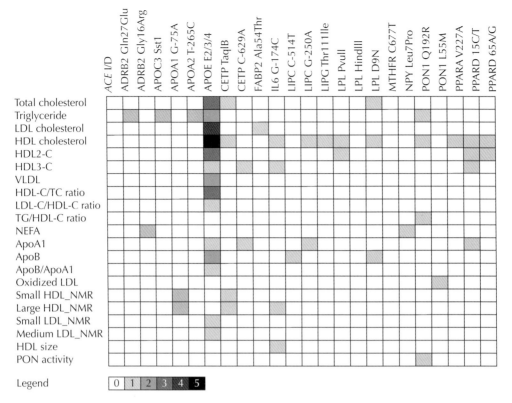

Figure 20.5 Overview of gene–exercise and physical activity interactions on blood lipid and lipoprotein phenotypes. Genes and polymorphisms that have been investigated are listed in columns. Lipid and lipoprotein phenotypes are listed in rows. Phenotypes were included if at least one published study reported a positive association with a polymorphism. The figure is derived from published candidate gene studies investigating the lipid–lipoprotein responses to acute or chronic exercise and/or physical activity. Most references can be found in Bray et al. (2009). The color intensity illustrates the number of times a specific gene–phenotype association was found. The legend indicates the number of reports with a positive association. TC, total cholesterol; TG, triglyceride; NEFA, non-esterified fatty acid; NMR, nuclear magnetic resonance; PON, paraoxonase.

(*ADRB2*), Apolipoprotein C-III (*APOC3*), and paraoxonase 1 (*PON1*). Second, a more important number of studies investigated the influence of gene polymorphisms on the response following a chronic (3–9 months) training regimen. The training programs vary substantially across studies, but generally consist of three to five sessions of aerobic exercise per week. The duration and intensity of exercise were almost always increased progressively throughout the programs with ranges of 30–50 min per session at an intensity corresponding to 50–85% $\dot{V}O_2max$. A number of polymorphisms located in the following genes were found to modulate the lipid and lipoprotein responses

to chronic exercise: *LPL, APOE, CETP*, endothelial lipase (*LIPG*), interleukin 6 (*IL6*), *LIPC, FABP2, APOA1*, peroxisome proliferator-activated receptor delta (*PPARD*), and *PON1*. Finally, a similar number of studies investigated gene–fitness or physical activity interactions on blood lipid and lipoprotein levels. These studies were conducted on a larger number of subjects whose level of physical activity or fitness was most often estimated using questionnaires. Significant gene–physical activity interactions were found with the following genes: *APOE, LPL, PON1, ADRB2, APOA2, LIPC, CETP* and peroxisome proliferator-activated receptor alpha (*PPARA*).

Genetics of Adiposity Phenotypes and Exercise

Physical activity and exercise play a crucial role in the prevention and treatment of obesity as well as in the maintenance of weight loss. Considerable individual differences in weight loss response are observed following an exercise training program. Family and twin studies suggest that a large part of this variability is attributable to genetic factors (Bouchard et al., 1994; Rice et al., 1999). Several studies were conducted to test the association between candidate genes and the response of adiposity phenotypes to regular exercise or physical activity (Bray et al., 2009). Figure 20.6 summarizes these findings. The most commonly reported positive associations were found with polymorphisms located in the adrenoreceptor genes (*ADRA2B, ADRB2, ADRB3*). As described in the previous section, β-adrenoreceptors stimulate lipolysis in adipose tissues (Figure 20.2). The Gln27Glu polymorphism in the *ADRB2* gene was associated with impaired receptor function in response to catecholamines (Green et al., 1995). Interestingly, a gene–physical activity interaction study has shown that subjects who are homozygotes for the Gln27 allele benefit more in term of weight loss from physical activity compared to carriers of the Glu27 allele (Meirhaeghe et al., 1999). This observation is concordant with a lower respiratory quotient among homozygotes for the Gln27 allele following a single bout of exercise (Macho-Azcarate et al., 2002), indicating a higher ability to oxidize fat in response to exercise. Similarly, white women carriers of the Gln27 allele exhibited a greater reduction in percent body fat following a 20-week standardized endurance-training program (Garenc et al., 2003). However, the effect was not observed in men or in black individuals, suggesting that the influence of this polymorphism is gender specific and population specific. A subsequent study indicates that Glu27 carriers had greater reduction in percent body fat following a 24 weeks of supervised aerobic exercise training (Phares et al., 2004). The Gln27Glu polymorphism illustrates well the current ambiguity regarding genetic association studies in the field. As shown in Figure 20.6, the following genes were associated with exercise training-induced changes in adiposity phenotypes: angiotensin I converting enzyme (*ACE*), catechol-*O*-methyltransferase (*COMT*), cytochrome P450, family 19, subfamily A, polypeptide 1 (*CYP19A1*), ectonucleotide pyrophosphatase/phosphodiesterase 1 (*ENPP1*), guanine nucleotide–binding protein (or G protein) (*GNAO1*), beta polypeptide 3 (*GNB3*), *LPL*, phenylethanolamine *N*-methyltransferase (*PNMT*), *PPARA*, peroxisome proliferator-activated receptor gamma (*PPARG*), and uncoupling protein 3 (*UCP3*). Overall, these results are intriguing, but need to be replicated in larger studies.

Conclusion

The utilization of lipids is a key determinant of the ability to sustain prolonged physical work. Lipids constitute the largest energy store in the human body. An innate ability to favor this energy substrate is likely to be a limiting factor for maximal performance in athletic endurance events. Animal models and naturally occurring mutations leading to Mendelian disorders have provided key insights on the molecular pathways that regulate: (1) fat storage and mobilization in adipose tissues; (2) lipid transport within the bloodstream, across cell membranes, and within intra- and extracellular spaces; and (3) lipid uptake and oxidation in skeletal muscles. It is anticipated that molecular adaptations occurring with training within these lipid processes contribute to the maximal achievable athletic capacity. A body of scientific data suggests that the interindividual response of health-related fitness and physical performance phenotypes is highly heterogeneous following a training program. Genetic epidemiology studies support the concept that these individual differences in trainability are largely determined by genetic factors. Research to identify the molecular causes of these differences is still in its infancy. Very little is known about DNA-genetic variants that impair or enhance lipid utilization and influence performance. Genes and molecular markers associated with training induce changes in lipid–lipoprotein profile, and adiposity phenotypes are being used as surrogates, but the results so far are either preliminary or controversial. Recent advances in genomic research are

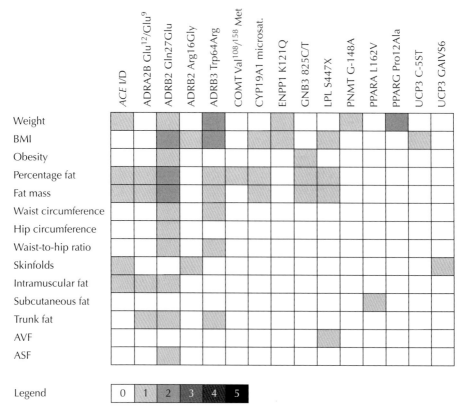

Figure 20.6 Overview of gene–exercise and physical activity interactions on adiposity indices. Visual representation is explained in Figure 20.5. All references can be found in Bray et al. (2009). AVF, abdominal visceral fat; ASF, abdominal subcutaneous fat.

offering unprecedented capabilities to interrogate the human genome and uncover the genetic complexity governing the person-to-person variability in training response. Intensive efforts, appropriate study designs, and large sample sizes will be required to achieve significant advances about the genetic factors influencing lipid utilization and athletic performance.

References

Bouchard, C., Tremblay, A., Despres, J.P., et al. (1994). The response to exercise with constant energy intake in identical twins. *Obesity Research* 2, 400–410.

Bray, M.S., Hagberg, J.M., Perusse, L., et al. (2009). The human gene map for performance and health- q fitness phenotypes: The 2006–2007 update. *Medicine and Science in Sports and Exercise* 41, 35–73.

Edwards, H.T., Margaria, R. & Dill, D.B. (1934). Metabolic rate, blood sugar and the utilization of carbohydrate. *American Journal of Physiology* 108, 203–209.

Fernandez, C., Hansson, O., Nevsten, P., Holm, C. & Klint, C. (2008). Hormone-sensitive lipase is necessary for normal mobilization of lipids during submaximal exercise. *American Journal of Physiology—Endocrinology and Metabolism* 295, E179–E186.

Garenc, C., Perusse, L., Chagnon, Y.C., et al. (2003). Effects of beta2-adrenergic receptor gene variants on adiposity: The HERITAGE Family Study. *Obesity Research* 11, 612–618.

Green, S.A., Turki, J., Hall, I.P. & Liggett, S.B. (1995). Implications of genetic variability of human beta 2-adrenergic receptor structure. *Pulmonary Pharmacology* 8, 1–10.

Helge, J.W., Watt, P.W., Richter, E.A., Rennie, M.J. & Kiens, B. (2001). Fat utilization during exercise: Adaptation to a fat-rich diet increases utilization of plasma fatty acids and very low density lipoprotein-triacylglycerol in humans. *Journal of Physiology* 537, 1009–1020.

Holloway, G.P., Luiken, J.J., Glatz, J.F., Spriet, L.L. & Bonen, A. (2008). Contribution of FAT/CD36 to the regulation of skeletal muscle fatty acid oxidation: An overview. *Acta Physiologica Oxford* 194, 293–309.

Horowitz, J.F. (2003). Fatty acid mobilization from adipose tissue during exercise. *Trends in Endocrinology and Metabolism* 14, 386–392.

Kiens, B. (2006). Skeletal muscle lipid metabolism in exercise and insulin resistance. *Physiological Reviews* 86, 205–243.

Kiens, B., Kristiansen, S., Jensen, P., Richter, E.A. & Turcotte, L.P. (1997). Membrane associated fatty acid binding protein (FABPpm) in human skeletal muscle is increased by endurance training. *Biochemical and Biophysical Research Communications* 231, 463–465.

Kiens, B., Roepstorff, C., Glatz, J.F., et al. (2004). Lipid-binding proteins and lipoprotein lipase activity in human skeletal muscle: Influence of physical activity and gender. *Journal of Applied Physiology* 97, 1209–1218.

Kim, J.K., Gimeno, R.E., Higashimori, T., et al. (2004). Inactivation of fatty acid transport protein 1 prevents fat-induced insulin resistance in skeletal muscle. *Journal of Clinical Investigation* 113, 756–763.

Lass, A., Zimmermann, R., Haemmerle, G., et al. (2006). Adipose triglyceride lipase-mediated lipolysis of cellular fat stores is activated by CGI-58 and defective in Chanarin-Dorfman syndrome. *Cell Metabolism* 3, 309–319.

Lefevre, C., Jobard, F., Caux, F., et al. (2001). Mutations in CGI-58, the gene encoding a new protein of the esterase/lipase/thioesterase subfamily, in Chanarin-Dorfman syndrome. *American Journal of Human Genetics* 69, 1002–1012.

Leon, A.S. & Sanchez, O.A. (2001). Response of blood lipids to exercise training alone or combined with dietary intervention. *Medicine and Science in Sports and Exercise* 33, S502–S515; discussion S528–S529.

Leon, A.S., Gaskill, S.E., Rice, T., et al. (2002). Variability in the response of HDL cholesterol to exercise training in the HERITAGE Family Study. *International Journal of Sports Medicine* 23, 1–9.

Luiken, J.J., Koonen, D.P., Coumans, W.A., et al. (2003). Long-chain fatty acid uptake by skeletal muscle is impaired in homozygous, but not heterozygous, heart-type-FABP null mice. *Lipids* 38, 491–496.

Lusis, A.J. & Pajukanta, P. (2008). A treasure trove for lipoprotein biology. *Nature Genetics* 40, 129–130.

Macho-Azcarate, T., Calabuig, J., Marti, A. & Martinez, J. (2002). A maximal effort trial in obese women carrying the beta2-adrenoceptor Gln27Glu polymorphism. *Journal of Physiology Biochemistry* 58, 103–108.

Martinez-Botas, J., Anderson, J.B., Tessier, D., et al. (2000). Absence of perilipin results in leanness and reverses obesity in Lepr(db/db) mice. *Nature Genetics* 26, 474–479.

Meirhaeghe, A., Helbecque, N., Cottel, D. & Amouyel, P. (1999). Beta2-adrenoceptor gene polymorphism, body weight, and physical activity. *Lancet* 353, 896.

Moro, C., Crampes, F., Sengenes, C., et al. (2004). Atrial natriuretic peptide contributes to physiological control of lipid mobilization in humans. *The FASEB Journal* 18, 908–910.

Nordby, P., Saltin, B. & Helge, J.W. (2006). Whole-body fat oxidation determined by graded exercise and indirect calorimetry: A role for muscle oxidative capacity? *Scandinavian Journal of Medicine and Science in Sports* 16, 209–214.

Phares, D.A., Halverstadt, A.A., Shuldiner, A.R., et al. (2004). Association between body fat response to exercise training and multilocus ADR genotypes. *Obesity Research* 12, 807–815.

Rice, T., Hong, Y., Perusse, L., et al. (1999). Total body fat and abdominal visceral fat response to exercise training in the HERITAGE Family Study: Evidence for major locus but no multifactorial effects. *Metabolism* 48, 1278–1286.

Rice, T., Despres, J.P., Perusse, L., et al. (2002). Familial aggregation of blood lipid response to exercise training in the health, risk factors, exercise training, and genetics (HERITAGE) Family Study. *Circulation* 105, 1904–1908.

Roepstorff, C., Vistisen, B., Roepstorff, K. & Kiens, B. (2004). Regulation of plasma long-chain fatty acid oxidation in relation to uptake in human skeletal muscle during exercise. *American Journal of Physiology—Endocrinology and Metabolism* 287, E696–E705.

Romijn, J.A., Coyle, E.F., Sidossis, L.S., Gastaldelli, A., Horowitz, J.F., Endert, E. & Wolfe, R.R. (1993). Regulation of endogenous fat and carbohydrate metabolism in relation to exercise intensity and duration. *American Journal of Physiology—Endocrinology and Metabolism* 265, E380–E391.

Saltin, B. & Astrand, P.O. (1993). Free fatty acids and exercise. *American Journal of Clinical Nutrition* 57, 752S–757S; discussion 757S–758S.

Seip, R.L., Mair, K., Cole, T.G. & Semenkovich, C.F. (1997). Induction of human skeletal muscle lipoprotein lipase gene expression by short-term exercise is transient. *American Journal of Physiology* 272, E255–E261.

Chapter 21

Genes, Exercise, and Glucose and Insulin Metabolism

LEONIDAS G. KARAGOUNIS AND JOHN A. HAWLEY

HIRi, School of Medical Sciences, RMIT University, Bundoora, VIC, Australia

Skeletal muscle displays a remarkable plasticity in adapting to numerous and varied external stimuli including contractile activity (loading and unloading), hormonal and substrate supply, and environmental factors (hypoxia and heat stress). If skeletal muscle is repeatedly exposed to specific stimuli, over time, it will adapt and alter its profile to meet the demands of its new environment (Figure 21.1). Repeated submaximal contractile activity (i.e., endurance exercise) and the concomitant increased flux of metabolic fuels for cellular oxidation lead to profound changes in the metabolic potential of muscle. These include mitochondrial biogenesis and an increase in the ability of the muscle to oxidize fat-based fuels in preference to carbohydrate (CHO)-based fuels, alterations in angiogenesis, neural adaptations, and fiber-type shifts. Knowledge of the genetic, molecular, and cellular events that regulate skeletal muscle plasticity can define the potential for adaptation in metabolism and performance, and may also lead to the discovery of novel pathways in common clinical disease states. This chapter describes the regulation of glucose metabolism in skeletal muscle. Focus is on the early genetic and molecular events that occur in response to contraction, and how the subsequent changes in the hormonal milieu and nutrient availability interact to modify cellular glucose metabolism.

The Energy for Skeletal Muscle Contraction

The energy for all forms of biological functions, including contractile activity, is provided chemically by an energy-rich molecule, adenosine triphosphate (ATP). Large amounts of free energy are liberated by the hydrolysis of ATP to adenosine diphosphate (ADP) and inorganic phosphate (P_i) or adenosine monophosphate (AMP) and pyrophosphate (PP_i) to yield energy for contractile activity. However, the small amount of ATP present within skeletal muscle fibers (25–30 mmol/g dm) is not quantitatively an important energy store but is considered an immediate donor of free energy. It is probably inappropriate to consider ATP as a metabolic fuel store due to its constant turnover during muscle contraction. In this regard, an ATP molecule is consumed within a minute following its formation, and whole-body ATP turnover in an adult human is about 40 kg of ATP per 24-h period. During strenuous exercise, the rate of ATP utilization may reach 0.5 kg/min; yet during maximal, intermittent sprinting exercise, ATP concentrations are conserved within a very narrow range (25–40% of resting levels) (McCartney et al., 1986). This is because even small declines in ATP content within muscle have a direct impact on the ability to sustain contractile force, making it imperative that under conditions of increased energy demand (which may increase up to 100-fold in the transition from rest to supramaximal exercise), the metabolic pathways responsible for synthesizing ATP must be able to rapidly respond and maintain an ATP equilibrium.

Genetic and Molecular Aspects of Sport Performance, 1st edition. Edited by Claude Bouchard and Eric P. Hoffman. Published 2011 by Blackwell Publishing Ltd.

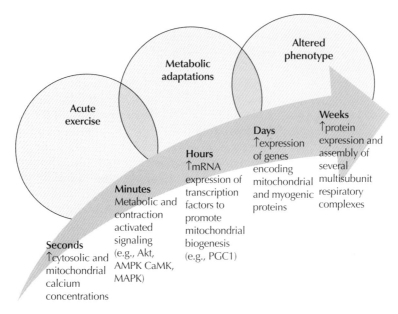

Figure 21.1 Schematic representation of the timescale of typical exercise, gene, and protein kinetics in skeletal muscle.

At rest, glucose availability exerts a dominant influence on the mix of oxidized fuels in healthy individuals. However, within 10–12 h of fasting, fatty acid (FA) oxidation accounts for ~80% of resting energy requirements. Strenuous exercise can lead to a 20-fold increase in the levels of whole-body energy metabolism, resulting in a depletion of endogenous CHO stores. During prolonged exercise such as a marathon run, there is a reliance on CHO-based fuels for oxidative metabolism and a resultant depletion of muscle and liver glycogen reserves that typically coincide with fatigue. Accordingly, diet–exercise strategies that increase pre-exercise glycogen stores along with the provision of endogenous CHO (i.e., glucose feeding) during exercise will result in increased exercise capacity. Training can also alter glucose metabolism by increasing glucose uptake by the trained musculature. Glucose uptake is higher in endurance-trained than in untrained human skeletal muscle when working at the same relative workload and in the face of low glycogen concentrations. This difference may be related to the higher muscle glucose transporter (GLUT) protein content in trained versus untrained muscle. The higher glucose utilization capacity in trained muscle probably explains why these subjects are able to utilize glucose at remarkably high rates

(~2 g/min) when fed CHO or infused with glucose during prolonged exercise when glycogen stores are low (Hawley et al., 1994).

Glucose metabolism is regulated by insulin and the counter-regulatory hormone glucagon. Systemic blood glucose concentrations are tightly regulated within a narrow range (4–5 mM in the healthy adult) despite large exercise-induced increases in the turnover of this pool. After a glucose load (i.e., postabsorptive), increased circulating levels of glucose trigger the release of insulin from the pancreatic beta cells within the Islets of Langerhans. Circulating insulin stimulates glucose transport and glycogen storage in both the muscle and liver. Muscle cells are not freely permeable to glucose, and therefore insulin is required to facilitate glucose uptake into the muscle by stimulating specific GLUTs (discussed subsequently). Unlike muscle, hepatic cell membranes are permeable to glucose and therefore do not require activation of putative GLUTs for uptake. In liver, insulin stimulates glycogen synthesis from the glucose absorbed by the small intestine. This is achieved firstly by activating the enzyme hexokinase, which phosphorylates glucose, trapping it within the cell. Coincidently, insulin acts to inhibit the activity of glucose-6-phosphatase (G6Pase). Insulin also activates several of the enzymes that are directly involved in glycogen synthesis, including phosphofructokinase and glycogen synthase. This chain of events leads to the storage of glycogen when glucose availability is high. In contrast, when blood glucose concentrations are low, glucagon is secreted by the alpha-pancreatic cells targeting the liver and stimulating hepatic glycogenolysis and gluconeogenesis while simultaneously inhibiting glycogen synthesis and thereby increasing the availability of blood glucose.

CHO-Based Fuels

CHOs are aldehyde or ketone compounds with multiple hydroxyl groups that have a ratio of one carbon atom to water ($C:H_2O$) and serve as energy stores, fuels, and metabolic intermediates. The most common forms of dietary CHOs are the monosaccharides: glucose, fructose, and galactose. Of these, glucose is the principal currency for CHO metabolism and is stored in polymer-form glycogen, which ranges in molecular weight from 50,000 to several million daltons. Most glucose units in glycogen are linked by alpha1,4-glycosidic bonds whereas the branches are formed by alpha1,6-glycosidic bonds. The branches serve to increase the terminal glucose sites, allowing for rapid degradation during exercise, and also rapid resynthesis following glycogen-depleting exercise. The energy oxidation equivalent for the terminal oxidation of glucose to CO_2 and H_2O is 5.1 kcal/l of oxygen.

The endogenous sources of CHO are muscle and liver glycogen, blood glucose, and blood, muscle, and liver lactate. The blood glucose pool contains ~25 g of glucose and is a constant state of turnover depending on the prevailing hormonal milieu and the energetic demands of insulin-sensitive tissues, such as skeletal muscle. The concentration of glycogen found in skeletal muscle can vary and is dependent upon the dietary and exercise habits of an individual (Hawley et al., 1997). Resting muscle glycogen concentrations in an untrained individual consuming a mixed diet are typically in the range of ~350 mmol/kg/dm and can reach ~500 mmol/kg/dm in endurance-trained athletes. Of note is that after several days of a high-CHO diet (>8 g of CHO/kg body mass) and an exercise taper, muscle glycogen content can be elevated to values exceeding 900 mmol/kg/dm (Hawley et al., 1997).

Aside from muscle, the liver, with an average weight of ~1.2 kg, contains about 100 g of glycogen. As with the muscle stores of CHO, the exact content of glycogen present within the liver varies according to the dietary and training status of the individual. The concentration of glycogen is greater in the liver than in skeletal muscle, although overall much more glycogen is stored in the muscle due to its greater mass. Results of early studies investigating the role of CHO availability on liver metabolism demonstrated that the liver is extremely sensitive to changes in dietary CHO intake. For instance, hepatic biopsies have shown that following 1 day of CHO restriction, hepatic glycogen stores were reduced almost 10-fold with further CHO restrictions leading to a release of hepatic glucose from gluconeogenesis.

Molecular Regulation of CHO Oxidation (Nutrient/Gene Interactions)

The intracellular signaling pathways that are activated by exercise in skeletal muscle can modulate the expression of genes involved in glucose and glycogen metabolism. Genetic information flows from DNA to RNA to protein, and the abundance of any gene product at any given time represents a balance between the rate of synthesis (i.e., translation of mRNA) and the rate of degradation of that protein. The rate of synthesis of a given protein is determined by the concentration of its corresponding mRNA which is a function of that mRNA's rate of transcription and its rate of degradation. Exercise activates a series of signal transduction cascades controlling glucose uptake, glycogen synthesis, gene expression, and protein synthesis. Indeed, a single bout of moderate-intensity exercise in humans increases GLUT4, hexokinase II, and glycogenin gene expression. When exercise is repeated over time (i.e., exercise training), these increases in gene expression lead to an increased expression of these proteins in skeletal muscle and contribute to the enhanced skeletal muscle glucose uptake, glycogen synthesis, and insulin action.

Glucose transport is a saturable process and occurs via passive ATP-independent mechanisms through a family of GLUTs. Primary facilitators of insulin-stimulated glucose transport across the plasma membrane are the muscle-specific GLUTs—GLUT1 and GLUT4. The GLUT1 isoform is believed to be involved in basal state glucose uptake as it is located in the plasma membrane, whereas the GLUT4 isoform is translocated from the cytosol to the plasma membrane in response to insulin-dependent kinase activation. In a normal healthy state, GLUT4 concentrations determine the capacity at which glucose transport can occur (Kern et al., 1990). In individuals with type 2 diabetes, insulin-dependent regulation of glucose uptake is impaired but contraction-stimulated regulation of glucose uptake remains intact. This highlights the fact that glucose transport (into muscle) is regulated by both insulin- and contraction-mediated mechanisms.

Because of the close temporal relationship between glucose metabolism and muscle contraction, it is perhaps not surprising that the regulation of glucose uptake into insulin-sensitive tissues involves several independent and interdependent intracellular signaling cascades (Figure 21.2).

Protein Kinase B (PKB/Akt) and Insulin Signaling

The specific cascades linking exercise and insulin with the genetic regulation of metabolic genes are multiple and function both independently and in tandem with each other. One of the best characterized pathways is the PKB/Akt pathway which is activated by both exercise and nutrient availability. Following the ingestion of a mixed meal, insulin is released by the pancreas to facilitate glucose uptake by the muscles and liver. Circulating insulin binds to the alpha subunits of the insulin receptors on the surface of muscle cells and activates a tyrosine (Tyr) kinase in the intracellular tails of the beta subunits which results in the autophosphorylation of the Tyr residues in the insulin receptors. Once the insulin receptor is rendered active, it phosphorylates insulin receptor substrate-1 (IRS-1) which activates phosphatidylinositol 3-kinase (PI3K). Activation of PI3K acts upon protein kinase B (PKB, also known as Akt), a key protein involved in this metabolic orchestration. Akt exists as three isoforms (Akt1, Akt2, and Akt3) of which Akt1 and Akt2 are the predominant isoforms expressed in skeletal muscle. *si*RNA-mediated gene silencing reveals

specificity in insulin signaling responses governing myotube formation and glucose/lipid metabolism. Insulin signaling pathways involving IRS-1 and Akt2 are required for myotube formation, phosphorylation of Akt and AS160, and glucose uptake. Conversely, insulin signaling pathways involving IRS-2 are necessary for lipid uptake and metabolism. The activation of the PI3K/Akt cascade is responsible for glycogen synthesis, GLUT4 translocation, and glucose transport, as well as the regulation of gene expression and both protein synthesis and degradation (Greenhaff et al., 2008; Turinsky & Damrau-Abney, 1999). Evidence that PI3K activation is directly involved in glucose transport comes from a study that chemically blocked PI3K activation and found a resultant inhibition of glucose transport (Lee et al., 1995).

Several Akt substrates have been identified as putative links between insulin signaling and muscle contraction and metabolic or gene-regulatory responses. Specifically, a novel 160-kDa signaling protein has been characterized as an Akt substrate (initially called AS160; now TBC1D4 GenBank designator) which is phosphorylated in rat skeletal muscle in response to contraction and insulin and also in human skeletal muscle by exercise (Deshmukh et al., 2006). AS160 contains six Akt phosphorylation sites, five of which are phosphorylated in response to insulin stimulation. However, overexpression of a construct where four of these sites are changed to alanines (AS160-4P) leads to the inhibition of GLUT4 translocation. AS160 also harbors a GTPase-activating protein (GAP) domain specific for Rab proteins that are responsible for the regulation of vesicular traffic. Therefore, the GAP domain of AS160 may lead to the negative regulation of one or more Rab isoforms in the absence of insulin. Insulin-stimulated AS160 phosphorylation may result in altered subcellular localization or induce conformational changes rendering its GAP domain inactive and therefore permitting the conversion of the Rabs to their guanosine triphosphate (GTP)-loaded active forms.

Reports from studies utilizing *in vivo* gene transfer techniques have shown that both insulin and contraction stimulation increased wild-type AS160 (AS160-WT) phosphorylation. Furthermore, the AS160-4P mutant (but not AS160-WT) led to the inhibition of insulin-stimulated glucose uptake that

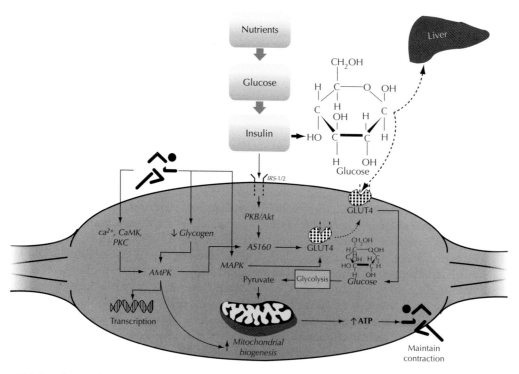

Figure 21.2 Regulation of glucose uptake by insulin and muscle contraction. Ingested nutrients (CHO) lead to increased levels of circulating glucose triggering insulin secretion. This stimulates glucose uptake by the liver by passive diffusion. In skeletal muscle, insulin stimulates IRS1/2 leading to the activation of AS160 via Akt. GLUT4 translocates to the plasma membrane to facilitate glucose uptake into the muscle. Once inside, the muscle glucose gets converted to pyruvate via the glycolytic pathway before being metabolized in the tricarboxylic acid cycle in the mitochondria, yielding increases in ATP. Muscle contraction depletes glycogen stores leading to the activation of AMPK. AMPK is able to alter gene expression resulting in increases in mitochondrial biogenesis as well as activating AS160 which ultimately leads to the translocation of GLUT4 to the plasma membrane facilitating glucose uptake.

required a GAP domain, whereas both AS160-WT and AS160-4P inhibited contraction-induced glucose uptake. These observations suggest that, although both insulin- and contraction-stimulated glucose uptake are AS160 dependent, contraction *per se* cannot fully repress AS160 activity. In studies that have used pharmacological agents (wortmannin) to abolish contraction-induced AS160 phosphorylation in isolated skeletal muscle, glucose uptake is not inhibited. Thus, while AS160 may be a critical point of convergence for insulin- and exercise-mediated glucose uptake in skeletal muscle, the precise role of the training-induced effects on AS160 signaling on glucose transport is not known.

The mechanism for selectivity between specific IRS and Akt isoforms in human skeletal muscle remains to be established. However, an emerging concept considers that the complexity of signaling networks and the integrated control of diverse cellular responses in insulin-sensitive tissues is under the control of essential mediators or "critical nodes" (Taniguchi et al., 2006). With this concept, isoform-specific functions at the level of IRS, PI3K, and Akt provide a framework for the diverse metabolic and gene-regulatory responses, thereby introducing specificity into the insulin signaling pathway.

AMPK: At the Crossroads of Exercise and Glucose Metabolism

The 5'-AMP-activated protein kinase (AMPK) is a member of a metabolite-sensing protein kinase family that functions as a metabolic "fuel gauge" in skeletal muscle (Hardie & Sakamoto, 2006). It is activated by increases in the cellular AMP:ATP ratio caused by metabolic stresses that either interfere with ATP production (e.g., deprivation for glucose or oxygen) or that accelerate ATP consumption (e.g., muscle contraction). Once activated, AMPK switches on catabolic pathways that generate ATP, while switching off ATP-consuming processes such as biosynthesis and cell growth and proliferation. AMPK exists as a heterotrimer, consisting of alpha, beta, and gamma subunits. The alpha subunit contains a specific threonine residue (Thr[172]) that functions as an activating phosphorylation site for one of several "upstream" AMPK kinases (the AMP-activated protein kinase kinase [AMPKK]). Activity of the AMPK alpha1 isoform has been linked to exercise training-induced adaptations and muscle glycogen storage. The beta subunit contains a glycogen-binding domain that allows AMPK to act as a glycogen sensor *in vivo*.

The first evidence that AMPK was involved in the coordinated regulation of skeletal muscle metabolism was provided by Winder and Hardie (1996). They observed a two- to threefold increase in AMPK activity in skeletal muscle of rats subjected to treadmill running within 5 min of the onset of exercise. This increase in AMPK was intensity dependent, and activity remained elevated for up to 30 min postexercise, suggesting that AMPK might also be involved during the postexercise recovery of muscle CHO reserves. With regard to the effects of exercise in humans, low- to moderate-intensity cycling in both untrained and trained individuals resulted in an isoform-specific and intensity-dependent increase in AMPK alpha2, but not alpha1-associated activity (Wojtaszewski et al., 2000). In contrast, alpha1 and alpha2 AMPK isoform activation occurred after maximal, short duration sprint-type exercise (Chen et al., 2000). It is likely that activation of the alpha1 isoform after short-term anaerobic exercise is related to the rate of fuel utilization rather than the degree of substrate depletion, as no change in AMPK alpha1 activity is observed after prolonged, continuous, low-intensity cycling leading to exhaustion and muscle glycogen depletion.

As previously noted, both muscle contraction and insulin increase glucose uptake through distinct mechanisms, yet AMPK and muscle contractions both increase glucose uptake via a common mechanism that is distinct from insulin. However, transgenic mice expressing an inactive (dominant-negative) form of AMPK display a 30% reduction in contraction-stimulated muscle glucose uptake, suggesting that although AMPK plays a role in glucose uptake, it does not fully account for the increased muscle glucose uptake in response to exercise/contraction. One possibility is that AMPK-generated increases in ATP during exercise occur through the promotion of FA oxidation, with enhanced CHO metabolism most likely a secondary effect of activation of this kinase. In skeletal muscle, both insulin and AMPK converge upon AS160, the phosphorylation of which is involved in the translocation of GLUT4 to the plasma membrane leading to increased glucose uptake. Because of the differing nature of insulin (anabolic) and AMPK (catabolic) signals, the fate of the glucose is not the same. In the case of insulin stimulation of AS160, glucose uptake leads to glycogen synthesis whereas AMPK-activated AS160 leads to glycolysis.

Calcium Signaling

At the onset of muscle contraction, sarcoplasmic Ca^{2+} levels increase rapidly. Such increases in Ca^{2+} concentrations play a fundamental role in excitation–contraction coupling, stimulation of ATP-generating metabolic pathways, as well as the activation of transcriptional responses. Calcium-dependent signaling pathways include Ca^{2+}/calmodulin-dependent kinase (CaMK), Ca^{2+}-dependent protein kinase C (PKC) isoforms, and the Ca^{2+}/calmodulin-dependent phosphatase calcineurin. CaMK II appears to be the dominant isoform expressed in skeletal muscle, although other CaMK isoforms have also been identified (Chin, 2005). Exercise stimulates increases in CaMK II activity in human skeletal muscle, with the magnitude of this increase being greater during high-intensity (~100% $\dot{V}O_2max$) than moderate-intensity (~75% $\dot{V}O_2max$) exercise. Early studies in rat muscle have demonstrated that PKC activity increases with electrical stimulation which is consistent with activation of conventional and novel PKC isoforms by contraction-induced increases in Ca^{2+} and diacylglycerol (Richter et al., 2003).

Calcium binds directly to and activates phosphorylase kinase. The phosphorylation of the kinase leads to the transformation of phosphorylase, rendering it active, and thereby contributing to enhanced glycogenolysis in skeletal muscle during exercise. In addition, further stimulation of glycogenolysis is provided by increases in P_i, AMP, and ADP. Calcium also increases glucose uptake, although the specific signaling intermediates are unresolved. For instance, inhibition of CaMK has been reported to lead to a ~50% reduction in glucose uptake during contraction *in vitro* (Wright et al., 2004), while CaMK inhibition resulted in complete abolishment of glucose uptake during tetanic contractions (Wright et al., 2005). The PKC inhibitor calphostin C, however, attenuates the contraction-induced increase in skeletal muscle glucose uptake (Richter et al., 2003). Increases in Ca^{2+} have also been reported to increase the transcriptional activation of numerous genes involved in mitochondrial biogenesis and muscle hypertrophy (Chin, 2005). All of the Ca^{2+}-dependent signaling pathways are believed to be involved in metabolic and gene-regulatory events to some extent. For example, CaMK inhibition abolishes Ca^{2+}-induced mitochondrial biogenesis (Ojuka et al., 2003), while PKC has been implicated in increased cytochrome *c* expression following Ca^{2+} stimulation (Chin, 2005), and calcineurin has a role in determining skeletal muscle fiber type (Bassel-Duby & Olson, 2006).

Mitogen-Activated Protein Kinase

Exercise is a powerful activator of the mitogen-activated protein kinase (MAPK) signaling cascade, and this pathway has been identified as a candidate system that may convert contraction-induced biochemical perturbations into appropriate intracellular responses (Hawley & Zierath, 2004). Muscular contraction rapidly activates several MAPK isoforms such as p38 MAPK and ERK 1/2 (extracellular signal-regulated kinases). The activity of the p38 MAPK isoform has been shown to increase in response to various muscle contractile models in both humans and rodents (Lee et al., 2002; Nader & Esser, 2001). The exact role of p38 MAPK is not clear, although evidence suggests it may play a role in skeletal muscle adaptation by promoting gene expression and specific coactivators involved in mitochondrial biogenesis and slow muscle fiber formation (Akimoto

et al., 2005) as well as contraction-induced activation of muscle glucose uptake (Richter et al., 2003). MAPK activation may also mediate gene expression by inducing the production of transcription factors as well as stimulating the activity of the translational stage during protein synthesis.

Effects of CHO Availability on Exercise-Induced Gene Responses

Manipulation of dietary macronutrient content is associated with marked changes in substrate stores, metabolic flux, and subsequent fuel oxidation. Changes in dietary intake alter the concentration of blood-borne nutrients and hormones and, via substrate availability, regulate the short-term macronutrient oxidative and storage profile of skeletal muscle. Perturbations in muscle and blood substrates alter the uptake and flux of these fuel-specific intermediates within related metabolic pathways. This immediate response serves to redirect enzymatic processes involved in substrate metabolism and the subsequent concentration of particular proteins critical for metabolic pathway function. Altering substrate availability impacts not only resting energy metabolism but also regulatory processes underlying gene expression (Arkinstall et al., 2004). To bring about such modifications, a number of highly coordinated processes occur, including gene transcription, RNA transport from the nucleus, protein synthesis, and, in some cases, posttranslational modification of the protein. However, the initiation of gene transcription is strongly related to changes in dietary intake and composition. The results of many studies have clearly demonstrated that exercise is capable of inducing the transcription of several genes with roles in substrate metabolism and mitochondrial biogenesis. But while it is well accepted that exercise results in a rapid (within several hours, see Figure 21.1) increase in the abundance of multiple mRNAs, the results from several recent investigations have demonstrated that CHO availability can modify the normal contraction-induced transcriptional response of many genes. For example, exercise with low muscle glycogen concentration has been shown to result in a greater transcriptional activation of pyruvate dehydrogenase kinase 4, hexokinase, and heat-shock protein 72 compared with high or normal muscle glycogen concentration at the start of exercise. We have recently reported that AMPK phosphorylation was greater when subjects commenced a single bout of intense exercise with low, compared with normal, muscle glycogen availability (Yeo et al., 2008), while others have shown that the isoform-specific activation of the alpha2 subunit under conditions of low muscle glycogen enhances AMPK alpha2 nuclear transloca-tion and increases GLUT4 mRNA expression. The role of (low) muscle glycogen levels in modulating the early gene–protein responses in these studies could be explained by the fact that several transcrip-tion factors include glycogen-binding domains. When muscle glycogen is low, these factors are released and become free to associate with different targeting pro-teins. Removal of the outer tier of glycogen by phos-phorylase releases about 30% of the available glucose and halves the number of nonreducing ends. The functional outcome of a reduction in the number of active nonreducing ends is that even a modest reduc-tion in glycogen content might cause release of sig-nificant quantities of AMPK from the polysaccharide so that more of the kinase becomes available to phos-phorylate targets involved in glucose uptake and/or insulin signaling. Such a scenario may help to explain why the increase in insulin-stimulated glucose uptake following a single bout of exercise is strongly corre-lated with the extent of glycogen depletion.

Glycogen Storage Diseases

Glycogen storage diseases (GSDs) represent a class of genetic disorders that result in diminished exer-cise performance and skeletal muscle ability to train. Some of these disorders are briefly introduced here to illustrate how a genetic defect associates with exer-cise intolerance. GSDs (also known as glycogenosis and dextrinosis) result in the defective processing of glycogen synthesis or breakdown within muscles and the liver. It is believed that the current incidence of GSDs is one case per 20,000–43,000 live births (see Ozen, 2007 for a detailed review of GSDs).

von Gierke Disease (GSD Type I)

von Gierke disease (GSD I) is brought about by a mutation in G6Pase, an enzyme involved in the hepatic regulation of blood glucose levels and the catalysis of the terminal step in both gluconeo-genesis and glycogenolysis. In more severe cases, this disease is characterized by nephromegaly,

hepatomegaly, obesity, and short stature. Long-term chronic complications may occur including, but not limited to, hepatic adenomas, gout, and renal fail-ure. There are four subtypes of this disease, depend-ent upon the abnormality of the G6Pase system.

McArdle Disease (GSD Type V)

Perhaps one of the most common of the diseases is McArdle disease (GSD type V). McArdle disease was first described in 1951 by McArdle in a patient who presented with exercise-induced myalgia and low blood lactate concentrations during ischemic forearm exercise. McArdle disease is the consequence of a mutation of the muscle glycogen phosphorylase gene, which is localized to 11q13. The glycogen phosphorylase enzyme is responsible for the catalysis of the rate-limiting step in the degrada-tion of glycogen releasing glucose-1-phosphate from the terminal alpha1,4-glycosidic bond. Therefore, a deficiency in glycogen phosphorylase leads to the inability to mobilize intramuscular glycogen stores during anaerobic exercise. Furthermore, oxidative phosphorylation is also inhibited due to an abnor-mally low flux of substrate through the tricarboxylic acid (TCA) cycle. To date, more than 65 mutations in the muscle isoform glycogen phosphorylase gene have been identified and associated with McArdle disease (Nogales-Gadea et al., 2007).

Summary

It is clear that the regulation of glucose metabolism is dependent upon several interacting signaling cas-cades that are responsible for regulating cellular and genetic alterations in response to fuel (i.e., CHO) availability, insulin level, and muscle contractions. To date, much of the research has focused on improving our understanding of the role of glucose ingestion on muscle metabolism during and after exercise. Recent advances in molecular biology techniques have allowed for the investigation of exercise–nutrient interactions on skeletal muscle gene expression and the early signaling responses to these interventions. The biggest challenge for exercise physiologists in the forthcoming years will be to directly link specific glucoregulatory signaling cascades to defined meta-bolic responses and specific changes in gene expres-sion in skeletal muscle that occur during and after exercise. This will be complicated by the fact that many of these pathways are not linear, but rather they constitute a complex network with a high degree of cross talk, feedback regulation, and transient activation. An increased understanding of these pathways through studies of comparative genomics between humans, rodents, and model organisms, coupled with greater knowledge of the functional relevance of the signaling intermediates, will facilitate the understanding of the mechanisms by which exercise and stress alters gene expression (Hawley & Zierath, 2004).

References

Akimoto, T., Pohnert, S.C., Li, P., et al. (2005). Exercise stimulates Pgc-1alpha transcription in skeletal muscle through activation of the p38 MAPK pathway. *Journal of Biological Chemistry* 280, 19587–19593.

Arkinstall, M.J., Tunstall, R.J., Cameron-Smith, D. & Hawley, J.A. (2004). Regulation of metabolic genes in human skeletal muscle by short-term exercise and diet manipulation. *American Journal of Physiology—Endocrinology and Metabolism* 287, E25–E31.

Bassel-Duby, R. & Olson, E.N. (2006). Signaling pathways in skeletal muscle remodeling. *Annual Review of Biochemistry* 75, 19–37.

Chen, Z.P., McConell, G.K., Michell, B.J., Snow, R.J., Canny, B.J. & Kemp, B.E. (2000). AMPK signaling in contracting human skeletal muscle: Acetyl-CoA carboxylase and NO synthase phosphorylation. *American Journal of Physiology—Endocrinology and Metabolism* 279, E1202–E1206.

Chin, E.R. (2005). Role of Ca2+ / calmodulin-dependent kinases in skeletal muscle plasticity. *Journal of Applied Physiology* 99, 414–423.

Deshmukh, A., Coffey, V.G., Zhong, Z., Chibalin, A.V., Hawley, J.A. & Zierath, J.R. (2006). Exercise-induced phosphorylation of the novel Akt substrates AS160 and filamin A in human skeletal muscle. *Diabetes* 55, 1776–1782.

Greenhaff, P.L., Karagounis, L.G., Peirce, N., et al. (2008). Disassociation between the effects of amino acids and insulin on signaling, ubiquitin ligases, and protein turnover in human muscle. *American Journal of Physiology—Endocrinology and Metabolism* 295, E595–E604.

Hardie, D.G. & Sakamoto, K. (2006). AMPK: A key sensor of fuel and energy status in skeletal muscle. *Physiology* 21, 48–60.

Hawley, J.A. & Zierath, J.R. (2004). Integration of metabolic and mitogenic signal transduction in skeletal muscle. *Exercise and Sport Science Reviews* 32, 4–8.

Hawley, J.A., Bosch, A.N., Weltan, S.M., Dennis, S.C. & Noakes, T.D. (1994). Glucose kinetics during prolonged exercise in euglycaemic and hyperglycaemic subjects. *Pflugers Archiv* 426, 378–386.

Hawley, J.A., Schabort, E.J., Noakes, T.D. & Dennis, S.C. (1997). Carbohydrate-loading and exercise performance. An update. *Sports Medicine* 24, 73–81.

Kern, M., Wells, J.A., Stephens, J.M., et al. (1990). Insulin responsiveness in skeletal muscle is determined by glucose transporter (Glut4) protein level. *Biochemical Journal* 270, 397–400.

Lee, A.D., Hansen, P.A. & Holloszy, J.O. (1995). Wortmannin inhibits insulin-stimulated but not contraction-stimulated glucose transport activity in skeletal muscle. *FEBS Letters* 361, 51–54.

Lee, J.S., Bruce, C.R., Spurrell, B.E. & Hawley, J.A. (2002). Effect of training on activation of extracellular signal-regulated kinase 1/2 and p38 mitogen-activated protein kinase pathways in rat soleus muscle. *Clinical and Experimental Pharmacology and Physiology* 29, 655–660.

McCartney, N., Spriet, L.L., Heigenhauser, G.J., Kowalchuk, J.M., Sutton, J.R. & Jones, N.L. (1986). Muscle power and metabolism in maximal intermittent exercise. *Journal of Applied Physiology* 60, 1164–1169.

Nader, G.A. & Esser, K.A. (2001). Intracellular signaling specificity in skeletal muscle in response to different modes of exercise. *Journal of Applied Physiology* 90, 1936–1942.

Nogales-Gadea, G., Arenas, J. & Andreu, A.L. (2007). Molecular genetics of McArdle's disease. *Current Neurology and Neuroscience Reports* 7, 84–92.

Ojuka, E.O., Jones, T.E., Han, D.H., Chen, M.A.Y. & Holloszy, J.O. (2003). Raising Ca2+ in L6 myotubes mimics effects of exercise on mitochondrial biogenesis in muscle. *FASEB Journal* 17, 675–681.

Ozen, H. (2007). Glycogen storage diseases: New perspectives. *World Journal of Gastroenterology* 13, 2541–2553.

Richter, E.A., Nielsen, J.N., Jorgensen, S.B., Frosig, C. & Wojtaszewski, J.F. (2003). Signalling to glucose transport in skeletal muscle during exercise. *Acta Physiologica Scandinavica* 178, 329–335.

Taniguchi, C.M., Emanuelli, B. & Kahn, C.R. (2006). Critical nodes in signalling pathways: Insights into insulin action. *Nature Reviews Molecular Cell Biology* 7, 85–96.

Turinsky, J. & Damrau-Abney, A. (1999). Akt kinases and 2-deoxyglucose uptake in rat skeletal muscles in vivo: Study with insulin and exercise. *American Journal of Physiology—Regulatory, Integrative and Comparative Physiology* 276, R277–R282.

Winder, W.W. & Hardie, D.G. (1996). Inactivation of acetyl-CoA carboxylase and activation of AMP-activated protein kinase in muscle during exercise. *American Journal of Physiology—Endocrinology and Metabolism* 270, E299–E304.

Wojtaszewski, J.F.P., Nielsen, P., Hansen, B.F., Richter, E.A. & Kiens, B. (2000). Isoform-specific and exercise intensity-dependent activation of 5′-AMP-activated protein kinase in human skeletal muscle. *The Journal of Physiology* 528, 221–226.

Wright, D.C., Hucker, K.A., Holloszy, J.O. & Han, D.H. (2004). Ca2+ and AMPK both mediate stimulation of glucose transport by muscle contractions. *Diabetes* 53, 330–335.

Wright, D.C., Geiger, P.C., Holloszy, J.O. & Han, D.H. (2005). Contraction- and hypoxia-stimulated glucose transport is mediated by a Ca2+-dependent mechanism in slow-twitch rat soleus muscle. *American Journal of Physiology—Endocrinology and Metabolism* 288, E1062–E1066.

Yeo, W.K., Lessard, S.J., Chen, Z.P., et al. (2008). Fat adaptation followed by carbohydrate restoration increases AMPK activity in skeletal muscle from trained humans. *Journal of Applied Physiology* 105, 1519–1526.

Chapter 22

Genes, Exercise, and Cardiovascular Phenotypes

JAMES M. HAGBERG

Department of Kinesiology, School of Public Health, University of Maryland, College Park, MD, USA

Human performance is a complex phenotype dependent on the integrated expression of numerous physiological systems. While the debate concerning the "limiting factors" for performance continues, the cardiovascular (CV) system is clearly of paramount importance in endurance events. Endurance athletes have an optimized CV profile that contributes to their performance capacities. At rest, this includes reduced heart rate (HR), increased stroke volume (SV) with unchanged cardiac output (Q), and generally unchanged blood pressure (BP) and total peripheral resistance. During maximal exercise, endurance athletes can have $\dot{V}O_2$max, Q, and SV values that are nearly double that of their sedentary peers, while their total peripheral resistance is lower and their BP and HR similar. While the training undergone by such athletes plays a major role in optimizing this phenotype, genetic factors also contribute to their enhanced CV phenotypes.

It is appealing to hypothesize that a specific profile containing a small number of gene variants, each independently affecting a critical physiological system, would optimize a person's CV phenotype. However, genome sequencing has taught us that humans and all higher life forms are highly complex relative to the genes underlying virtually any phenotype, as all of our physiological functions are regulated by numerous redundant signals, proteins, and genes. Thus, a potentially very complex genetic variation profile across numerous pathways impacts the final phenotype. This complexity is highlighted further by the current estimate of about 10 million single-nucleotide polymorphisms (SNPs) across the human genome (Frazer et al., 2007). While a number of Mendelian (monogenic) diseases are caused by single mutations in single genes, the most common public health problems, such as CV disease, are believed to be the result of numerous genetic variants (SNPs) dispersed across potentially equally numerous genes, and their interaction with the environment (e.g., diet and exercise). Thus, our initial hope of finding a small number of genetic variations with substantial independent effects on any phenotype, including CV disease, athletic performance, or rest or exercise CV phenotypes, is likely simplistic.

Our purpose is to review, in the context of these initial misdirected and bound-to-fail expectations, what we know relative to genetic variations and CV phenotypes that relate to endurance performance. To best accomplish this, we would review a vast body of literature that quantified the impact of genetic variations on CV phenotypes of endurance athletes at rest and during exercise. However, very few such data have been published. What we do have are studies assessing the relationship between genotypes and CV phenotypes in populations other than endurance athletes.

Heritability of Resting CV Phenotypes

The initial step in assessing the relationship between genetics and any phenotype is to determine whether

Genetic and Molecular Aspects of Sport Performance, 1st edition.
Edited by Claude Bouchard and Eric P. Hoffman.
Published 2011 by Blackwell Publishing Ltd.

the phenotype is heritable. Over the past 20 years, numerous studies have assessed the heritability of various resting CV phenotypes. These studies have generally found heritability for systolic, diastolic, and mean BP to be in the range of 0.20–0.50 (Adeyemo et al., 2009; de Oliveira et al., 2008), indicating that 20–50% of the interindividual variations in BP can be attributed to heritable factors. Resting HR has also been found to be highly heritable, with genetics accounting for 0.30–0.80 of the interindividual differences in resting HR (Gombojav et al., 2008; de Oliveira et al., 2008).

The heritability of CV structural phenotypes has also been assessed. For example, in 1987, left ventricular mass (LVM) was reported to be heritable in male twins, although minimally after accounting for body mass (Fagard et al., 1987). Recently, LVM heritability was found to be 0.27 in a large Chinese population (Chien et al., 2006) and ~0.47 in an even larger population of Caucasians and African-Americans (de Simone et al., 2007). Carotid intimal-medial thickness heritability in male twin pairs was 0.69, and remained high after adjusting for covariates (Zhao et al., 2008). A report from the classic Framingham study found that reflected wave amplitude ($h = 0.48$), forward wave amplitude ($h = 0.21$), and carotid-femoral pulse wave velocity (PWV) ($h = 0.40$) were also heritable (Mitchell et al., 2005).

Fewer studies have assessed the heritability of CV functional phenotypes. For example, a recent study reported that flow-mediated dilation was heritable in male twin pairs ($h = ~0.40$) (Zhao et al., 2007). In Framingham, Benjamin and coworkers (2004) reported heritability of 0.14 for flow-mediated dilation. Another Framingham study reported heritability of 0.30–0.52 for echocardiographic and brachial artery function measures (Vasan et al., 2007). A number of studies have shown that heart rate variability (HRV), an index of cardiac autonomic function, has heritability of up to 0.65 (Sinnreich et al., 1999; Wang et al., 2009), although Singh and coworkers (1999) reported heritabilities of 0.13–0.23 for HRV measures in the Framingham population (Singh et al., 1999).

Thus, over the past three decades, numerous studies have demonstrated substantial heritability for a range of resting CV hemodynamic, structural, and functional phenotypes. After establishing that a phenotype is heritable, which these data clearly do, the next step is to determine the chromosomal loci that underlie the phenotype.

Genome-Wide Linkage and Association Studies of Resting CV Phenotypes

One method to identify the chromosomal loci underlying a phenotype is to perform a genome-wide linkage or association study. Two or three decades ago, such studies assessed linkage in families between specific phenotypes and general chromosomal loci using 400–800 polymorphic markers across the genome. Now 500k/1M SNP chips are used for this purpose, allowing genome-wide association studies (GWAS) to be performed in large populations. Numerous studies in humans have identified chromosomal loci linked or associated with resting CV phenotypes. For example, a 2004 study from Framingham reported linkage on chromosomes 5, 7, 15, and 19 for pulse pressure (DeStefano et al., 2004). A recent study in African-Americans assessed >800,000 polymorphic markers; the 10 best SNP associations had P-values $<1.03 \times 10^{-5}$ for hypertension, $<1.12 \times 10^{-6}$ for systolic BP, and $<4.47 \times 10^{-6}$ for diastolic BP (Adeyemo et al., 2009). Another recent study assessed 2.5 million SNPs in ~34,000 Europeans and replicated the findings in three additional cohorts totaling ~110,000 individuals (Newton-Cheh et al., 2009). They found eight loci associated with either systolic or diastolic BP with P-values $<1 \times 10^{-8}$. While the results of these three examples are very promising, it is also important to note that the first two GWAS for BP-related traits (Saxena et al., 2007; Wellcome Trust Case Control Consortium, 2007) did not find any significant relationships.

A small number of genome-wide linkage and association studies have assessed CV functional measures as their phenotypes. One study quantified linkage of HRV in families and sib pairs from Framingham (Singh et al., 2002). Their two genetic loci demonstrating the highest linkage were on chromosomes 2 and 15. Plausible candidate genes located in these regions included one coding for a cholinergic receptor and another for a K$^+$ channel. Another genome-wide scan from Framingham assessed linkage with PWV and reflected pulse wave amplitude (Mitchell et al., 2005). They found two strong loci for reflected wave amplitude on chromosomes 4 and 8, and one for forward wave amplitude on chromosome 7. A more recent Framingham study used arterial stiffness as the functional phenotype (Levy et al., 2007). Based on more than 70,000 SNPs, they found two markers in the MADS box transcription enhancer factor 2, polypeptide C (*MEF2C*) gene, a regulator of cardiac morphogenesis, that were highly significant: one with the reflected wave ($P = 2.21 \times 10^{-6}$) and a second with carotid-brachial PWV ($P = 2.5 \times 10^{-6}$). Another Framingham GWAS found 10 SNPs associated with CV phenotypes (nine for echocardiographic dimensions, one for brachial artery endothelial function) with *P*-values for the top eight relationships all $<1.13 \times 10^{-5}$ (Vasan et al., 2007).

Thus, genome-wide linkage and association studies have identified several chromosomal loci related to resting CV hemodynamic, structural, and functional measures. However, in general, these studies have not identified novel genes that play a powerful role in determining these CV phenotypes.

Candidate Gene Association Studies of Resting CV Phenotypes

The majority of studies relating genes to resting CV phenotypes are candidate gene association studies. The first of these studies was published ~20 years ago and capitalized on the recent development, at that time, of relatively time- and cost-efficient assays for isolated SNPs. Thus, it actually became possible to genotype a small number of SNPs in previously established cohorts with already measured CV phenotypes. Genes and SNPs were selected for study as "candidates" because they could plausibly be proposed to impact the phenotype of interest. Examples of such candidate genes would be those in the renin–angiotensin system (RAS) for BP or CV phenotypes.

The first large-scale studies that assessed candidate genes and CV phenotypes investigated hypertension relative to the angiotensin-converting enzyme (*ACE*) genotype. In spontaneously hypertensive rats, Jacob and coworkers mapped a locus containing the *ACE* gene that was linked to hypertension (Jacob et al., 1991), and Nara and coworkers found that a microsatellite in the *ACE* gene co-segregated with BP (Nara et al., 1991). Following Zee and coworkers (1992), Harrap and coworkers (1993) reported that a 287 base pair (bp) insertion (I)/deletion (D) in the *ACE* gene is associated with hypertension in humans. Although this variant is intronic, it alters plasma ACE activity levels and thus must play some functional role (Tiret et al., 1992). The *ACE* gene has remained a high-profile CV candidate gene as evidenced by an October 2009 PubMed search on *ACE* genotype and hypertension resulting in 716 citations. While the results are not completely consistent, this variant has frequently been shown to be related to a wide range of CV phenotypes, including physiological, structural, and pathological outcomes.

A Met → Thr substitution at bp 235 in the angiotensinogen (*AGT*) gene appears to be the second genetic variant studied relative to CV phenotypes. Jeunemaitre and colleagues (1992) reported a highly significant association between this variant and hypertension in Caucasians. This was followed, in 1993–1994, by similar reports by Bennett and colleagues (1993) in severe familial hypertension, Hata and colleagues (1994) in Japanese hypertensives, and Hegele and coworkers (1994) in a Canadian isolate group, the Hutterites. The Met → Thr AGT 235 variant has also been shown to affect plasma AGT levels (Jeunemaitre et al., 1992) and, again, a recent PubMed search generated 261 citations for AGT genotype and hypertension.

In addition to these two candidate CV genes that have been studied intensely, many other candidate genes and SNPs have been investigated relative to CV phenotypes. The extent of this literature is easily demonstrated by two facts. First, an October 2009 PubMed search of candidate genes and CV system generated 971 citations, clearly a substantial body of literature. Second, two high-profile journals (*Hypertension* and *Circulation*) in just 1 year (2008) published 40–50 genetic studies related to CV phenotypes. Thus, the literature in this area is already substantial and is expanding rapidly.

Heritability of CV Responses to Acute Interventions

While the genetics underlying resting CV pheno-types are important to understand, physiologists would much rather study the genetics underlying the response of CV phenotypes to interventions. Recently, a few researchers have begun to quantify the heritability of CV responses to acute interventions. Gu et al. (2007) assessed the heritability of BP responses to alterations in Na^+ and K^+ intake. Heritability for baseline BP was 0.31–0.34. The heritability of all BP measures increased after the interventions, with the values ranging from 0.47 to 0.53, and the heritability of the responses was also relatively high, ranging from 0.20 to 0.33. Roy-Gagnon and coworkers (2008) studied extended Amish families and found heritability of 0.14–0.27 for systolic and diastolic BP responses to and during recovery from a cold pressor test. Wang and coworkers (2009) assessed the response of HRV to three stressors in European and African-American twins. They found the heritability of baseline HRV to be 0.48–0.59 and that it increased somewhat, though not significantly, in response to the stressors. However, the heritability of the HRV response to the stressors ranged from 0.11 to 0.23.

Thus, these examples and numerous other studies provide strong evidence that the responses of CV phenotypes to various acute interventions are also heritable. The logical next question would be whether the CV hemodynamic responses to acute exercise are heritable.

Heritability of Acute Exercise CV Response Phenotypes

Unfortunately, relatively few studies have assessed the heritability of CV responses to acute exercise. In one of the first studies, Klissouras (1971) reported heritability of maximal HR and $\dot{V}O_2$max to be very high ($h = 0.81$–0.93) in 25 male twin pairs. In another early study, Slattery and coworkers (1988) found that $\dot{V}O_2$max and recovery HR in male twin pairs had heritability of 0.75 and 0.99, respectively (Slattery et al., 1988). However, another early study reported, at best, minimal heritability for maximal

HR and $\dot{V}O_2$max (Lesage et al., 1985). Hunt and coworkers (1989) found heritability of 0.32 and 0.42 for systolic and diastolic BP, respectively, during cycle ergometer exercise, and 0.19 and 0.35 for systolic and diastolic BP, respectively, during isometric handgrip (Hunt et al., 1989). In a study of male twin pairs, Bielen et al. (1991a) found that the heritability of the increase in LV end-diastolic dimension and fractional shortening from rest to exercise were 0.24 and 0.47, respectively. However, later Bielen and coworkers (1991b) found BP and CV hemodynamics during supine cycle ergometer exercise to be only minimally heritable.

Bouchard and colleagues (1998) from Health, Risk Factors, Exercise Training, and Genetics (HERITAGE) reported a heritability for $\dot{V}O_2$max in the sedentary state of 0.50. An and coworkers (2000) from the white HERITAGE families reported heritability of 0.41–0.46 for SV and Q during sub-maximal exercise. Ingelsson and coworkers (2007) from Framingham found that HR, systolic BP, and diastolic BP responses during exercise and recovery had heritability of 0.32–0.34, 0.15–0.25, and 0.13–0.26, respectively. In a smaller subset of Framingham, heritability was 0.16–0.40 for HR and systolic and diastolic BP during and following an exercise treadmill test (Vasan et al., 2007).

These data represent the entire body of literature on the heritability of CV responses to acute exercise. These data clearly do not quantify the heritable aspects of anywhere near all of the CV responses to the numerous forms of acute exercise. However, they do generally consistently show that the CV hemodynamic and functional responses that have been assessed are moderately heritable. Obviously, substantially more studies are necessary to quantify the heritability of the numerous CV phenotype responses to the wide range of forms of acute exercise individuals undergo.

Heritability of CV Responses to Exercise Training

The next logical question in this progression is whether the CV responses to exercise training are heritable. Unfortunately, again, very few studies have addressed this issue. In reality, only the

HERITAGE Family Study directed by Claude Bouchard and colleagues has an appropriate sample size and study design to assess the heritability of CV responses to endurance exercise training. They implemented a highly standardized 20-week endurance exercise training intervention in sedentary individuals in 113 black and 99 white two-generation families (Bouchard et al., 1995). In 1999, they reported that the response of $\dot{V}O_2$max to this training was 2.5 times more variable between than within Caucasian families (Bouchard et al., 1999), translating into a heritability of 0.47. Rice et al. (2002a) from HERITAGE found that systolic and diastolic BP and HR changes at rest with training had heritabilities of 0.14–0.24 (Rice et al., 2002a). However, in normotensives, only diastolic BP ($h = 0.19$) responses to training were heritable, contrasting with the elevated BP group, in which systolic BP ($h = 0.20$) and HR ($h = 0.36$) responses to training were heritable. An and colleagues (2003) also reported heritability in the white HERITAGE families of 0.29–0.34 for the training response of HR and 0.22 for systolic BP training response during submaximal exercise. In another study, An and colleagues (2000) found heritability for the submaximal exercise SV and Q changes with training to be 0.32–0.40, values that were generally similar to the heritability of these responses to exercise prior to training.

Thus, very little is known about the heritability of the CV responses to endurance training. Clearly, more studies are required to provide the data necessary to verify that all CV responses to training are heritable. However, the available data consistently demonstrate that, at least for the limited CV phenotypes studied to date, these responses are generally moderately heritable. This provides a rationale to pursue more detailed studies to identify the specific genetic loci underlying this heritability.

Genome-Wide Linkage and Association Studies of Acute Exercise and Exercise Training CV Responses

Very few studies have assessed genome-wide linkage or association with CV responses to exercise training, and all the data are from HERITAGE. In an early HERITAGE report, Gagnon and colleagues found no linkage between seven markers on chromosome 22 and $\dot{V}O_2$max response to training (Gagnon et al., 1997); chromosome 22, the shortest autosomal chromosome, was studied first because it required fewer markers to assess linkages. Rankinen and colleagues (2001) presented the first complete genome-wide linkage or association study for CV responses to training. They found one locus linked to the training-induced response of submaximal exercise systolic BP in whites (8q21), while in blacks the training-induced change in submaximal exercise diastolic BP was linked to 10q21-q23. Numerous linkages with "suggestive" evidence were found for the training-induced changes in submaximal exercise BP. Rankinen and coworkers (2002a) found in whites one promising linkage for the change in submaximal exercise SV with training (10p11.2) and one suggestive linkage for the change in submaximal exercise SV and Q with training (2q31.1). They did not find any linkages in blacks for these phenotypes. Rice and coworkers (2002b) found very few suggestive or promising linkages for the resting BP change with training. Rico-Sanz and coworkers (2004) found in blacks one promising linkage (1p31) and two suggestive linkages (16q22, 20q13.1) for training-induced change in $\dot{V}O_2$max; in whites they identified three suggestive linkages for this phenotype (4q27, 7q34, and 13q12). An and colleagues (2006) found two "potentially interesting" linkages in whites (1q42.2 and 21q22.3) and five in blacks (3p14.2, 3p21.2, 3p14.1, 20p11.23, and 21q21.1) for the resting HR change with training. Spielmann and colleagues (2007) quantified linkage with the training-induced change in submaximal exercise HR responses. In whites, they found two "promising" linkages (18q21-q22 and 2q33.3) but none in blacks.

The next step would be to fine-map these specific loci exhibiting linkage to identify potentially novel genes of functional significance for the phenotype. In the first study to do this relative to exercise and CV phenotypes, HERITAGE investigators further mapped the 2q31-q32 locus they found to be linked to the submaximal exercise SV response to exercise training (Rankinen et al., 2002a) by typing another 12 microsatellite markers over this region

(Rankinen et al., 2003). They found strong evidence of linkage with two markers near the titin (*TTN*) gene. They proposed that this gene could be a key modulator of cardiomyocyte elasticity and involved in the Frank–Starling mechanism, although no specific variant has since been identified to functionally account for this relationship.

The same earlier study from HERITAGE also found linkage for the submaximal exercise SV response to training at 10p11.2 (Rankinen et al., 2002a). Argyropoulos and coworkers (2009) further explored this locus and found the best linkage with the kinesin heavy chain (*KIF5B*) gene. They sequenced *KIF5B* and found numerous polymorphisms; their follow-up functional studies substantiated this gene's potential to alter the submaximal exercise SV response to training.

Thus, only a few studies have assessed genome-wide linkage for CV responses to exercise training. Two of these chromosomal loci have been fine mapped; and relatively novel candidate genes (*TTN*, *KIF5B*) were identified that appear to play strong roles in determining some of the CV responses to exercise training.

Candidate Gene Studies of CV Responses to Acute Exercise

Candidate gene studies provide most of our knowledge relative to the genetics of CV responses to acute exercise. In fact, 35 studies have addressed this issue and have reported significant results, and the rate at which new data are generated has increased substantially recently (Bray et al., 2009). Just as for candidate gene studies for resting CV phenotypes, the *ACE* gene variant is the most frequently investigated and the first polymorphism studied relative to CV responses to exercise (Bray et al., 2009). Clearly, because of its central role in the RAS, *ACE* variants would be highly plausible candidates to affect CV phenotypes, especially since the *ACE* I/D genotype affects plasma ACE activity (Jeunemaitre et al., 1992). Friedl and coworkers (1996) found that diastolic BP during and following exercise varied by *ACE* genotype with DD, ID, and II individuals having maximal exercise diastolic BP of 93 ± 10, 85 ± 10, and

82 ± 8 mmHg, respectively, with similar significant differences evident during recovery. This group reported that *ACE* DD individuals increased plasma atrial natriuretic peptide (ANP) levels more with exercise than *ACE* II individuals (Friedl et al., 1998).

Currently, 14 studies have found significant effects of this variant on CV responses to acute exercise (Bray et al., 2009). The CV phenotypes studied range from pulmonary hemodynamic measures to HR and BP during submaximal and maximal exercise. However, eight of the nine studies in which the independent effects of the *ACE* genotype on CV responses during exercise could be quantified found that *ACE* DD individuals had the worst CV responses in terms of higher maximal exercise diastolic BP, plasma ANP and lactate levels, pulmonary artery pressures, and pulmonary vascular resistance, and reduced pulmonary O_2 delivery (Friedl et al., 1996, 1998; Hagberg et al., 1998; Kanazawa et al., 2000, 2002, 2003a, b, 2004). The only study not consistent with this trend is from our laboratory (Hagberg et al., 2002) in which *ACE* II postmenopausal women had about 10 bpm higher submaximal exercise HR than ID or DD women.

Relative to the other five studies assessing *ACE* genotype and acute exercise CV responses, Ashley et al. (2006) reported that following a 24-h running race, *ACE* II athletes had a larger decrease in LV function than other genotypes. Two of the remaining studies (Blanchard et al., 2006; Pescatello et al., 2007) assessed ambulatory BP following exercise in hypertensives and found varied effects of the *ACE* and other genotypes, depending on exercise intensity and Ca^{++} intake. The final two studies, from Framingham (Ingelsson et al., 2007; Vasan et al., 2007), found associations between the *ACE* genotype and HR and BP during exercise and recovery, but the direction and magnitude of the effects could not be ascertained.

Additional candidate gene studies have assessed the effect of other RAS gene variants on CV responses to acute exercise. Five studies have reported significant impacts of the *AGT* Met235Thr variant on CV responses to exercise (Bray et al., 2009). In the first study, Krizanova and coworkers (1998) found that diastolic BP during exercise was lower in *AGT* TT individuals, but there was no effect of the variant on systolic BP, HR, or plasma pressor hormone levels. Two HERITAGE studies by Rankinen et al. (1999, 2000a) reported no overall effect of *AGT* Met235Thr genotype on exercise BP. However, in white men maximal exercise diastolic BP was significantly higher, by 6–9 mmHg, in *AGT* TT individuals (Rankinen et al., 2000a). We found that TT genotype white women had 11–14 bpm higher HR during maximal and submaximal exercise (McCole et al., 2002). In two studies from Framingham (Ingelsson et al., 2007; Vasan et al., 2007), significant associations were found between *AGT* genotypes and BP during exercise and recovery, but it was not possible to quantify the effect. Also within the RAS, one study assessed the impact of an AT1R variant on ambulatory BP following exercise; however, they did not consistently find an effect of this variant independent of exercise intensity and Ca^{++} intake (Blanchard et al., 2006; Pescatello et al., 2007).

Another plausible mechanism that could affect CV responses to exercise would be the sympathetic nervous system, especially the adrenergic receptor (*ADR*) genes. In fact, genetic variations in six *ADR* genes have been assessed in this regard (Bray et al., 2009). However, one study assessed the impact of an *ADRA2B* variant, three investigated an *ADRB1* variant, and one study assessed *ADRA1A*, *ADRA1B*, and *ADRA1D* variants. Also, while another six studies quantified the effect of an *ADRB2* variant on acute exercise CV responses, they assessed the impact of three different *ADRB2* variants. Furthermore, four of these studies assessed the impact of these *ADRB2* variants on CV responses to isometric handgrip exercise. Again, the two studies from Framingham (Ingelsson et al., 2007; Vasan et al., 2007) found significant associations between *ADRA1A*, *ADRA1B*, *ADRA1D*, and *ADRB2* genetic variations and HR and BP during exercise and recovery.

The nitric oxide (NO) pathway is another mechanism that regulates CV responses to exercise. In 2006, we reported that in postmenopausal white women, endothelial NO synthase (*NOS3*) 894G homozygotes had higher submaximal and maximal exercise

HR, by about 7–8 bpm, than 894T allele carriers; however, 894 T allele carriers had higher submaximal and a somewhat higher maximal exercise SV, so that submaximal and maximal exercise BP and Q levels did not differ between genotypes (Hand et al., 2006). We found no effect of the *NOS3* T-786C variant on any CV phenotypes during submaximal or maximal exercise. Also, Kim and coworkers (2007) reported that Korean 894T allele carriers were significantly less likely to develop a hypertensive response to maximal exercise (Kim et al., 2007).

Variations in the endothelin-1 (*EDN1*) signaling pathway would also be candidates to affect CV responses to exercise. The only study to assess this relationship (Tiret et al., 1999) found that *EDN1* G198T TT genotype men had ~8 mmHg higher maximal exercise systolic BP than GG genotype men, while the difference was ~3 mmHg in women. However, no genotype-dependent relationship existed in normal weight individuals but there was a significant difference of ~9 mmHg in maximal exercise systolic BP in overweight homozygotes.

The remainder of the genes studied found to be significantly related to the CV responses to acute exercise are from single studies that have not been replicated (Bray et al., 2009). These genes include adenosine monophosphate deaminase 1 (*AMPD1*), adducin 1 (*ADD1*), hemochromatosis (*HFE*), cholinergic receptor muscarinic 2 (*CHRM2*), guanine nucleotide binding protein beta polypeptide 3 (*GNB3*), angiogenin ribonuclease (*ANG*), transforming growth factor beta 1 (*TGFB1*), GNAS complex locus (*GNAS*), and cytochrome 450 family 2 subfamily D polypeptide 6 (*CYP2D6*) genes. These genes clearly span a wide array of plausible pathways that could affect CV responses to acute exercise. However, since each of these genes has been assessed only in a single study and the results have not been replicated, it is clearly premature to draw any strong conclusions from them at this point.

Thus, in summary, the *ACE* DD genotype appears to have a negative impact on CV responses to acute exercise. These results are consistent with the higher rates of CV disease pathologies generally reported in *ACE* DD genotype individuals. The *AGT* TT genotype may also confer adverse

CV responses to acute exercise. However, this conclusion is not completely consistent across all the data published to date. While a number of studies assessed the impact of genetic variations within *ADR* genes on CV responses to acute exercise, there are virtually no replicated data on which to base conclusions. Perhaps the only conclusion that can be drawn is from three papers from the Joyner group (Mayo Clinic) which consistently found that *ADRB2* Gly16 homozygotes have larger HR and BP responses to isometric handgrip (Eisenach et al., 2004, 2005; Snyder et al., 2006).

Candidate Gene Studies of Exercise Training CV Phenotypes

In our last update of the Human Gene Map for Performance and Health-Related Fitness Phenotypes, there were only 23 studies that had reported significant genotype-dependent training response effects (Bray et al., 2009). Just as for other candidate gene studies relative to CV phenotypes, the *ACE* variant has been investigated most frequently in regard to CV responses to exercise training with seven studies (Bray et al., 2009). The first study by Montgomery and coworkers (1997) found that military training increased LVM 36−38% in ID/DD individuals, while no increase in LVM occurred in II individuals.

A number of studies reporting significant effects of *ACE* genotype on CV responses to training were published from 1999 to 2002. Rankinen and coworkers (2000a) from HERITAGE reported that white *ACE* DD men reduced their submaximal exercise diastolic BP more with training than *ACE* I allele carriers (−4.4 vs. −2.6 mmHg). Fatini and coworkers (2000) also reported that over a soccer season, *ACE* D allele carriers increased LVM more than *ACE* II homozygotes. Myerson and colleagues (2001) also found that LVM increased more with military training in *ACE* DD compared to II men (12 vs. 5 g).

In a small population of hypertensives, we reported that *ACE* I allele carriers reduced diastolic BP more with training than *ACE* DD men (−10 vs. −1 mmHg) (Hagberg et al., 1999). Zhang and coworkers (2002) found that systolic and diastolic BP decreased with training in *ACE* I allele

carrier Japanese hypertensives, but did not change in *ACE* DD individuals. We also found that African-American hypertensive *ACE* I homozygotes increased Na$^+$ excretion with exercise training, while *ACE* D allele carriers did not (Jones et al., 2006).

Also consistent with other candidate gene studies for CV phenotypes, the second most studied polymorphism relative to training-induced CV phenotype changes is the *AGT* Met235Thr variant. Rankinen and coworkers (2000a) found that in the white HERITAGE men, *AGT* Met235Thr M allele carriers reduced submaximal exercise diastolic BP with training by ~3.5 mmHg, whereas the reduction was less in T homozygotes (−0.4 mmHg). They also reported a significant interaction as individuals carrying the *AGT* TT genotype and the *ACE* D allele did not reduce submaximal exercise diastolic BP with training. Rauramaa and coworkers (2002) found that, over 6 years, *AGT* Met235Thr M homozygotes undergoing exercise training decreased their resting systolic BP by 1 mmHg, while MM genotype controls increased their systolic BP by ~15 mmHg. We found differences in resting BP responses to strength training as a function of two RAS genotypes (*AGT* A-20C genotype for systolic BP, *AGTR1* A1166C genotype for diastolic BP).

Rankinen and coworkers (2000b) showed in HERITAGE that *NOS3* G894 homozygotes had ~3 mmHg greater reduction in submaximal exercise diastolic BP with training than T homozygotes. This genotype-dependent response was similar in men and women; Glu homozygotes also tended to reduce submaximal exercise systolic BP response and submaximal exercise rate-pressure product more with training. Erbs and colleagues (2003) found that in patients with CV disease, the endothelium-dependent response of arterial average peak velocity increase with exercise training was dependent on *NOS3* T-786C, but not on G894T, genotype (Erbs et al., 2003).

Two studies also found an effect of lipoprotein lipase (*LPL*) variants on CV responses to training. We reported that in a small number of hypertensives, the *LPL* HindIII + allele carriers decreased resting systolic and diastolic BP more than − allele homozygotes (−10 vs. +3 mmHg, −9 vs. +2 mmHg)

(Hagberg et al., 1999). Flavell and coworkers (2006) later reported that with military training young *LPL* X447 carrier men increased LVM less (2.2% vs. 5.8%), but had a greater decrease in systolic BP (−6 vs. +2 mmHg).

Rankinen and coworkers (2002b) also found effects of the *GNB3* C825T variant on CV responses to training in HERITAGE. They found that black CC genotype individuals exhibited greater training-induced reductions in submaximal exercise HR than TT individuals, with a similar but nonsignificant trend evident in whites. Also, CC genotype black women reduced resting systolic and diastolic BP more than TT black women with training.

Recently, Rankinen and coworkers (2007) published a groundbreaking paper relative to *EDN1* genotype, physical activity, and CV phenotypes. In the HERITAGE population, they first found that *EDN1* haplotypes did not relate to any BP and HR phenotypes in the baseline state prior to exercise training. However, a number of *EDN1* haplotypes were significantly associated with various CV responses to exercise training. Then, in another cohort, they again found that *EDN1* genotypes did not differ between hypertension cases and normotensive controls. However, two *EDN1* SNPs interacted significantly with CV fitness levels to impact risk for developing hypertension, with both SNPs exerting their effects only on the risk of developing hypertension in low-fit individuals with the risk roughly doubling in individuals carrying the risk allele or genotype.

The remaining evidence for genotype-dependent effects on CV responses to exercise training consists of only single studies that found significant effects of different candidate genes, including *AMPD1*, nuclear factor of kappa light polypeptide gene enhancer in beta cells (*NFKB1*), fatty acid binding protein-2 (*FABP2*), *EDN1*, *CHRM2*, potassium voltage-gated channel KQT-like subfamily member 1 (*KCNQ1*), *HBB*, bradykinin receptor B2 (*BDKRB2*), apolipoprotein E (*APOE*), and peroxisome proliferator-activated receptor alpha (*PPARA*) genes (Bray et al., 2009). The CV responses to training in these studies include reactive hyperemic blood flow, systolic and diastolic BP at rest and during submaximal and maximal exercise, SV and Q during submaximal exercise, and LVM, among others. Thus, clearly more studies are necessary to assess the impact of the numerous potential candidate genes that could be involved in regulating the different CV phenotype responses to exercise training.

One conclusion, at least at this point in time, appears to be that the *ACE* D allele confers advantageous CV responses to exercise training in healthy individuals, whereas the *ACE* I allele may be beneficial in terms of responses to training in hypertensives. However, this conclusion is based on a limited number of studies and a minimal range of CV phenotypes, and this conclusion is not consistent with the general finding that there is an excess of *ACE* I alleles in elite endurance athletes (Myerson et al., 1999). The two studies that assessed the effect of *AGT* variants concluded that Met235 carriers may have a better training-induced BP response. It also appears that common variants in the *NOS3*, *LPL*, *GNB3*, and *EDN1* genes may affect CV responses to endurance exercise training.

Limitations, Conclusions, and the Future

More data are needed before strong conclusions can be drawn relative to the genetics underlying CV responses to acute exercise and exercise training. A small number of studies have been published relative to this issue, and they pale in comparison to the number of candidate genes and CV phenotypes that must be studied. It does appear, even with the minimal data currently available, that the *ACE* I/D genotype may affect CV responses to both acute exercise and exercise training. In addition, *AGT* Met235Thr genotype appears to affect CV responses to exercise training. Some data indicate that a small number of other genes may also affect the CV responses to endurance exercise training. Therefore, we are beginning to develop a profile of a small number of common genetic variants that may be important in determining interindividual differences in CV responses to acute exercise and exercise training.

One major limitation relative to obtaining more data concerning the genetics underlying CV responses to acute exercise and exercise training is the fact that most genetic studies currently being published in other fields incorporate very large populations (>1000 individuals) with replication within the same study also in relatively large populations. Such requirements are especially difficult to implement relative to the genetics underlying CV responses to acute exercise or exercise training because (a) of the large number of potential candidate genes to be investigated, (b) the wide range of CV phenotypes to be assessed, (c) the absolutely essential need to utilize highly standardized CV phenotype measurements, and (d) to impose similarly highly standardized acute exercise or exercise training interventions. Thus, given these constraints, it is difficult to see how definitive, replicated evidence will be generated to robustly quantify the impact of common genetic variations on CV responses to acute exercise and exercise training that are critical to endurance performance without a new, larger HERITAGE Family Study.

References

Adeyemo, A., Gerry, N., Chen, G., et al. (2009). A genome-wide association study of hypertension and blood pressure in African Americans. *PLoS Genetics* 5, e1000564.

An, P., Rice, T., Gagnon, J., et al. (2000). Familial aggregation of stroke volume and cardiac output during submaximal exercise: The HERITAGE Family Study. *International Journal of Sports Medicine* 21, 566–572.

An, P., Perusse, L., Rankinen, T., et al. (2003). Familial aggregation of exercise heart rate and blood pressure in response to 20 weeks of endurance training: The HERITAGE family study. *International Journal of Sports Medicine* 24, 57–62.

An, P., Rice, T., Rankinen, T., et al. (2006). Genome-wide scan to identify quantitative trait loci for baseline resting heart rate and its response to endurance exercise training: The HERITAGE Family Study. *International Journal of Sports Medicine* 27, 31–36.

Argyropoulos, G., Stutz, A.M., Ilnytska, O., et al. (2009). *KIF5B* gene sequence variation and response of cardiac stroke volume to regular exercise. *Physiological Genomics* 36, 79–88.

Ashley, E.A., Kardos, A., Jack, E.S., et al. (2006). Angiotensin-converting enzyme genotype predicts cardiac and autonomic responses to prolonged exercise. *Journal of the American College of Cardiology* 48, 523–531.

Benjamin, E.J., Larson, M.G., Keyes, M.J., et al. (2004). Clinical correlates and heritability of flow-mediated dilation in the community: The Framingham Heart Study. *Circulation* 109, 613–619.

Bennett, C.L., Schrader, A.P. & Morris, B.J. (1993). Cross-sectional analysis of Met235 → Thr variant of angiotensinogen gene in severe, familial hypertension. *Biochemical and Biophysical Research Communications* 197, 833–839.

Bielen, E.C., Fagard, R.H. & Amery, A.K. (1991a). Inheritance of acute cardiac changes during bicycle exercise: An echocardiographic study in twins. *Medicine and Science in Sports and Exercise* 23, 1254–1259.

Bielen, E.C., Fagard, R.H. & Amery, A.K. (1991b). Inheritance of blood pressure and haemodynamic phenotypes measured at rest and during supine dynamic exercise. *Journal of Hypertension* 9, 655–663.

Blanchard, B.E., Tsongalis, G.J., Guidry, M.A., et al. (2006). RAAS polymorphisms alter the acute blood pressure response to aerobic exercise among men with hypertension. *European Journal of Applied Physiology* 97, 26–33.

Bouchard, C., Leon, A.S., Rao, D.C., Skinner, J.S., Wilmore, J.H. & Gagnon, J. (1995). The HERITAGE Family Study. Aims, design, and measurement protocol. *Medicine and Science in Sports and Exercise* 27, 721–729.

Bouchard, C., Daw, E.W., Rice, T., et al. (1998). Familial resemblance for V̇O$_2$max in the sedentary state: The HERITAGE Family Study. *Medicine and Science in Sports and Exercise* 30, 252–258.

Bouchard, C., An, P., Rice, T., et al. (1999). Familial aggregation of V̇O$_2$max response to exercise training: Results from the HERITAGE Family Study. *Journal of Applied Physiology* 87, 1003–1008.

Bray, M.S., Hagberg, J.M., Perusse, L., et al. (2009). The human gene map for performance and health-related fitness phenotypes: The 2006–2007 update.

Medicine and Science in Sports and Exercise 41, 35–73.

Chien, K.L., Hsu, H.C., Su, T.C., Chen, M.F. & Lee, Y.T. (2006). Heritability and major gene effects on left ventricular mass in the Chinese population: A family study. *BMC Cardiovascular Disorders* 6, 37.

Delmonico, M.J., Ferrell, R.E., Meerasahib, A., et asl. (2005). Blood pressure response to strength training may be influenced by angiotensinogen A-20C and angiotensin II type I receptor A1166C genotypes in older men and women. *Journal of the American Geriatric Society* 53, 204–210.

DeStefano, A.L., Larson, M.G., Mitchell, G.F., et al. (2004). Genome-wide scan for pulse pressure in the National Heart, Lung and Blood Institute's Framingham Heart Study. *Hypertension* 44, 152–155.

Eisenach, J.H., McGuire, A.M., Schwingler, R.M., Turner, S.T. & Joyner, M.J. (2004). The Arg16/Gly beta2-adrenergic receptor polymorphism is associated with altered cardiovascular responses to isometric exercise. *Physiological Genomics* 16, 323–328.

Eisenach, J.H., Barnes, S.A., Pike, T.L., et al. (2005). Arg16/Gly beta2-adrenergic receptor polymorphism alters the cardiac output response to isometric exercise. *Journal of Applied Physiology* 99, 1776–1781.

Erbs, S., Baither, Y., Linke, A., et al. (2003). Promoter but not exon 7 polymorphism of endothelial nitric oxide synthase affects training-induced correction of endothelial dysfunction. *Arteriosclerosis, Thrombosis, and Vascular Biology* 23, 1814–1819.

Fagard, R., Van Den Broeke, C., Bielen, E. & Amery, A. (1987). Maximum oxygen uptake and cardiac size and function in twins. *American Journal of Cardiology* 60, 1362–1367.

Fatini, C., Guazzelli, R., Manetti, P., et al. (2000). RAS genes influence exercise-induced left ventricular hypertrophy: An elite athletes study. *Medicine and Science in Sports and Exercise* 32, 1868–1872.

Flavell, D.M., Wootton, P.T., Myerson, S.G., et al. (2006). Variation in the lipoprotein lipase gene influences exercise-induced left ventricular growth. *Journal of Molecular Medicine* 84, 126–131.

Frazer, K.A., Ballinger, D.G., Cox, D.R., et al. (2007). A second generation human haplotype map of over 3.1 million SNPs. *Nature* 449, 851–861.

Friedl, W., Krempler, F., Sandhofer, F. & Paulweber, P. (1996). Insertion/deletion polymorphism in the angiotensin-converting-enzyme gene and blood pressure during ergometry in normal males. *Clinical Genetics* 50, 541–544.

Friedl, W., Mair, J., Pichler, M., Paulweber, B., Sandhofer, F. & Puschendorf, B. (1998). Insertion/deletion polymorphism in the angiotensin-converting enzyme gene is associated with atrial natriuretic peptide activity after exercise. *Clinica Chimica Acta* 274, 199–211.

Gagnon, J., Ho-Kim, M.A., Chagnon, Y.C., et al. (1997). Absence of linkage between $\dot{V}O_2$max and its response to training with markers spanning chromosome 22. *Medicine and Science in Sports and Exercise* 29, 1448–1453.

Gombojav, B., Park, H., Kim, J.I., et al. (2008). Heritability and linkage study on heart rates in a Mongolian population. *Experimental & Molecular Medicine* 40, 558–564.

Gu, D., Rice, T., Wang, S., et al. (2007). Heritability of blood pressure responses to dietary sodium and potassium intake in a Chinese population. *Hypertension* 50, 116–122.

Hagberg, J.M., Ferrell, R.E., McCole, S.D., Wilund, K.R. & Moore, G.E. (1998). $\dot{V}O_2$max is associated with ACE genotype in postmenopausal women. *Journal of Applied Physiology* 85, 1842–1846.

Hagberg, J.M., Ferrell, R.E., Dengel, D.R. & Wilund, K.R. (1999). Exercise training-induced blood pressure and plasma lipid improvements in hypertensives may be genotype dependent. *Hypertension* 34, 18–23.

Hagberg, J.M., McCole, S.D., Brown, M.D., et al. (2002). ACE insertion/deletion polymorphism and submaximal exercise hemodynamics in postmenopausal women. *Journal of Applied Physiology* 92, 1083–1088.

Hand, B.D., McCole, S.D., Brown, M.D., et al. (2006). NOS3 gene polymorphisms and exercise hemodynamics in postmenopausal women. *International Journal of Sports Medicine* 27, 951–958.

Harrap, S.B., Davidson, H.R., Connor, J.M., et al. (1993). The angiotensin I converting enzyme gene and predisposition to high blood pressure. *Hypertension* 21, 455–460.

Hata, A., Namikawa, C., Sasaki, M., et al. (1994). Angiotensinogen as a risk factor for essential hypertension in Japan. *Journal of Clinical Investigation* 93, 1285–1287.

Hegele, R.A., Brunt, J.H. & Connelly, P.W. (1994). A polymorphism of the angiotensinogen gene associated with variation in blood pressure in a genetic isolate. *Circulation* 90, 2207–2212.

Hunt, S.C., Hasstedt, S.J., Kuida, H., Stults, B.M., Hopkins, P.N. & Williams, R.R. (1989). Genetic heritability and common environmental components of resting and stressed blood pressures, lipids, and body mass index in Utah pedigrees and twins. *American Journal of Epidemiology* 129, 625–638.

Ingelsson, E., Larson, M.G., Vasan, R.S., et al. (2007). Heritability, linkage, and genetic associations of exercise treadmill test responses. *Circulation* 115, 2917–2924.

Jacob, H.J., Lindpaintner, K., Lincoln, S.E., et al. (1991). Genetic mapping of a gene causing hypertension in the stroke-prone spontaneously hypertensive rat. *Cell* 67, 213–224.

Jeunemaitre, X., Soubrier, F., Kotelevtsev, Y.V., et al. (1992). Molecular basis of human hypertension: Role of angiotensinogen. *Cell* 71, 169–180.

Jones, J.M., Park, J.J., Johnson, J., et al. (2006). Renin-angiotensin system genes and exercise training-induced changes in sodium excretion in African American hypertensives. *Ethnicity & Disease* 16, 666–674.

Kanazawa, H., Okamoto, T., Hirata, K. & Yoshikawa, J. (2000). Deletion polymorphisms in the angiotensin converting enzyme gene are associated with pulmonary hypertension evoked by exercise challenge in patients with chronic obstructive pulmonary disease. *American Journal of Respiratory and Critical Care Medicine* 162, 1235–1238.

Kanazawa, H., Otsuka, T., Hirata, K. & Yoshikawa, J. (2002). Association between the angiotensin-converting enzyme gene polymorphisms and tissue oxygenation during exercise in patients with COPD. *Chest* 121, 697–701.

Kanazawa, H., Hirata, K. & Yoshikawa, J. (2003a). Effects of captopril administration on pulmonary haemodynamics and tissue oxygenation during exercise in ACE gene subtypes in patients with COPD: A preliminary study. *Thorax* 58, 629–631.

Kanazawa, H., Hirata, K. & Yoshikawa, J. (2003b). Influence of oxygen administration on pulmonary haemodynamics and tissue oxygenation during exercise in COPD patients with different ACE genotypes. *Clinical Physiology and Functional Imaging* 23, 332–336.

Kanazawa, H., Tateishi, Y. & Yoshikawa, J. (2004). Acute effects of nifedipine administration in pulmonary haemodynamics and oxygen delivery during exercise in patients with chronic obstructive pulmonary disease: Implication of the angiotensin-converting enzyme gene polymorphisms. *Clinical Physiology and Functional Imaging* 24, 224–228.

Kim, J.S., Cho, J.R., Park, S., et al. (2007). Endothelial nitric oxide synthase Glu298Asp gene polymorphism is associated with hypertensive response to exercise in well-controlled hypertensive patients. *Yonsei Medical Journal* 48, 389–395.

Klissouras, V. (1971). Heritability of adaptive variation. *Journal of Applied Physiology* 31, 338–344.

Krizanova, O., Koska, J., Vigas, M. & Kvetnansky, R. (1998). Correlation of Met235Thr DNA polymorphism with cardiovascular and endocrine responses during physical exercise in healthy subjects. *Physiological Research* 47, 81–88.

Lesage, R., Simoneau, J.A., Jobin, J., Leblanc, J. & Bouchard, C. (1985). Familial resemblance in maximal heart rate, blood lactate and aerobic power. *Human Heredity* 35, 182–189.

Levy, D., Larson, M.G., Benjamin, E.J., et al. (2007). Framingham Heart Study 100K Project: Genome-wide associations for blood pressure and arterial stiffness. *BMC Medical Genetics* 8(Suppl. 1), S3.

McCole, S.D., Brown, M.D., Moore, G.E., et al. (2002). Angiotensinogen Met235Thr polymorphism associates with exercise hemodynamics in postmenopausal women. *Physiological Genomics* 10, 63–69.

Mitchell, G.F., DeStefano, A.L., Larson, M.G., et al. (2005). Heritability and a genome-wide linkage scan for arterial stiffness, wave reflection, and mean arterial pressure: The Framingham Heart Study. *Circulation* 112, 194–199.

Montgomery, H.E., Clarkson, P., Dollery, C.M., et al. (1997). Association of angiotensin-converting enzyme gene

I/D polymorphism with change in left ventricular mass in response to physical training. *Circulation* 96, 741–747.

Myerson, S., Hemingway, H., Budget, R., Martin, J., Humphries, S. & Montgomery, H. (1999). Human angiotensin I-converting enzyme gene and endurance performance. *Journal of Applied Physiology* 87, 1313–1316.

Myerson, S.G., Montgomery, H.E., Whittingham, M., et al. (2001). Left ventricular hypertrophy with exercise and *ACE* gene insertion/deletion polymorphism: A randomized controlled trial with losartan. *Circulation* 103, 226–230.

Nara, Y., Nabika, T., Ikeda, K., Sawamura, M., Endo, J. & Yamori, Y. (1991). Blood pressure cosegregates with a microsatellite of angiotensin I converting enzyme (*ACE*) in F2 generation from a cross between original normotensive Wistar-Kyoto rat (WKY) and stroke-prone spontaneously hypertensive rat (SHRSP). *Biochemical and Biophysical Research Communications* 181, 941–946.

Newton-Cheh, C., Johnson, T., Gateva, V., et al. (2009). Genome-wide association study identifies eight loci associated with blood pressure. *Nature Genetics,* May 10 [Epub a head of print].

de Oliveira, C.M., Pereira, A.C., de Andrade, M., Soler, J.M. & Krieger, J.E. (2008). Heritability of cardiovascular risk factors in a Brazilian population: Baependi Heart Study. *BMC Medical Genetics* 9, 32.

Pescatello, L.S., Turner, D., Rodriguez, N., et al. (2007). Dietary calcium intake and renin angiotensin system polymorphisms alter the blood pressure response to aerobic exercise: A randomized control design. *Nutrition & Metabolism* 4, 1.

Rankinen, T., Gagnon, J., Perusse, L., et al. (1999). Body fat, resting and exercise blood pressure and the angiotensinogen Met235Thr polymorphism: The HERITAGE Family Study. *Obesity Research* 7, 423–430.

Rankinen, T., Gagnon, J., Perusse, L., et al. (2000a). AGT Met235Thr and ACE ID polymorphisms and exercise blood pressure in the HERITAGE Family Study. *American Journal of Physiology Heart Circulation Physiology* 279, H368–H374.

Rankinen, T., Rice, T., Perusse, L., et al. (2000b). NOS3 Glu298Asp genotype and blood pressure response to endurance training: The HERITAGE family study. *Hypertension* 36, 885–889.

Rankinen, T., An, P., Rice, T., et al. (2001). Genomic scan for exercise blood pressure in the Health, Risk Factors, Exercise Training and Genetics (HERITAGE) Family Study. *Hypertension* 38, 30–37.

Rankinen, T., An, P., Perusse, L., et al. (2002a). Genome-wide linkage scan for exercise stroke volume and cardiac output in the HERITAGE Family Study. *Physiological Genomics* 10, 57–62.

Rankinen, T., Rice, T., Leon, A.S., et al. (2002b). G protein beta 3 polymorphism and hemodynamic and body composition phenotypes in the HERITAGE Family Study. *Physiological Genomics* 8, 151–157.

Rankinen, T., Rice, T., Boudreau, A., et al. (2003). Titin is a candidate gene for stroke volume response to endurance training: The HERITAGE Family Study. *Physiological Genomics* 15, 27–33.

Rankinen, T., Church, T., Rice, T., et al. (2007). Effect of endothelin 1 genotype on blood pressure is dependent on physical activity or fitness levels. *Hypertension* 50, 1120–1125.

Rauramaa, R., Kuhanen, R., Lakka, T.A., et al. (2002). Physical exercise and blood pressure with reference to the angiotensinogen Met235Thr polymorphism. *Physiological Genomics* 10, 71–77.

Rice, T., An, P., Gagnon, J., et al. (2002a). Heritability of HR and BP response to exercise training in the HERITAGE Family Study. *Medicine and Science in Sports and Exercise* 34, 972–979.

Rice, T., Rankinen, T., Chagnon, Y.C., et al. (2002b). Genomewide linkage scan of resting blood pressure: HERITAGE Family Study. Health, Risk Factors, Exercise Training, and Genetics. *Hypertension* 39, 1037–1043.

Rico-Sanz, J., Rankinen, T., Rice, T., et al. (2004). Quantitative trait loci for maximal exercise capacity phenotypes and their responses to training in the HERITAGE Family Study. *Physiological Genomics* 16, 256–260.

Roy-Gagnon, M.H., Weir, M.R., Sorkin, J.D., et al. (2008). Genetic influences on blood pressure response to the cold pressor test: Results from the Heredity and Phenotype Intervention Heart Study. *Journal of Hypertension* 26, 729–736.

Saxena, R., Voight, B.F., Lyssenko, V., et al. (2007). Genome-wide association analysis identifies loci for type 2 diabetes and triglyceride levels. *Science* 316, 1331–1336.

de Simone, G., Tang, W., Devereux, R.B., et al. (2007). Assessment of the interaction of heritability of volume load and left ventricular mass: The HyperGEN offspring study. *Journal of Hypertension* 25, 1397–1402.

Singh, J.P., Larson, M.G., O'Donnell, C.J., Tsuji, H., Evans, J.C. & Levy, D. (1999). Heritability of heart rate variability: The Framingham Heart Study. *Circulation* 99, 2251–2254.

Singh, J.P., Larson, M.G., O'Donnell, C.J., Tsuji, H., Corey, D. & Levy, D. (2002). Genome scan linkage results for heart rate variability (the Framingham Heart Study). *American Journal of Cardiology* 90, 1290–1293.

Sinnreich, R., Friedlander, Y., Luria, M.H., Sapoznikov, D. & Kark, J.D. (1999). Inheritance of heart rate variability: The kibbutzim family study. *Human Genetics* 105, 654–661.

Slattery, M.L., Bishop, D.T., French, T.K., Hunt, S.C., Meikle, A.W. & Williams, R.R. (1988). Lifestyle and blood pressure levels in male twins in Utah. *Genetic Epidemiology* 5, 277–287.

Snyder, E.M., Beck, K.C., Dietz, N.M., et al. (2006). Arg16Gly polymorphism of the beta2-adrenergic receptor is associated with differences in cardiovascular function at rest and during exercise in humans. *The Journal of Physiology* 571, 121–130.

Spielmann, N., Leon, A.S., Rao, D.C., et al. (2007). Genome-wide linkage scan for submaximal exercise heart rate in the HERITAGE family study. *American Journal of Physiology Heart Circulation Physiology* 293, H3366–H3371.

Tiret, L., Rigat, B., Visvikis, S., et al. (1992). Evidence, from combined segregation and linkage analysis, that a variant of the angiotensin I-converting enzyme (ACE) gene controls plasma ACE levels. *American Journal of Human Genetics* 51, 197–205.

Tiret, L., Poirier, O., Hallet, V., McDonagh, T.A., Morrison, C., McMurray, J.J., et al. (1999). The Lys198Asn polymorphism in the endothelin-1 gene is associated with blood pressure in overweight people. *Hypertension* 33, 1169–1174.

Vasan, R.S., Larson, M.G., Aragam, J., et al. (2007). Genome-wide association of echocardiographic dimensions, brachial artery endothelial function and treadmill exercise responses in the Framingham Heart Study. *BMC Medical Genetics* 8(Suppl. 1), S2.

Wang, X., Ding, X., Su, S., et al. (2009). Genetic influences on heart rate variability at rest and during stress. *Psychophysiology* 46, 458–465.

Wellcome Trust Case Control Consortium. (2007). Genome-wide association study of 14,000 cases of seven common diseases and 3,000 shared controls. *Nature* 447, 661–678.

Zee, R.Y., Lou, Y.K., Griffiths, L.R. & Morris, B.J. (1992). Association of a polymorphism of the angiotensin I-converting enzyme gene with essential hypertension. *Biochemical and Biophysical Research Communications* 184, 9–15.

Zhang, B., Sakai, T., Miura, S., et al. (2002). Association of angiotensin-converting-enzyme gene polymorphism with the depressor response to mild exercise therapy in patients with mild to moderate essential hypertension. *Clinical Genetics* 62, 328–333.

Zhao, J., Cheema, F.A., Reddy, U., et al. (2007). Heritability of flow-mediated dilation: A twin study. *Journal of Thrombosis and Haemostasis* 5, 2386–2392.

Zhao, J., Cheema, F.A., Bremner, J.D., et al. (2008). Heritability of carotid intima-media thickness: A twin study. *Atherosclerosis* 197, 814–820.

Chapter 23

Genes, Exercise, and Protein Metabolism

MARK A. TARNOPOLSKY

Department of Pediatrics and Medicine, McMaster University Medical Center, Hamilton, ON, Canada

Athletes and coaches have traditionally believed that high-dietary protein intakes were required for optimal sports performance. Eventually, the primacy of carbohydrate (CHO) and fat oxidation during exercise was demonstrated (Cathcart, 1925). There is still a widely held belief that high levels of dietary protein are needed to maximize strength and mass gains for resistance-trained/power athletes in the general population. An ongoing debate remains as to whether exercise has a negative impact on net protein balance and increase dietary protein requirements (Phillips, 2006; Tarnopolsky, 2004; Tipton & Witard, 2007; Tipton & Wolfe, 2004).

This chapter will evaluate the molecular regulation of protein turnover as it pertains to resistance and endurance exercise. The main emphasis will be on gene expression and regulation with a brief overview of signaling molecules and the control of protein translation.

Protein Metabolism

Proteins serve important structural (i.e., dystrophin, sarcoglycan, desmin, and collagen) and regulatory (i.e., citrate synthase and lactate dehydrogenase) functions. Proteins are comprised of amino acids containing an amino ($-NH_2$), carboxylic acid ($-COOH$), and a radical (different for each amino acid) component. There are 20 amino acids, with

nine considered essential (histidine, isoleucine, leucine, lysine, methionine, phenylalanine, threonine, tryptophan, and valine), and one considered conditionally essential (arginine) (Pellet, 1990). Essential amino acids come from the diet and/or from endogenous protein breakdown. A detailed examination of protein metabolism would require a dedicated textbook; consequently, I shall focus on the basics of protein turnover at the molecular level to allow for a conceptual understanding of protein metabolism and nutrition during exercise.

Protein Degradation in Skeletal Muscle

There are three main pathways for protein degradation in human skeletal muscle including lysosomal (cathepsin) and nonlysosomal (ubiquitin and calpain) pathways. The two nonlysosomal pathways in human skeletal muscle include the adenosine triphosphate (ATP)-dependent ubiquitin pathway (Lecker et al., 2004; Mitch & Goldberg, 1996) and the calcium-activated neutral protease (calpain) pathway (Bartoli & Richard, 2005; Belcastro et al., 1998). Skeletal muscle protein breakdown is elevated for about 36 h following resistance exercise but is of lesser magnitude and duration compared to the increase in muscle synthetic rates (~48 h) (Phillips et al., 1997). The increase in muscle breakdown rates following exercise occurs in untrained but not in trained men (Phillips et al., 1999). Following a period of resistance exercise training, there is an increase in resting fed-state fractional muscle protein breakdown but no acute exercise-induced increase in breakdown (Phillips et al., 2002). The latter observations are consistent with

Genetic and Molecular Aspects of Sport Performance, 1st edition.
Edited by Claude Bouchard and Eric P. Hoffman.
Published 2011 by Blackwell Publishing Ltd.

a training-induced adaptation such that the muscle is more resistant to acute physiological stress following training and consistent with the rapid adaptations seen even after a single bout of eccentric exercise termed the "second bout phenomena" (Stupka et al., 2001). In contrast to the data on resistance exercise, there is little information regarding acute or chronic changes in muscle protein breakdown rates in skeletal muscle following endurance exercise.

Lysosomal pathways degrade endocytosed proteins, some cytosolic proteins, hormones, and immune modulators (Mitch & Goldberg, 1996). The lysosomal pathways are generally considered to not contribute significantly to human skeletal muscle protein degradation in healthy humans (Lowell et al., 1986), but are involved in pathological states such as muscular dystrophy (Tidball & Spencer, 2000). One study found that cathepsin B + L activity was elevated 14 days after a single bout of lengthening (eccentric) contractions (Feasson et al., 2002). The latter study did not look at the time course of cathepsin mRNA changes; however, cathepsin mRNA species are not seen at +3 h or +48 h following eccentric (Mahoney et al., 2008), or endurance (Mahoney et al., 2005), exercise using microarray analysis. It is likely that the cathepsin pathway is more important in mediating skeletal muscle atrophy compared to exercise adaptations as they seem to play a role in Fox03-mediated autophagy (Zhao et al., 2007).

The calpain pathway has traditionally been considered to play some role in skeletal muscle proteolysis following exercise (Belcastro et al., 1998), although the evidence for such a role in humans is lacking. The calpains are a group of nonlysosomal cysteine proteases activated by intracellular calcium. The two main forms in skeletal muscle are calpain 1 (μ, "mu," responding to micromolar [Ca^{++}]) and calpain 2 (m, responding to millimolar [Ca^{++}]), and they are inhibited by the endogenous protein calpastatin. Calpain 3 (p94) mutations cause limb girdle muscular dystrophy type 2A; however, a role in exercise proteolysis is only just emerging (Murphy & Lamb, 2009). It has been suggested that the fatigue of long duration that occurs after some forms of exhausting exercise could

be due to exercise-induced activation of calpain 1 (mu) and subsequent damage to the excitation–contraction coupling apparatus (Verburg et al., 2009). It remains unclear whether calpain 1 activation plays any role at all in the acute postexercise proteolysis activation; Murphy and colleagues (2007) found no evidence of activation following eccentric exercise in humans. Although the role of calpain 3 in exercise-induced muscle damage or repair is unclear, calpain 3 activation does occur 24 h following eccentric exercise (Murphy et al., 2007), at a time when calpain 3 mRNA is lower (Feasson et al., 2002). To date, there are no reports of calpain 1, 2, or 3 activation following endurance exercise; however, with the minimal changes in calpain 1 activation seen even after eccentric exercise (Murphy et al., 2007), a significant role in proteolysis activation following exercise (other than calpain 3 following eccentric) seems unlikely.

Calpain 1 mRNA abundance tends to decrease, while calpain 2 mRNA abundance tends to increase acutely following endurance exercise (Table 23.1); yet, they are usually unchanged following acute resistance exercise (Table 23.2). There does not appear to be an effect of endurance or resistance training on steady-state mRNA abundance for any of the calpains as determined by microarray (Tables 23.3 and 23.4).

The ubiquitin pathway has been considered to be a major player in models of muscle atrophy and wasting, especially during pathological conditions such as cancer cachexia and sepsis, following the discovery of atrogenes, atrogin-1/MAFbx and MuRF-1 (Bodine et al., 2001; Gomes et al., 2001; Lecker et al., 2004). Activation of the ubiquitin–proteosomal pathway is first initiated by targeting proteins for degradation with polyubiquitin tags, followed by proteolysis within the core proteosome. Ubiquitin molecules are linked to lysine residues through a series of pathways catalyzed by three enzymes, termed E1, E2, and E3, and are then degraded by the 26S proteosome into peptides (Mitch & Goldberg, 1996). This pathway is activated during starvation and muscle atrophy (Bodine et al., 2001; Gomes et al., 2001; Jagoe et al., 2002), and maybe involved in remodeling of skeletal muscle (Murton et al., 2008). It appears that a major level of

Table 23.1 Changes in mRNA species related to protein turnover following endurance exercise

References	Study design	Gene	Direction of change	Timepoint
Coffey et al. (2006)	Acute, 60 min cycling at 70% VO$_2$peak in endurance-trained (ET) or strength-trained (ST) subjects	Atrogin-1	Increase	3 h in ST and ET
Feasson et al. (2002)	Acute, 30 min downhill running at 80% VO$_2$max, VL	Calpain 1	Decrease	24 h
		Calpain 2	Increase	24 h
		Calpain 3	Decrease	Immediately post and 24 h
Harber et al. (2009)	Acute, 45 min level running at 75% VO$_2$max, VL and SOL	Atrogin-1	Unchanged	
		MuRF-1	Increase	VL: 4 h
		FOXO3A	Increase	SOL: 24 h
Louis et al. (2007)	Acute, 30 min level running at 75% VO$_2$max, LG	Atrogin-1	Increase	1, 2, and 4 h
		MuRF-1	Increase	1, 2, 4, and 24 h
		FOXO3A	Increase	1 h
		Calpain 1	Decrease	1 h
		Calpain 2	Increase	24 h

VL, vastus lateralis; SOL, soleus; LG, lateral gastrocnemius.

Table 23.2 Changes in mRNA species related to protein turnover following resistance exercise

References	Study design	Gene	Direction of change	Timepoint
Drummond et al. (2009)	Acute RT (8 sets of 10 reps at 70% 1RM + EAA) in young (29 years) and old (70 years)	REDD1	Decrease	Post 6 h, old
		TSC1	Decrease	Post 6 h, old
		REDD2	Decrease	Post 3 and 6 h, old and young
		TSC2	Decrease	Post 3 and 6 h, old
		Rheb	Increase	Post 3 and 6 h, young
		cMyc	Increase	Post 6 h, old and young
		TCTP	Unchanged	
		mTOR	Unchanged	
		S6K1	Unchanged	
		hVPS34	Unchanged	
		MAP4K3	Unchanged	
Drummond et al. (2008)	Acute RT with or without blood flow restrictions (75 repetitions over 4 sets at 20% 1RM), crossover design	HIF-1α	Increase	Post 3 h
		REDD1	Decrease	
		mTOR	Unchanged	
		S6K	Unchanged	
		Atrogin-1	Unchanged	
		MuRF-1		

EAA = essential amino acid.

Table 23.2 (Continued)

References	Study design	Gene	Direction of change	Timepoint
Lamas et al. (2009)	Chronic RT (8 weeks, 3 days/week) of either strength (ST; 4–10RM) or power (PT; 30–60% 1RM) training in young males (ST = 22.5 years, PT = 24.2 years)	*RAPTOR*	Increase	Post 48 h ST and PT
		mTOR	Increase	ST and PT
		RICTOR	Increase	ST and PT
		4EBP-1	Decrease	ST and PT groups pooled
		Calcineurin	Unchanged	
		Calcipressin	Increase	ST and PT
Kubica et al. (2005)	Acute RT, rats lifting bodyweight plus weight vest, four sessions over 8 days	*eIF2Bε*	Increase	Post 16 h
		eIF2Bδ	Unchanged	
Zanchi et al. (2009)	Chronic RT (12 weeks), rats 12 weeks (48 sessions), concentric contractions 80–95% voluntary max	*GSK3β*	Unchanged	Post 24 h
		4EBP-1	Unchanged	
		eIF2Bε	Unchanged	
		MuRF-1	Decrease	
		Atrogin-1	Decrease	
Lamon et al. (2009)	Chronic RT 8 weeks, 3 days/week, low (3–5RM) or high (20–28RM), groups collapsed for analysis	STARS	Increase	Post 48–72 h (groups pooled)
		MRTF-A	Increase	
		MRTF-B	Increase	
		SRF	Increase	
Mascher et al. (2008)	Acute RT, two bouts 48 h apart (4 sets of 10RM at 80% 1RM)	Atrogin-1	Decrease	Post 2 h each bout
		MuRF-1	Increase	Post 2 h each bout
Raue et al. (2007)	Acute RT in old (85 years) and young (23 years) women (3 sets of 10 reps at 70% 1RM)	*FOXO3A*	Unchanged	Post 4 h
		TNFα	Unchanged	
		Atrogin-1	Increase	Post 4 h, old
		MuRF-1		Post 4 h, old and young
Coffey et al. (2006)	Acute RT (8 sets of 5RM, leg extensions) in endurance-trained (ET) or strength-trained (ST) subjects	Atrogin-1	Unchanged	Post 3 h, ET and ST
Yang et al. (2006)	Acute RT, 3 sets of 10 reps at 65%1RM, MHC1 versus MHC2a	Calpain 1	Unchanged	
		Calpain 2	Unchanged	
		MuRF-1	Increased	Post 4 h
		Atrogin-1	Decreased	Post 24 h, MHC1
Churchley et al. (2007)	Acute RT (8 sets of 5 reps at 80% 1RM) in normal or glycogen-depleted state	Atrogin-1	Decrease	Post 3 h, normal
		MuRF	Unchanged	
Louis et al. (2007)	Acute RT (3 sets of 10 reps at 70% 1RM)	Calpain 1	Unchanged	
		Calpain 2	Unchanged	
		MuRF-1	Increase	Post 1, 2, 4 h
		FOXO3A	Decrease	Post 8, 12 h (trend)
		Atrogin-1	Decrease	Post 8, 12 h

(Continued)

Table 23.2 (Continued)

References	Study design	Gene	Direction of change	Timepoint
Deldicque et al. (2008)	Acute RT (10 sets of 10 repetitions at 80% 1RM) creatine or placebo	Calpain 1	Increase	Post 72 h
		Atrogin-1	Increase	Immediately post
		Atrogin-1	Decrease	Post 24 h
Leger et al. (2006)	Chronic RT 8 weeks, 3 days/week, low (3–5RM) or high (20–28RM), groups collapsed for analysis	Atrogin-1	Increase	Post 48–72 h
		MuRF	Increase	
		Ubiquitin	Increase	
		TNFα	Unchanged	
		Proteosome 20 s α-1	Increased	

RT, resistance training; 1RM = 1 repetition maximal strength; EAA, essential amino acid; MHC = myosin heavy chain.

control is activation of atrogin-1 and MuRF-1 (both E3 ligases) (Bodine et al., 2001; Lecker et al., 2004). Caspase-3 activation appears to be a prerequisite for initiation of ubiquitin–proteosome-mediated proteolysis of myofibrillar proteins (Du et al., 2004).

Human exercise literature evaluating the role of the ubiquitin–proteosome pathway has been limited from a lack of specific antibodies for MuRF-1 and atrogin-1 protein content. Consequently, most studies have quantified the mRNA abundance of MuRF-1, atrogin-1, and other components of the ubiquitin–proteosome pathway in response to acute and chronic exercise. In general, atrogin-1 mRNA abundance decreases, while MuRF-1 mRNA abundance increases following an acute bout of resistance exercise (Table 23.2). Similarly, MuRF-1 increases and atrogin-1 either did not change or increased following acute endurance exercise (Table 23.2). MuRF-1 or atrogin-1 mRNA was not represented in any of the acute or chronic microarray studies involving either endurance or resistance exercise (Tables 23.3 and 23.4). Although a number of other components of the ubiquitin–proteosome pathway were seen in the microarray studies, there was no consistency in the changes (Tables 23.3 and 23.4), in contrast to that seen in the wasting/atrophy literature (Bodine et al., 2001; Lecker et al., 2004).

Amino Acid Oxidation in Skeletal Muscle

Although skeletal muscle can oxidize eight amino acids (alanine, asparagine, aspartate, glutamate, isoleucine, leucine, lysine, and valine) (Goldberg & Chang, 1978), the branched-chain amino acids (BCAAs: isoleucine, leucine, and valine) are preferentially oxidized during acute endurance exercise (Goldberg & Chang, 1978; Wolfe et al., 1984). Studies have shown that acute endurance exercise increases leucine oxidation (Bowtell et al., 2000; Hamadeh et al., 2005; Lamont et al., 1995, 2001a; McKenzie et al., 2000; Phillips et al., 1993), and that it is intensity dependent (Lamont et al., 2001b). Acute increases in leucine oxidation during endurance exercise return to baseline in the first hour after exercise (Phillips et al., 1993). In contrast, we have shown that leucine oxidation is minimal with acute resistance exercise (Tarnopolsky et al., 1991). Leucine oxidation was lower in women as compared to men during endurance exercise (Phillips et al., 1993), and women had lower leucine oxidation at rest and during exercise before and after a 38-day endurance exercise training program (McKenzie et al., 2000). 17-β-estradiol partly mediates this effect (Hamadeh et al., 2005); however, it is likely due to greater lipid oxidation leading to sparing of CHO and amino acid oxidation (Tarnopolsky, 2008). These findings suggest that dietary protein recommendations (Table 23.5) maybe 10–20% higher for men versus women endurance athletes.

BCAAs are transaminated to their keto acids by branched-chain aminotransferase and are then oxidized by the rate-limiting branched-chain oxo-acid dehydrogenase enzyme (BCKD) (Boyer & Odessey, 1991; Khatra et al., 1977). The BCKD enzyme is a

Table 23.3 Endurance exercise and protein turnover mRNA abundance: Summary of microarray results

References	Study design	Microarray	Gene			Function		Fold change
Stepto et al. (2009)	Training—Young endurance-trained cyclists (~8 years of cycling 250–600 km/week; n = 7) versus control (n = 7)—resting biopsy = cross-sectional endurance training evaluation	Human 8K, glass microarray, WEHI, Melbourne, Australia	LIM domain 7	LMO7	H22826	Protein ubiquitination	UP	2.94
			Glutamic-oxaloacetic transaminase 1, soluble	GOT1	H22856	Amino acid metabolism	UP	11.21
			Branched-chain keto acid dehydrogenase E1, beta polypeptide	BCKDHB	AA427739	Branched-chain family amino acid catabolism	UP	1.43
			Metaxin 1	MTX1	AA670347	Protein transport	UP	1.24
			Granzyme M (lymphocyte met-ase1)	GZMM	AI124941	Proteolysis	UP	1.4
			Kallikrelin-related peptidase 11	KLK11	AA477283	Proteolysis	UP	1.52
			Methionine sulfoxide reductase A	MSRA	AI253074	Protein metabolism; methionine metabolism	UP	1.64
			6-Pyruvoyltetrahydro protein synthase	PTS	AA877347	Amino acid metabolism; L-phenylalanine catabolism	UP	1.26
			Small optic lobes homolog (Drosophila)	SOLH	AI796687	Proteolysis; calpain activity	UP	1.48
			Ubiquitin carboxyl-terminal esterase L3 (ubiquitin thioesterase)	UCHL3	N27190	Ubiquitin-dependent protein catabolism	UP	1.8
			Eukaryotic translation initiation factor 3, subunit 1 alpha, 35 kDa	EIF3S1	AA455070	Protein biosynthesis	UP	2.02
			Cathepsin G	CTSG	W92603	Proteolysis	DOWN	0.75
			Aminoacylase 1	ACY1	AA402915	Proteolysis; amino acid metabolism	DOWN	0.85

(Continued)

Table 23.3 (Continued)

References	Study design	Microarray	Gene		Function		Fold change
			Proteasome subunit, alpha type, 5	PSMA5	AA598815	Ubiquitin proteolysis	DOWN	0.59
			Ribosomal protein L31	RPL31	W15277	Protein biosynthesis	DOWN	0.55
			Ribosomal protein L35a	RPL35A	AA873351	Protein biosynthesis	DOWN	0.56
			Ribosomal protein L7a	RPL7A	H23422	Protein biosynthesis	DOWN	0.55
			Ribosomal protein S14	RPS14	H73727	Protein translation	DOWN	0.62
			Ribosomal protein S7	RPS7	AW005693	Protein biosynthesis	DOWN	0.62
			Ubiquitin-activating enzyme E1-like	UBE1L	N23454	Ubiquitin pathway	DOWN	0.75
			Ubiquitin-conjugating enzyme E2L3	UBE2L3	AA669526	Ubiquitin-dependent protein catabolism	DOWN	0.73
			EMG1 nucleolar protein homolog (S. cerevisiae)	EMG1	AA701981	Translational elongation	DOWN	0.61
Schmutz et al. (2006)	Acute—6 young untrained men—30 min high-intensity endurance cycle—biopsies pre, +1, +8, +24 h postexercise biopsies	Custom-designed low-density Atlas® cDNA expression arrays (BD Biosciences)	Human metalloproteinase-2 inhibitor	TIMP2	J05593	Matrix degradation	UP at +8 h	2.6
			Human metalloproteinase-2 inhibitor	TIMP3	U14394	Matrix degradation	UP at +8 h	1.42
			Membrane-type matrix metalloproteinase-2	MMP15	Z48482	Matrix degradation	UP at +8 h	1.18
			Membrane-type matrix metalloproteinase-2	MMP14	D26512	Matrix degradation	UP at +8 h	2.56
			Membrane-type matrix metalloproteinase-8	MMP8	J05556	Matrix degradation	UP at +8 h	1.11
			Human ubiquitin	UBC	M26880	Protein breakdown	UP at +8 h	1.64
			Ribosomal protein L13a	RPL13A	X56932	Protein breakdown	UP at +8 h	1.66

Mahoney et al. (2005)	Acute—14 young men biopsied before and after (+3 h and +48 h) exhaustive (~70 min) cycling ergometry	Custom DNA microarrays were prepared from a Research Genetics human cDNA library	Human ribosomal protein S9	RPS9	U14971	Protein breakdown	UP at +8 h	2.34
			CTP:phosphocholine cytidylyltransferase 1	CCT1	X55282	Protein breakdown	UP at +8 h	1.36
			Forkhead transcription factor O1A	FOXO1A	AA134749	Activates PDK4 and PGC1alpha		5.2 (+3 h), 1.1 (+48 h)
			Mitochondrial ribosomal protein L2	MRPL2	N94366	Mitochondrial protein translation		3.4 (+3 h), 5.5 (+48 h)

Training—these are training studies with samples taken before and after training OR in cross-sectional studies of untrained versus trained (indicated). Acute—there are acute single-bout studies with samples taken before and at various times after exercise.

Table 23.4 Resistance exercise and protein turnover mRNA abundance: Summary of microarray results

References	Study design	Microarray	Gene			Function		Fold change
Stepto et al. (2009)	Training—Young strength-trained power lifters (>8 years of training; n = 6) versus control (n = 7); men—resting biopsy = cross-sectional resistance training evaluation	Human 8K, glass microarray, WEHI, Melbourne, Australia	*LMO7*	LIM domain 7	H22826	Protein ubiquitination	UP	4.46
			GOT1	Glutamic-oxaloacetic transaminase 1, soluble (aspartate aminotransferase 1)	H22856	Amino acid metabolism	UP	3.76
			BCKDHB	Branched-chain keto acid dehydrogenase E1, beta polypeptide	AA427739	Branched-chain amino acid catabolism	UP	1.23
			MTX1	Metaxin 1	AA670347	Protein transport	UP	1.83
			GZMM	Granzyme M (lymphocyte met-ase1)	AI124941	Proteolysis	UP	2.14
			KLK11	Kallikrein-related peptidase 11	AA477283	Proteolysis	UP	2.2
			MSRA	Methionine sulfoxide reductase A	AI253074	Protein metabolism	UP	1.84
			PTS	6-Pyruvoyltetrahydroprotein synthase	AA877347	Amino acid metabolism	UP	1.45
			SOLH	Small optic lobes homolog (*Drosophila*)	AI796687	Proteolysis; calpain activity	UP	1.58
			UCHL3	Ubiquitin carboxyl-terminal esterase L3 (ubiquitin thioesterase)	N27190	Ubiquitin degradation	UP	2.3
			EIF3S1	Eukaryotic translation initiation factor 3, subunit 1 alpha, 35 kDa	AA455070	Protein biosynthesis	UP	3.17
			CTSG	Cathepsin G	W92603	Proteolysis	DOWN	0.69
			ACY1	Aminoacylase 1	AA402915	Amino acid metabolism	DOWN	0.75
			PSMA5	Proteasome subunit, alpha type, 5	AA598815	Protein ubiquitination	DOWN	0.46
			RPL31	Ribosomal protein L31	W15277	Protein biosynthesis	DOWN	0.5
			RPL35A	Ribosomal protein L35a	AA873351	Protein biosynthesis	DOWN	0.44

Study	Protocol	Platform	Gene name	Symbol	Accession	Function	Direction	Fold
			Ribosomal protein L7a	*RPL7A*	H23422	Protein biosynthesis	DOWN	0.53
			Ribosomal protein S14	*RPS14*	H73727	Protein translation	DOWN	0.6
			Ribosomal protein S7	*RPS7*	AW005693	Protein biosynthesis	DOWN	0.44
			Ubiquitin-activating enzyme E1-like	*UBE1L*	N23454	Ubiquitin cycle	DOWN	0.67
			Ubiquitin-conjugating enzyme E2L 3	*UBE2L3*	AA669526	Protein ubiquitination	DOWN	0.77
			EMG1 nucleolar protein homolog (*S. cerevisiae*)	*EMG1*	AA701981	Translational elongation	DOWN	0.44
Melov et al. (2007)	Training—Healthy older adults (*n* = 14; 6 men and 8 women); whole-body resistance training (3 sets of 10 repetitions; twice a week for 26 weeks; progressed from one set of each exercise at 50% of the initial 1RM to 3 sets at 80% of new 1RM over the training period); pre versus post vastus lateralis biopsy obtained 72 h after last bout of exercise	Human Ref-8 BeadChips, Illumina	Mitochondrial ribosomal protein S33 (MRPS33)	*MRPS33*	NM_053035.1	Mitochondrial protein translation	UP	1.31
			Mitochondrial ribosomal protein L34 (MRPL34)	*MRPL34*	NM_023937.2	Mitochondrial protein translation	UP	1.21
			Eukaryotic translation initiation factor 2C, 4	*EIF2C4*	NM_017629.2	Mitochondrial protein translation	UP	1.08
			Mitochondrial ribosomal protein L41	*MRPL41*	NM_032477.1	Mitochondrial protein translation	UP	1.26
			Mitochondrial ribosomal protein L35	*MRPL35*	NM_145644.1	Mitochondrial protein translation	UP	1.14
			Mitochondrial ribosomal protein L13	*MRPL13*	NM_014078.4	Mitochondrial protein translation	UP	1.30
			Mitochondrial ribosomal protein S28	*MRPS28*	NM_014018.2	Mitochondrial protein translation	UP	1.20
			Mitochondrial ribosomal protein S16	*MRPS16*	NM_016065.2	Mitochondrial protein translation	UP	1.24
			Mitochondrial ribosomal protein S5	*MRPS5*	NM_031902.3	Mitochondrial protein translation	UP	1.20

(Continued)

Table 23.4 (Continued)

References	Study design	Microarray	Gene		Function	Fold change
			Mitochondrial ribosomal protein L49	MRPL49	NM_004927.2	Mitochondrial protein translation	DOWN 0.89
			Mitochondrial translation optimization 1 homolog	MTO1	NM_012123.1	Mitochondrial protein translation	DOWN 0.93
Mahoney et al. (2008)	Acute—Young healthy men (n = 3, n = 15 RT-PCR)—eccentric contraction (30 sets of 10 repetitions separated by 1 min); pre versus post +3 and +48 h vastus lateralis biopsy	Custom cDNA microarrays, Buck Institute of Age Research, Novato, CA, USA	Glutamyl-prolyl-tRNA synthetase	EPRS	Hs. 497788	Participates in translation	UP at +48 h 2.60
			Mitochondrial ribosomal protein L32	MRPL32	Hs. 50252	Mitochondrial protein synthesis	UP at +3 h 3.30
			Mitochondrial ribosomal protein L2	MRPL2	Hs.55041	Mitochondrial translation	UP at +3 h and +48 h 6.8 (+3 h) and 5.0 (+48 h)
			Ubiquitin-specific protease 2	USP2	Hs.524085	E4 ubiquitin recycling enzyme	UP at +3 h 7.40
			Cullin 1	CUL1	Hs.146806	E3 ubiquitin ligase	UP at +3 h 4.40
			Ubiquitin-activating enzyme E1-like	UBE1L2	Hs.16695	UPS-mediated proteolysis	UP at +48 h 2.30
			Proteasome 26S subunit 3	PSMC3	Hs.12970	Regulatory subunit of the 26 proteasome	DOWN at +3 h 0.40
			Ubiquitin-specific protease 1	USP1	Hs.35086	Protein degradation	DOWN at +3 h 0.30
			E2 ubiquitin-conjugating enzyme	UBE2N	Hs.529420	Ubiquitin ligase	DOWN at +48 h 0.30
Kostek et al. (2007)	Acute—Young healthy men (n = 3); eccentric contractions (one leg) and concentric contractions; pre versus post +3, +6, and +24 h biopsies	Custom-made Affymetrix MuscleChip, Santa Clara, CA, USA	Ribosomal protein S20	RPS20	NM_001023.2	Protein translation	UP at +24 h 1.4
			Ribosomal protein SA	RPSA	NM_001012321	Protein biosynthesis	UP at +3 h 1.3
			Ribosomal protein S17	RPS17	NM_001021	Protein biosynthesis	UP at +24 h 1.2

Study	Methods	Gene name	Symbol	Accession	Function	Response	Fold
		Ribosomal protein L15	RPL15	NM_002948	Protein biosynthesis	UP at +24 h	1.4
		Ribosomal protein S24	RPS24	NM_001026	Protein biosynthesis	UP at +3 h	1.6
		F-box protein 32	FBXO32	NM_058229	Protein breakdown	DOWN at +6 h	−3.4
		Ribosomal protein S27	RPS27	NM_001030	Protein biosynthesis	DOWN at +6 h	−1.2
		Ribosomal protein S28	RPS28	NM_001031	Protein biosynthesis	DOWN at +6 h	−1.3
Chen et al. (2003)	Acute—Young healthy men ($n = 3$); 300 eccentric contractions; pre versus post vastus lateralis biopsy + 4, + 6 h 12,000 gene microarray; Affymetrix U95Av2 Santa Clara, CA, USA	Cardiac ankyrin-repeated protein	CARP	NM_01439	Skeletal muscle hypertrophy	UP	7.5

Training—these are training studies with samples taken before and after training OR in cross-sectional studies of untrained versus trained (indicated). Acute—there are acute single-bout studies with samples taken before and at various times after exercise.

Table 23.5 Estimated protein requirements for athletes and nonathletes

Group	Protein intake (g/kg/day)
Sedentary men and women	0.80–1.0 or ~10% of dietary energy
Elite male endurance athletes	1.5–1.7
Moderate-intensity endurance athletes[a]	1.1–1.4
Recreational endurance athletes[b]	0.80–1.0 or ~10% of dietary intake
Football, power sports	1.5–1.8
Resistance athletes (early training)	1.5–1.7
Resistance athletes (steady state)	1.1–1.4
Women athletes[c]	~10–20% lower than men athletes

[a] Exercising approximately four/five times per week for 45–60 min.
[b] Exercising four to five times per week for 30 min at <55% VO$_2$peak.
[c] Estimated from Phillips et al. (1993).

supramolecular complex (metabolom) consisting of dihydrolipoyl transacylase subunits (E2), branched-chain keto acid decarboxylase/dehydrogenase (E1), dihydrolipoamide dehydrogenase (E3), BCKD kinase, and BCKD phosphatase (Islam et al., 2007). The complex is less active when phosphorylated by BCKD kinase at a serine residue in the E1 complex (Ser-292) (Islam et al., 2007). About 5–8% of skeletal muscle BCKD is active at rest. This increases to 20–25% with acute endurance exercise (McKenzie et al., 2000; Wagenmakers et al., 1989) and is correlated with leucine oxidation during acute endurance exercise (McKenzie et al., 2000). The activation of BCKD by acute exercise is thought to be related to a lower ATP/ADP ratio, higher intramuscular acidity muscle glycogen depletion (Kasperek, 1989; Wagenmakers et al., 1991). The BCKD metabolom will not form in the presence of high mitochondrial energy charge (NADH/NAD$^+$ ratio) (Islam et al., 2007), suggesting that with lower energy during exercise, the BCKD complex will become functionally competent and increase BCAA oxidation. We measured

BCKD and BCKD kinase mRNA before and immediately after 90 min of endurance exercise in men and women and did not see an acute effect (Fu et al., 2009), nor did we see any changes 3 h after acute endurance exercise using a microarray approach (Mahoney et al., 2005). The above data indicate that the acute exercise-induced activation of BCAA oxidation is at the posttranslational level in skeletal muscle.

The total amount of BCKD protein and total activity increases following endurance exercise training (Howarth et al., 2007; McKenzie et al., 2000), which may increase the total "capacity" for amino acid oxidation in highly trained endurance athletes. We evaluated the effect of 38 days of endurance exercise training on acute BCKD activation following an acute bout in men and women (McKenzie et al., 2000). Leucine oxidation during exercise was lower following training, and this was co-temporal with lesser BCKD activation (McKenzie et al., 2000). Another group also reported that whole-body amino acid oxidation was lower following endurance exercise training (Gaine et al., 2005). Resting BCKD protein is higher following endurance training, and the attenuation of acute activation was associated with higher BCKD kinase protein content (Howarth et al., 2007). Taken together, these data are consistent with the fact that endurance training attenuates amino acid oxidation during exercise. However, with the greater total BCKD activity and content seen following endurance exercise training (Howarth et al., 2007; McKenzie et al., 2000), it is possible that highly trained athletes training for long hours at a high relative intensity could flux enough amino acids through this pathway to negatively impact dietary protein requirements. To date, there is no data on the steady-state mRNA abundance for either BCKD or BCKD kinase following endurance exercise training (see Table 23.4).

An emerging area of interest to exercise physiologists is the potential for microRNAs (miRNAs) to regulate gene translation. miRNAs are small noncoding molecules that are highly conserved (Grimson et al., 2008). They can be transcribed from noncoding intronic regions of the gene that they regulate or from their own gene as primary transcripts cleaved by Drosha in the nucleus. The pre-miRNAs are exported to the cytosol by Exportin-5 where they are cleaved into 21–24 nucleotide miRNAs by the enzyme Dicer. The miRNAs are part of a RNA-silencing complex (RISC) where the miRNA can bind to the 3′ UTR of the mRNA and either block translation at the ribosome (imperfect miRNA/target RNA match) or target the mRNA for degradation (perfect miRNA match). It has been shown that human miR29b targets the mRNA for the dihydrolipoyl branched-chain acyltransferase core (E2) of the BCKD complex and prevents its translation (Mersey et al., 2005). We have recently shown that acute endurance exercise does alter the abundance of several miRNAs in rodent skeletal muscle (Safdar et al., 2009); however, we did not look at miR29b in that study.

Regulation of Protein Synthesis in Response to Exercise in Skeletal Muscle

Protein synthesis starts with a signal initiating mRNA transcription from DNA that is eventually translated into a functional protein. The basic aspects of nuclear and mitochondrial gene transcription and translation have been covered in previous chapters. I shall focus on how myofibrillar and mitochondrial protein synthesis is influenced by acute resistance and endurance exercise and the response to training. Once in the cytosol, the mRNA is translated into a protein through the process of translation by the ribosomes that exist free in the cytosol or bound to rough endoplasmic reticulum. The process of translation requires that transfer RNAs (tRNAs) be combined with their respective amino acids via specific tRNA synthases to form amino-acyl-tRNA complexes (e.g., leucyl-tRNA and histidinyl-tRNA). Within the ribosome scaffolding, mRNA codons are read by the specific tRNA anticodons and deliver the specific amino acids. The process of translation requires three steps: initiation, elongation, and termination.

Endurance exercise training leads to an increase in mitochondrial biogenesis (Holloszy & Booth, 1976), and resistance training leads to skeletal muscle hypertrophy (Lemon et al., 1992). In apparent contrast, studies using mixed muscle protein synthesis have shown similar increases in the protein synthetic rate following a single bout of both endurance-like (Miller et al., 2005) and resistance (Chesley et al., 1992; Miller et al., 2003; Phillips et al., 1997, 1999; Tipton et al., 2001) exercise. More recent data are more in-line with the phenotypic responses with myofibrillar and mitochondrial fraction analysis showing that acute resistance exercise promotes synthesis of both fractions, while acute endurance exercise only stimulates mitochondrial protein synthesis (Wilkinson et al., 2008).

The increase in muscle protein synthesis following an acute bout of resistance exercise is measurable after 3–4 h and persists for up to 72 h (Chesley et al., 1992; Phillips et al., 1997). Resistance exercise training results in an increase in the basal mixed muscle protein synthetic rate with an attenuation of the "pulse" amplitude and duration of the response (Phillips et al., 2002), with the myofibrillar protein synthesis rate still showing similar postexercise increases following resistance training (Kim et al., 2005). The acute increase in mixed and myofibrillar protein synthesis that occurs in the hours following exercise is likely mediated at the level of translation.

Adenosine monophosphate-activated protein kinase (AMPK) is generally considered to be a signaling molecule that senses cellular energy charge and is activated (phosphorylation) under conditions of low cellular energy charge (e.g., high AMP content). Activated AMPK can increase lipid oxidation and glucose transport, increase mitochondrial biogenesis (Hardie, 2004; Jorgensen et al., 2006; Richter & Ruderman, 2009), and reduce protein and lipid synthesis (Matsakas & Patel, 2009; Richter & Ruderman, 2009). In general, AMPK pathway activation is felt to be an important mediator of the adaptive response to endurance exercise training.

The mTOR pathway increases protein synthesis with mTOR phosphorylation being activated by upstream kinases including the insulin signaling pathway (Matsakas & Patel, 2009). In general, it appears that the mTOR pathway and the AMPK pathway work in opposite directions, and activation of the AMPK pathway can inhibit mTOR signaling (Deldicque et al., 2005). mTOR is activated by amino acids (primarily leucine) or insulin that leads to activation of downstream kinases including p70S6K1 and its downstream target, rpS6 (Bolster et al., 2003). Activation of this pathway leads to a reduction in the phosphorylation of epsilon subunit of eukaryotic initiation factor 2B (eIF2Bε) and activates translation.

Both resistance and endurance exercise activate the Akt-mTOR-p70S6K1 pathway following exercise. This effect is enhanced, at least for resistance exercise, in the presence of protein/amino acids (Glover et al., 2008; Hulmi et al., 2009). One study used electrical stimulation of rat muscle *in situ* to mimic an endurance pattern (low frequency) and a resistance exercise pattern (high frequency), finding AMPK-phosphorylation to increase with the low-frequency paradigm, while p70S6k, 4E-BP1, eIF2B, AMPK, and eEF2 responded in a prolonged fashion to the high-frequency stimulation (Atherton et al., 2005). The latter group coined the term "AMPK-PKB switch" to indicate a critical pathway distinction leading to hypertrophy (PKB-TSC2-mTOR) or mitochondrial biogenesis (AMPK-PGC-1α) (Atherton et al., 2005). Others have found that p70S6k activation is less robust and more transient with endurance versus resistance exercise (Benziane et al., 2008; Mascher et al., 2007; Wilkinson et al., 2008).

Although the balance of evidence indicates that acute changes in signaling molecule activation is the major factor mediating the changes in protein synthesis following exercise, mRNA abundance also changes. Microarray data indicate that a few mRNA species of relevance to protein synthesis are induced in the 3–24 h following eccentrically biased resistance-type exercise (Table 23.4). Using targeted analysis Reverse transcriptase-polymerase chain reaction (RT-PCR), it appears that mTOR mRNA abundance is not induced, nor are there any directionally consistent changes in transcript abundance for other relevant mRNA species (e.g., eIF2B) (Table 23.3). Using a microarray approach reveals that some transcripts are acutely induced following endurance exercise directionally consistent with an increase in mitochondrial biogenesis (e.g., FOXO1A and MRPL2), but none related to the general protein synthetic machinery (Tables 23.1 and 23.3).

At the overall protein synthesis level, the response to differing modes of contraction in specific protein fractions becomes more specialized

with exercise training. The acute stimulation of myofibrillar, but not mitochondrial, protein synthesis is retained after resistance exercise training, while endurance exercise training continues to stimulate mitochondrial protein synthesis exclusively (Wilkinson et al., 2008). Paradoxically, the activation of the mTOR-Akt-p70 pathway and AMPK did not display any exercise mode-specific effects following training, although resistance exercise-induced activation of p70s6k does become more transient (Wilkinson et al., 2008). The baseline (resting) phosphorylation levels of some translational signaling molecules (e.g., Akt, GSK3-β, eIF4E, and mTOR) are enhanced following both modes of training (Leger et al., 2006; Wilkinson et al., 2008).

In contrast to the apparent lack of pre-translational control over the acute protein synthetic response following endurance or resistance exercise, directional changes in mRNA abundance following training are more consistent with the phenotypic response. The mRNA content for mTOR, RAPTOR, and RICTOR mRNA abundance are higher following resistance training (Lamas et al., 2009) (Table 23.2). Components of the electron transport chain were induced in men following endurance exercise training; however, the specific gene identifiers were not obtainable from that paper (Timmons et al., 2005). Microarray data regarding training-induced changes in transcripts encoding components of the translational machinery are limited, but data from a recent study of resistance and endurance-trained young subjects suggest that transcripts for ribosomal proteins are actually lower relative to control subjects for both exercise groups (Tables 23.3 and 23.4; Stepto et al., 2009). Older adults may be unique in showing less contraction-specific genetic and phenotypic response to training in that they exhibit an increase in mitochondrial enzyme activity following resistance training (Parise et al., 2005), and a robust increase in mRNA transcript abundance for many components of mitochondrial structure/function (Melov et al., 2007). In the latter study, there were also increases in transcripts consistent with the hypertrophy phenotype commonly observed following resistance exercise training including ribsomal proteins and eIF transcripts (Melov et al., 2007) (Table 23.4).

In spite of the extensive amount of research evaluating signaling molecules and the control of protein translation during different types of exercise, many questions remain. One of the most important is to delineate the role of the satellite cells in the hypertrophic response to resistance exercise. Satellite cells represent about 3–5% of the myonuclei in mature human skeletal muscle and are activated following resistance exercise (Olsen et al., 2006). All of the signaling molecules and RNA work cited earlier have been completed on mixed muscle homogenates, and it is impossible to delineate the contribution from satellite cells. The latter fact that is important for animal studies have clearly shown that compensatory hypertrophy of skeletal muscle in response to overload does not occur in the absence of satellite cells (Rosenblatt et al., 1994). Another important area for future research is to evaluate the role of miRNA in the regulation of muscle hypertrophy following resistance exercise. MicroRNA 1, 133a, and 206 are all expressed in skeletal muscle (miR 206 exclusively), important in muscle differentiation (Anderson et al., 2006), and decrease (miR 1 and miR 133a) following muscle hypertrophy (McCarthy & Esser, 2007). We have recently shown that a number of miRNAs change following acute endurance exercise in mice and that the changes in miRNA are directionally consistent with corresponding changes in predicted mRNA targets (e.g., lower miR 23 and higher PCG-1α) (Safdar et al., 2009).

Putting It All Together—Dietary Protein Requirements

The athlete is interested in how exercise training and acute exercise influence protein metabolism and what effect this has on training and dietary strategies. The determinants of protein requirements are almost infinite and include exercise intensity, duration, environmental factors, timing of postexercise nutrition, training status, sex, and likely by a variety of inherited (single nucleotide polymorphisms, SNPs) and other genetic factors. Understanding the molecular regulation of protein

turnover allows one to adjust protein requirement levels and dietary strategies for a given athlete. For the endurance athlete, the ultimate goal is to maximize cardiac output and mitochondrial biogenesis, which are the two main determinants of maximal aerobic power ($\dot{V}O_2$max = delivery of oxygen $\times$ peripheral oxygen uptake), while maintaining a lean body composition for optimal biomechanical advantage. In strength athletes such as power lifters, heavyweight wrestlers, sumo wrestlers, or linemen in American football or rugby, the goal is for maximal muscle mass strength irrespective of body size. For optimal performance in athletes who engage in short sprints of high intensity, such as running backs, sprinters, team sports, and jumpers, it is desirable to have high strength and power with relatively low percent body fat for optimal performance.

The optimal protein requirement for an athlete would provide the optimal amount of amino acids for protein synthesis, yet not an excessive amount that would lead to excessive oxidation and result in the displacement of other macronutrients, such as CHO, which are important in muscle glycogen storage. Protein requirements for humans have traditionally been determined using a "black box" approach with the nitrogen balance (NBAL) technique (Pellet, 1990). The premise of NBAL is that all of the nitrogen coming into a human via food and all of the nitrogen leaving the human body (urine, feces, sweat, and miscellaneous losses) are quantified, and the protein requirement is that where net balance is zero ([intake − output] × 2 SD safety margin). NBAL provides a "safe" intake value required to achieve a zero NBAL without providing any information on how that balance was achieved at the molecular level.

Protein Requirements for Resistance Athletes

We evaluated dietary protein requirements in varsity team sport players who are also completing several bouts of weekly resistance exercise and found that a dietary protein intake at 2.8 g/kg/day versus 1.4 g/kg/day was associated with higher amino acid oxidation and no change in protein synthesis (Tarnopolsky et al., 1992). The same athletes completed a NBAL study, and our protein intake estimate based on NBAL experiments for these athletes was a safe estimate and was found to be 1.76 g/kg/day (Tarnopolsky et al., 1992). We also evaluated NBAL after initiation of a heavy weight training program (6 days/wk, 2 h/day) and used NBAL to determine a safe protein intake of 1.65 g/kg/day (Lemon et al., 1992). Together, these studies demonstrated that athletes during the early phase of a very intensive weight training program, that is those who were doing three to four sessions of weight training plus varsity level football and rugby drills required ~1.7 gPRO/kg/day, and that the stabile isotope tracer methods agreed nicely with the NBAL data (Table 23.5).

There is also evidence for an adaptation to steady-state resistance training since we have shown that bodybuilders who have resistance trained for many years required only 1.2 gPRO/kg/day (Tarnopolsky et al., 1988), and others have shown that those participating in fairly recreational endurance and mixed endurance resistance protocols required between 1.2 gPRO/kg/day and 1.4 gPRO/kg/day of mixed source protein (Consolazio et al., 1975; Torun et al., 1977). There is no experimental evidence to support that protein intakes greater than 2 g/kg/day are any more effective at increasing or maintaining muscle mass or strength in resistance athletes versus the earlier recommended intakes.

Protein Requirements for Endurance Athletes

Three studies have used NBAL technology to evaluate protein requirements in top sport endurance-trained athletes using NBAL (Brouns et al., 1989; Friedman & Lemon, 1989; Tarnopolsky et al., 1988). The estimated safe protein requirement from the three aforementioned studies ranged from 1.49 g/kg/day to 1.8 g/kg/day. These are likely the highest dietary intake levels required for any endurance athlete given that one study was completed as a simulated *Tour de France* cycling study (Brouns et al., 1989), and in another, six elite male endurance athletes with a mean $\dot{V}O_2$max of 76.2 ml/kg/min were evaluated (Tarnopolsky et al., 1988). Athletes performing moderate-intensity endurance

activities require only marginally more protein intake than US/Canadian/European requirements with values ranging from 1.0 gPRO/kg/day to 1.2 gPRO/kg/day (Forslund et al., 1998; Meredith et al., 1989; Phillips et al., 1993).

A summary of the estimated safe protein requirements is provided in Table 23.5; however, it is critical to note that these are mean values on a mixed diet, and a wide variety of factors will increase or decrease the requirement. In general, protein requirements are *positively* impacted by (lower requirement): female sex, high CHO intake, high energy intake (egg white, milk), high biological value proteins, timing of consumption (early postexercise); and *negatively* impacted by (higher requirement): low CHO intake, low energy intake (e.g., weight loss or esthetic sports), low biological value proteins, higher intensity and/or duration than seen in the published studies, during a period of increased training and intensity (i.e., training camp), and dehydration. It is important to note that most endurance and resistance athletes meet the protein requirements outlined in Table 23.5 with their normal diet.

Conclusions and Future Directions

Protein is an important dietary component since proteins are involved in almost every structural and functional component of the human body. Protein turnover involves protein synthesis, protein breakdown, and amino acid oxidation. Acute endurance exercise may negatively impact on the need for dietary protein by increasing the oxidation of amino acids, especially leucine. The acute regulation of amino acid oxidation during exercise is primarily at the posttranslational level through dephosphorylation of BCKD. Acute resistance exercise increases both protein synthesis and degradation for up to 48 h, with degradation returning to baseline more rapidly. The activation of protein degradation may involve the ubiquitin–proteosome pathway, but much more work is required in this area. The recent discovery of calpain 3 activation following eccentric exercise deserves further evaluation. The coadministration of protein (amino acids) during or after acute resistance exercise enhances the anabolic response via posttranslational control of signaling molecule activation (phosphorylation/dephosphorylation).Proteinrequirements follow the general rules of physiological adaptation in response to exercise training in that pulses of stress lead to compensatory changes at the molecular/cellular level that attenuate the threat to muscle homeostasis. Moderate-intensity endurance exercise training attenuates the acute exercise-induced increase in leucine oxidation in association with lesser activation of BCKD in response to acute exercise. Following endurance exercise training, the total capacity of the body to oxidize amino acids is increased; however, only in the elite athlete would this capacity ever be taxed and lead to a significant negative impact upon dietary protein requirements. Consequently, the maximal dietary PRO requirement for top sport endurance athletes would be in the order of ~1.6 g/kg/day. The maximal dietary PRO requirement for resistance-trained athletes either in the early stages of very intensive program or with concomitant sport practices (e.g., rugby and football) is at most 1.7 g/kg/day. A dietary PRO of 10–15% of total energy intake with an energy-sufficient diet should cover the PRO requirements for nearly all resistance and endurance athletes.

There are still many unanswered questions regarding the influence of different types of exercise on protein turnover both acutely and in response to chronic training. The microarray data are based on a very small number of subjects and, with newer oligonucleotide arrays covering most of the human genome, much information can be mined using such technology. Along these lines, there are no studies looking at the acute response of mRNA or signaling molecules and linking them to the phenotypic outcomes in large cohorts. There is an emerging understanding of the role of miRNA in the regulation of mRNA abundance and translational control, and this is an area ripe for future investigation. Other novel areas on the horizon include the role of SNPs in determining the acute and chronic (training) response to different modes of exercise. Emerging concepts such as the possibility that riboswitches may regulate metabolism through metabolite–ribozyme interactions is very exciting (Klein & Ferre-D'Amare, 2006).

References

Anderson, C., Catoe, H. & Werner, R. (2006). MIR-206 regulates connexin43 expression during skeletal muscle development. *Nucleic Acids Research* 34, 5863–5871.

Atherton, P.J., Babraj, J., Smith, K., Singh, J., Rennie, M.J. & Wackerhage, H. (2005). Selective activation of AMPK-PGC-1alpha or PKB-TSC2-mTOR signaling can explain specific adaptive responses to endurance or resistance training-like electrical muscle stimulation. *FASEB Journal* 19, 786–788.

Bartoli, M. & Richard, I. (2005). Calpains in muscle wasting. *International Journal of Biochemistry & Cell Biology* 37, 2115–2133.

Belcastro, A.N., Shewchuk, L.D. & Raj, D.A. (1998). Exercise-induced muscle injury: A calpain hypothesis. *Molecular and Cellular Biochemistry* 179, 135–145.

Benziane, B., Burton, T.J., Scanlan, B., et al. (2008). Divergent cell signaling after short-term intensified endurance training in human skeletal muscle. *American Journal of Physiology—Endocrinology and Metabolism* 295, E1427–E1438.

Bodine, S.C., Latres, E., Baumhueter, S., et al. (2001). Identification of ubiquitin ligases required for skeletal muscle atrophy. *Science* 294, 1704–1708.

Bolster, D.R., Kubica, N., Crozier, S.J., et al. (2003). Immediate response of mammalian target of rapamycin (mTOR)-mediated signalling following acute resistance exercise in rat skeletal muscle. *The Journal of Physiology* 553, 213–220.

Bowtell, J.L., Leese, G.P., Smith, K., et al. (2000). Effect of oral glucose on leucine turnover in human subjects at rest and during exercise at two levels of dietary protein. *The Journal of Physiology* 525(Pt 1), 271–281.

Boyer, B. & Odessey, R. (1991). Kinetic characterization of branched chain ketoacid dehydrogenase. *Archives of Biochemistry and Biophysics* 285, 1–7.

Brouns, F., Saris, W.H., Stroecken, J., et al. (1989). Eating, drinking, and cycling. A controlled Tour de France simulation study, Part I. *International Journal of Sports Medicine* 10(Suppl. 1), S32–S40.

Cathcart, E.P. (1925). Influence of muscle work on protein metabolism. *Physiological Reviews* 5, 225–243.

Chen, Y.W., Hubal, M.J., Hoffman, E.P., Thompson, P.D. & Clarkson, P.M. (2003). Molecular responses of human muscle to eccentric exercise. *Journal of Applied Physiology* 95, 2485–2494.

Chesley, A., MacDougall, J.D., Tarnopolsky, M.A., Atkinson, S.A. & Smith, K. (1992). Changes in human muscle protein synthesis after resistance exercise. *Journal of Applied Physiology* 73, 1383–1388.

Churchley, E.G., Coffey, V.G., Pedersen, D.J., et al. (2007). Influence of preexercise muscle glycogen content on transcriptional activity of metabolic and myogenic genes in well-trained humans. *Journal of Applied Physiology* 102, 1604–1611.

Coffey, V.G., Shield, A., Canny, B.J., Carey, K.A., Cameron-Smith, D. & Hawley, J.A. (2006). Interaction of contractile activity and training history on mRNA abundance in skeletal muscle from trained athletes. *American Journal of Physiology—Endocrinology and Metabolism* 290, E849–E855.

Consolazio, C.F., Johnson, H.L., Nelson, R.A., Dramise, J.G. & Skala, J.H. (1975). Protein metabolism during intensive physical training in the young adult. *American Journal of Clinical Nutrition* 28, 29–35.

Deldicque, L., Theisen, D. & Francaux, M. (2005). Regulation of mTOR by amino acids and resistance exercise in skeletal muscle. *European Journal of Applied Physiology* 94, 1–10.

Deldicque, L., Atherton, P., Patel, R., et al. (2008). Effects of resistance exercise with and without creatine supplementation on gene expression and cell signaling in human skeletal muscle. *Journal of Applied Physiology* 104, 371–378.

Drummond, M.J., Fujita, S., Abe, T., Dreyer, H.C., Volpi, E. & Rasmussen, B.B. (2008). Human muscle gene expression following resistance exercise and blood flow restriction. *Medicine and Science in Sports and Exercise* 40, 691–698.

Drummond, M.J., Miyazaki, M., Dreyer, H.C., et al. (2009). Expression of growth-related genes in young and older human skeletal muscle following an acute stimulation of protein synthesis. *Journal of Applied Physiology* 106, 1403–1411.

Du, J., Wang, X., Miereles, C., et al. (2004). Activation of caspase-3 is an initial step triggering accelerated muscle proteolysis in catabolic conditions. *The Journal of Clinical Investigation* 113, 115–123.

Feasson, L., Stockholm, D., Freyssenet, D., et al. (2002). Molecular adaptations of neuromuscular disease-associated proteins in response to eccentric exercise in human skeletal muscle. *The Journal of Physiology* 543, 297–306.

Forslund, A.H., Hambraeus, L., Olsson, R.M., El-Khoury, A.E., Yu, Y.M. & Young, V.R. (1998). The 24-h whole body leucine and urea kinetics at normal and high protein intakes with exercise in healthy adults. *The American Journal of Physiology* 275, E310–E320.

Friedman, J.E. & Lemon, P.W. (1989). Effect of chronic endurance exercise on retention of dietary protein. *International Journal of Sports Medicine* 10, 118–123.

Fu, M.H., Maher, A.C., Hamadeh, M.J., Ye, C. & Tarnopolsky, M.A. (2009). Exercise, sex, menstrual cycle phase and 17{beta}-estradiol influence metabolism related genes in human skeletal muscle. *Physiological Genomics* 22:4(7), e6335.

Gaine, P.C., Viesselman, C.T., Pikosky, M.A., et al. (2005). Aerobic exercise training decreases leucine oxidation at rest in healthy adults. *The Journal of Nutrition* 135, 1088–1092.

Glover, E.I., Oates, B.R., Tang, J.E., Moore, D.R., Tarnopolsky, M.A. & Phillips, S.M. (2008). Resistance exercise decreases eIF2Bepsilon phosphorylation and potentiates the feeding-induced stimulation of p70S6K1 and rpS6 in young men. *American Journal of Physiology—Regulatory, Integrative and Comparative Physiology* 295, R604–R610.

Goldberg, A.L. & Chang, T.W. (1978). Regulation and significance of amino acid metabolism in skeletal muscle. *Federation Proceedings* 37, 2301–2307.

Gomes, M.D., Lecker, S.H., Jagoe, R.T., Navon, A. & Goldberg, A.L. (2001). Atrogin-1, a muscle-specific F-box protein highly expressed during muscle atrophy. *Proceedings of the National Academy of Sciences of the United States of America* 98, 14440–14445.

Grimson, A., Srivastava, M., Fahey, B., et al. (2008). Early origins and evolution of microRNAs and Piwi-interacting RNAs in animals. *Nature* 455, 1193–1197.

Hamadeh, M.J., Devries, M.C. & Tarnopolsky, M.A. (2005). Estrogen supplementation reduces whole body leucine and carbohydrate oxidation and increases lipid oxidation in men during endurance exercise. *The Journal of Clinical Endocrinology and Metabolism* 90, 3592–3599.

Harber, M.P., Crane, J.D., Dickinson, J.M., et al. (2009). Protein synthesis and the expression of growth-related genes are altered by running in human vastus lateralis and soleus muscles. *American Journal of Physiology—Regulatory, Integrative and Comparative Physiology* 296, R708–R714.

Hardie, D.G. (2004). AMP-activated protein kinase: A key system mediating metabolic responses to exercise. *Medicine and Science in Sports and Exercise* 36, 28–34.

Holloszy, J.O. & Booth, F.W. (1976). Biochemical adaptations to endurance exercise in muscle. *Annual Review of Physiology* 38, 273–291.

Howarth, K.R., Burgomaster, K.A., Phillips, S.M. & Gibala, M.J. (2007). Exercise training increases branched-chain oxoacid dehydrogenase kinase content in human skeletal muscle. *American Journal of Physiology—Regulatory, Integrative and Comparative Physiology* 293, R1335–R1341.

Hulmi, J.J., Tannerstedt, J., Selanne, H., Kainulainen, H., Kovanen, V. & Mero, A.A. (2009). Resistance exercise with whey protein ingestion affects mTOR signaling pathway and myostatin in men. *Journal of Applied Physiology* 106(5), 1720–1729.

Islam, M.M., Wallin, R., Wynn, R.M., et al. (2007). A novel branched-chain amino acid metabolon. Protein-protein interactions in a supramolecular complex. *The Journal of Biological Chemistry* 282, 11893–11903.

Jagoe, R.T., Lecker, S.H., Gomes, M. & Goldberg, A.L. (2002). Patterns of gene expression in atrophying skeletal muscles: Response to food deprivation. *FASEB Journal* 16, 1697–1712.

Jorgensen, S.B., Richter, E.A. & Wojtaszewski, J.F. (2006). Role of AMPK in skeletal muscle metabolic regulation and adaptation in relation to exercise. *The Journal of Physiology* 574, 17–31.

Kasperek, G.J. (1989). Regulation of branched-chain 2-oxo acid dehydrogenase activity during exercise. *The American Journal of Physiology* 256, E186–E190.

Khatra, B.S., Chawla, R.K., Sewell, C.W. & Rudman, D. (1977). Distribution of branched-chain alpha-keto acid dehydrogenases in primate tissues. *The Journal of Clinical Investigation* 59, 558–564.

Kim, P.L., Staron, R.S. & Phillips, S.M. (2005). Fasted-state skeletal muscle protein synthesis after resistance exercise is altered with training. *The Journal of Physiology* 568, 283–290.

Klein, D.J. & Ferre-D'Amare, A.R. (2006). Structural basis of glmS ribozyme activation by glucosamine-6-phosphate. *Science* 313, 1752–1756.

Kostek, M.C., Chen, Y.W., Cuthbertson, D.J., et al. (2007). Gene expression responses over 24 h to lengthening and shortening contractions in human muscle: Major changes in CSRP3, MUSTN1, SIX1, and FBXO32. *Physiological Genomics* 31, 42–52.

Kubica, N., Bolster, D.R., Farrell, P.A., Kimball, S.R. & Jefferson, L.S. (2005). Resistance exercise increases muscle protein synthesis and translation of eukaryotic initiation factor 2Bepsilon mRNA in a mammalian target of rapamycin-dependent manner. *The Journal of Biological Chemistry* 280, 7570–7580.

Lamas, L., Aoki, M.S., Ugrinowitsch, C., et al. (2010). Expression of genes related to muscle plasticity after strength and power training regimens. *Scandinavian Journal of Medicine & Science in Sports* 20(2):216–225.

Lamon, S., Wallace, M.A., Leger, B. & Russell, A.P. (2009). Regulation of STARS and its downstream targets suggest a novel pathway involved in human skeletal muscle hypertrophy and atrophy. *The Journal of Physiology* 587, 1795–1803.

Lamont, L.S., McCullough, A.J. & Kalhan, S.C. (1995). Beta-adrenergic blockade heightens the exercise-induced increase in leucine oxidation. *The American Journal of Physiology* 268, E910–E916.

Lamont, L.S., McCullough, A.J. & Kalhan, S.C. (2001a). Gender differences in leucine, but not lysine, kinetics. *Journal of Applied Physiology* 91, 357–362.

Lamont, L.S., McCullough, A.J. & Kalhan, S.C. (2001b). Relationship between leucine oxidation and oxygen consumption during steady-state exercise. *Medicine and Science in Sports and Exercise* 33, 237–241.

Lecker, S.H., Jagoe, R.T., Gilbert, A., et al. (2004). Multiple types of skeletal muscle atrophy involve a common program of changes in gene expression. *FASEB Journal* 18, 39–51.

Leger, B., Cartoni, R., Praz, M., et al. (2006). Akt signalling through GSK-3beta, mTOR and Foxo1 is involved in human skeletal muscle hypertrophy and atrophy. *The Journal of Physiology* 576, 923–933.

Lemon, P.W., Tarnopolsky, M.A., MacDougall, J.D. & Atkinson, S.A. (1992). Protein requirements and muscle mass/strength changes during intensive training in novice bodybuilders. *Journal of Applied Physiology* 73, 767–775.

Louis, E., Raue, U., Yang, Y., Jemiolo, B. & Trappe, S. (2007). Time course of proteolytic, cytokine, and myostatin gene expression after acute exercise in human skeletal muscle. *Journal of Applied Physiology* 103, 1744–1751.

Lowell, B.B., Ruderman, N.B. & Goodman, M.N. (1986). Evidence that lysosomes are not involved in the degradation of myofibrillar proteins in rat skeletal muscle. *The Biochemical Journal* 234, 237–240.

Mahoney, D.J., Parise, G., Melov, S., Safdar, A. & Tarnopolsky, M.A. (2005). Analysis of global mRNA expression in human skeletal muscle during recovery from endurance exercise. *FASEB Journal* 19, 1498–1500.

Mahoney, D.J., Safdar, A., Parise, G., et al. (2008). Gene expression profiling in

human skeletal muscle during recovery from eccentric exercise. *American Journal of Physiology—Regulatory, Integrative and Comparative Physiology* 294, R1901–R1910.

Mascher, H., Andersson, H., Nilsson, P.A., Ekblom, B. & Blomstrand, E. (2007). Changes in signalling pathways regulating protein synthesis in human muscle in the recovery period after endurance exercise. *Acta Physiologica (Oxford)* 191, 67–75.

Mascher, H., Tannerstedt, J., Brink-Elfegoun, T., Ekblom, B., Gustafsson, T. & Blomstrand, E. (2008). Repeated resistance exercise training induces different changes in mRNA expression of MAFbx and MuRF-1 in human skeletal muscle. *American Journal of Physiology—Endocrinology and Metabolism* 294, E43–E51.

Matsakas, A. & Patel, K. (2009). Intracellular signalling pathways regulating the adaptation of skeletal muscle to exercise and nutritional changes. *Histology and Histopathology* 24, 209–222.

McCarthy, J.J. & Esser, K.A. (2007). MicroRNA-1 and microRNA-133a expression are decreased during skeletal muscle hypertrophy. *Journal of Applied Physiology* 102, 306–313.

McKenzie, S., Phillips, S.M., Carter, S.L., Lowther, S., Gibala, M.J. & Tarnopolsky, M.A. (2000). Endurance exercise training attenuates leucine oxidation and BCOAD activation during exercise in humans. *American Journal of Physiology—Endocrinology and Metabolism* 278, E580–E587.

Melov, S., Tarnopolsky, M.A., Beckman, K., Felkey, K. & Hubbard, A. (2007). Resistance exercise reverses aging in human skeletal muscle. *PLoS One* 2, e465.

Meredith, C.N., Zackin, M.J., Frontera, W.R. & Evans, W.J. (1989). Dietary protein requirements and body protein metabolism in endurance-trained men. *Journal of Applied Physiology* 66, 2850–2856.

Mersey, B.D., Jin, P. & Danner, D.J. (2005). Human microRNA (miR29b) expression controls the amount of branched chain alpha-ketoacid dehydrogenase complex in a cell. *Human Molecular Genetics* 14, 3371–3377.

Miller, S.L., Tipton, K.D., Chinkes, D.L., Wolf, S.E. & Wolfe, R.R. (2003). Independent and combined effects of amino acids and glucose after resistance exercise. *Medicine and Science in Sports and Exercise* 35, 449–455.

Miller, B.F., Olesen, J.L., Hansen, M., et al. (2005). Coordinated collagen and muscle protein synthesis in human patella tendon and quadriceps muscle after exercise. *The Journal of Physiology* 567, 1021–1033.

Mitch, W.E. & Goldberg, A.L. (1996). Mechanisms of muscle wasting. The role of the ubiquitin-proteasome pathway. *New England Journal of Medicine* 335, 1897–1905.

Murphy, R.M. & Lamb, G.D. (2009). Endogenous calpain-3 activation is primarily governed by small increases in resting cytoplasmic [Ca2+] and is not dependent on stretch. *The Journal of Biological Chemistry* 284, 7811–7819.

Murphy, R.M., Goodman, C.A., McKenna, M.J., Bennie, J., Leikis, M. & Lamb, G.D. (2007). Calpain-3 is autolyzed and hence activated in human skeletal muscle 24 h following a single bout of eccentric exercise. *Journal of Applied Physiology* 103, 926–931.

Murton, A.J., Constantin, D. & Greenhaff, P.L. (2008). The involvement of the ubiquitin proteasome system in human skeletal muscle remodelling and atrophy. *Biochimica Biophysica Acta* 1782, 730–743.

Olsen, S., Aagaard, P., Kadi, F., et al. (2006). Creatine supplementation augments the increase in satellite cell and myonuclei number in human skeletal muscle induced by strength training. *The Journal of Physiology* 573, 525–534.

Parise, G., Brose, A.N. & Tarnopolsky, M.A. (2005). Resistance exercise training decreases oxidative damage to DNA and increases cytochrome oxidase activity in older adults. *Experimental Gerontology* 40, 173–180.

Pellet, P.L. (1990). Protein requirements in humans. *American Journal of Clinical Nutrition* 51, 723–737.

Phillips, S.M. (2006). Dietary protein for athletes: From requirements to metabolic advantage. *Applied Physiology, Nutrition, and Metabolism* 31, 647–654.

Phillips, S.M., Atkinson, S.A., Tarnopolsky, M.A. & MacDougall, J.D. (1993). Gender differences in leucine kinetics and nitrogen balance in endurance athletes. *Journal of Applied Physiology* 75, 2134–2141.

Phillips, S.M., Tipton, K.D., Aarsland, A., Wolf, S.E. & Wolfe, R.R. (1997). Mixed muscle protein synthesis and breakdown after resistance exercise in humans. *The American Journal of Physiology* 273, E99–E107.

Phillips, S.M., Tipton, K.D., Ferrando, A.A. & Wolfe, R.R. (1999). Resistance training reduces the acute exercise-induced increase in muscle protein turnover. *The American Journal of Physiology* 276, E118–E124.

Phillips, S.M., Parise, G., Roy, B.D., Tipton, K.D., Wolfe, R.R. & Tamopolsky, M.A. (2002). Resistance-training-induced adaptations in skeletal muscle protein turnover in the fed state. *Canadian Journal of Physiology and Pharmacology* 80, 1045–1053.

Raue, U., Slivka, D., Jemiolo, B., Hollon, C. & Trappe, S. (2007). Proteolytic gene expression differs at rest and after resistance exercise between young and old women. *Journals of Gerontology Series A: Biological Sciences and Medical Sciences* 62, 1407–1412.

Richter, E.A. & Ruderman, N.B. (2009). AMPK and the biochemistry of exercise: Implications for human health and disease. *The Biochemical Journal* 418, 261–275.

Rosenblatt, J.D., Yong, D. & Parry, D.J. (1994). Satellite cell activity is required for hypertrophy of overloaded adult rat muscle. *Muscle & Nerve* 17, 608–613.

Safdar, A., Abadi, A., Akhtar, M., Hettinga, B.P. & Tarnopolsky, M.A. (2009). miRNA in the regulation of skeletal muscle adaptation to acute endurance exercise in C57Bl/6J male mice. *PLoS One* 4, e5610.

Schmutz, S., Dapp, C., Wittwer, M., Vogt, M., Hoppeler, H. & Fluck, M. (2006). Endurance training modulates the muscular transcriptome response to acute exercise. *Pflugers Archiv* 451, 678–687.

Stepto, N.K., Coffey, V.G., Carey, A.L., et al. (2009). Global gene expression in skeletal muscle from well-trained strength and endurance athletes. *Medicine and Science in Sports and Exercise* 41, 546–565.

Stupka, N., Tarnopolsky, M.A., Yardley, N.J. & Phillips, S.M. (2001). Cellular adaptation to repeated eccentric exercise-induced muscle damage. *Journal of Applied Physiology* 91, 1669–1678.

Tarnopolsky, M. (2004). Protein requirements for endurance athletes. *Nutrition* 20, 662–668.

Tarnopolsky, M.A. (2008). Sex differences in exercise metabolism and the role of 17-beta estradiol. *Medicine and Science in Sports and Exercise* 40, 648–654.

Tarnopolsky, M.A., MacDougall, J.D. & Atkinson, S.A. (1988). Influence of protein intake and training status on nitrogen balance and lean body mass. *Journal of Applied Physiology* 64, 187–193.

Tarnopolsky, M.A., Atkinson, S.A., MacDougall, J.D., Senor, B.B., Lemon, P.W. & Schwarcz, H. (1991). Whole body leucine metabolism during and after resistance exercise in fed humans. *Medicine and Science in Sports and Exercise* 23, 326–333.

Tarnopolsky, M.A., Atkinson, S.A., MacDougall, J.D., Chesley, A., Phillips, S. & Schwarcz, H.P. (1992). Evaluation of protein requirements for trained strength athletes. *Journal of Applied Physiology* 73, 1986–1995.

Tidball, J.G. & Spencer, M.J. (2000). Calpains and muscular dystrophies. *The International Journal of Biochemistry & Cell Biology* 32, 1–5.

Timmons, J.A., Larsson, O., Jansson, E., et al. (2005). Human muscle gene expression responses to endurance training provide a novel perspective on Duchenne muscular dystrophy. *FASEB Journal* 19, 750–760.

Tipton, K.D., Rasmussen, B.B., Miller, S.L., et al. (2001). Timing of amino acid-carbohydrate ingestion alters anabolic response of muscle to resistance exercise. *American Journal of Physiology—Endocrinology and Metabolism* 281, E197–E206.

Tipton, K.D. & Wolfe, R.R. (2004). Protein and amino acids for athletes. *Journal of Sports Science* 22, 65–79.

Tipton, K.D. & Witard, O.C. (2007). Protein requirements and recommendations for athletes: Relevance of ivory tower arguments for practical recommendations. *Clinics in Sports Medicine* 26, 17–36.

Torun, B., Scrimshaw, N.S. & Young, V.R. (1977). Effect of isometric exercises on body potassium and dietary protein requirements of young men. *American Journal of Clinical Nutrition* 30, 1983–1993.

Verburg, E., Murphy, R.M., Richard, I. & Lamb, G.D. (2009). Involvement of calpains in Ca2+-induced disruption of excitation-contraction coupling in mammalian skeletal muscle fibers. *American Journal of Physiology Cell Physiology* 296(5):C1115–C1122.

Wagenmakers, A.J., Brookes, J.H., Coakley, J.H., Reilly, T. & Edwards, R.H. (1989). Exercise-induced activation of the branched-chain 2-oxo acid dehydrogenase in human muscle. *European Journal of Applied Physiology and Occupational Physiology* 59, 159–167.

Wagenmakers, A.J., Beckers, E.J., Brouns, F., et al. (1991). Carbohydrate supplementation, glycogen depletion, and amino acid metabolism during exercise. *The American Journal of Physiology* 260, E883–E890.

Wilkinson, S.B., Phillips, S.M., Atherton, P.J., et al. (2008). Differential effects of resistance and endurance exercise in the fed state on signalling molecule phosphorylation and protein synthesis in human muscle. *The Journal of Physiology* 586, 3701–3717.

Wolfe, R.R., Wolfe, M.H., Nadel, E.R. & Shaw, J.H. (1984). Isotopic determination of amino acid-urea interactions in exercise in humans. *Journal of Applied Physiology* 56, 221–229.

Yang, Y., Jemiolo, B. & Trappe, S. (2006). Proteolytic mRNA expression in response to acute resistance exercise in human single skeletal muscle fibers. *Journal of Applied Physiology* 101, 1442–1450.

Zanchi, N.E., de Siqueira Filho, M.A., Lira, F.S., et al. (2009). Chronic resistance training decreases MuRF-1 and Atrogin-1 gene expression but does not modify Akt, GSK-3beta and p70S6K levels in rats. *European Journal of Applied Physiology* 106, 415–423.

Zhao, J., Brault, J.J., Schild, A., et al. (2007). FoxO3 coordinately activates protein degradation by the autophagic/lysosomal and proteasomal pathways in atrophying muscle cells. *Cell Metabolism* 6, 472–483.

Chapter 24

The Regulation of Physical Activity by Genetic Mechanisms: Is There a Drive to Be Active?

NISHAN SUDHEERA KALUPAHANA[1], NAIMA MOUSTAID-MOUSSA[1],
JUNG HAN KIM[1,2], BRYNN H. VOY[1], DAVID BASSETT[1], MOLLY S. BRAY[3]
AND J. TIMOTHY LIGHTFOOT[4]

[1]University of Tennessee Obesity Research Center, Knoxville, TN, USA
[2]Department of Pharmacology, Physiology, and Toxicology, Marshall University, Huntington, WV, USA
[3]Department of Epidemiology, Heflin Center for Genomic Sciences, University of Alabama at Birmingham, Birmingham, AL, USA
[4]Department of Health and Kinesiology, Texas A&M University, College Station, TX, USA

Regular physical activity is associated with longevity (Landi et al., 2008; Lanza et al., 2008) and a lowered risk for obesity (Healy et al., 2008) and chronic diseases such as cardiovascular disease (Richardson et al., 2004), type 2 diabetes mellitus (Hu et al., 2004; Jeon et al., 2007), and site-specific cancers such as breast and colon cancer (Irwin et al., 2008; Slattery et al., 1997). Although regular physical activity is inversely associated with obesity, several studies have shown that the beneficial effects of physical activity in terms of risk reduction for chronic diseases are independent of its effects on body weight and obesity (Crespo et al., 2002). In fact, some studies have shown that normal-weight physically inactive individuals have a higher risk for cardiovascular disease than overweight physically active individuals (Sui et al., 2007). Further, there is evidence that the longevity and reduced incidence of cardiovascular mortality associated with regular physical activity is not due to genetic selection (Carlsson et al., 2007). Based on

this evidence, the Centers for Disease Control and Prevention (CDC), the American College of Sports Medicine (ACSM), and the American Heart Association (AHA) recommend that adults engage in a minimum of 150 min of moderate-intensity physical activity each week for improving and maintaining of health (Haskell et al., 2007; Pate et al., 1995). Recently, the government of the United States issued similar recommendations for the general public (Physical Activity Guidelines Advisory Committee, 2008). Despite public awareness of the beneficial effects of regular physical activity, less than 50% of adult Americans meet these guidelines (Haskell et al., 2007). Further, the duration of most physical activity interventions is short lived (van der Bij et al., 2002). In this context, it is important to understand the factors affecting the level of physical activity in an individual. While there is an extensive literature base considering the effects of environmental factors on physical activity levels, a growing body of literature suggests that the drive to be active or inactive may arise from genetic factors. Although the impact of genetics on fitness and physical performance has been studied in great detail, the impact of genetics on physical activity levels is yet to be explored.

Genetic and Molecular Aspects of Sport Performance, 1st edition.
Edited by Claude Bouchard and Eric P. Hoffman.
Published 2011 by Blackwell Publishing Ltd.

Evidence for a Genetic Basis of Physical Activity in Animal Models

Other chapters in this volume have extensively covered the associations discovered in human models between heritability and physical activity. Supporting this data, there is a large amount of similar research using animal models. Since most genes are conserved between mouse and human genomes (Waterston et al., 2002), the function of genes discovered in mouse and other rodent models can be easily translated to humans. For example, several single-gene mutations resulting in obesity were first identified using animal models (Chua et al., 1996; Pelleymounter et al., 1995). The ease of manipulation of environmental and breeding conditions allows the study of animal models to provide important information not otherwise discoverable in human studies. The two main approaches in these animal studies have been to compare different inbred strains of mice and mice selectively bred for specific physical activity characteristics (i.e., phenotype). Using the first approach, several investigators have compared the differences in various measures of physical activity between strains of inbred mice to estimate the broad-sense heritability by means of calculation of interclass correlation and coefficient of genetic determination. Festing (1977) compared the wheel-running activity (revolutions per 24 h) between 26 inbred strains of mice for 2 and 7 days and found that the intraclass correlations were 0.29 and 0.26, respectively. Lightfoot et al. (2004) found similar intraclass correlations (0.30) for distance run in an experiment comparing the wheel-running activity between 13 inbred strains of mice with differing heritability estimates by sex. Lerman et al. (2002) also compared the wheel-running activity in seven inbred strains of mice and found a broad-sense heritability estimate of 0.39 for the total duration. Thus, the broad-sense heritability for overall wheel-running activity is generally between 18% and 58% depending on the index of activity used.

Other investigators have used a selective inbreeding strategy to generate mice with a high degree of physical activity (Swallow et al., 1998). Koteja and colleagues used outbred, genetically variable Hsd: ICR house mice and selectively bred four lines for high wheel-running activity. They found that in the 13th generation, the high-active mice ran 2.2 times more revolutions per day than control mice (Koteja et al., 1999). By the 17th generation, this difference was 2.7-fold (Gomes et al., 2009). Interestingly, in addition to being more physically active, some of these selectively bred mice had lower hind limb muscle mass suggestive of increased aerobic capacity (Gomes et al., 2009). They also had lower percent body fat and plasma leptin levels, increased insulin-stimulated glucose uptake, and increased levels of glucose transporter 4 (GLUT4) in the skeletal muscle (Girard et al., 2007).

Thus, the evidence from human twin and family studies plus selectively and inbred rodent studies suggests that the propensity for physical activity is heritable.

Identifying the Genes Regulating the Level of Physical Activity

The two conventional approaches to identify genes associated with a trait of interest are association studies and linkage analyses. Association studies are based on candidate genes selected by their physiological relevance to the trait of interest. Statistical associations between the genotype/allele frequency and the phenotype are made to identify possible associations. Linkage studies, on the other hand, investigate whether a gene locus is transmitted to the next generation with the trait of interest. The assumption is that the gene loci can be identified and quantified—usually termed *quantitative trait loci* (QTL). This allows for focused, hypothesis-driven investigation on the known genes in that QTL. Alternatively, positional cloning allows for gene identification without prior knowledge of genes in that locus. To offset the possibility of a false-positive QTL, the standard of statistical proof in genetic linkage studies is higher with alpha values of 0.05 only being designated as "suggestive," and alpha values of 0.01 and 0.001 being reserved for the terms *significant* and *highly significant*, respectively. Thus, the finding of a significant QTL confidently narrows the genomic area in which to look for genes that may be involved in regulating daily activity levels.

QTLs Linked to Physical Activity Phenotypes

Linkage analysis is often used as a first tool to identify genomic regions potentially harboring functional genes related to the trait of interest prior to analysis of individual genes. This approach is considered a *hypothesis generating* statistical method to narrow the search for candidate genes in a systematic and unbiased way. Two linkage studies of physical activity and/or inactivity have been conducted to date in humans (Table 24.1). The Quebec Family Study (QFS) includes a large cohort of French Canadian families who were extensively phenotyped for body size measures and obesity-related traits (Borecki et al., 1991). Using both single-point and multipoint linkage

analysis, Simonen and colleagues identified a region on chromosome 2p22-p16 that was strongly linked to physical inactivity. Putative candidate genes in this region include hereditary spastic paraplegia type 4, striatin, and member 1 of the solute carrier family 8 (Simonen et al., 2003b). This study also identified a linkage region on 13q22-q31 for both total daily physical activity and moderate to strenuous physical activity phenotypes, and the insulin-like growth factor binding protein 1 (*IGFBP1*) region on chromosome 7p11.2 was linked to both inactivity scores as well as moderate to strenuous physical activity. Additional linkage signals were also reported on chromosomes 9, 11, 15, and 20 (Simonen et al., 2003b).

Table 24.1 Summary of QTL linked to physical activity phenotypes

Study	Study group	Phenotype	Chromosome/ cM (CI)	*P*-value	Nearest marker/gene
Simonen et al. (2003b)	Human (*n* = 767)	Physical inactivity	2/50.1–59.6	0.0013	D2S2347/ D2S2305
			7/72.2	0.021	IGFBPf
			20/57.6	0.012	PLC1
		Time spent on PA	11/15	0.0036	C11P15_3
			15/19.1	0.012	D15S165
		Total daily PA	13/70.6	0.031	D13S317
		Moderate to strenuous PA	9/99.5	0.028	D9S938
Cai et al. (2006)	Human (*n* = 1030)	Sedentary activity	18/63–71	<0.01	D18S1102/ D18S474
Lightfoot et al. (2008)	Mouse/C57/LJ X C3H/HeJ	Wheel running			
		Duration	13/11 (1–15)	<0.01	Rs6329684
		Distance	13/11 (1–15)	<0.01	Rs6329684
		Speed	9/7 (1–15)	<0.01	Rs13480073
			13/9 (1–17)	<0.01	Rs6329684
Leamy et al. (2008)	Mouse/C57/LJ X C3H/HeJ	Wheel running	Epistasis		
		Duration	1/107 and 8/74	<0.05	
			3/68 and 10/41	<0.05	
			4/110 and 14/32	<0.05	
			5/19 and 11/54	<0.05	
			12/17 and 14/15	<0.05	
		Distance	3/62 and 10/34	<0.05	
			6/80 and 15/4	<0.05	
		Speed	10/80 and 11/56	<0.05	
			12/25 and 15/8	<0.05	
Kas et al. (2008)	Mouse/C57BL/ 6J X A/J (chromosome substitution)	Horizontal distance traveled (home cage)	1	<0.01	A830043J08Rik

PA, physical activity.

Linkage analysis of energy expenditure and physical activity was also performed in a large study of obesity in Hispanic families called the VIVA LA FAMILIA Study. A total of 1030 children, aged 4–19 years, from 319 families were utilized for these analyses. Free-living physical activity was monitored continuously for 3 days using Actiwatch accelerometers (Mini Mitter, Bend, OR), and the percentage of awake time spent in sedentary, light, moderate, and vigorous activity was computed using predetermined thresholds for each category (Cai et al., 2006). A maximum likelihood variance components method was used to estimate the heritability and identify putative linkage regions for the physical activity traits. This study identified a significant linkage region on chromosome 18q for sedentary activity, which was also linked to light and moderate physical activity. The gastrin-releasing peptide and melanocortin 4 receptor genes are putative candidate genes lying proximal to this linkage region.

As in human models, there are limited physical activity genetic linkage studies using animal models. The first study to perform a comprehensive QTL analysis on physical activity traits of mice in a home cage environment was by Lightfoot and colleagues in 2008 (Lightfoot et al., 2008). They studied the length, duration, and speed of voluntary wheel-running activity of F_2 mice derived from highly physically active C57L/J and low-active C3H/HeJ inbred strains. They found a significant QTL located on chromosome 13 which was associated with all three phenotypes. In addition, there was a significant QTL on Chr 9, which was associated with the speed of wheel running. While the Chr 13 QTL accounted for 6% of variability of physical activity in each phenotype, the Chr 9 QTL accounted for 11% of the variation. While fundamentally a different phenotype, Kas et al. (2008) found a significant QTL on Chr 1 linked with the total horizontal distance traveled by mice in home cage environment (Table 24.1). However, one of the general issues with the available QTL studies is that they only consider the effect of single QTL on the variance of the concerned phenotype. As with most complex traits, the effect of a single QTL on physical activity has generally been shown to account

for only a few percent of the phenotypic variance (e.g., 1–6%; Lightfoot, 2008).

Thus, since direct-effect QTL were noted to only partially account for the heritability levels of physical activity (14–34% of the total variance; Lightfoot, 2008), the interaction of each pair of QTL in different chromosomes (epistasis) was investigated by Leamy et al. (2008) in a subsequent study on the same population of mice investigated by Lightfoot's group. In this first-ever study of epistatic interactions in any exercise behavior, they showed that epistatic interactions accounted for nearly 26% of the total variation in the three physical activity traits (Table 24.1). There were one to three significant epistatic pairs for each activity index (e.g., Chr 3, 62 cM x Chr 10, 34 cM for distance) as well as several suggestive epistatic pairs (e.g., $n = 8$ for distance). Most important was that the total variance explained by the combination of the direct-effect and the epistatic QTL approached the total variance explained by genetics indicating that the genetic influence on activity characteristics was a mixture of both direct-effect and epistatic loci (Leamy et al., 2008). Further, all effects of the QTL were sex independent, that is, there were no sex-linked QTL. This is a surprising finding considering that there have been reports of differences in habitual physical activity by sex (Lightfoot, 2008); however, Lightfoot and colleagues have attributed most of the sex-related variation in physical activity to downstream regulators including sex hormones.

Of the five linkage studies performed to date for physical activity traits, there is little to no overlap in the findings. This may not be surprising, since the studies have included subjects of different species and ethnic groups. In addition, differences in measuring or estimating physical activity can lead to difficulty in generating consistent results using gene discovery approaches for physical activity. Ultimately, linkage studies require additional follow-up to fine map the linkage regions in order to identify functional variants that influence physical activity behavior. Until specific candidate genes are identified, the mechanisms linking genetic variation to physical activity behavior remain unknown.

Candidate Genes for Habitual Physical Activity

It remains uncertain how the mechanisms arising from genetic variation may influence physical activity behavior. Like the mouse models bred for physical activity behavior, genetic variation in humans may influence the physical capacity for exercise and activity by altering motivation, metabolism, and/or endurance parameters. As such, genetic variants may also alter the production or reuptake of neuropeptides that influence pain, pleasure, or other neural reward systems associated with physical activity. Most genetic association studies of candidate genes for physical activity have focused on these types of mechanisms in selecting genes for study.

Since motivation for physical activity is thought to be an important factor in determining the level of physical activity in an individual, the dopaminergic system, which is involved with regulation of motor movement and motivation, has been studied as a candidate for the regulation of physical activity level. More specifically, the midbrain dopaminergic system has been shown to be important in mediating motor control and reward mechanisms (Salamone, 1992) with suggestions that D1 receptors are involved in activity regulation (Rhodes & Garland, 2003). Recently, Knab et al. investigated the changes in expression of seven genes of the dopaminergic system in male high-active C57/LJ versus low-active C3H/Hej mice (Knab et al., 2009). Half the mice from each strain were given access to running wheels for 21 days, during which their wheel-running activity was measured. After analyzing mRNA from the nucleus accumbens and striatum of the midbrain, they found that of the seven dopamine genes studied (dopamine receptors D1, D2, D3, D4, and D5, tyrosine hydroxylase, and dopamine transporter) two (dopamine receptor D1 and tyrosine hydroxylase) had lower expression in the high-active C57/LJ mice when compared to the low-active C3H/Hej mice. Further, the expression of these genes was not affected by having access to a running wheel. Thus, the findings of Knab et al. showed that dopamine receptor D1 expression levels were not dependent upon physical activity and, given the observed expression differences

between high- and low-active mice as well as the pharmacological work of Rhodes and Garland (2003) appear to be involved in the regulation of physical activity in inbred mice.

While human studies have also suggested the involvement of the dopaminergic system in the regulation of activity, unlike the available animal studies, there have been associations shown between dopamine receptor D2 (DRD2) and physical activity level. Simonen and colleagues studied the association between a marker in the DRD2 gene (a fragment length polymorphism in exon 6) and physical activity in individuals who took part in the QFS and the Health, Risk Factors, Exercise Training, and Genetics (HERITAGE) study (Simonen et al., 2003a). The frequency of the minor allele among whites was 28% in the former study, while it was 30% in the latter. In the former study, among white women, the TT homozygotes had 25% and 34% lower physical activity levels when compared to the CC homozygotes and CT heterozygotes, respectively. Similar changes were observed in the HERITAGE study with 29% and 38% lower sports index (graded activity levels and frequency in an increasing order) in the TT homozygotes versus CC homozygotes and CT heterozygotes, respectively. The relationship between dopamine D2 receptors and activity has been suggested previously in Drd2 knockout mice (Kelly et al., 1998). These mice were reported to have only 50% of the horizontal activity when compared to wild-type mice. However, a clear confounding factor in these mice was that their motor coordination was also impaired, which could have impacted their voluntary activity level. Interestingly, in an animal study using a different strategy, a similar association between Drd2 and physical activity was reported subsequently (Bronikowski et al., 2004). In that study, Bronikowski and colleagues selectively inbred mice for high wheel-running activity and compared the gene expression in the hippocampal region of the brain in those mice with a control group. Using high-density oligonucleotide microarrays, they identified 30 differentially regulated genes between the two groups, one of which was Drd2, which had 20% higher expression in the mice with high voluntary wheel-running activity.

Other studies have shown that exercise activates dopaminergic motivation systems such as reward-related motivation (Chen et al., 2008). In fact, it was recently shown that long-term exercise reduces the rewarding efficacy of 3,4-methylenedioxymethamphetamine (MDMA), a strong psychostimulant, in C57BL/6J mice (Chen et al., 2008). Thus, gene polymorphisms in the dopaminergic motivation systems have been implicated in physical inactivity of individuals.

The transcription factor nescient helix loop helix 2 (*Nhlh2*) is another gene that has been studied with regard to regulation of physical activity level. Good and colleagues (1997) first reported that targeted neuronal deletion of this transcription factor in mice (*Nhlh2*-KO) results in adult-onset obesity. Subsequently, Coyle et al. (2002) characterized the pattern of weight gain in these animals and showed that reduced voluntary wheel-running activity, and not increased food intake or lower body temperature, preceded their adult-onset obesity. Since there were no disturbances in the circadian rhythm, balance, or motor control, these investigators concluded that the lower physical activity in the *Nhlh2*-KO mice was the result of a behavioral change. Supporting this, other investigators have shown that *Nhlh2* may be involved in the regulation of alpha-melanocyte stimulating hormone (MSH) and thyrotropin-releasing hormone (TRH) production in the hypothalamic arcuate and paraventricular nuclei, via transcriptional regulation of prohormone convertase 1 and 2 (Jing et al., 2004). Further, there is evidence that *Nhlh2* may be a target of leptin, a regulator of food intake and energy expenditure (Vella et al., 2007).

While there exists more work on dopamine receptors and nescient helix-loop-helix, other potential gene candidates whose participation in activity regulation has been suggested by early, but nonconclusive evidence. Table 24.2 provides an overview of the association studies for specific genes that have been performed for physical activity traits. Genes that have been significantly associated with physical activity include those related to neural signaling (*BSX, DRD2, LEPR, MC4R, MCH1R*), genes related to cardiovascular function (*ACE*), and genes related to specific peripheral

metabolic functions, such as bone formation (*CASR*), estrogen biosynthesis (*CYP19A1*), or glucose transport (*Glut4*). Each of the genes examined represent single association studies without replication; thus, no consensus of gene association can be made across multiple studies/populations. Although some studies include adjustment for covariates such as age and gender, no studies consider either the joint effects or the interaction between psychological and social variables and genetic variation in determining physical activity outcomes. Within the human studies, only one study (Stefan et al., 2002) does not use some kind of questionnaire-based instrument to assess the physical activity. Although physical activity has been reported to moderate the effects of the *FTO* gene on obesity and diabetes-related outcomes, the *FTO* gene does not appear to influence physical activity levels directly.

Brain-specific homeobox transcription factor (*Bsx*) is another hypothalamic factor like *Nhlh2* which has been found to be necessary for normal locomotor activity, with mice lacking *Bsx* exhibiting hypolocomotion (Sakkou et al., 2007). Deacon et al. (2006) found that mice lacking adenosine triphosphate (ATP)-sensitive potassium channel subunit Kir6.2 were more active in the home cage environment than their wild-type counterparts. Melanin-concentrating hormone receptor-1 (*Mch1r*)-deficient mice were also reported to exhibit increased wheel-running activity (Zhou et al., 2005).

As opposed to the potential central genes that may regulate activity (e.g., *Drd1* and *Nhlh2*), Tsao et al. (2001) found that mice overexpressing glucose transporter 4 (*Glut4*) in fast-twitch skeletal muscle exhibited a fourfold increase in voluntary wheel-running activity. This potential peripheral mechanism regulating activity was supported by the finding that mice selectively bred for high wheel-running activity also showed higher increases in skeletal muscle *Glut4* in response to physical activity, when compared to controls (Gomes et al., 2009).

While several studies have shown rather robust associations between candidate genes and physical activity level, most of them have not been further validated by gene knockout or overexpression studies to show causality. Thus, more studies of the

Table 24.2 Association studies of physical activity

References	Gene/marker	Variant	Location (human genome)	N Subjects/sib-pairs	Method of PA assessment	Covariates	Phenotype	P-values
Winnicki et al. (2004)	*ACE*	I/D	17q23	265 men, 90 women (33 ± 9 years) with untreated stage 1 hypertension	Questionnaire	Age, sex, marital status, profession, coffee, and alcohol consumption	Sedentary versus active	0.001
Loos et al. (2005)	*AGRP, CART, MC3R, NPY, NPYY1R, POMC*	Multiple		669 subjects (parents: 52 ± 3.4 years; offspring; 28 ± 8.7 years)	3-day PA diary; habitual activity via questionnaire	Age, gender, and BMI	Moderate to strenuous activity; inactivity	NS
Lorentzon et al. (2001)	*CASR*	A986S (G/T)	3q21–q24	97 healthy Caucasian girls (16.9 ± 1.2 years)	Questionnaire	None	Weight-bearing physical activity	0.01
Salmen et al. (2003)	*CYP19A1*	TTTA repeat	15q21.1	331 early postmenopausal Caucasian women (52.7 ± 2.3 years)	Questionnaire	None	Physical activity	0.039
Knab et al. (2009)	*Drd1*	—	5q35.1	17 high-active C57L/J male mice; 20 low-active C3H/HeJ male mice	Wheel-running activity	None	Distance run/day, duration of activity, speed of running	0.0001
Simonen et al. (2003a)	*DRD2*	C/T	11q23	402 white women (40.1 ± 14.2 years)	Questionnaire	None	Work index	0.004
Simonen et al. (2003a)	*DRD2*	C/T	11q23	256 white women (35.9 ± 14.6 years)	Questionnaire	BMI	Physical activity	0.016
Simonen et al. (2003a)	*DRD2*	C/T	11q23	256 white women (35.9 ± 14.6 years)	Questionnaire	BMI	Sports index	0.023
Simonen et al. (2003b)	*DRD2*	C/T	11q23	275 AA men and women; 241 white males	Questionnaire	None	Physical activity; sports index; work index	NS

(Continued)

Table 24.2 (Continued)

References	Gene/marker	Variant	Location (human genome)	N Subjects/sib-pairs	Method of PA assessment	Covariates	Phenotype	P-values
Hakanen et al. (2009)	FTO			438 healthy boys and girls (15 years)	Questionnaire	None	PA index	NS
Berentzen et al. (2008)	FTO	Rs9939609	16q12.2	234 obese and 323 nonobese men (median age, 48 years)	Direct measures; questionnaire	Age	REE, GIT, $\dot{V}O_2$ max, or LTPA	NS
Stefan et al. (2002)	LEPR			268 Pima men and women (~27 years)	Respiratory chamber	Age, sex, body composition, energy balance, and degree of Pima Indian heritage	Total PA (24-h EE/sleeping EE)	0.008
Loos et al. (2005)	MC4R	C-2745T	18q22	669 subjects (parents: 52 ± 3.4 years; offspring: 28 ± 8.7 years)	3-day PA diary; habitual activity via questionnaire	Age, gender, and BMI	Moderate to strenuous activity; inactivity	0.005, 0.01[a]
Coyle et al. (2002)	Nhlh2	—	1p12-p11	Nhlh2 knockout males and females versus wild-type males and females	Wheel-running activity	None	Wheel rotations/day	0.001–0.047
De Moor et al. (2009)	PAPSS2	37 novel SNPs	10q23.2	1644 Dutch and 978 American adults	Questionnaire	Sex, age, and BMI	Exercise participation	6.26×10^{-6}
Tsao et al. (2001)	Slc2a4 (Glut4)	—	17p13	Overexpression transgenic male mice	Wheel-running activity	None	Distance run/day	<0.05

[a]Two separate analyses.
PA, physical activity; BMI, body mass index; NS, not significant; REE, resting energy expenditure; GIT, glucose-induced thermogenesis; LTPA, leisure-time physical activity; EE, energy expenditure; SNPs, single-nucleotide polymorphisms.

latter category and association studies with high power (e.g., genome-wide association scans) are needed to identify specific genes regulating physical activity level. Recently, De Moor and colleagues (2009) reported the first genome-wide association study of exercise behavior in unrelated adult cohorts from the Netherlands and the United States. The study included 2622 individuals of European ancestry, and physical activity was characterized using a detailed self-report instrument from which subjects were classified as being nonexercisers or regular exercisers (equivalent to performing a minimum of 4 MET-hours/week) (De Moor et al., 2009). More than 1.6 million single-nucleotide polymorphisms (SNPs) were examined for association to exerciser status. The study identified both novel SNPs in the *PAPSS2* gene as well as two intergenic regions on chromosomes 2q33.1 and 18p11.32 that were associated with exercise participation (pooled *P*-values $<1.0 \times 10$) (De Moor et al., 2009). None of the previously reported associations or linkage results for physical activity were replicated in the genome-wide scan, with the exception of a modest signal for *LEPR*. Nevertheless, most of the previous association studies used very different means to characterize physical activity behavior; thus, nonreplication of these findings is not surprising. With the availability of technologies capable of scanning the entire genome for association to behavioral traits, the potential to discover specific genes and pathways that influence physical activity and exercise behavior is quite exciting.

Gene–Environment Interaction and Epigenetics

In recent times, environmental conditions have become less favorable for physical activity due to reductions in occupation, transportation, and domestic work-related energy expenditure, while the same conditions have become more favorable for sedentary behaviors like watching television. It is possible that there is an interaction of these genes with other environmental factors (e.g., diet) in determining the final activity phenotype. Recently, Le Galliard and colleagues showed that although locomotor performance of the common lizard is highly heritable, only the high-endurance neonates with dietary restriction continued their superior physical performance over time (Le Galliard et al., 2004).

Epigenetics could also be important in determining the final physical activity phenotype. Epigenetics refers to changes in gene expression due to chemical modifications of DNA or histone proteins without changes in DNA sequence in genes (Jaenisch & Bird, 2003). There is evidence that these changes occur during fetal life and could continue even after birth (Weaver et al., 2004). Supporting this interesting hypothesis is data from Vickers and colleagues who found that offspring of female Wistar rats that were subjected to limited food intake during pregnancy exhibited lower locomotor activity than controls (Vickers et al., 2003). Interestingly, the sedentary behavior of the offspring was further accentuated when they were given a high-fat diet.

Further work investigating the interaction of environment and genetics awaits confirmation of the genetic factors associated with physical activity regulation.

Conclusions

It is clear from the available data that voluntary physical activity is heritable. Although several QTL and candidate genes associated with both physical activity and inactivity have been identified, no definitive biomarker of sedentary behavior has been identified to date. It is likely that multiple QTL with individual small effects might be interacting with each other to bring about the large degree of variability in physical activity. Further research in this area is certainly warranted, especially in the further identification and confirmation of the genetic factors that are associated with and regulate physical activity. Additionally, since several gene polymorphisms associated with obesity have been identified, and since physical activity is a contributing factor to obesity, association studies between these markers and physical activity are warranted, an example of which is the work of Coyle and colleagues (2002) relating *Nhlh2* to physical activity regulation. Although the propensity for physical activity is distinct from exercise endurance and physical fitness phenotypes, it is interesting to speculate whether the former and genetic factors associated with it are related to sports performance, based on its potential impact on motivation for prolonged training. While the applied impact of the identification of the genetic factors associated with activity regulation is hard to predict, identification of biomarkers of physical activity could lead to the development of potential customizable approaches to encouraging physical activity in those predisposed to inactivity.

References

Berentzen, T., Kring, S.I., Holst, C., et al. (2008). Lack of association of fatness-related FTO gene variants with energy expenditure or physical activity. *The Journal of Clinical Endocrinology and Metabolism* 93(7), 2904–2908.

Borecki, I.B., Rice, T., Bouchard, C. & Rao, D.C. (1991). Commingling analysis of generalized body mass and composition measures: The Quebec Family Study. *International Journal of Obesity* 15, 763–773.

Bronikowski, A.M., Rhodes, J.S., Garland, T., Jr., Prolla, T.A., Awad, T.A. & Gammie, S.C. (2004). The evolution of gene expression in mouse hippocampus in response to selective breeding for increased locomotor activity. *Evolution* 58, 2079–2086.

Cai, G., Cole, S.A., Butte, N., et al. (2006). A quantitative trait locus on chromosome 18q for physical activity and dietary intake in Hispanic children. *Obesity (Silver Spring)* 14, 1596–1604.

Carlsson, S., Andersson, T., Lichtenstein, P., Michaelsson, K. & Ahlbom, A. (2007). Physical activity and mortality: Is the association explained by genetic selection? *American Journal of Epidemiology* 166, 255–259.

Chen, H.I., Kuo, Y.M., Liao, C.H., et al. (2008). Long-term compulsive exercise reduces the rewarding efficacy of 3,4-methylenedioxymethamphetamine. *Behavioural Brain Research* 187, 185–189.

Chua, S.C., Jr., Chung, W.K., Wu-Peng, X.S., et al. (1996). Phenotypes of mouse diabetes and rat fatty due to mutations in the OB (leptin) receptor. *Science* 271, 994–996.

Coyle, C.A., Jing, E., Hosmer, T., Powers, J.B., Wade, G. & Good, D.J. (2002). Reduced voluntary activity precedes adult-onset obesity in Nhlh2 knockout mice. *Physiology & Behavior* 77, 387–402.

Crespo, C.J., Palmieri, M.R., Perdomo, R.P., et al. (2002). The relationship of physical activity and body weight with all-cause mortality: Results from the Puerto Rico Heart Health Program. *Annals of Epidemiology* 12, 543–552.

Deacon, R.M., Brook, R.C., Meyer, D., et al. (2006). Behavioral phenotyping of mice lacking the K ATP channel subunit Kir6.2. *Physiology & Behavior* 87, 723–733.

De Moor, M.H., Liu, Y.J., Boomsma, D.I., et al. (2009). Genome-wide association study of exercise behavior in Dutch and American adults. *Medicine and Science in Sports and Exercise* 41, 1887–1895.

Festing, M.F. (1977). Wheel activity in 26 strains of mouse. *Laboratory Animals* 11, 257–258.

Girard, I., Rezende, E.L. & Garland, T., Jr. (2007). Leptin levels and body composition of mice selectively bred for high voluntary locomotor activity. *Physiological and Biochemical Zoology* 80, 568–579.

Gomes, F.R., Rezende, E.L., Malisch, J.L., et al. (2009). Glycogen storage and muscle glucose transporters (GLUT- 4) of mice selectively bred for high voluntary wheel running. *The Journal of Experimental Biology* 212, 238–248.

Good, D.J., Porter, F.D., Mahon, K.A., Parlow, A.F., Westphal, H. & Kirsch, I.R. (1997). Hypogonadism and obesity in mice with a targeted deletion of the Nhlh2 gene. *Nature Genetics* 15, 397–401.

Hakanen, M., Raitakari, O.T., Lehtimäki, T., et al. (2009). *FTO* genotype is associated with body mass index after the age of seven years but not with energy intake or leisure-time physical activity. *The Journal of Clinical Endocrinology and Metabolism* 94(4), 1281–1287.

Haskell, W.L., Lee, I.M., Pate, R.R., et al. (2007). Physical activity and public health: Updated recommendation for adults from the American College of Sports Medicine and the American Heart Association. *Medicine and Science in Sports Exercise* 39, 1423–1434.

Healy, G.N., Wijndaele, K., Dunstan, D.W., et al. (2008). Objectively measured sedentary time, physical activity, and metabolic risk: The Australian Diabetes, Obesity and Lifestyle Study (AusDiab). *Diabetes Care* 31, 369–371.

Hu, G., Lindstrom, J., Valle, T.T., et al. (2004). Physical activity, body mass index, and risk of type 2 diabetes in patients with normal or impaired glucose regulation. *Archives of Internal Medicine* 164, 892–896.

Irwin, M.L., Smith, A.W., McTiernan, A., et al. (2008). Influence of pre- and postdiagnosis physical activity on mortality in breast cancer survivors: The health, eating, activity, and lifestyle study. *Journal of Clinical Oncology* 26, 3958–3964.

Jaenisch, R. & Bird, A. (2003). Epigenetic regulation of gene expression: How the genome integrates intrinsic and environmental signals. *Nature Genetics* 33(Suppl.), 245–254.

Jeon, C.Y., Lokken, R.P., Hu, F.B. & van Dam, R.M. (2007). Physical activity of moderate intensity and risk of type 2 diabetes: A systematic review. *Diabetes Care* 30, 744–752.

Jing, E., Nillni, E.A., Sanchez, V.C., Stuart, R.C. & Good, D.J. (2004). Deletion of the Nhlh2 transcription factor decreases the levels of the anorexigenic peptides alpha melanocyte-stimulating hormone and thyrotropin-releasing hormone and implicates prohormone convertases I and II in obesity. *Endocrinology* 145, 1503–1513.

Kas, M.J., de Mooij-van Malsen, A., de Krom, M., et al. (2008). High resolution genetic mapping of mammalian motor activity levels in mice. *Genes, Brain and Behavior* 8, 13–22.

Kelly, M.A., Rubinstein, M., Phillips, T.J., et al. (1998). Locomotor activity in D2 dopamine receptor-deficient mice is determined by gene dosage, genetic background, and developmental adaptations. *The Journal of Neuroscience* 18, 3470–3479.

Knab, A.M., Bowen, R.S., Hamilton, A.T., Gulledge, A.A. & Lightfoot, J.T. (2009). Altered dopaminergic profiles: Implications for the regulation of voluntary physical activity. *Behavioural Brain Research* 204(1), 147–152.

Koteja, P., Garland, T., Jr., Sax, J.K., Swallow, J.G. & Carter, P.A. (1999). Behaviour of house mice artificially selected for high levels of voluntary wheel running. *Animal Behaviour* 58, 1307–1318.

Landi, F., Russo, A., Cesari, M., et al. (2008). Walking one hour or more per day prevented mortality among older persons: Results from ilSIRENTE study. *Preventive Medicine* 47, 422–426.

Lanza, I.R., Short, D.K., Short, K.R., et al. (2008). Endurance exercise as a countermeasure for aging. *Diabetes* 57, 2933–2942.

Le Galliard, J.F., Clobert, J. & Ferriere, R. (2004). Physical performance and Darwinian fitness in lizards. *Nature* 432, 502–505.

Leamy, L.J., Pomp, D. & Lightfoot, J.T. (2008). An epistatic genetic basis for physical activity traits in mice. *The Journal of Heredity* 99, 639–646.

Lerman, I., Harrison, B.C., Freeman, K., et al. (2002). Genetic variability in forced and voluntary endurance exercise performance in seven inbred mouse strains. *Journal of Applied Physiology* 92, 2245–2255.

Lightfoot, J.T. (2008). Sex hormones' regulation of rodent physical activity: A review. *International Journal of Biological Sciences* 4, 126–132.

Lightfoot, J.T., Turner, M.J., Daves, M., Vordermark, A. & Kleeberger, S.R. (2004). Genetic influence on daily wheel running activity level. *Physiological Genomics* 19, 270–276.

Lightfoot, J.T., Turner, M.J., Pomp, D., Kleeberger, S.R. & Leamy, L.J. (2008). Quantitative trait loci for physical activity traits in mice. *Physiological Genomics* 32, 401–408.

Loos, R.J., Rankinen, T., Tremblay, A., Pérusse, L., Chagnon, Y. & Bouchard, C. (2005). Melanocortin-4 receptor gene and physical activity in the Québec Family Study. *International Journal of Obesity (London)* 29(4), 420–428.

Lorentzon, M., Lorentzon, R., Lerner, U.H. & Nordström, P. (2001). Calcium sensing receptor gene polymorphism, circulating calcium concentrations and bone mineral density in healthy adolescent girls. *European Journal of Endocrinology* 144(3), 257–261.

Pate, R.R., Pratt, M., Blair, S.N., et al. (1995). Physical activity and public health. A recommendation from the Centers for Disease Control and Prevention and the American College of Sports Medicine. *Journal of the American Medical Association* 273, 402–407.

Pelleymounter, M.A., Cullen, M.J., Baker, M.B., et al. (1995). Effects of the obese gene product on body weight regulation in ob/ob mice. *Science* 269, 540–543.

Physical Activity Guidelines Advisory Committee (2008). *Physical Activity Guidelines Advisory Committee Report*. U.S. Department of Health and Human Services, Washington, DC.

Rhodes, J.S. & Garland, T. (2003). Differential sensitivity to acute administration of Ritalin, apomorphine, SCH 23390, but not raclopride in mice selectively bred for hyperactive wheel-running behavior. *Psychopharmacology (Berlin)* 167, 242–250.

Richardson, C.R., Kriska, A.M., Lantz, P.M. & Hayward, R.A. (2004). Physical activity and mortality across cardiovascular disease risk groups. *Medicine and Science in Sports Exercise* 36, 1923–1929.

Sakkou, M., Wiedmer, P., Anlag, K., et al. (2007). A role for brain-specific homeobox factor Bsx in the control of hyperphagia and locomotory behavior. *Cell Metabolism* 5, 450–463.

Salamone, J.D. (1992). Complex motor and sensorimotor functions of striatal and accumbens dopamine: Involvement in instrumental behavior processes. *Psychopharmacology (Berlin)* 107, 160–174.

Salmen, T., Heikkinen, A.M., Mahonen, A., et al. (2003). Relation of aromatase gene polymorphism and hormone replacement therapy to serum estradiol levels, bone mineral density, and fracture risk in early postmenopausal women. *Annals of Medicine* 35(4), 282–288.

Simonen, R.L., Rankinen, T., Perusse, L., et al. (2003a). A dopamine D2 receptor gene polymorphism and physical activity in two family studies. *Physiology & Behavior* 78, 751–757.

Simonen, R.L., Rankinen, T., Perusse, L., et al. (2003b). Genome-wide linkage scan for physical activity levels in the Quebec Family study. *Medicine and Science in Sports Exercise* 35, 1355–1359.

Slattery, M.L., Edwards, S.L., Ma, K.N., Friedman, G.D. & Potter, J.D. (1997). Physical activity and colon cancer. A public health perspective. *Annals of Epidemiology* 7, 137–145.

Stefan, N., Vozarova, B., Del Parigi, A., et al. (2002). The Gln223Arg polymorphism of the leptin receptor in Pima Indians: Influence on energy expenditure, physical activity and lipid metabolism. *International Journal of Obesity and Related Metabolic Disorders* 26, 1629–1632.

Sui, X., LaMonte, M.J., Laditka, J.N., et al. (2007). Cardiorespiratory fitness and adiposity as mortality predictors in older adults. *Journal of the American Medical Association* 298, 2507–2516.

Swallow, J.G., Carter, P.A. & Garland, T., Jr. (1998). Artificial selection for increased wheel-running behavior in house mice. *Behaviour Genetics* 28, 227–237.

Tsao, T.S., Li, J., Chang, K.S., et al. (2001). Metabolic adaptations in skeletal muscle overexpressing GLUT4: Effects on muscle and physical activity. *FASEB Journal* 15, 958–969.

van der Bij, A.K., Laurant, M.G. & Wensing, M. (2002). Effectiveness of physical activity interventions for older adults: A review. *American Journal of Preventive Medicine* 22, 120–133.

Vella, K.R., Burnside, A.S., Brennan, K.M. & Good, D.J. (2007). Expression of the hypothalamic transcription factor Nhlh2 is dependent on energy availability. *Journal of Neuroendocrinology* 19, 499–510.

Vickers, M.H., Breier, B.H., McCarthy, D. & Gluckman, P.D. (2003). Sedentary behavior during postnatal life is determined by the prenatal environment and exacerbated by postnatal hypercaloric nutrition. *American Journal of Physiology— Regulatory, Integrative and Comparative Physiology* 285, R271–R273.

Waterston, R.H., Lindblad-Toh, K., Birney, E., et al. (2002). Initial sequencing and comparative analysis of the mouse genome. *Nature* 420, 520–562.

Weaver, I.C., Cervoni, N., Champagne, F.A., et al. (2004). Epigenetic programming by maternal behavior. *Nature Neuroscience* 7, 847–854.

Winnicki, M., Accurso, V., Hoffmann, M., et al. (2004). Physical activity and angiotensin-converting enzyme gene polymorphism in mild hypertensives. HARVEST Study Group. *American Journal of Medical Genetics. Part A* 125A(1), 38–44.

Zhou, D., Shen, Z., Strack, A.M., Marsh, D.J. & Shearman, L.P. (2005). Enhanced running wheel activity of both Mch1r- and Pmch-deficient mice. *Regulatory Peptides* 124, 53–63.

Chapter 25

Genes, Exercise, and Psychological Factors

ECO J.C. DE GEUS AND MARLEEN H.M. DE MOOR

Department of Biological Psychology, VU University Amsterdam, Amsterdam, The Netherlands

Regular exercisers, defined as individuals who voluntarily seek out moderate to vigorous physical activity on a weekly or even daily basis, have a different psychological profile than non-exercisers. Regular exercisers score lower on neuroticism and higher on extraversion, conscientiousness, and sensation-seeking (De Moor et al., 2006; Hoyt et al., 2009; Rhodes & Smith, 2006). These stable differences in psychological traits are reflected in assessments of ongoing mood and one of the most compelling descriptions of the psychological profile of regular exercisers is that of the "iceberg profile" in Morgan's mental health state model (Morgan & Johnson, 1977). Regular exercisers have higher vigor (the peak of the iceberg) but lower levels of anxiety, tension, apprehension, depression, and fatigue compared to the average of these mood states in non-exercisers.

Originally, the interest of sports psychologists in the psychological profile of exercisers was driven by the idea that personality and mood could help predict athletic performance in elite athletes. This did not turn out to be a very useful addition to the coach's toolbox. A large meta-analysis, for instance, showed that Profile of Mood States scores accounted for less than 1% of athletic success (Rowley et al., 1995). However, Morgan's research facilitated the growing interest in the relation between exercise and psychology in a different field, that of preventive medicine. The core idea there is that exercise (not necessarily at the elite level) increases psychological well-being and can attenuate or even prevent the development of anxiety and depressive disorders in the population at large. This converges with folk wisdom that exercise "makes you feel better" and can help combat stress. In the last two decades, this has become a dominant angle in sports psychology.

Exercise and Psychological Well-Being

When compared to sedentary individuals, regular exercisers have been found to have higher self-esteem (Sonstroem & Morgan, 1989), life satisfaction and happiness (Stubbe et al., 2006b), better perceived health, and quality of life (De Moor et al., 2007b; Klavestrand & Vingard, 2009). Most importantly, regular exercisers have lower levels of anxious and depressive symptoms in both adolescents and adults (Brown et al., 2005; Camacho et al., 1991; Cooper-Patrick et al., 1997; De Moor et al., 2006; Farmer et al., 1988; Kritz-Silverstein et al., 2001; McKercher et al., 2009; Rhodes & Smith, 2006; Stephens, 1988; Strawbridge et al., 2002; van Gool et al., 2003; Weiss et al., 2008; Weyerer, 1992; Wise et al., 2006). However, co-existence of a healthy mind and a healthy body does not prove that the healthy mind is a consequence of the healthy body (such causality was never suggested by Juvenal, the Roman satirist who originated the adage of "mens sana in corpore sano"—he just prayed for both). The association could as easily

Genetic and Molecular Aspects of Sport Performance, 1st edition.
Edited by Claude Bouchard and Eric P. Hoffman.
Published 2011 by Blackwell Publishing Ltd.

reflect a reversed causality, where emotionally well-adjusted, agreeable, and self-confident individuals with low levels of stress are simply more attracted to sports and exercise, or that only such persons have the necessary energy and self-discipline to maintain an exercise regime. In short, the favorable psychological profile of exercisers may reflect self-selection.

To resolve causality, various studies have addressed the association in longitudinal designs (Brown et al., 2005; Camacho et al., 1991; Cooper-Patrick et al., 1997; Farmer et al., 1988; Kritz-Silverstein et al., 2001; Strawbridge et al., 2002; van Gool et al., 2003; Weyerer, 1992; Wise et al., 2006). Most of these longitudinal studies reported that regular exercise at baseline was associated with less depression and anxiety at follow-up. Some studies, however, did not find evidence for a longitudinal association (Cooper-Patrick et al., 1997; Kritz-Silverstein et al., 2001; Weyerer, 1992), and one study found a longitudinal association in White women but not in men and Black women (Farmer et al., 1988). More importantly, observational longitudinal studies cannot truly resolve causality. Higher psychological well-being in exercisers may reflect the operation of common underlying influences on both exercise behavior and psychological well-being.

These "third factors" may consist of environmental influences like low socio-economic status and poor social support networks that affect both exercise behavior and psychological well-being. Alternatively the association between regular exercise behavior and the psychological profile may be partly due to underlying genetic factors that have a favorable effect on both traits. This means that the genetic variation that leads to, for example, increased levels of depression may also influence voluntary exercise behavior. This phenomenon, where low-level biological variation has effects on multiple complex traits at the organ and behavioral level, is called genetic *pleiotropy*. If present in a time-lagged form, that is, when genetic effects on exercise behavior precede effects of the same genes on psychological health at a later time point, this phenomenon can

cause longitudinal correlations that mimic the causal effects of exercise. To test for the presence of underlying genetic "third factors," genetically informative designs are needed, for instance, a study in twins.

Twin studies can directly decompose familial resemblance into genetic and shared environmental influences by comparing the resemblance in exercise behavior between monozygotic (MZ) and dizygotic (DZ) twins. When twins are reared together, they share part of their environment, and this sharing of the family environment is postulated to be the same for MZ and DZ twins. The important difference between MZ and DZ twins is that the former share (close to) all of their genotypes, whereas the latter share on average only half of the genotypes segregating in that family. If the resemblance in exercise behavior within MZ pairs is larger than in DZ pairs, this suggests that genetic factors influence exercise behavior. If the resemblance in exercise behavior is as large in DZ twins as it is in MZ twins, this points to shared environmental factors as the cause of family resemblance (Boomsma et al., 2002). As opposed to parent–offspring family designs, a twin study estimates heritability within members of the same generation, which avoids dilution of within-family resemblance by cohort effects.

In a twin study, four possible components and their interactions and correlations are thought to contribute to the total variance in a trait: unique environmental factors ("E"), shared environmental factors ("C"), additive genetic factors ("A"), and dominant genetic factors ("D"). Shared environmental factors ("C"), and additive ("A") and dominant ("D") genetic factors can cause twin resemblance, whereas the extent to which twins do not resemble each other is ascribed to the unique (or non-shared) environmental factors. These include all unique experiences like differential jobs or lifestyle, accidents or other life events, and in childhood, differential treatment by the parents, and non-shared peers. Using twins, we can only estimate three components of variance at the same time (A, C, and E, or A, D, and E). One solution is to add parents or offspring of twins to the design (Keller et al., 2009). If data on parents are not available, one needs to make the assumption that either C or D is absent. The presence of dominance can be inferred from the pattern of twin correlations because it yields DZ correlations that are much lower than half the MZ correlation. In contrast, the presence of shared environmental effects yields DZ correlations that are much higher than half the MZ correlations.

Twin researchers typically use structural equation modeling to estimate the relative contribution of A, D/C, and E to the individual differences in the trait. In structural equation modeling, the relationships between several latent unobserved variables (e.g., genetic and environmental factors) and observed variables are summarized by a series of equations. Additional equations can specify the correlation between the latent genetic and environmental factors if these are known. It is possible to derive the variance–covariance matrix implied by the total set of equations (*the model*) through the use of covariance algebra. When the complexity and number of the equations increase, the structural equation model can be formulated more easily by application of path-tracing rules on the complete representation of all relationships between observed and unobserved variables in a so-called *path diagram*. An example is depicted in Figure 25.1 where exercise behavior has been measured in DZ and MZ twin pairs. In this example inspection of the twin correlations had suggested that dominance does not play a role and that all the genetic variance in exercise behavior is additive genetic variance. Hence only the latent factors A, C, and E are used in the diagram, and the latent D factor was omitted.

Using maximum likelihood estimation, we can iteratively test the fit of the expected covariances/variances to the actual observed covariances/variances in a sample of hundreds or thousands of twins over a range of possible values for the path coefficients. From the best fitting model, we take the estimates for the path coefficients (e.g., a, c, and e) and determine the relative contribution of the latent factors to the total variance in leisure-time exercise behavior. Heritability of this behavior,

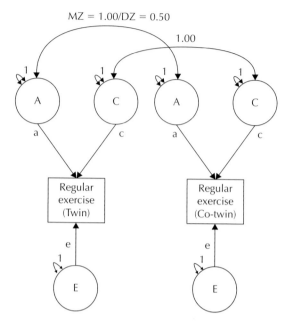

Figure 25.1 Path diagram depicting correlated latent additive genetic (A) and shared environmental (C) factors that can cause twin resemblance in exercise behavior, as well as unique environmental (E) factors that are uncorrelated in the twins.

defined as the relative proportion of the total variance explained by genetic factors, is obtained as the ratio of $a^2/(a^2 + e^2 + c^2)$. The heritability can also be expressed as a percentage by multiplying this ratio by a hundred.

Heritability of Exercise Behavior

Using the twin design, it has been established that genetic factors importantly contribute to individual differences in exercise participation and measures of exercise frequency, duration and/or intensity. An overview of these twin studies is given in Table 25.1. We focused on voluntary leisure-time exercise behavior and it is important to note that different heritability estimates may be obtained for total physical activity that includes strenuous activities at the workplace or commuting by bicycle. Indeed, the factors underlying total physical activity need not overlap with the factors that underlie leisure-time

Table 25.1 Twin studies on exercise behavior

Study	Sample	Phenotype(s)	Results
Carlsson et al. (2006)	5334 MZ and 8028 DZ pairs (aged 14–46 years)	Leisure-time PA	$h^2 = 64\%$; $c^2 = 7\%$ for males 14–28 years $h^2 = 51\%$; $c^2 = 15\%$ for females 14–28 years $h^2 = 40\%$; $c^2 = 0\%$ for males 29–46 years $h^2 = 41\%$; $c^2 = 0\%$ for females 29–46 years
Eriksson et al. (2006)	1022 Swedish male twin pairs (aged 19–29 years)	Total PA, leisure-time PA incl. and excl. sport, sport during leisure-time, occupational PA	$h^2 = 40$–65%; $c^2 = 0\%$
Stubbe et al. (2006a)	13,676 MZ and 23,375 DZ pairs from seven different countries in Europe and Australia (aged 19–40 years)	Leisure-time exercise participation	$h^2 = 27$–67%; $c^2 = 0$–37% for males $h^2 = 48$–71%; $c^2 = 0\%$ for females
Stubbe et al. (2005a)	2628 Dutch twin pairs (aged 13–14, 15–16, 17–18, 19–20 years)	Leisure-time exercise participation	$h^2 = 0\%$; $c^2 = 84\%$ for 13- to 14-year-old twins $h^2 = 0\%$; $c^2 = 78\%$ for 15- to 16-year-old twins $h^2 = 36\%$; $c^2 = 47\%$ for 17- to 18-year-old twins $h^2 = 85\%$; $c^2 = 0\%$ for 19- to 20-year-old twins
Beunen & Thomis (1999)	92 male and 91 female Belgium twin pairs (aged 15 years)	Number of hours spent on sports each week	$h^2 = 83\%$; $c^2 = 0\%$ for males $h^2 = 44\%$; $c^2 = 54\%$ for females
de Geus et al. (2003)	157 adolescent (aged 13–22 years) and 208 middle-aged Dutch twin pairs (aged 35–62 years)	Weekly METs for vigorous leisure-time exercise	$h^2 = 79\%$; $c^2 = 0\%$ for adolescent twins $h^2 = 41\%$; $c^2 = 0\%$ for middle-aged twins
Frederiksen & Christensen (2003)	616 MZ and 642 same-sex DZ twin pairs (aged 45–68 years)	Leisure-time exercise participation in any of 11 activities	$h^2 = 49\%$; $c^2 = 0\%$ for males and females
Kujala et al. (2002)	Data on both members of 1772 MZ and 3551 DZ same-sex twin pairs (aged 24–60 years)	Participation in vigorous PA	$h^2 = 56\%$; $c^2 = 4\%$ for vigorous activity
Maia et al. (2002)	411 Portuguese twin pairs (aged 12–25 years)	Sports participation	$h^2 = 68\%$; $c^2 = 20\%$ for males $h^2 = 40\%$; $c^2 = 26\%$ for females
Aarnio et al. (1997)	3254 twins at age 16, their parents and grandparents	Engagement in five types of PA	$h^2 = 54\%$; $c^2 = 18\%$ for males $h^2 = 46\%$; $c^2 = 18\%$ for females
Lauderdale et al. (1997)	3344 male twin pairs of the Vietnam Era Twin Registry (aged 33–51 years)	Regular participation in five types of sports	$h^2 = 53\%$ for jogging $h^2 = 48\%$; $c^2 = 4\%$ for racquet sports $h^2 = 30\%$; $c^2 = 17\%$ for strenuous sports $h^2 = 58\%$ for bicycling $h^2 = 8\%$; $c^2 = 31\%$ for swimming

MZ, monozygotic; DZ, dizygotic; PA, physical activity; h^2, heritability; c^2, shared environmental factors.

activity, or they may even be negatively correlated as high-energy expenditure demands at work or during commuting may countermand voluntary exercise behavior.

A striking finding in these studies is that the genetic architecture of leisure-time exercise behavior is vastly different across the life span. Studies conducted in adult twins aged between 19 and 68 years (Beunen & Thomis, 1999; Carlsson et al., 2006; Eriksson et al., 2006; Frederiksen & Christensen, 2003; Kujala et al., 2002; Lauderdale et al., 1997; Stubbe et al., 2006a) show that variation in exercise behavior is accounted for by genetic and non-shared environmental factors, with heritability estimates ranging between 35% and 83% of the variance in exercise. Twin studies in adolescence (Aarnio et al., 1997; Carlsson et al., 2006; Maia et al., 2002; Stubbe et al., 2005a) show that variation in adolescent leisure-time exercise behavior is explained by a combination of genetic, shared environmental, and unique environmental factors. In a study of adolescent twins aged 13–20 years (Stubbe et al., 2005a), it was observed that exercise behavior in young adolescents (up to 16 years) is largely determined by shared environmental factors. The influence of these factors rapidly wanes when adolescents become young adults and genetic factors start to appear. After age 18, heritability estimates for exercise behavior are as high as 80% (de Geus et al., 2003; Stubbe et al., 2005a).

Bivariate Heritability of Exercise Behavior and Psychological Well-Being

We now deal with the question of whether the genetic factors underlying exercise behavior also play a role in well-being. First it is important to note that a number of (twin) family studies have demonstrated that general well-being and aspects of it, such as life satisfaction and happiness, are influenced by genetic factors just like anxiety and depression (Hettema et al., 2001; Lykken & Tellegen, 1996; Nes et al., 2006; Stubbe et al., 2005b; Sullivan et al., 2000). Similarly, significant genetic influences have been reported for self-rated health

and quality of life (De Moor et al., 2007b; Svedberg et al., 2005). To test for an overlap in the genetic factors for exercise behavior and well-being we used a bivariate extension of the structural equation model for twin data (Neale & Cardon, 1992). This allows a whole new set of hypotheses to be tested because the observed information now includes all possible cross-twin cross-trait correlations, for instance the correlation of exercise behavior in a twin with the psychological well-being of his or her co-twin.

Using this addition to the covariance matrix we can test the extent to which the heritability of these traits is caused by common genetic factors that influence all of the traits, as well as the extent to which heritability is caused by genetic factors that are specific to each trait. As shown in Figure 25.2, there are now two genetic factors: common genetic factor A1 influences regular exercise as well as psychological well-being, whereas specific genetic factor A2 influences well-being only. When multivariate models are depicted in a path diagram, the number of arrows (and path loadings) can become overwhelming so

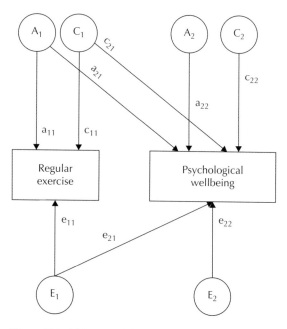

Figure 25.2 A bivariate ACE model for exercise behavior and well-being.

Figure 25.2 differs from Figure 25.1 in that the part of only one twin is depicted (the other twin and non-twin sibling is still used in the analyses with their latent factors correlated as before; they are just no longer drawn).

Path coefficient a_{11} quantifies the effect of genetic factor A_1 on regular exercise; a_{21} quantifies the pleiotropic genetic effect of A_1 on well-being. Coefficient a_{22} quantifies the effect of specific genetic factor A_2 on well-being. If a_{21} is zero and a_{22} is significantly different from zero, the association between exercise and well-being does not derive from the same genetic factor and there is no evidence of pleiotropy. In a similar way, path coefficients e_{11}, c_{11}, e_{21}, c_{21}, e_{22}, and c_{22} quantify the effects of common and specific factors E and C on exercise and well-being. If c_{21} is non-significant but c_{22} significantly differs from zero, the association between exercise and well-being does not derive from the same common environmental factor.

Alternatively, the bivariate twin model in Figure 25.2 can be used to decompose a phenotypic correlation between two traits into its three possible sources: overlapping genetic, overlapping shared environmental, or overlapping unique environmental factors. The word *overlapping* can be defined more precisely as the correlation between the latent genetic (R_g), shared environmental (R_c), or unique environmental (R_e) factors influencing regular exercise (A_1, C_1, E_1) and psychological well-being (A_2, C_2, E_2). More generally, R_g between two traits is derived as the genetic covariance between the traits divided by the square root of the product of their genetic variances ($R_g = (a_{11}*a_{21})/\sqrt{(a_{11}^2)} * \sqrt{(a_{21}^2 + a_{22}^2)}$). If the genetic correlation is close to 1, the two genetic factors (A_1 and A_2) overlap completely. In this case, a_{22} will be zero and there are no genes that emerge specifically for the second trait. If the genetic correlation is significantly less than 1, there may be overlap in the genetic factors (A_1 and A_2) that influence both traits but the overlap is imperfect, and a_{22} will be non-zero. If the genetic correlation is close to 0, there is no overlap in the genetic factors that influence both traits and the heritability of the second trait is determined completely by a_{22}. Analogously, the environmental correlations R_g and R_c between two traits are derived as the

environmental covariance divided by the square root of the product of the environmental variances of the two traits ($R_c = (c_{11}*c_{21})/\sqrt{(c_{11}^2)} * \sqrt{(c_{21}^2 + c_{22}^2)}$; $R_e = (e_{11}*e_{21})/\sqrt{(e_{11}^2)} * \sqrt{(e_{21}^2 + e_{22}^2)}$).

Computation of genetic and environmental correlations in longitudinal data across 2-, 4-, 7-, 9-, and 11-year follow-up periods from 8558 twins and their family members was used to test a crucial prediction from the causal hypothesis (De Moor et al., 2008). If exercise causally influences symptoms of anxiety and depression, all genetic and environmental factors that influence variance in exercise behavior will, through the causal chain, also influence the variance in these symptoms. Translating this to the structural equation models used on twin family data, this means that, if A, C, and E contribute to both traits, the genetic (R_g) and environmental (R_e, R_c) correlations between exercise and symptoms must *both* be significantly different from zero. Furthermore, this should apply to the cross-sectional cross-trait correlations as well as the longitudinal cross-trait correlations (De Moor et al., 2008). In spite of sufficient power, we found only the genetic correlation to be significant (ranging between -0.16 and -0.44 for different symptom scales and different time-lags). Environmental correlations, however, were essentially zero. This means that the environmental factors that cause a person to take up exercise do not cause lower anxiety or depressive symptoms in that person, currently or at any future time point. In contrast, the genetic factors that cause a person to take up exercise also cause lower anxiety or depressive symptoms in that person at the present and all future time points.

We have also addressed the association between exercise and well-being in a sample of 5140 Dutch adult twins and their non-twin siblings from 2831 families using self-rated health as an index of well-being (De Moor et al., 2007b). Bivariate genetic models tested the contribution of genetic and environmental factors to the observed correlation between exercise participation and self-rated health. We showed that the genetic factors influencing exercise participation and self-rated health partially overlap ($R_g = 0.36$) and, importantly, this overlap fully explains their association. Again this argues in favor of genetic pleiotropy and suggests that the well-being of genetically identical individuals shows a high resemblance, even if one is a fervent exerciser and the other is a couch potato.

Training Studies

Based on the work described above, we conclude that, in the population at large, exercise participation is associated with higher levels of perceived health, life satisfaction, and happiness, as well as lower levels of anxiety and depression, largely through genetic factors that influence both exercise behavior and psychological well-being. At first sight, these pleiotropic effects would suggest that recruitment of "genetic non-exercisers" in exercise activities would do little to change well-being because exercising would not change their genotype. However, it is important to recognize that we assessed leisure-time exercise behavior that is completely initiated and maintained by individuals on a voluntary basis. Although the association between such voluntary self-chosen exercise behavior and well-being does not reflect a causal effect, it still does not rule out the possibility that successful recruitment of sedentary individuals, for instance as part of a (cardiac) rehabilitation or psychiatric therapy program, might have an effect on well-being.

To test the potential of exercise to change the well-being of individuals who do not voluntarily engage in exercise, an experimental manipulation of exercise behavior in these individuals is needed. Several well-designed training studies have reported improved mood and coping behavior or reduction in depression and anxiety after a program of aerobic exercise training in comparison to control manipulations (Barbour & Blumenthal, 2002; Brosse et al., 2002; Steptoe et al., 1989). Unfortunately, many others have failed to replicate these training effects (de Geus et al., 1993; King et al., 1989). Without denying the potential of exercise to change well-being in subsets of individuals, critical reviews on this topic express only cautious optimism about the use of exercise for the enhancement of well-being in the population at large (Dunn et al., 2005; Lawlor & Hopker, 2001).

The most promising evidence comes from studies in individuals who had low initial levels of well-being at the start of the exercise program, like depression patients. Various randomized controlled trials showed that regular exercise can be used as a treatment to relieve symptoms in these patients (Babyak et al., 2000; Blumenthal et al., 2007). These studies reported beneficial psychological effects of exercise that match or even exceed those of pharmacological treatment. Although these results are promising it should be kept in mind that these studies examined the effects of prescribed and externally monitored exercise treatments in select subgroups. Whether these individuals are able to maintain exercise behavior in the long run is a serious concern, since up to 50% of individuals recruited in exercise programs may drop out within the first few months (Dishman & Ickes, 1981). Moreover, due to the strong societal beliefs about the efficacy of this intervention, severe methodological problems cling to even the most well-designed randomized controlled trial (Ekkekakis, 2008).

At present, to assume that regular exercise may prevent depression in a non-clinical population sample, because it has been successfully used as a therapy in subsets of clinically depressed patients, is still a leap of faith. What is needed is a better understanding of why non-exercisers chose *not* to exercise in spite of the well-advertised benefits in many domains. Below we argue that increased knowledge of the genetic factors determining the acute psychological response to exercise may be a powerful way forward.

Genes for Exercise Behavior

What types of genes are relevant to exercise behavior? As with any other behavior, for exercise behavior to be repeated time and again, the net rewarding effects of exercise would need to outweigh the net aversive effects. Although emphasis traditionally is put on "feeling good" after exercise, the reality of sedentariness dictates that many individuals may be "feeling bad" after exercise. Figure 25.3 proposes that individuals for whom the aversive effects are stronger than the rewarding effects will become non-exercisers. In contrast,

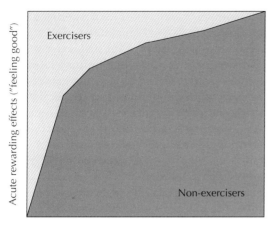

Figure 25.3 Hypothesized relation between the acute psychological effects of exercise and long-term exercise behavior.

individuals for whom the rewarding effects are dominant will repeat the behavior and become regular lifetime exercisers. The figure proposes that, to become an exerciser, the rewarding effects must outweigh the aversive effects to a substantial degree. It also proposes that this may be the case for a modest percentage of the population, in keeping with the low percentage of individuals (17% of the Dutch population) that regularly engage in vigorous exercise (De Moor et al., 2007b).

Whether the rewarding or punishing effects of a behavior affect the future frequency of the behavior strongly depends on the contingency in time of behavior and reinforcement. Hence, the psychological effects during and shortly after exercising, often termed the *acute effects* of exercise will be especially important. Among the hypothesized factors that determine the acute psychological effects of exercise, a homeostatic "need for activity" should probably be first on the list. The motivation to exercise may be a real drive in the classical Hullian sense, not different from sex drive, hunger, or thirst (Rowland, 1998). Hypothalamic and brain stem control systems are likely biased to food intake over energy expenditure, but large individual differences in this homeostatic system may exist with some individual intrinsically driven to more exercise than others. Fulfillment of this homeostatic

drive may be directly coupled to dopaminergic midbrain reward systems, not unlike eating and drinking.

Additional activation of the rewarding system may come from two well-known social psychological mechanisms. The first may be specific to individuals with above-average exercise abilities. People generally like doing what they are good at, and will pursue those activities in leisure time as much as possible. Taken the strong positive cultural attitudes toward exercise ability, individuals who notice that they gain more in performance than others (that nonetheless follow the same exercise regime) will experience strong feelings of competence and mastery. A second mechanism may be limited to individuals who have strong self-regulatory skills, including self-discipline. These individuals may experience reward by the sheer accomplishment of an exercise challenge, even when exercising has strong aversive effects.

Among the aversive effects of exercise, the first one to list is that the person may need to abort other ongoing activities. Put otherwise, exercise has to compete with other, potentially more pleasurable, activities available to the individual. But the aversive effects of exercise may also be direct through feelings of pain, fatigue, or even exertion. According to Ekkekakis and colleagues (Ekkekakis et al., 2005, 2008), intensity is an important determinant of the aversive responses to exercise. At intensities that exceed the individuals' ventilatory threshold, strong interoceptive cues, inherently charged with negative affect, may present a major obstacle to perceiving pleasure. Tolerance for aversive interoceptive cues may depend on the same serotonergic punishment systems in the brain that are also involved in the pathogenesis of depression and anxiety. Many novice exercisers may lack the skills to adequately regulate the intensity of exercise at the optimal level and fail to use their felt discomfort as a meaningful guideline, because they may have been brainwashed to push harder by their coaches, parents, or peers ("no pain, no gain").

In the less able exercisers, the aversive effects may also stem from the reverse variant of the psychosocial mechanisms that make exercise rewarding for those who have above-average exercise abilities.

Put otherwise, people generally *don't* like doing things they are *not* good at. Individuals who achieve low levels of performance, even after substantial training, may feel disappointment and even shame, particularly when the exercise is performed in a competitive context. In this regard, it is perhaps not striking that the highest heritabilities of exercise behavior are achieved during adolescence (Stubbe & de Geus, 2009), when the sensitivity to one's own relative ranking among peers may be the largest.

We have previously hypothesized that the acute aversive and rewarding effects of exercise show strong individual differences and that these differences may be strongly genetically determined (de Geus & De Moor, 2008). That is, there are gene-by-exercise interaction effects for the acute psychological effects of exercise. Furthermore, in keeping with the finding that regular exercisers report greater acute exercise-induced mood enhancement than non-exercisers (Hoffman & Hoffman, 2008), we predicted that genes determining individual differences in the acute psychological effects of exercise to a large extent explain the heritability of voluntary exercise behavior as tabulated in Table 25.1. More specifically we may predict that the heritability of exercise behavior reflects genes influencing individual differences in the immediate rewarding and aversive effects, which may directly depend on genetic variation in the homeostatic exercise drive, exercise-induced engagement of the opioid and dopamine reward systems, exercise-induced engagement of punishment systems or, indirectly, on exercise ability.

These predictions can be most optimally tested in a training study in twins by showing significant genetic correlation between the exercise participation and individual differences in the acute psychological response to exercise, for instance as assessed by a change in mood. To our knowledge, no such studies have been performed to date. Below, we follow an alternative strategy by examining the genes currently implicated in variation in exercise behavior and see whether they fit our theoretical model. In total, nine genes have been found to be associated with exercise behavior, or more broadly defined physical activity phenotypes, in six candidate gene

association studies (Loos et al., 2005; Lorentzon et al., 2001; Salmen et al., 2003; Simonen et al., 2003a; Stefan et al., 2002; Winnicki et al., 2004), three genome-wide linkage studies (Cai et al., 2006; De Moor et al., 2007a; Simonen et al., 2003b), and one genome-wide association study (GWAS) (De Moor et al., 2009).

Two genes were identified that are highly expressed in the hypothalamic area and may fit our idea of genetic variation in the "need for activity." The leptin receptor (*LEPR*) gene was associated with total physical activity in Pima Indians (Stefan et al., 2002) and was also associated with leisure-time exercise participation in a sample of American Caucasians (De Moor et al., 2009), in both studies independent of BMI. The melanocortin-4 receptor (*MC4R*) gene was associated with daily physical activity levels in a combined sample of adult men and women (Loos et al., 2005), and the same region on chromosome 18 was also identified in a linkage study on daily physical activity in children (Cai et al., 2006).

Two of the exercise genes identified could influence the reward or aversive experience of exercise directly. In a candidate gene study, the dopamine 2 receptor (*DRD2*) gene was found to be associated with physical activity, sports participation, and occupational physical activity in adult females (Simonen et al., 2003a). In the GWAS, the *GABRG3* gene was associated with exercise participation in a combined sample of 2622 Dutch and American individuals (De Moor et al., 2009) and one of the linkage studies had also pointed to the 15q12-13 region that harbors the *GABRG3* gene (Simonen et al., 2003b). The *GABRG3* gene may be involved in the aversive effects of exercise-induced fatigue as shown in an elegant study on physical exercise-associated gene expression that found higher expression levels of the *GABRG3* gene after a bout of exhaustive exercise but not after aerobic exercise at 60% $\dot{V}O_2$max (Kawai et al., 2007).

Three of the exercise genes identified may specifically act to increase or decrease exercise ability. In a study of adolescent females, the calcium-sensing receptor (*CASR*) gene was associated with hours spent on physical activities per week (Lorentzon et al., 2001). The GWAS further strongly pointed to the *PAPSS2* gene on chromosome 10q23 (De Moor et al., 2009). This gene is widely expressed in skeletal and smooth muscles and the brain and has previously been linked to maximal exercise capacity in a genome-wide linkage study of 453 sib pairs (Rico-Sanz et al., 2004). Finally, in a study of mild male and female hypertensives, the angiotensin-converting enzyme (*ACE*) gene accounted for 21% of the variation in leisure-time physical activity (Winnicki et al., 2004). The regular exercisers had the same genotype (I/I) that has been associated with high exercise ability and, in particular, an increased responsiveness to training (Williams et al., 2000; Woods et al., 2000).

Conclusion

In the population at large, regular leisure-time exercise is associated with better mental health largely through genetic pleiotropic effects. Genetic factors that increase the chances that an individual engages in voluntary exercise behavior also influence the risk for anxiety and depression. These genetic factors may in part reflect a differential sensitivity to the acute psychological effects of exercise. For exercisers, exercising may be associated with a strong "feel good" experience and constitute an excellent short-term coping strategy that helps to unwind more rapidly from daily pressures experienced in the school, job, or home environment, up to a point where stopping exercise could lead to loss of well-being and self-esteem. In fact, most of the popular ideas on the psychological benefits of exercise may stem from exercisers themselves. For persistent sedentary individuals the aversive effects of exercise, at least in the forms that they tried so far, may greatly overwhelm the rewarding effects, causing them to drop out. Increased understanding of the genetic factors influencing the psychological response to exercise may help create better exercise programs to re-engage these individuals in some form of regular exercise.

References

Aarnio, M., Winter, T., Kujala, U.M. & Kaprio, J. (1997). Familial aggregation of leisure-time physical activity—A three generation study. *International Journal of Sports Medicine* 18, 549–556.

Babyak, M., Blumenthal, J.A., Herman, S., et al. (2000). Exercise treatment for major depression: Maintenance of therapeutic benefit at 10 months. *Psychosomatic Medicine* 62, 633–638.

Barbour, K.A. & Blumenthal, J.A. (2002). Exercise and depression in older adults. *Neurobiology of Aging* 26, S118–S123.

Beunen, G. & Thomis, M. (1999). Genetic determinants of sports participation and daily physical activity. *International Journal of Obesity* 23, S55–S63.

Blumenthal, J.A., Babyak, M.A., Doraiswamy, P.M., et al. (2007). Exercise and pharmacotherapy in the treatment of major depressive disorder. *Psychosomatic Medicine* 69, 587–596.

Boomsma, D., Busjahn, A. & Peltonen, L. (2002). Classical twin studies and beyond. *Nature Reviews Genetics* 3, 872–882.

Brosse, A.L., Sheets, E.S., Lett, H.S. & Blumenthal, J.A. (2002). Exercise and the treatment of clinical depression in adults—Recent findings and future directions. *Sports Medicine* 32, 741–760.

Brown, W.J., Ford, J.H., Burton, N.W., Marshall, A.L. & Dobson, A.J. (2005). Prospective study of physical activity and depressive symptoms in middle-aged women. *American Journal of Preventive Medicine* 29, 265–272.

Cai, G.W., Cole, S.A., Butte, N., et al. (2006). A quantitative trait locus on chromosome 18q for physical activity and dietary intake in Hispanic children. *Obesity Research* 14, 1596–1604.

Camacho, T.C., Roberts, R.E., Lazarus, N.B., Kaplan, G.A. & Cohen, R.D. (1991). Physical activity and depression: Evidence from the Alameda County Study. *American Journal of Epidemiology* 134, 220–231.

Carlsson, S., Andersson, T., Lichtenstein, P., Michaelsson, K. & Ahlbom, A. (2006). Genetic effects on physical activity: Results from the Swedish twin registry. *Medicine and Science in Sports and Exercise* 38, 1396–1401.

Cooper-Patrick, L., Ford, D.E., Mead, L.A., Chang, P.P. & Klag, M.J. (1997). Exercise and depression in midlife: A prospective study. *American Journal of Public Health* 87, 670–673.

De Moor, M.H.M., Beem, A.L., Stubbe, J.H., Boomsma, D.I. & de Geus, E.J.C. (2006). Regular exercise, anxiety, depression and personality: A population-based study. *Preventive Medicine* 42, 273–279.

De Moor, M.H.M., Posthuma, D., Hottenga, J.J., Willemsen, A.H.M., Boomsma, D.I. & de Geus, E.J.C. (2007a). Genome-wide linkage scan for exercise participation in Dutch sibling pairs. *European Journal of Human Genetics* 15, 1252–1259.

De Moor, M.H.M., Stubbe, J.H., Boomsma, D.I. & de Geus, E.J.C. (2007b). Exercise participation and self-rated health: Do common genes explain the association? *European Journal of Epidemiology* 22, 27–32.

De Moor, M.H.M., Boomsma, D.I., Stubbe, J.H., Willemsen, G. & de Geus, E.J.C. (2008). Testing causality in the association between regular exercise and symptoms of anxiety and depression. *Archives of General Psychiatry* 65, 897–905.

De Moor, M.H.M., Liu, Y.J., Boomsma, D.I., et al. (2009). Genome-wide association study of exercise behavior in Dutch and American adults. *Medicine and Science in Sports and Exercise* 41, 1887–1895.

Dishman, R.K. & Ickes, W. (1981). Self-motivation and adherence to therapeutic exercise. *Journal of Behavioral Medicine* 4, 421–438.

Dunn, A.L., Trivedi, M.H., Kampert, J.B., Clark, C.G. & Chambliss, H.O. (2005).

Exercise treatment for depression—Efficacy and dose response. *American Journal of Preventive Medicine* 28, 1–8.

Ekkekakis, P. (2008). The genetic tidal wave finally reached our shores: Will it be the catalyst for a critical overhaul of the way we think and do science? *Mental Health and Physical Activity* 1, 47–52.

Ekkekakis, P., Hall, E.E. & Petruzzello, S.J. (2005). Some like it vigorous: Measuring individual differences in the preference for and tolerance of exercise intensity. *Journal of Sport & Exercise Psychology* 27, 350–374.

Ekkekakis, P., Hall, E.E. & Petruzzello, S.J. (2008). The relationship between exercise intensity and affective responses demystified: To crack the 40-year-old nut, replace the 40-year-old nutcracker! *Annals of Behavioral Medicine* 35, 136–149.

Eriksson, M., Rasmussen, F. & Tynelius, P. (2006). Genetic factors in physical activity and the equal environment assumption—The Swedish young male twins study. *Behavior Genetics* 36, 238–247.

Farmer, M.E., Locke, B.Z., Moscicki, E.K., Dannenberg, A.L., Larson, D.B. & Radloff, L.S. (1988). Physical activity and depressive symptoms: The NHANES I epidemiologic follow-up study. *American Journal of Epidemiology* 128, 1340–1351.

Frederiksen, H. & Christensen, K. (2003). The influence of genetic factors on physical functioning and exercise in second half of life. *Scandinavian Journal of Medicine & Science in Sports* 13, 9–18.

de Geus, E.J.C. & De Moor, M.H.M. (2008). A genetic perspective on the association between exercise and mental health. *Mental Health and Physical Activity* 1, 53–61.

de Geus, E.J., van Doornen, L.J. & Orlebeke, J.F. (1993). Regular exercise and aerobic fitness in relation to psychological make-up and physiological stress reactivity. *Psychosomatic Medicine* 55, 347–363.

de Geus, E.J.C., Boomsma, D.I. & Snieder, H. (2003). Genetic correlation of exercise with heart rate and respiratory sinus arrhythmia. *Medicine and Science in Sports and Exercise* 35, 1287–1295.

Hettema, J.M., Neale, M.C. & Kendler, K.S. (2001). A review and meta-analysis of the genetic epidemiology of anxiety disorders. *The American Journal of Psychiatry* 158, 1568–1578.

Hoffman, M.D. & Hoffman, D.R. (2008). Exercisers achieve greater acute exercise-induced mood enhancement than nonexercisers. *Archives of Physical Medicine and Rehabilitation* 89, 358–363.

Hoyt, A.L., Rhodes, R.E., Hausenblas, H. A. & Giacobbi, P.R. (2009). Integrating five-factor model facet-level traits with the theory of planned behavior and exercise. *Psychology of Sport and Exercise* 10, 565–572.

Kawai, T., Morita, K., Masuda, K., et al. (2007). Physical exercise-associated gene expression signatures in peripheral blood. *Clinical Journal of Sport Medicine* 17, 375–383.

Keller, M.C., Medland, S.E., Duncan, L.E., et al. (2009). Modeling extended twin family data I: Description of the Cascade model. *Twin Research and Human Genetics* 12, 8–18.

King, A.C., Taylor, C.B., Haskell, W.L. & Debusk, R.F. (1989). Influence of regular aerobic exercise on psychological health—A randomized, controlled trial of healthy middle-aged adults. *Health Psychology* 8, 305–324.

Klavestrand, J. & Vingard, E. (2009). The relationship between physical activity and health-related quality of life: A systematic review of current evidence. *Scandinavian Journal of Medicine & Science in Sports* 19, 300–312.

Kritz-Silverstein, D., Barrett-Connor, E. & Corbeau, C. (2001). Cross-sectional and prospective study of exercise and depressed mood in the elderly—The Rancho Bernardo Study. *American Journal of Epidemiology* 153, 596–603.

Kujala, U.M., Kaprio, J. & Koskenvuo, M. (2002). Modifiable risk factors as predictors of all-cause mortality: The roles of genetics and childhood environment. *American Journal of Epidemiology* 156, 985–993.

Lauderdale, D.S., Fabsitz, R., Meyer, J.M., Sholinsky, P., Ramakrishnan, V. & Goldberg, J. (1997). Familial determinants of moderate and intense physical activity: A twin study. *Medicine and Science in Sports and Exercise* 29, 1062–1068.

Lawlor, D.A. & Hopker, S.W. (2001). The effectiveness of exercise as an intervention in the management of depression: Systematic review and meta-regression analysis of randomised controlled trials. *British Medical Journal* 322, 763–767.

Loos, R.J.F., Rankinen, T., Tremblay, A., Perusse, L., Chagnon, Y. & Bouchard, C. (2005). Melanocortin-4 receptor gene and physical activity in the Quebec Family Study. *International Journal of Obesity* 29, 420–428.

Lorentzon, M., Lorentzon, R., Lerner, U.H. & Nordstrom, P. (2001). Calcium sensing receptor gene polymorphism, circulating calcium concentrations and bone mineral density in healthy adolescent girls. *European Journal of Endocrinology* 144, 257–261.

Lykken, D. & Tellegen, A. (1996). Happiness is a stochastic phenomenon. *Psychological Science* 7, 186–189.

Maia, J.A.R., Thomis, M. & Beunen, G. (2002). Genetic factors in physical activity levels—A twin study. *American Journal of Preventive Medicine* 23, 87–91.

McKercher, C.M., Schmidt, M.D., Sanderson, K.A., Patton, G.C., Dwyer, T. & Venn, A.J. (2009). Physical activity and depression in young adults. *American Journal of Preventive Medicine* 36, 161–164.

Morgan, W.P. & Johnson, R.W. (1977). Psychologic characterization of elite wrestler—A mental health model. *Medicine and Science in Sports and Exercise* 9, 55–56.

Neale, M.C. & Cardon, L.R. (1992). *Methodology for Genetic Studies of Twins and Families.* Kluwer Academic Publishers, Dordrecht.

Nes, R.B., Roysamb, E., Tambs, K., Harris, J.R. & Reichborn-Kjennerud, T. (2006). Subjective well-being: Genetic and environmental contributions to stability and change. *Psychological Medicine* 36, 1033–1042.

Rhodes, R.E. & Smith, N.E.I. (2006). Personality correlates of physical activity: A review and meta-analysis. *British Journal of Sports Medicine* 40, 958–965.

Rico-Sanz, J., Rankinen, T., Rice, T., et al. (2004). Quantitative trait loci for maximal exercise capacity phenotypes and their responses to training in the HERITAGE Family Study. *Physiological Genomics* 16, 256–260.

Rowland, T.W. (1998). The biological basis of physical activity. *Medicine and Science in Sports and Exercise* 30, 392–399.

Rowley, A.J., Landers, D.M., Kyllo, L.B. & Etnier, J.L. (1995). Does the iceberg profile discriminate between successful and less successful athletes—A meta-analysis. *Journal of Sport & Exercise Psychology* 17, 185–199.

Salmen, T., Heikkinen, A.M., Mahonen, A., et al. (2003). Relation of aromatase gene polymorphism and hormone replacement therapy to serum estradiol levels, bone mineral density, and fracture risk in early postmenopausal women. *Annals of Medicine* 35, 282–288.

Simonen, R.L., Rankinen, T., Perusse, L., et al. (2003a). A dopamine D2 receptor gene polymorphism and physical activity in two family studies. *Physiology & Behavior* 78, 751–757.

Simonen, R.L., Rankinen, T., Perusse, L., et al. (2003b). Genome-wide linkage scan for physical activity levels in the Quebec family study. *Medicine and Science in Sports and Exercise* 35, 1355–1359.

Sonstroem, R.J. & Morgan, W.P. (1989). Exercise and self-esteem—Rationale and model. *Medicine and Science in Sports and Exercise* 21, 329–337.

Stefan, N., Vozarova, B., Del Parigi, A., et al. (2002). The Gln223Arg polymorphism of the leptin receptor in Pima Indians: Influence on energy expenditure, physical activity and lipid metabolism. *International Journal of Obesity* 26, 1629–1632.

Stephens, T. (1988). Physical-activity and mental-health in the United States and Canada—Evidence from 4 population surveys. *Preventive Medicine* 17, 35–47.

Steptoe, A., Edwards, S., Moses, J. & Mathews, A. (1989). The effects of exercise training on mood and perceived coping ability in anxious adults from the general population. *Journal of Psychosomatic Research* 33, 537–547.

Strawbridge, W.J., Deleger, S., Roberts, R.E. & Kaplan, G.A. (2002). Physical activity reduces the risk of subsequent depression for older adults. *American Journal of Epidemiology* 156, 328–334.

Stubbe, J.H. & de Geus, E.J. (2009). Genetics of exercise behavior. In: Y.K. Kim (ed.) *Handbook of Behavior Genetics*. Springer Verlag, Berlin.

Stubbe, J.H., Boomsma, D.I. & de Geus, E.J.C. (2005a). Sports participation during adolescence: A shift from environmental to genetic factors. *Medicine and Science in Sports and Exercise* 37, 563–570.

Stubbe, J.H., Posthuma, D., Boomsma, D.I. & de Geus, E.J.C. (2005b). Heritability of life satisfaction in adults: A twin-family study. *Psychological Medicine* 35, 1581–1588.

Stubbe, J.H., Boomsma, D.I., Vink, J.M., et al. (2006a). Genetic influences on exercise participation: A comparative study in adult twin samples from seven countries. *PLoS ONE* 1, e22.

Stubbe, J.H., De Moor, M.H.M., Boomsma, D. & de Geus, E.J. (2006b). The association between exercise participation and well-being: A co-twin study. *Preventive Medicine* 44, 148–152.

Sullivan, P.F., Neale, M.C. & Kendler, K.S. (2000). Genetic epidemiology of major depression: Review and meta-analysis. *The American Journal of Psychiatry* 157, 1552–1562.

Svedberg, P., Gatz, M., Lichtenstein, P., Sandin, S. & Pedersen, N.L. (2005). Self-rated health in a longitudinal perspective: A 9-year follow-up twin study. *Journals of Gerontology Series B—Psychological Sciences and Social Sciences* 60, S331–S340.

van Gool, C.H., Kempen, G.I.J.M., Penninx, B.W.J.H., Deeg, D.J.H., Beekman, A.T.F. & van Eijk, J.T.M. (2003). Relationship between changes in depressive symptoms and unhealthy lifestyles in late middle aged and older persons: Results from the Longitudinal Aging Study Amsterdam. *Age and Ageing* 32, 81–87.

Weiss, A., Bates, T.C. & Luciano, M. (2008). Happiness is a personal(ity) thing—The genetics of personality and well-being in a representative sample. *Psychological Science* 19, 205–210.

Weyerer, S. (1992). Physical inactivity and depression in the community. Evidence from the Upper Bavarian Field Study. *International Journal of Sports Medicine* 13, 492–496.

Williams, A.G., Rayson, M.P., Jubb, M., et al. (2000). Physiology—The *ACE* gene and muscle performance. *Nature* 403, 614.

Winnicki, M., Accurso, V., Hoffmann, M., et al. (2004). Physical activity and angiotensin-converting enzyme gene polymorphism in mild hypertensives. *American Journal of Medical Genetics* 125A, 38–44.

Wise, L.A., Adams-Campbell, L.L., Palmer, J.R. & Rosenberg, L. (2006). Leisure time physical activity in relation to depressive symptoms in the Black Women's Health Study. *Annals of Behavioral Medicine* 32, 68–76.

Woods, D.R., Humphries, S.E. & Montgomery, H.E. (2000). The *ACE* I/D polymorphism and human physical performance. *Trends in Endocrinology and Metabolism* 11, 416–420.

PART 4

SYSTEMS BIOLOGY OF EXERCISE AND TRAINING

Chapter 26

A Primer on Systems Biology, as Applied to Exercise Physiology and Metabolism

JAMES A. TIMMONS[1] AND JORN W. HELGE[2]

[1]*Lifestyle Research Group, Royal Veterinary College, University of London, London, UK*
[2]*Center for Healthy Ageing, Section of Systems Biology Research, Panum Institute, University of Copenhagen, Copenhagen, Denmark*

Systems biology is a new approach to biomedical research that shares some common principles with physiology. Physiology, having survived the onslaught of molecular biology (financially and politically), seems to have picked a quarrel over the genealogy of systems biology (Auffray & Noble, 2009). William Harvey, postulated as a physiologist who epitomized and (by inference) practiced systems biology, used logic and arithmetic to deduce that the blood circulates around the body, powered by the contractility of the heart. He worked with units of volume and time combined with a logical interpretation of the unidirectional function of the valves he could visualize with the naked eye, to conclude that it was highly unlikely that the blood was formed *de novo* from nutritional intake during each heartbeat. Rather, he concluded that blood flow seems unidirectional and sufficiently rapid that it must be recycled, passing through the heart and lungs during this process (Auffray & Noble, 2009). We provide a perspective on systems biology as defined by data integration and phenotype prediction. We describe the promise of systems biology for the field of physiology and metabolism, and then provide some specific examples of where progress is being made.

Genetic and Molecular Aspects of Sport Performance, 1st edition.
Edited by Claude Bouchard and Eric P. Hoffman.
Published 2011 by Blackwell Publishing Ltd.

Systems Biology: Raising Expectations?

It has been stated of the functional interactions between the key components of cells, organs, and systems that:

> *"This information resides neither in the genome nor even in the individual proteins that genes code for. It lies at the level of protein interactions within the context of subcellular, cellular, tissue, organ, and system structures. There is therefore no alternative to copying nature and computing these interactions to determine the logic of healthy and diseased states."* (Noble, 2002)

This implies that one cannot draw reliable predictions or conclusions about a physiological system or its ability to respond to physiological perturbation from merely cataloging the sequence of the genome, the abundance of the product of transcription (RNA transcriptome), or the consequential products of translation (proteome) from a single tissue. Indeed, modeling biology should require such data to be placed in the correct cellular or anatomical context. Unlike the situation that William Harvey found himself in, where the function of a vessel could be deduced by physical occlusion, today's biologists are confronted with millions of individual observations, which cannot be assigned a physiological function or cellular location because we simply have no definitive idea what each protein or sequence of DNA does. Thus, rather than applying logic to the molecular variables (as Harvey did to his anatomical variables), the aim of systems biology is to focus on statistical rigor and experimental design, as we cannot perturb each "gene" *in situ* (as Harvey occluded veins) in humans to generate a scenario open to logical deduction alone.

How will systems biology be applied to exercise physiology? Consider generating a genome-wide RNA expression data set from a blood sample. Imagine if some information within that data were able to predict who would respond to an exercise training program and who would not. That is, using a mathematically derived composite score derived from 30 of 50,000 RNA measurements, we could then tell who will demonstrate the largest improvement in their aerobic performance (a complex integrated multiorgan capacity) despite all subjects having similar baseline physiological characteristics. It is important to point out that many of these 30 genes will not code for proteins, as many RNA fragments never form proteins, but rather form bioactive noncoding RNA molecules (Mattick, 2004). Such molecules can provide feedback (Nagano et al., 2008), control activity from the genome, or determine which allele is transcribed, and thus which of your parents contributes to your transcriptome.

You have just visualized one theoretical example of systems biology being applied to study human physiology, one that required no knowledge of what those 30 genes do or where they act in the system. Systems biology is not simply integrated molecular biology. What is critical is that the reproducibility of such an observation is established, demonstrating that such data has an inherent quality to inform about a functional response (physiology). At this stage, one may then investigate how the individual genes function, how the predictor may respond to a novel situation but with sobering thought in mind, that it may infrequently be possible to genuinely assign function to a gene, independent of its precise context. Thus, transcriptomics, proteomics, and metabolomics should not be seen as attempts to produce massive parts catalogs. The function of a collection (network) of genes

or molecules can be defined by how they predict a system property, and thus the individual function of the network members will always be, obviously, context dependent. It has been stated that no particular scale has a privileged relevance for systems biology (Kohl & Noble, 2009). This appears to justify computational approaches to study isolated subsystems which may potentially repeat the flaws of molecular reductionism. From the perspective of medicine, we wish to understand human physiology and disease; thus, investment in systems biology approaches should focus on working with integrated biological systems (defined by disease or physiology), which represent the minimum level of complexity, yet still faithfully reproduce or reflect the characteristics of the human, *in vivo*.

Systems biology and physiology are highly distinct research approaches, both of which will drive understanding of human disease. It remains critical that the physiological paradigm studied is genuinely representative of the system and question being considered. Establishment of a *reproducible* molecular signature which captures biological variance in a physiological system, under a variety of conditions, means that you have somehow linked molecules to a system property. While subsequent hypothesis-driven research may propose detailed explanation, the key is to put this initial molecular signature to practical use. In the following sections, we are going to provide some case examples in which omic technologies combined with systems biology have been implemented for the study of human physiology and metabolism. In particular, we will focus on metabolic flexibility in the context of insulin resistance, endurance exercise, and the role of coding and noncoding RNA. We will speculate how such approaches can be used in the future to guide athlete selection to a particular sport and tailor nutritional advice.

The System Biology Technologies

Generating a model from high-density data to discover molecules responsible for disease or physiological function usually involves a stage in which the researcher assumes or attempts to establish causality. Originally, a gene was the hidden cause of a trait, such as color or behavior—in short, it was a concept. Like all aspects of language, definitions evolve. Now a gene can be thought as a stable piece of inherited DNA sequence that codes for a functional molecule (protein or RNA) that is tasked to represent the function of that DNA sequence ("gene"). DNA sequence may not be the only inheritable factor (Noble, 2008); accessibility of a gene to be read (epigenetic) can be inherited to a point and other organic matter (e.g., cell components) passed from mother to offspring, but these can be categorized as early environmental parameters that may not be stable beyond a couple of generations.

The concept that variations in the DNA sequence can cause changes in biological phenotype led to a multitude of single-gene manipulation technologies (transgenic or knockout [KO] mouse), aiming to define the role of each gene within a physiological process. This basic concept reflects the central dogma of molecular biology and can be challenged for many reasons. It is beyond the purpose of this chapter to examine this issue in detail, but, for example, one can consider the point of view that rarely do single-gene manipulations ever yield a reproducible phenotype (Calvo et al., 2008; Choi et al., 2008), probably reflecting the intuitive idea that, in isolation, a single gene is not responsible for any complex phenotype. True, mitochondrial uncoupling protein 1 appears essential for nonshivering thermogenesis (Feldmann et al., 2009) but no more so than the presence or absence of cold or substrate. If we consider the definition of a gene as the DNA sequence that yields a bioactive protein or RNA molecule, and we accept that most complex physiological phenotypes are the result of the interaction of these gene products (foreshortened to the idea of "gene networks"; Keller et al., 2007), then the aim of systems biology is to capture these interactions and the interactions between metabolites and protein/RNA status to yield a "fingerprint."

It is also a fact that not all systems biology technologies (transcriptomics, proteomics, and metabolomics) are ready to deliver unbiased phenome-wide data for human physiology studies. The primary aim is to yield a fingerprint that captures as much of the variance as possible. Thus, deep sequencing of an individual's genome will

yield >1 million sequence data points; the most advanced expression technologies will give the RNA abundance for all of the genome. Global proteomics gives at this time a chemically biased and relatively small snapshot of the protein level of the tissue. Metabolomics, when targeted to a subset of biochemical intermediates, appears promising for profiling specific areas of metabolism (Shaham et al., 2008) and a useful phenotyping tool. Such factors influence which technology is used to generate a systems biology fingerprint and how informative it will be. One thing is certain, though, which is that it is not useful or accurate to think of an RNA profile, for example, simply as a surrogate for protein changes or indeed protein measurement without spatial context.

The Molecular Basis for Metabolic Inflexibility

Regular physical activity on average improves insulin sensitivity and vascular function and, when combined with other lifestyle changes, can halve the progression from impaired glucose tolerance to Type II diabetes (T2DM) (Knowler et al., 2002; Tuomilehto et al., 2001). Lack of cardiorespiratory (aerobic) capacity and low habitual physical activity promotes premature death by enhancing the progression of cardiovascular disease (Blair et al., 1989, 1996; Wisloff et al., 2005). Consistent with these or similar observations, the metabolic syndrome (or syndrome X) was presented as a means to identify individuals at higher risk of cardiovascular or other metabolic disease conditions (Reaven, 2003). The metabolic syndrome reflects the development of metabolic imbalance and there does not appear to be a unifying single molecular mechanism that can fully explain this. However, in order to further understand the sequence of events leading to metabolic imbalance, the concept of metabolic inflexibility may be a useful framework for discussion (Storlien et al., 2004). Metabolic inflexibility is a concept initially coined by Kelley and colleagues as an inadequate ability to switch substrate utilization from predominantly fat oxidation to carbohydrate oxidation, when studied using a euglycemic hyperinsulinemic clamp (Kelley et al., 1999). Metabolic inflexibility has since been further outlined (Storlien et al., 2004) as the occurrence of one or more of the following three metabolic traits (Figure 26.1):

1. impaired cephalic phase insulin secretion
2. impaired ability to transition between lipid (post-prandial) and carbohydrate (prandial) oxidation in skeletal muscle
3. impaired transition from fatty acid efflux to fatty acid storage after a meal.

One of the many consequences of this inflexibility is the exacerbation of the amplitude, as well as the duration of the substrate fluxes in the bloodstream followed by promotion of ectopic fat storage and disruption of the vascular system. Metabolic *inflexibility* is reduced by physical activity while a concurrent improvement in metabolic fitness is induced (Helge et al., 2008). Metabolic fitness is not easily defined but is largely conceived as healthy values of four of the five parameters, excluding central obesity, that currently define the metabolic syndrome (Figure 26.1). The challenge for systems biology is to take these metabolic concepts and provide, firstly, reproducible molecular fingerprints to help identify those at risk or the particular stages of metabolic syndrome, and then secondly molecular targets for therapy development (in the absence of adherence to lifestyle solutions). At this stage, the metabolite profiling approaches remain to be harmonized, such that the framework for cause–effect association remains to be reproducibly defined (Skovbro et al., 2008).

Transcriptomics and Skeletal Muscle Adaptation to Aerobic Training

There have been many studies documenting the gene-by-gene acute metabolic and molecular responses to aerobic exercise (Baar et al., 2002; Hawley et al., 2006; Pedersen et al., 2004; Widegren et al., 1998). The idea has been postulated (Schmutz et al., 2006) that acute responses to endurance training somehow add up over time to allow the muscle phenotype to drift toward a slower oxidative type. Others have taken a more systems approach, profiling the response of the genome to acute exercise (Mahoney et al., 2005) and found changes

suggesting that acute molecular responses simply reflect the acute energy and ionic homeostasis challenges. It is also unclear whether array analysis of athletes versus nonathletes (cross-sectional analysis) is informative (Wittwer et al., 2004) of molecular processes responsible for training adaptation or whether it largely reflects baseline genetic, epigenetic, or environmental differences incidental to the influence of aerobic training. Nevertheless, more studies are required to address all of these important questions.

Humans demonstrate a heterogeneous ability to improve their aerobic fitness, with some individuals unable to improve at all (Bouchard & Rankinen, 2001; Kohrt et al., 1991; Lortie et al., 1984; Timmons et al., 2005a). In recent years, genomics technologies have provided insight into the molecular responses to long-term aerobic training and have demonstrated that the switching on of gene products varies according to physiological adaptation (Timmons et al., 2005a, b). So far, cause-and-effect relationships have not been established with such an approach. Note that acute studies cannot prove that the acute perturbation leads to physiological adaptation such that both acute and chronic responses to exercise should be studied, longitudinally, in the same subjects.

Mutation or knockout of candidate genes (identified previously from canonical pathway analysis studies in cells), such as AMP-activated protein kinase (*AMPK*), Ca^{2+}/calmodulin-dependent protein kinase (*CaMK*), and peroxisome proliferator-activated receptor-gamma coactivator-1α (*PGC-1α*), have attempted to prove that such genes are causal regulators of muscle adaptation following exercise (Akimoto et al., 2004; Handschin & Spiegelman, 2008; Jorgensen et al., 2007; Leick et al., 2008). In fact, none of these much-studied (Hardie, 2008) genes appear critical for exercise-induced endurance training adaptation in terms of muscle biochemistry or performance. Furthermore, there is little evidence that *AMPK* or *PGC-1α* directly regulate the large number of molecular changes noted following endurance changes in humans (Hittel et al., 2003, 2005; Keller et al., 2007; Radom-Aizik et al., 2005; Timmons et al., 2005b). Thus, the much-argued master regulators identified by these traditional gene-by-gene approaches

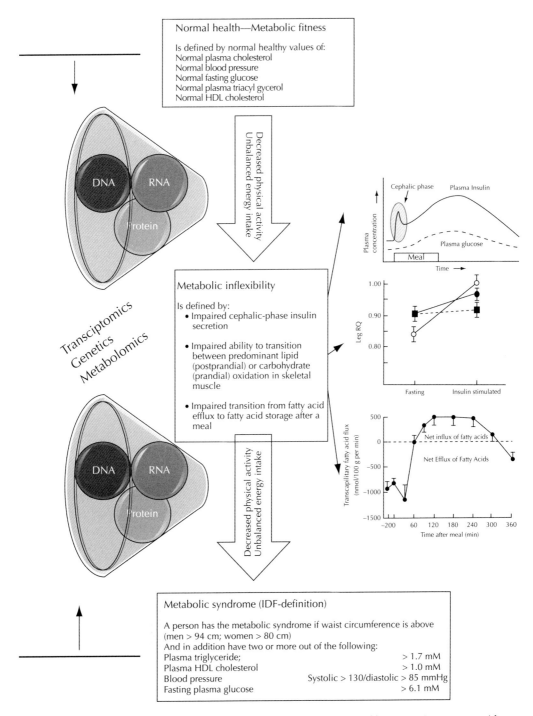

Figure 26.1 A schematic representation of the transition from normal physiological homeostasis to system-wide metabolic dysfunction. The "eyepiece" objects are diagrammatic representations of the process of taking a snapshot of the transcriptome, proteome/metabolome, and genome. Examination of these "omes" at the various stages for the development of metabolic syndrome would allow for a systems biology approach to modeling disease development via the establishment of a "time course" of gene network perturbation. Identification of biomarkers that reliably inform about disease progression rate or severity would be one tangible outcome of such studies. (The three embedded right-hand side figures are reproduced from Storlien et al. (2004), with permission from *Proceedings of the Nutrition Society*.)

have not identified the global mechanisms directly responsible for the chronic adaptation of human skeletal muscle in response to aerobic exercise.

How can systems biology deliver understanding of the biological processes that underpin aerobic training-induced adaptation in human skeletal muscle and systemic metabolism? Again, we return to the theme of focusing on the correct physiological model. So far, murine models of endurance capacity lack evidence that they are relevant to the human situation. Indeed, few molecular mechanisms translate from an experimental model to enhanced endurance functional capacity in the clinic (Calvert et al., 2008; Hespel et al., 2001; Timmons et al., 1996, 1998). Those that have were based on knowledge that was consistent with human metabolism in the first instance. If we can generate global molecular and metabolite profiles to study how exercise training induces improved muscle efficiency and maximal capacity directly in humans, then this may yield more reliable knowledge.

A second strategy is to utilize the fact that, despite undergoing identical training programs, many humans do not improve their physiological capacity or their performance (Bouchard & Rankinen, 2001; Timmons et al., 2005a; Vollaard et al., 2009). Recently, we have been able to make progress in generating a molecular predictor that forecasts who will be a low or high responder to training. This was achieved using a pretraining global profile of >20,000 RNA molecules expressed in each subject's muscle tissue. One group acted as a test group to generate a predictor set, and a second group acted as a blind validation group. This analysis not only yielded a tissue-based predictor but is offering new opportunities to find underlying genetic variants that contribute to the extent of the training-induced response (Noble, 2002). Using this new knowledge, to provide personalized health advice is a plausible outcome.

Transcriptomics and Skeletal Muscle Insulin Resistance

When insulin fails to rapidly promote the removal of glucose from the bloodstream, the target tissue (collectively; muscle cells, extracellular matrix, and vascular cells) is considered to be insulin resistant.

The cause of this is unclear. It could be as simple as the lack of turnover of the muscle carbohydrate stores (glycogen) due to physical inactivity, providing physiological feedback to block additional glucose flux into the cell (Babraj et al., 2009). This would be mediated by alteration of the proteins that regulate glucose to glycogen metabolism or the spatial distribution of the glycogen (Prats et al., 2009). Manipulation of related genes in nonphysiological cell models (where metabolic flux is not physiologically coupled) or thermally stressed mice models (Feldmann et al., 2009) does increase the flux of glucose into the muscle. However, this provides little clue to the underlying *causal* pathophysiology (nor a potential treatment). Applying a system-wide approach may help to identify unique features of insulin-resistant skeletal muscle that directly contribute to an alteration in physiological status over time. This could involve dysfunction of the vascular and muscle components of insulin-mediated peripheral glucose uptake.

RNA expression studies of muscle insulin resistance, an early feature of T2DM, have been carried out. Four small gene-chip studies, relying on partial coverage of the human transcriptome, have taken a systems biology approach to establish the basis of insulin resistance in human skeletal muscle (Mootha et al., 2003; Patti et al., 2003; Sreekumar et al., 2002; Yang et al., 2002). In one prominent study (Patti et al., 2003), muscle samples from a small group of subjects were studied using the Affymetrix HuGeneFL array platform (representing about 15% of the RNA transcriptome). The 10 control subjects and five subjects with T2DM in this study were not matched for body mass index or age, making data interpretation challenging. Subset analysis of about 500 mitochondrial mRNAs is a powerful statistical approach to detecting *coordinated* changes in mitochondrial function *in vivo*. In a larger study, a group of older subjects (about 66 years) were profiled using a microarray platform that provides greater coverage of the transcriptome (about 20,000 sequences) (Mootha et al., 2003). The authors presented evidence that there was a statistically significant downregulation of a group of genes involved in oxidative metabolism (*OXPHOS*) in skeletal muscle of subjects with T2DM. The loss of *OXPHOS* mRNA expression was attributed to an intrinsic defect in peroxisome proliferator-activated receptor-gamma coactivator-1α (*PGC-1α*) activity in the muscle, and this was considered the underlying cause of skeletal muscle insulin resistance (Kim et al., 2008; Morino et al., 2006; Spiegelman, 2007). However, in the study by Mootha, the muscle RNA was obtained following exposure to several hours of pharmacological hyperinsulinemia. As patients with T2DM will respond differently to pharmacological levels of insulin infusion compared to control subjects (Sreekumar et al., 2002), this creates an artificial (pharmacological) difference in gene expression. Despite continuing debate in the field (Rabol et al., 2006; Timmons et al., 2006), the concept of an *OXPHOS* gene-expression impairment (Handschin & Spiegelman, 2008; Kim et al., 2008; Petersen & Shulman, 2006) continues to guide diabetes research and provide an explanation as to why physical activity might help prevent the progression of insulin resistance to T2DM (as exercise activates this pathway).

How will a systems approach deliver understanding of insulin resistance? It is usually informative to study the response of a physiological system to a change in demand. This is especially true in humans, where responses are heterogeneous. Bouchard and colleagues identified gene-expression signatures that differed between subjects who had either a good or a poor response to an endurance training program for improvements in glucose tolerance (Teran-Garcia et al., 2005). Such differentially regulated genes provide clues to the gene networks within the muscle tissue (and beyond) that associate with the *magnitude* of the improved insulin action. Profiling of muscle expression phenotype prior to the intervention, coupled with independent validation, can allow for the development of a molecular predictor (Knudsen & Knudsen, 2004), that identifies a set of values of gene abundance which contains information relevant for the adaptive potential of the insulin signaling physiological system. Such signatures could even be generated in the blood, using metabolite profiling. The key aim is to utilize tools that integrate genetic variation to produce a signature and robust statistical validation. So far, classic genome-wide analysis,

focused on the highest ranking genetic variants in large population studies, has failed to deliver such information (Snyder et al., 2009). A combination of all the above approaches may actually be the best approach. To then determine whether the genes and metabolites are directly and causally related to insulin resistance, one would wish to study what happens to these marker molecules under multiple physiological conditions known to alter insulin sensitivity. This level of proof will not be widely accepted within the molecular biology community as they would expect direct gene manipulation to yield data that proves a cause–effect role. However, as discussed earlier, such hypothesis-driven reductionist strategies are theoretically flawed (Koch & Britton, 2008) and are rarely reproducible across multiple murine/genetic backgrounds.

What Is the Role of Nonprotein-Coding RNA?

The short answer is that there is insufficient human data to demonstrate an important role of nonprotein-coding RNA in the physiological regulation of human muscle metabolism during exercise. There are many types of noncoding RNA (Baek et al., 2008; Timmons & Good, 2006). In the context of skeletal muscle, recent studies have demonstrated that noncoding antisense RNA may influence mitochondrial gene splicing *in vivo* when physical activity is increased or decreased (Scheele et al., 2007). Furthermore, under conditions of extreme human muscle dysfunction, such as muscular dystrophy (Eisenberg et al., 2007) or sepsis (Fredriksson et al., 2008), there are selective changes in microRNA (miRNA) expression. miRNAs, emerging regulators of mammalian cell phenotype (Baek et al., 2008; Grimson et al., 2007; Selbach et al., 2008), are short RNA posttranscriptional molecules able to regulate the translation of protein-coding genes (Lewis et al., 2005). miRNAs play important roles in cellular development and differentiation (Esau et al., 2004; Rao et al., 2006). The expression of selected miRNAs changes with age in human skeletal muscle (Drummond et al., 2008). We have found that about 25 miRNAs are modulated in response to endurance training, including downregulation of miRNA-92a. Recently, miRNA-92a was shown in a variety of *in vivo* models to negatively regulate angiogenesis (Bonauer et al., 2009), and thus it would appear that at least one of the endurance training-induced miRNAs in humans regulates biological networks relevant for functional gains in aerobic capacity (Timmons et al., 2005b). However, the impact of a single miRNA *in vivo* in humans is likely to be modest; rather, we believe that groups of miRNAs, perhaps modulated only to a small degree, will combine to regulate muscle protein changes. The reason for claims made of single miRNAs (e.g., miRNA-92a; Bonauer et al., 2009) is that in the model systems studied, extreme changes in miRNA expression are used to generate the phenotype. During physiological adaptation in human skeletal muscle, we have never observed large shifts in miRNA concentration. However, a role for the various noncoding RNAs in human biology is a recent development (Faghihi & Wahlestedt, 2009), and both the rules of action and the relative importance to human physiology have still to be established.

Future Directions

In medical research, there is an overreliance on single-gene cellular and *in vivo* molecular approaches to provide key proof about metabolic properties and disease processes despite the fact that, over the longer term, the phenotype of such models rarely reflects the complexity of the outbred human. As an alternative to animal models, gene chips are a powerful molecular tool that readily provides a fingerprint of the molecular events following physiological or hormonal interventions in humans (Fredriksson et al., 2008; Mahoney et al., 2005; Stump et al., 2003; Timmons et al., 2005b). Such data can be combined with network analysis (Calvano et al., 2005) and detailed physiological profiling, to infer with high statistical rigor, the molecular basis of physiological traits, or human disease. Such studies provide drug targets and biomarkers for independent clinical verification, while successful development of a novel treatment based on such data should be considered as the gold standard for final validation. This alternative translational research strategy will only supersede reliance on reductionist laboratory model systems if the clinical studies are both adequately powered, efficiently run to deliver timely results and, of course, reproduced across labs. The same rules apply to the application of any system biology tool and biomarker generation for the refinement of athletic training and exercise rehabilitation programs.

References

Akimoto, T., Ribar, T.J., Williams, R.S. & Yan, Z. (2004). Skeletal muscle adaptation in response to voluntary running in Ca2+/calmodulin-dependent protein kinase IV-deficient mice. *American Journal of Physiology—Cell Physiology* 287, C1311–C1319.

Auffray, C. & Noble, D. (2009). Origins of systems biology in William Harvey's masterpiece on the movement of the heart and the blood in animals. *International Journal of Molecular Sciences* 10, 1658–1669.

Baar, K., Wende, A.R., Jones, T.E., et al. (2002). Adaptations of skeletal muscle to exercise: Rapid increase in the transcriptional coactivator PGC-1. *The FASEB Journal* 16, 1879–1886.

Babraj, J.A., Vollaard, N.B., Keast, C., Guppy, F.M., Cottrell, G. & Timmons, J.A. (2009). Extremely short duration high intensity interval training substantially improves insulin action in young healthy males. *BMC Endocrine Disorders* 9, 3.

Baek, D., Villen, J., Shin, C., Camargo, F.D., Gygi, S.P. & Bartel, D.P. (2008). The impact of microRNAs on protein output. *Nature* 455, 64–71.

Blair, S.N., Kohl, H.W., III, Paffenbarger, R.S., Jr., Clark, D.G., Cooper, K.H. & Gibbons, L.W. (1989). Physical fitness and all-cause mortality. A prospective study of healthy men and women. *Journal of the American Medical Association* 262, 2395–2401.

Blair, S.N., Kampert, J.B., Kohl, H.W., III, et al. (1996). Influences of cardiorespiratory fitness and other precursors on cardiovascular disease and all-cause mortality in men and women. *Journal of the American Medical Association* 276, 205–210.

Bonauer, A., Carmona, G., Iwasaki, M., et al. (2009). MicroRNA-92a controls angiogenesis and functional recovery of ischemic tissues in mice. *Science* 324, 1710–1713.

Bouchard, C. & Rankinen, T. (2001). Individual differences in response to regular physical activity. *Medicine and Science in Sports and Exercise* 33, S446–S451.

Calvano, S.E., Xiao, W., Richards, D.R., et al. (2005). A network-based analysis of systemic inflammation in humans. *Nature* 437, 1032–1037.

Calvert, L.D., Shelley, R., Singh, S.J., et al. (2008). Dichloroacetate enhances performance and reduces blood lactate during maximal cycle exercise in chronic obstructive pulmonary disease. *American Journal of Respiratory and Critical Care Medicine* 177, 1090–1094.

Calvo, J.A., Daniels, T.G., Wang, X., et al. (2008). Muscle-specific expression of PPARgamma coactivator-1alpha improves exercise performance and increases peak oxygen uptake. *Journal of Applied Physiology* 104, 1304–1312.

Choi, C.S., Befroy, D.E., Codella, R., et al. (2008). Paradoxical effects of increased expression of PGC-1alpha on muscle mitochondrial function and insulin-stimulated muscle glucose metabolism. *Proceedings of the National Academy of Sciences of the United States of America* 105, 19926–19931.

Drummond, M.J., McCarthy, J.J., Fry, C.S., Esser, K.A. & Rasmussen, B.B. (2008). Aging differentially affects human skeletal muscle microRNA expression at rest and after an anabolic stimulus of resistance exercise and essential amino acids. *American Journal of Physiology—Endocrinology and Metabolism* 295, E1333–E1340.

Eisenberg, I., Eran, A., Nishino, I., et al. (2007). Distinctive patterns of microRNA expression in primary muscular disorders. *Proceedings of the National Academy of Sciences of the United States of America* 104, 17016–17021.

Esau, C., Kang, X., Peralta, E., et al. (2004). MicroRNA-143 regulates adipocyte differentiation. *The Journal of Biological Chemistry* 279, 52361–52365.

Faghihi, M.A. & Wahlestedt, C. (2009). Regulatory roles of natural antisense transcripts. *Nature Reviews Molecular Cell Biology* 10, 637–643.

Feldmann, H.M., Golozoubova, V., Cannon, B. & Nedergaard, J. (2009). UCP1 ablation induces obesity and abolishes diet-induced thermogenesis in mice exempt from thermal stress by living at thermoneutrality. *Cell Metabolism* 9, 203–209.

Fredriksson, K., Tjader, I., Keller, P., et al. (2008). Dysregulation of mitochondrial dynamics and the muscle transcriptome in ICU patients suffering from sepsis induced multiple organ failure. *PLoS ONE* 3, e3686.

Grimson, A., Farh, K.K., Johnston, W.K., Garrett-Engele, P., Lim, L.P. & Bartel, D.P. (2007). MicroRNA targeting specificity in mammals: Determinants beyond seed pairing. *Molecular Cell* 27, 91–105.

Handschin, C. & Spiegelman, B.M. (2008). The role of exercise and PGC1alpha in inflammation and chronic disease. *Nature* 454, 463–469.

Hardie, D.G. (2008). AMPK: A key regulator of energy balance in the single cell and the whole organism. *International Journal of Obesity (London)* 32(Suppl. 4), S7–S12.

Hawley, J.A., Hargreaves, M. & Zierath, J.R. (2006). Signalling mechanisms in skeletal muscle: Role in substrate selection and muscle adaptation. *Essays in Biochemistry* 42, 1–12.

Helge, J.W., Damsgaard, R., Overgaard, K., et al. (2008). Low-intensity training dissociates metabolic from aerobic fitness. *Scandinavian Journal of Medicine & Science in Sports* 18, 86–94.

Hespel, P., Op't eijnde, B., Van Leemputte, M., et al. (2001). Oral creatine supplementation facilitates the rehabilitation of disuse atrophy and alters the expression of muscle myogenic factors in humans. *The Journal of Physiology* 536, 625–633.

Hittel, D.S., Kraus, W.E. & Hoffman, E.P. (2003). Skeletal muscle dictates the fibrinolytic state after exercise training in overweight men with characteristics of metabolic syndrome. *The Journal of Physiology* 548, 401–410.

Hittel, D.S., Kraus, W.E., Tanner, C.J., Houmard, J.A. & Hoffman, E.P. (2005). Exercise training increases electron and substrate shuttling proteins in muscle of overweight men and women with the metabolic syndrome. *Journal of Applied Physiology* 98, 168–179.

Jorgensen, S.B., Jensen, T.E. & Richter, E.A. (2007). Role of AMPK in skeletal muscle gene adaptation in relation to exercise. *Applied Physiology Nutrition and Metabolism* 32, 904–911.

Keller, P., Vollaard, N.J.B., Babraj, J., Ball, D., Sewell, D.A. & Timmons, J.A. (2007). Using systems biology to define the essential biological networks responsible for adaptation to endurance exercise training. *Biochemical Society Transactions* 35, 1306–1309.

Kelley, D.E., Goodpaster, B., Wing, R.R. & Simoneau, J.A. (1999). Skeletal muscle fatty acid metabolism in association with insulin resistance, obesity, and weight loss. *American Journal of Physiology—Endocrinology and Metabolism* 277, E1130–E1141.

Kim, J.A., Wei, Y. & Sowers, J.R. (2008). Role of mitochondrial dysfunction in insulin resistance. *Circulation Research* 102, 401–414.

Knowler, W.C., Barrett-Connor, E., Fowler, S.E., et al. (2002). Reduction in the incidence of type 2 diabetes with lifestyle intervention or metformin. *The New England Journal of Medicine* 346, 393–403.

Knudsen, S. & Knudsen, S. (2004). *Guide to Analysis of DNA Microarray Data.* Wiley-Liss, Hoboken, NJ.

Koch, L.G. & Britton, S.L. (2008). Development of animal models to test the fundamental basis of gene-environment interactions. *Obesity (Silver Spring)* 16(Suppl. 3), S28–S32.

Kohl, P. & Noble, D. (2009). Systems biology and the virtual physiological human. *Molecular Systems Biology* 5, 292.

Kohrt, W.M., Malley, M.T., Coggan, A.R., et al. (1991). Effects of gender, age, and fitness level on response of $\dot{V}O_2$max to training in 60–71 yr olds. *Journal of Applied Physiology* 71, 2004–2011.

Leick, L., Wojtaszewski, J.F., Johansen, S.T., et al. (2008). PGC-1alpha is not mandatory for exercise- and training-induced adaptive gene responses in mouse skeletal muscle. *American Journal of Physiology—Endocrinology and Metabolism* 294, E463–E474.

Lewis, B.P., Burge, C.B. & Bartel, D.P. (2005). Conserved seed pairing, often flanked by adenosines, indicates that thousands of human genes are microRNA targets. *Cell* 120, 15–20.

Lortie, G., Simoneau, J.A., Hamel, P., Boulay, M.R., Landry, F. & Bouchard, C. (1984). Responses of maximal aerobic power and capacity to aerobic training. *International Journal of Sports Medicine* 5, 232–236.

Mahoney, D.J., Parise, G., Melov, S., Safdar, A. & Tarnopolsky, M.A. (2005). Analysis of global mRNA expression in human skeletal muscle during recovery from endurance exercise. *The FASEB Journal* 19, 1498–1500.

Mattick, J.S. (2004). RNA regulation: A new genetics? *Nature Reviews Genetics* 5, 316–323.

Mootha, V.K., Lindgren, C.M., Eriksson, K.F., et al. (2003). PGC-1alpha-responsive genes involved in oxidative phosphorylation are coordinately downregulated in human diabetes. *Nature Genetics* 34, 267–273.

Morino, K., Petersen, K.F. & Shulman, G.I. (2006). Molecular mechanisms of insulin resistance in humans and their potential links with mitochondrial dysfunction. *Diabetes* 55(Suppl. 2), S9–S15.

Nagano, T., Mitchell, J.A., Sanz, L.A., et al. (2008). The air noncoding RNA epigenetically silences transcription by targeting G9a to chromatin. *Science* 322, 1717–1720.

Noble, D. (2002). Modeling the heart—From genes to cells to the whole organ. *Science* 295, 1678–1682.

Noble, D. (2008). Genes and causation. *Philosophical Transactions of the Royal Society* 366, 3001–3015.

Patti, M.E., Butte, A.J., Crunkhorn, S., et al. (2003). Coordinated reduction of genes of oxidative metabolism in humans with insulin resistance and diabetes: Potential role of PGC1 and NRF1. *Proceedings of the National Academy of Sciences of the United States of America* 100, 8466–8471.

Pedersen, B.K., Steensberg, A., Fischer, C., et al. (2004). The metabolic role of IL-6 produced during exercise: is IL-6 an exercise factor? *Proceedings of the Nutrition Society* 63, 263–267.

Petersen, K.F. & Shulman, G.I. (2006). Etiology of insulin resistance. *The American Journal of Medicine* 119, S10–S16.

Prats, C., Helge, J.W., Nordby, P., et al. (2009). Dual regulation of muscle glycogen synthase during exercise by activation and compartmentalization. *The Journal of Biological Chemistry* 284(23), 15692–15700.

Rabol, R., Boushel, R. & Dela, F. (2006). Mitochondrial oxidative function and type 2 diabetes. *Applied Physiology, Nutrition, and Metabolism* 31, 675–683.

Radom-Aizik, S., Hayek, S., Shahar, I., Rechavi, G., Kaminski, N. & Ben-Dov, I. (2005). Effects of aerobic training on gene expression in skeletal muscle of elderly men. *Medicine and Science in Sports and Exercise* 37, 1680–1696.

Rao, P.K., Kumar, R.M., Farkhondeh, M., Baskerville, S. & Lodish, H.F. (2006). Myogenic factors that regulate expression of muscle-specific microRNAs. *Proceedings of the National Academy of Sciences of the United States of America* 103, 8721–8276.

Reaven, G.M. (2003). The insulin resistance syndrome. *Current Atherosclerosis Reports* 5, 364–371.

Scheele, C., Petrovic, N., Faghihi, M.A., et al. (2007). The human PINK1 locus is regulated in vivo by a non-coding natural antisense RNA during modulation of mitochondrial function. *BMC Genomics* 8, 74.

Schmutz, S., Dapp, C., Wittwer, M., Vogt, M., Hoppeler, H. & Fluck, M. (2006). Endurance training modulates the muscular transcriptome response to acute exercise. *Pflugers Archiv* 451, 678–687.

Selbach, M., Schwanhausser, B., Thierfelder, N., Fang, Z., Khanin, R. & Rajewsky, N. (2008). Widespread changes in protein synthesis induced by microRNAs. *Nature* 455, 58–63.

Shaham, O., Wei, R., Wang, T.J., et al. (2008). Metabolic profiling of the human response to a glucose challenge reveals distinct axes of insulin sensitivity. *Molecular Systems Biology* 4, 214.

Skovbro, M., Baranowski, M., Skov-Jensen, C., et al. (2008). Human skeletal muscle ceramide content is not a major factor in muscle insulin sensitivity. *Diabetologia* 51, 1253–1260.

Snyder, M., Weissman, S. & Gerstein, M. (2009). Personal phenotypes to go with personal genomes. *Molecular Systems Biology* 5, 273.

Spiegelman, B.M. (2007). Transcriptional control of mitochondrial energy metabolism through the PGC1 coactivators. *Novartis Foundation Symposium* 287, 60–63.

Sreekumar, R., Halvatsiotis, P., Schimke, J.C. & Nair, K.S. (2002). Gene expression profile in skeletal muscle of type 2 diabetes and the effect of insulin treatment. *Diabetes* 51, 1913–1920.

Storlien, L., Oakes, N.D. & Kelley, D.E. (2004). Metabolic flexibility. *Proceedings of the Nutrition Society* 63, 363–368.

Stump, C.S., Short, K.R., Bigelow, M.L., Schimke, J.M. & Nair, K.S. (2003). Effect of insulin on human skeletal muscle mitochondrial ATP production, protein synthesis, and mRNA transcripts. *Proceedings of the National Academy of Sciences of the United States of America* 100, 7996–8001.

Teran-Garcia, M., Rankinen, T., Koza, R.A., Rao, D.C. & Bouchard, C. (2005). Endurance training-induced changes in insulin sensitivity and gene expression. *American Journal of Physiology— Endocrinology and Metabolism* 288, E1168–E1178.

Timmons, J.A. & Good, L. (2006). Does everything now make (anti)sense? *Biochemical Society Transactions* 34, 1148–1150.

Timmons, J.A., Poucher, S.M., Constantin-Teodosiu, D., et al. (1996). Increased acetyl group availability enhances contractile function of canine skeletal muscle during ischemia. *Journal of Clinical Investigation* 97, 879–883.

Timmons, J.A., Gustafsson, T., Sundberg, C.J., Jansson, E. & Greenhaff, P.L. (1998). Muscle acetyl group availability is a major determinant of oxygen deficit in humans during submaximal exercise. *The American Journal of Physiology* 274, E377–E380.

Timmons, J.A., Jansson, E., Fischer, H., et al. (2005a). Modulation of extracellular matrix genes reflects the magnitude of physiological adaptation to aerobic exercise training in humans. *BMC Biology* 3, 19.

Timmons, J.A., Larsson, O., Jansson, E., et al. (2005b). Human muscle gene expression responses to endurance training provide a novel perspective on Duchenne muscular dystrophy. *The FASEB Journal* 19, 750–760.

Timmons, J.A., Norrbom, J., Scheele, C., Thonberg, H., Wahlestedt, C. & Tesch, P. (2006). Expression profiling following local muscle inactivity in humans provides new perspective on diabetes-related genes. *Genomics* 87, 165–172.

Tuomilehto, J., Lindstrom, J., Eriksson, J.G., et al. (2001). Prevention of type 2 diabetes mellitus by changes in lifestyle among subjects with impaired glucose tolerance. *The New England Journal of Medicine* 344, 1343–1350.

Vollaard, N.B., Constantin-Teodosiu, D., Fredriksson, K., et al. (2009). Systematic analysis of adaptations in aerobic capacity and submaximal energy metabolism provides a unique insight into determinants of human aerobic performance. *Journal of Applied Physiology* 106, 1479–1486.

Widegren, U., Jiang, X.J., Krook, A., et al. (1998). Divergent effects of exercise on metabolic and mitogenic signaling pathways in human skeletal muscle. *The FASEB Journal* 12, 1379–1389.

Wisloff, U., Najjar, S.M., Ellingsen, O., et al. (2005). Cardiovascular risk factors emerge after artificial selection for low aerobic capacity. *Science* 307, 418–420.

Wittwer, M., Billeter, R., Hoppeler, H. & Fluck, M. (2004). Regulatory gene expression in skeletal muscle of highly endurance-trained humans. *Acta Physiologica Scandinavica* 180, 217–227.

Yang, X., Pratley, R.E., Tokraks, S., Bogardus, C. & Permana, P.A. (2002). Microarray profiling of skeletal muscle tissues from equally obese, non-diabetic insulin-sensitive and insulin-resistant Pima Indians. *Diabetologia* 45, 1584–1593.

Chapter 27

Systems Biology Through Time Series Data— A Strength of Muscle Remodeling

MONICA HUBAL, ZUYI WANG AND ERIC P. HOFFMAN

Department of Integrative Systems Biology, George Washington University School of Medicine, Research Center for Genetic Medicine, Children's National Medical Center, Washington, DC, USA

Systems biology is the effort to integrate diverse and complex data sets into mathematical models able to predict a biological response. Given enough data, one should be able to predict an individual's response to a specific bout of training, with knowledge of each individual's personal genetics (polymorphisms), genomics [responses of muscle mRNA and microRNA (miRNA)], proteomics, and environmental conditions (training, diet). As discussed in Chapter 26, data integration alone can provide dense and potentially informative correlations (snapshots of complex biology), but infrequently yield the mechanistic "cause and effect" data considered important for robust systems biology models. Here, we extend the discussion in Chapter 26 to describe "time" as a key element in systems biology modeling efforts. Most simply, time series data can inform cause/effect relationships between genes, proteins, and tissue remodeling, where the early molecular events occurring after a muscle stimulus (e.g., exercise training bout), often within seconds or minutes, are "upstream" of the later events of muscle remodeling (measured in hours or days). By assessing molecular, proteomic, and physiological changes as a function of time, there is the potential to connect all the dots and assign mathematical models to muscle adaptation. These responses can

also be modeled as a function of underlying genetic sequence variability.

Skeletal muscle may prove the tissue system of choice for driving technologies needed for systems biology models. Muscle has an impressive ability to remodel based upon environmental stimuli such as exercise training, the tissue is relatively simple and highly conserved between species, and human studies often obtain multiple tissue samples as a function of time after an intervention. In this chapter, we describe time series applications to muscle remodeling from the perspective of cellular processes (stretch, energy), modifiers of cellular processes (training, genetics), and the muscle functional level (fiber type, hypertrophy, metabolism). Additionally, we provide a description of the complex bioinformatics techniques that are needed to build true systems biology models, and special considerations for applying these models.

The Building Blocks of Muscle: Parallel Assessments of DNA, RNA, and Protein

Of the three types of large molecules driving biology (DNA, RNA, and protein), DNA is relatively static and simple. DNA is a code, and this code, like a book or encyclopedia, is not particularly dynamic, although epigenetic modifications of DNA can change the activity of genes. DNA is made more variable by polymorphisms, but within a single individual even these polymorphisms are

Genetic and Molecular Aspects of Sport Performance, 1st edition.
Edited by Claude Bouchard and Eric P. Hoffman.
Published 2011 by Blackwell Publishing Ltd.

319

unchanging during the life of the person. RNA is more dynamic; different cells turn on different genes, and often use different parts of the gene in different ways (e.g., alternative splicing of exons). Time becomes an important variable, with the temporal sequence of gene activation driving cellular responses. The intrinsic variability afforded by gene regulation, alternative splicing, temporal sequences of activation/repression, and the different types of RNA (miRNA, mRNA, and others) greatly increases the complexity and flexibility of the systems a cell, tissue, and organism can use. Variations in protein production, activity, and localization take this level of complexity to new orders of magnitude. The timescale in which protein changes (such as phosphorylation) occur can range from milliseconds to hours or days. Proteins can associate and disassociate with each other, changing the function of themselves or their binding partners. Cells can communicate with each other through secreted proteins, such that a single protein can have one role within the cell expressing it, and various other roles within cells distant to its original production site.

The goal of systems biology is to understand and be able to model this complexity for a specific response to a stimulus, such as a bout of muscle training. These models would enable the prediction of responses to given stimuli and could be manipulated to predict responses given a set of input variables. Systems biology research demands robust multilevel data to inform its models. Research technologies have evolved to become highly parallel, providing much of the raw data needed to begin to develop systems biology models. As described in an earlier chapter in this volume, one million polymorphisms can be screened in a single individual with a single experiment, and these polymorphisms have been associated with specific features of athletic ability, such as *ACTN3* and endurance running (see Chapter 18). To add to genetic variation data, muscle biopsies can be taken over time following muscle training or injury, and all the genes of the genome can be monitored as they turn on and off (mRNA and miRNA transcriptomic profiling). Proteomic profiling is evolving to reach the level where thousands of muscle proteins can

be monitored following training or other stimuli that affect skeletal muscle (such as glucocorticoids; Reeves et al., 2009).

Static Versus Dynamic Models: Resolving Biological Responses into Event-Driven Pathways

Putting all this information together, and then deriving mathematical models of the response of muscle to training is becoming tangible, but the field is in its infancy. Most studies to date have looked at only one molecular level at a time; for example, mRNA profiling as a function of muscle damage, training, or repair. But mRNA is only 5% of the picture, and polymorphisms, miRNAs, and proteomics need to be generated on hundreds or thousands of individuals before the data is adequate to generate true systems biology models. Most important is that multiple levels from the same individuals are needed to power these models and then more subjects added to increase the dynamic range and predictive abilities of the models.

Where do we stand in the systems biology of muscle and its responses to stimuli? There are two key approaches that can aid in the interpretation of this highly dimensional data: time series and knowledge networks. Knowledge networks are computer databases and associated interfaces that assemble preexisting biological knowledge into a tool that can accept genomics data sets (e.g., mRNA and miRNA expression profiles) and help interpret the data within the context of preexisting knowledge. There are many types of knowledge networks (Fang et al., 2006). One group of networks consists of gene ontology databases where genes and encoded proteins are categorized into biochemical- or sequence-defined groups. Another approach is to code existing literature into biologically relevant canonical pathways and networks, and then overlay genomics data sets onto these networks. While many types of knowledge networks have been developed, the two most commonly utilized applications are Ingenuity Pathways Analysis (IPA; www.ingenuity.com) and GenMapp (www.genmapp.org). An example is provided in Figure 27.1, where muscle biopsies were taken from gastic bypass surgery patients both before and after surgery (Park et al., 2006). Three gene transcripts were found to be significantly decreased in muscle by the surgery-associated weight loss: Growth factor receptor-bound protein 14 (GRB14), glycerol-3-phosphate dehydrogenase 1 (GPD1), and growth differentiation factor 8 (GDF8; myostatin). These three proteins could be connected into the same network using computer programs that query pre-existing literature (Ingenuity; www.ingenuity.com). This provides a candidate network for molecular pathways in muscle influenced by the obese vs. lean state (Figure 27.1).

Network analysis software is a highly useful and practical tool to condense genome-wide data sets into biologically relevant pathways deserving further study, although the sensitivity and specificity

of these packages must be currently considered tentative at best. Pathways and networks are highly dependent on the cells, tissues, and organisms under study, and these packages generally assume that an interaction described in the literature between two proteins in mouse eye are relevant to the cancer cells. Also, they must be assumed to be a relatively blunt instrument (insensitive), as only a very small proportion of "biological truth" in terms of networks and pathways is currently known and published (perhaps 1%).

The second tool to assist in interpretation of genome-wide data is the "time series." The behavior of a gene or protein as a function of time allows an assessment of biological plausibility that can help reduce false positives. Most microarray

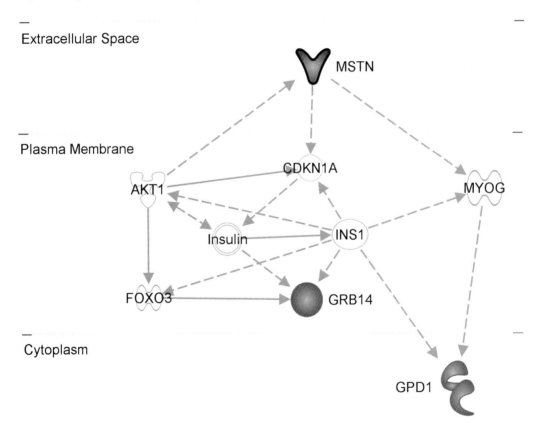

Figure 27.1 Ingenuity network analysis of skeletal muscle mRNA transcripts differentially regulated by the obesity state (from Park et al., 2006).

and proteomics experiments are "snap shots" of main effects (like group) at only a single point in time (e.g., affected versus unaffected subjects; see Figure 27.1 as typical example, with "affected" subjects being the obese group). It is impossible to determine any cause/effect of responses to obesity in the data in Figure 27.1. However, if serial muscle biopsies were taken as a function of time (perhaps after a gastric bypass operation or strict diet), then some of the transcriptional responses seen in Figure 27.1 could be assigned to an earlier time point than others, leading to models more suggestive of cause and effect, and leading to better systems biology models.

To provide an example of the value of time series data in supporting cause/effect relationships, we have described a 27 time point muscle regeneration series, where muscle damage was invoked in mouse models by cardiotoxin, and muscle samples

taken as a function of time during recovery from the muscle damage (Figure 27.2) (Zhao & Hoffman, 2004; Zhao et al., 2002, 2003). Two transcription factors, myogenin and MyoD, are both strongly transcriptionally induced around day 3 during recovery. Close inspection of the time series shows the peak of myogenin to be half a day later than the peak of MyoD; thus these results can support a hypothesis that MyoD induces myogenin (e.g., MyoD is upstream of myogenin). A cause/effect relationship between these two proteins can thus be established.

Future directions in biological computing should focus on combining data from multiple data sets, integrating both snap shot (cross-sectional) and time series projects done, with data cross-referenced and validated in different human and animal models. As an example, we used a multiple project multiple species approach to define molecular networks

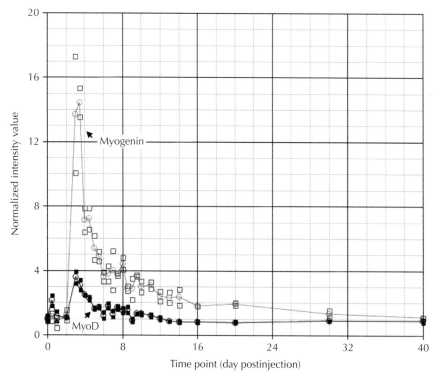

Figure 27.2 Time series data in muscle regeneration *in vivo*. MyoD shows a peak of expression at 3.0 days, while myogenin shows peak expression 0.5 day later, suggesting that MyoD is upstream of myogenin. Data is from Zhao et al. (2002, 2003).

disrupted due to mutations of lamin A/C (*LMNA*), a component of the nuclear envelope, and cause of premature aging, heart disease, muscular dystrophy, and lipodystrophies in human families (Bakay et al., 2006; Melcon et al., 2006). The first step utilized a cross-sectional study, where about 125 patient muscle biopsies from 13 disease groups were assessed for mRNA transcriptome profiles. Use of knowledge networks led to a relational tree of the different disorders (Figure 27.3). The disease of interest was EDMD (Emery Dreifus muscular dystrophy), where patients show mutations of components of the nuclear envelope; however, the molecular pathophysiology of EDMD was very poorly understood.

Transcripts (genes) that distinguished the 6-EDMD node in the tree from others (diagnostic genes) were then queried in a mouse degeneration/regeneration time series (see Figure 27.2). Based on the expression

of network members at different times in muscle regeneration, we defined a cause/effect transcriptional regulatory pathway that included many of the differentially expressed EDMD-specific transcripts (Figure 27.4) (Bakay et al., 2006). This model for a biochemical block in differentiation of muscle caused by mutations in the Lamin A/C (or interacting protein, Emerin) defined a model for disease pathogenesis (failure of events during muscle regeneration) that was then tested and validated in a mouse model of EDMD (Melcon et al., 2006).

These examples illustrate the types of approaches that can be utilized to begin to develop systems biology models of muscle. The remaining sections of this chapter give the status of data generation toward systems biology models using exercise-induced muscle damage and muscle adaptation to resistance exercise training as examples. This is followed by a discussion of current and future

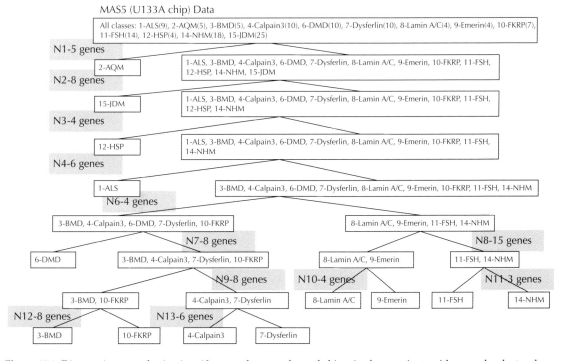

Figure 27.3 Diagnostic gene selection in a 13 group data set of muscle biopsies from patients with muscular dystrophy (see Bakay et al., 2006). The disorders separating at the top of the figure are most dissimilar to the other groups, whereas those near the bottom are closely related. The number of genes needed for accurate diagnosis of each group from microarray data is shown at the nodes.

(a) Normal development and regeneration

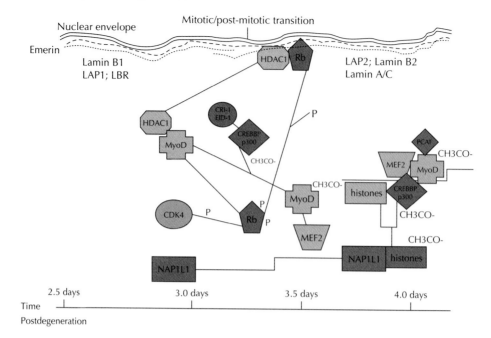

(b) EDMD regeneration

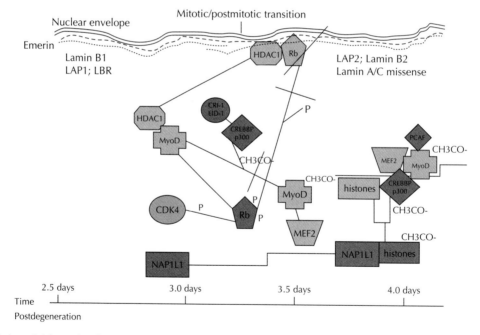

Figure 27.4 A model for molecular pathophysiology of muscular dystrophy involving mutations of the nuclear envelope constructed from time series data. (a) Normal muscle regeneration. (b) Disruption of regeneration in Emery Dreifus muscular dystrophy (EDMD). The X-axis (bottom) shows time series data from mouse regeneration. The remainder of the figure shows timed induction of transcriptional pathways during muscle regeneration, around the time of the transition from mitotically active cells to postmitotic differentiated myotubes (mitotic/postmitotic transition). The proteins boxed are those differentially regulated in patients with EDMD, while the cross-hatches indicate blocks in this molecular pathway due to nuclear envelope mutations (from Bakay et al., 2006).

bioinformatics trends in cause/effect time series data integration.

Exercise-Induced Muscle Damage Mechanisms Elucidated via Time Series Data

An example of data integration between expression profiling data, protein localizations, and DNA variations can be seen across a series of studies investigating the molecular mechanisms underlying exercise-induced muscle damage (Chen et al., 2003; Hubal et al., 2008, 2010). All these studies utilize a human maximal eccentric exercise muscle damage model that typically produces a moderate amount of muscle dysfunction following exercise (~50–70% decreases in voluntary strength with recovery to baseline between 3 and 7 days postexercise).

The earliest of this series (Chen et al., 2003) examined a single bout of eccentric exercise in the leg, presenting transcriptional data at 6 h postexercise along with quantitative reverse-transcription polymerase chain reaction (qRT-PCR) validation, Western blots, and immunohistochemistry for a subset of genes of interest. Functional and biochemical data (creatine kinase activity) were also collected at pre-exercise and several time points after exercise (immediately postexercise and 1, 3, 5, and 7 days postexercise). Results of this study highlighted the upregulation in a set of inflammatory factors in the hours following exercise.

Given that inflammation is known to play many roles in skeletal muscle during damage and repair, Hubal et al. (2008) examined this set of inflammatory factors following two bouts of eccentric exercise (using the same exercise model as Chen et al.), to determine the role of these factors in muscle adaptation. The rapid adaptation of muscle to eccentric exercise is manifested by lower indices of damage, known collectively as the "repeated bout effect." This study utilized qRT-PCR to assess transcriptional changes in eight inflammatory genes with further immunohistochemistry to localize protein products. Of the eight genes tested, five demonstrated upregulation with the first bout of exercise with no differences between bouts. The three remaining genes [monocyte chemoattractant protein 1 (*CCL2*), zinc

finger protein 36 (*ZFP36*), and CCAAT-enhancing binding protein delta (*CEBPD*)] demonstrated upregulation following the first exercise bout and then further significant upregulation following the second exercise bout.

The further upregulation of these three inflammatory factors seems to conflict with literature showing decreased overall inflammation with the repeated bout effect (reviewed in Peake et al., 2005), although studies of specific inflammatory factors were fairly mixed with regard to a repeated bout effect. Hubal et al. (2008) co-localized the CCL2 protein with both macrophages and muscle precursor cells (satellite cells) within the exercised muscle tissue, with little expression in myonuclei. Based on cellular studies that indicate CCL2 is a vital chemokine enabling cross talk between macrophages and satellite cells (Chazaud et al., 2003), it has been proposed that CCL2 plays an important role in muscle repair following exercise (Hubal et al., 2008). This is supported by data from the *CCR2* knockout mouse, showing little difference in the freeze-injury response in *CCR2* versus control animals at 3 days postinjury, but impaired regeneration, increased inflammation, and increased fibrosis at 14 days postinjury (Warren et al., 2005).

Most recently, Hubal et al. (2010) used time series study to track genetic influences of *CCL2* and *CCR2* variations across a 10-day period following eccentric exercise-induced damage on muscle function and biochemical measures. A key element within this analysis was the use of post hoc testing of significant repeated measures data to pinpoint times within the larger time series at which effects are strongest. Significant overall genotype–time interactions in strength values were found for several specific variants in *CCL2* and *CCR2*, and post hoc testing revealed that these differences were specific to the 4–10 days postexercise time period. These data support a role for CCL2 in muscle repair, rather than a role in the primary generation of muscle damage.

This progression of studies highlights the insight that can be gained from integrating multiple levels of genetic, cellular, molecular, and functional data, especially when time series data are utilized.

Muscle Adaptations to Resistance Exercise Training

The ability of exercise training to remodel skeletal muscle is well documented, eliciting specific adaptations depending on what type of exercise intervention is used. Chapter 26 of this book discussed primarily metabolic changes caused by endurance-type or aerobic exercise training. Here, we address primarily resistance-type training, which elicits remodeling changes in both muscle structure and function, the most obvious being increased muscle strength and size (hypertrophy).

On a functional level, a great deal of time series data has been collected in populations utilizing resistance exercise interventions. Environmental and biological factors such as sex, age, disease state, diet, and others have been tested for effects on traits such as muscle strength and fatigability before, during, and following acute or chronic exercise training of various designs. However, at the molecular level, time series data investigating muscle adaptation are more rare (especially in human studies using muscle biopsies from volunteers), and data sets are typically disparate with regard to the type of data generated (i.e., genomic, proteomic, metabolomic, and so on).

Using mostly traditional molecular biology techniques (Western blotting, real-time PCR, etc.), significant work has been done toward determining key pathways, such as the AKT/mTOR pathway, that regulate muscle adaptation via hypertrophy/atroph balance (see Hoffman & Nader, 2004; Phillips, 2009). Various animal and human models have been used to identify hypertrophy/atrophy pathway elements and regulation by factors such as mechanical stimuli, growth factors, and inflammatory cytokines, as well as the effects of exercise, diet, and pharmacological interventions. While this work has helped us begin to understand molecular pathways underlying adaptation, the limited focus of many studies to a few genes or pathways makes it difficult to fully model a very complex adaptation process.

High-throughput studies have been implemented in response to this limitation, and several studies have examined transcriptional profiles following exercise training. Recent investigations have just begun to explore proteomic and metabolomic responses to training, with the first publications exploring only aerobic exercise interventions. miRNA and epigenetics level data have yet to be utilized in examining muscle adaptations to training. Although many studies have examined genomic variant associations with muscle traits such as muscle strength or size (Bray et al., 2009), only a few high-throughput genetic association studies (genome-wide association studies or GWAS) have looked at exercise-related traits (i.e., maximal oxygen consumption, exercise heart rate, and so on), and to date no GWAS studies have addressed exercise-induced change in muscle traits.

Bioinformatics of Integrating Cause/ Effect Time Series Data to Model Skeletal Muscle Adaptations

Given the large numbers of participating components and processes at all biological levels with various space and time, the muscle remodeling system is organizationally very complex. The study of the cause/effect of such a system requires a multi-scale systems biology approach that may identify an appropriate modularization of the system and draw results through integrating the modules (Figure 27.5). The modularization part of systems biology aims at identifying specific roles of processes and signals in smaller, interdependent and fully regulated modules, and characterizing the function of each module in adequate detail. Whereas the integration counterpart computes what would happen if these modules were lacking or functioning in a different way. It is evident that the muscle remodeling system is hierarchical and modular, insights into an array of smaller modules may provide a foundation on a deeper understanding of the functionality of large-scale integrated systems. In addition, large-scale analyses without appropriate modularization often create mathematical complications and fail to capture subtle features in dynamics, regulation, and adaptation of each biological module (Figure 27.5).

When designing a modulated systems biology approach, it is not only important to carefully partition the system into modules, but also critical

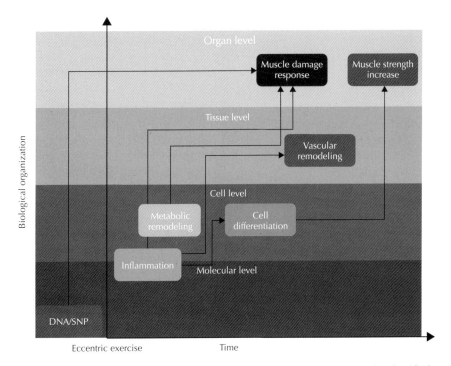

Figure 27.5 Schematic illustration of the modularization of the muscle remodeling system. The identified modules are positioned spatially in the bands representing the biological level where the modules mainly function, and they are also placed in sequence along the time axis to represent when the modules function. The integration of the modules aims to analyze timing and signal passing between different modules.

to evaluate the risks of partitioning incorrectly. On the one hand, modularization is the answer to managing complexity; on the other hand, inappropriate modularization may mislead the study and further increase the complexity. As an example for describing the modular nature of systems biology, we present here a modular architecture for analyzing the mechanism of muscle remodeling in response to exercise. The following description of effort not only illustrates the difficulty of designing an appropriate modular architecture of a complex system but also how much of a trial and error process it is at this stage.

Understanding the mechanism of muscle remodeling involves the analysis of large molecular networks and protein–protein interactions, the mapping of metabolic pathway systems, and determining the interactions of all participating subsystems at different levels of biological organization. As we can see in the above discussion, this system contains a number

of fundamental interdependent elements (such as inflammation, vascular remodeling, etc.) that interact in a convoluted way. As a first step in examining these questions, it is necessary to carefully examine over- and under-modularity; the key of realizing appropriate modular designs is to group strongly interacting elements or parts together and separate weakly interacting ones. The complexity often rises from the unknown nature and magnitude of interactions between different parts of the system. The nature of the interaction between two parts may be positive, negative, or unrelated. As a result, overall system response can exhibit highly nonlinear and/or nonmonotonic behavior in response to changes in one or more parts.

An example of a previous successful approach to integrating multiple omics data sets from parallel tissue samples can be seen in the work of Dobrin et al. (2009) or Yang et al. (2009). In both instances, the investigators were able to infer new information or reinforce biological theories about causal pathways and tissue–tissue subnetworks underlying the obesity trait by using an integrated approach to data analysis. This group has pioneered and extensively validated the likelihood-based causality model selection (LCMS) procedure (Schadt, 2009). Briefly, this method is applied to data on multiple levels [e.g., microarray expression profiles, clinical measures, loci of single-nucleotide polymorphisms (SNPs), database results] to generate likelihoods, or probabilities, that certain candidate data objects (e.g., eQTLs and mRNAs) lie on various nodes within the network of disease function acting as either causal factors, reactive factors, or independent factors within the organism or model of disease. Specific equations model interactions between nodes.

The LCMS procedure can uncover the relationships among many layers of data and provides probabilities that each component of the model plays either a causal, reactive, or interactive role in the development of disease. Models for all possible biological structures can be constructed, and then the statistical likelihood of each model evaluated based on the experimental data. The model with the highest likelihood given

in the data then provides insight into the underlying mechanisms of disease. Increasing levels of data that can be integrated and assessed by the statistical procedure will provide greater levels of insight, and increased confidence about that insight, regarding the mechanisms driving muscle adaptation.

Conclusion

The ultimate goal of systems biology is to model complex biological phenomenon, where the models are accurate in predicting response to environmental variables, including individual variability (SNPs) in the response. Muscle adaptation may emerge as the tissue of choice for developing systems biology models in human tissue remodeling. Developing such models will require large numbers of data sets, many involving time series data to define upstream versus downstream remodeling (cause/effect models). Initial data sets are emerging in muscle damage, and strength training, although hundred more data sets and considerable bioinformatic and computational research will be needed before this vision becomes a reality.

References

Bakay, M., Wang, Z., Melcon, G., et al. (2006). Nuclear envelope dystrophies show a transcriptional fingerprint suggesting disruption of Rb-MyoD pathways in muscle regeneration. *Brain* 129, 996–1013.

Bray, M.S., Hagberg, J.M., Pérusse, L., et al. (2009). The human gene map for performance and health-related fitness phenotypes: The 2006–2007 update. *Medicine and Science in Sports and Exercise* 41, 35–73.

Chazaud, B., Sonnet, C., Lafuste, P., et al. (2003). Satellite cells attract monocytes and use macrophages as a support to escape apoptosis and enhance muscle growth. *The Journal of Cell Biology* 163, 1133–1143.

Chen, Y.W., Hubal, M.J., Hoffman, E.P., Thompson, P.D. & Clarkson, P.M. (2003). Molecular responses of human muscle to eccentric exercise. *Journal of Applied Physiology* 95, 2485–2494.

Dobrin, R., Zhu, J., Molony, C., et al. (2009). Multi-tissue coexpression networks

reveal unexpected subnetworks associated with disease. *Genome Biology* 10, R55.

Fang, Z., Yang, J., Li, Y., Luo, Q. & Liu, L. (2006). Knowledge guided analysis of microarray data. *Journal of Biomedical Informatics* 39, 401–411.

Hoffman, E.P. & Nader, G.A. (2004). Balancing muscle hypertrophy and atrophy. *Nature Medicine* 10, 584–585.

Hubal, M.J., Chen, T.C., Thompson, P.D. & Clarkson, P.M. (2008). Inflammatory gene changes associated with the repeated-bout effect. *American Journal of Physiology—Regulatory, Integrative and Comparative Physiology* 294, R1628–R1637.

Hubal, M.J., Devaney, J.M., Hoffman, E. P., et al. (2010). CCL2 and CCR2 polymorphisms are associated with markers of exercise-induced muscle damage. *Journal of Applied Physiology* 108, 1651–1658.

Melcon, G., Kozlov, S., Cutler, D.A., et al. (2006). Loss of emerin at the nuclear

envelope disrupts the Rb1/E2F and MyoD pathways during muscle regeneration. *Human Molecular Genetics* 15, 637–651.

Park, J.J., Berggren, J.R., Hulver, M.W., Houmard, J.A. & Hoffman, E.P. (2006). GRB14, GPD1, and GDF8 as potential network collaborators in weight loss-induced improvements in insulin action in human skeletal muscle. *Physiological Genomics* 27, 114–121.

Peake, J., Nosaka, K. & Suzuki, K. (2005). Characterization of inflammatory responses to eccentric exercise in humans. *Exercise Immunology Review* 11, 64–85.

Phillips, S.M. (2009). Physiologic and molecular bases of muscle hypertrophy and atrophy: Impact of resistance exercise on human skeletal muscle (protein and exercise dose effects). *Applied Physiology, Nutrition, and Metabolism* 34(3), 403–410.

Reeves, E.K., Gordish-Dressman, H., Hoffman, E.P. & Hathout, Y. (2009).

Proteomic profiling of glucocorticoid-exposed myogenic cells: Time series assessment of protein translocation and transcription of inactive mRNAs. *Proteome Science* 7, 26.

Schadt, E.E. (2009). Molecular networks as sensors and drivers of common human diseases. *Nature* 461(7261), 218–223.

Warren, G.L., Hulderman, T., Mishra, D., et al. (2005). Chemokine receptor CCR2 involvement in skeletal muscle regeneration. *The FASEB Journal* 19(3), 413–415.

Yang, X., Deignan, J.L., Qi, H., et al. (2009). Validation of candidate causal genes for obesity that affect shared metabolic pathways and networks. *Nature Genetics* 41(4), 415–423.

Zhao, P. & Hoffman, E.P. (2004). Embryonic myogenesis pathways in muscle regeneration. *Developmental Dynamics* 229, 380–392.

Zhao, P., Iezzi, S., Carver, E., et al. (2002). Slug is a novel downstream target of MyoD. Temporal profiling in muscle regeneration. *The Journal of Biological Chemistry* 277, 30091–30101.

Zhao, P., Seo, J., Wang, Z., Wang, Y., Shneiderman, B. & Hoffman, E.P. (2003). In vivo filtering of in vitro expression data reveals MyoD targets. *Comptus Rendus Biologies* 326, 1049–1065.

Chapter 28

Proteomics in Exercise Training Research

DUSTIN S. HITTEL

Roger Jackson Center for Health and Wellness, Human Performance Laboratory, Faculty of Kinesiology, University of Calgary, Calgary, AB, Canada

Muscle remodeling in response to training occurs due to changes in proteins making up the muscle tissue. The changes in proteins then change the functional abilities of the myofiber, vasculature, and connective tissue, enabling greater functional capacities after training. To some extent, the remodeling of muscle proteins can be monitored by the changes in gene expression; specific genes turn off and on after bouts of training, leading to altered transcriptome, and translation of the transcriptome into proteins directly involved in the muscle remodeling. However, mRNA profiling is a very blunt instrument, and is able to detect only a small percentage of the protein changes causing greater fitness. In addition, many of the protein changes do not involve changes in gene transcription, but instead involve changes in protein modifications (phosphorylation, glycosylation), and protein localizations (e.g., movement from one subcellular compartment to another), or protein–protein interactions.

Recent advances in mass spectrometry (MS) permit monitoring global changes in proteins, an emerging field known as proteomics, and applications of proteomics to exercise biology. In Chapter 5 of this volume, the physics and chemistry underlying proteomic profiling technologies are described. In this chapter, we will review and expand upon the practical application of proteomics for investigating the physiological adaptations associated with exercise

training and human performance. The number of studies remains quite limited, yet proteomics is certain to advance our understanding of exercise physiology over the next decade.

Why Study the Proteome?

The sequencing and annotation of the human and rodent genomes has produced genome-wide expression-profiling technologies such as DNA microarrays. Although initially welcomed with great fanfare, gene expression (mRNA) levels alone have proven to be poor predictors of the abundance and activities of their corresponding protein products (Hittel et al., 2005, 2007; Yi et al., 2008). Regrettably, this has limited the acceptance of molecular profiling technologies by the greater exercise physiology community. The root of this problem is an incomplete understanding of the central dogma of biochemistry; that genetic information from the genome is transcribed into mRNA, then into proteins (Figure 28.1). For instance, the underlying assumption that a twofold increase in the abundance of an mRNA transcript always translates into a twofold increase in the corresponding protein product is valid only for prokaryotes (bacteria) and cannot be generalized to all organisms. By necessity and design, multicellular organisms regulate the expression of many genes at the posttranscriptional level using a variety of elegant control mechanisms: mRNA structure, alternative splicing, and microRNAs, to name a few. For example, the abundance of many proteins (i.e., mitochondrial enzymes) in muscle do not temporally correlate with the abundance of their

Genetic and Molecular Aspects of Sport Performance, 1st edition. Edited by Claude Bouchard and Eric P. Hoffman. Published 2011 by Blackwell Publishing Ltd.

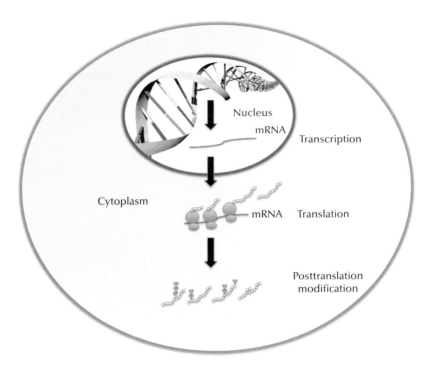

Figure 28.1 Life depends upon the information flow in a cell and the ability to change genetic information into functional proteins. This is referred to as the central dogma of biochemistry and is regulated at three levels of control: transcription, translation, and posttranslational modification. While there are subtle differences in these processes between eukaryotic and prokaryotic organisms, the basic steps of transcription and translation are conserved.

corresponding mRNA transcripts (Hittel et al., 2005; Yi et al., 2008). Further, many proteins expressed in skeletal muscle are not consistently detectable at the mRNA level [i.e., myosin light chain kinase (MLCK) and creatine kinase (CK)] because of tight transcriptional control (Yi et al., 2008).

A true understanding of muscle remodeling following training will require the modeling of all concurrent processes in a cell, tissue, or organism by global measurements (Smith & Figeys, 2006). Consequently, this systems approach toward exercise research is possible only if all biomolecular interactions are evaluated. Much of the recent progress in the field of systems biology has in fact been driven by MS-based proteomics (Hittel et al., 2007).

Posttranslational Modifications

The posttranslational modification (PTM) of proteins is an important (but often overlooked) contributor to the cellular adaptations with exercise training (Flueck, 2009; Riedmaier et al., 2009; Sheehan, 2006; Smith & Figeys, 2006). During protein synthesis, there are 20 different amino acids, each with their own unique chemical side chains, which confer the unique structure, function, and charge characteristics of a protein (Figure 28.2). After translation, PTMs can extend the range of function of a protein through the addition of biochemical groups such as phosphates, lipids, and carbohydrates or can produce structural changes, like the formation of disulfide bridges. From a practical perspective, PTMs have allowed anti-doping organizations to detect the abuse of recombinant forms of erythropoietin (EPO; a hematocrit boosting peptide hormone) in urine and blood samples from athletes (Lasne & de Ceauriz, 2000). Endogenous EPO has a unique pattern of glycosylation (carbohydrate side chains) that can be differentiated from the recombinant form by isoelectric focusing (IEF), a technique that separates

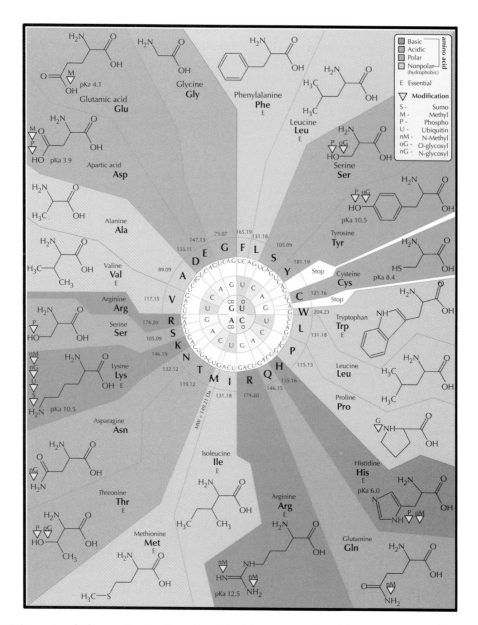

Figure 28.2 A genetic code diagram showing the amino acid residues as target of modification. During protein synthesis, 20 different amino acids can be incorporated in proteins. After translation, the posttranslational modification of amino acids extends the range of functions of the protein by attaching to it other biochemical functional groups such as acetate, phosphate, various lipids, and carbohydrates, by changing the chemical nature of an amino acid, or by making structural changes, like the formation of disulfide bridges. Image modified from http://en.wikipedia.org/wiki/File:GeneticCode22.svg.

proteins by charge. Initially, recombinant EPO was produced in a cell line derived from hamsters and thus exhibited an IEF "fingerprint" dissimilar from human EPO, making it easy to identify in urine and blood (Figure 28.3). More recently, some 2008 *Tour de France* competitors mistakenly believed that a new form of EPO known as CERA (continuous erythropoietin receptor activator) was undetectable

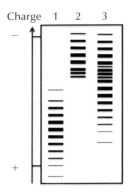

Charge 1   2   3

Figure 28.3 Representative isoelectric focusing gels of normal excreted urinary EPO. (1) Purified CERA, (2) urine containing a mix of endogenous EPO and CERA, (3) from an athlete who was caught doping.

Table 28.1 Boolean search of the Pubmed database for publications containing the words "genome" or "proteome" in combination with exercise.

Year	2006–2009	2003–2006	2000–2003	1990–2000
Genome and exercise	251	267	194	183
Proteome and exercise	22	7	5	2

because of a synthetic chemical modification that was added to make the drug more potent. In fact, this modification made the detection of EPO doping easier because the IEF fingerprint of CERA is quite distinct from both endogenous and recombinant EPO (Figure 28.3).

Exposure to oxidative stress also produces a wide range of reversible and irreversible PTMs (Sheehan, 2006). Increasingly, oxidative stress is believed to act as cellular signal mediating exercise adaptation and adverse metabolic changes, such as insulin resistance. Given the extremely short half-life of oxygen radicals, it is hypothesized that PTMs such as protein carbonylation and nitrosylation may function as signaling intermediates of oxidative stress by targeting key regulatory proteins for degradation (Scheele et al., 2009; Sheehan, 2006). One way in which proteomics can be used to study oxidative stress is through the identification of oxidative PTMs that change the charge characteristics of proteins.

Proteomics and Exercise Training: Techniques and Practical Applications

Quantitative proteomic profiling has lagged behind other gene profiling techniques, in part, due to the highly technical nature of protein preparation and MS (Hittel et al., 2007). However, in recent years, the number of proteomics publications has increased

considerably, a reflection perhaps of the improved accessibility and affordability of MS equipment and data analyses. Further, while exercise-related genomics publications have plateaued, the numbers of corresponding proteomics publications are on the rise (Table 28.1). As with the early days of genome sequencing and transcriptional profiling, proteomics has faced technical challenges, such as the complexity of cellular proteins (PTMs, alternate splicing isoforms), the low abundance and vast dynamic range of cellular proteins, and the small sample size of most clinical samples. Unlike DNA- and RNA-based techniques, currently there is no straightforward way to amplify protein samples; however, similar problems were overcome during large-scale sequencing projects by unforeseen technical advances (i.e., PCR) (Hittel et al., 2007). The purpose of this section is to review and provide a practical perspective on publications that have adopted a proteomics approach to study exercise adaptation.

Two-Dimensional Gel-Based Proteomics

The overarching goal of proteomics is to (a) identify and (b) differentially quantify proteins in cells, tissues, and organs under different physiological states. Unlike mRNA profiling, proteins cannot be identified in a simple one-step process. Instead, proteins must first be fractionated by a combination of differential extraction and gel electrophoresis. In addition, current instrumentation limits the mass window for most MS-based techniques, which require that proteins be digested into peptides. Accordingly, the workflow for "classic" proteomic profiling begins with protein separation by two-dimensional gel

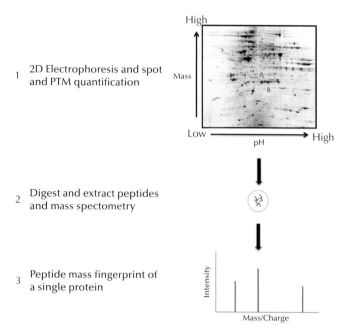

1   2D Electrophoresis and spot and PTM quantification

2   Digest and extract peptides and mass spectometry

3   Peptide mass fingerprint of a single protein

Figure 28.4 A "classic" 2D gel proteomics experimental workflow. (1) Proteins are first quantified in gels by densitometry. (2) Differentially expressed protein spots are then excised and digested into peptides that are analyzed by mass spectrometry. (3) The peptide mass fingerprint is then compared against a database of peptide masses to uniquely identify the protein and potential PTMs.

electrophoresis (2DGE) (Figure 28.4). Each protein spot is then quantified based on staining intensity and next compared between two or more gels. This technique pairs software-assisted pattern-matching algorithms with a researcher's observational skills and has been the standard method for quantitative proteomics for more than 20 years (Flueck, 2009; Hittel et al., 2007).

Skeletal muscle is central to the performance and metabolic adaptations associated with exercise training. Furthermore, proteomics is the ultimate means to resolve fine molecular adjustments that become manifest as structural and metabolic adaptations (Flueck, 2009; Hittel et al., 2005). In a recent study, proteomics was used to study adaptations in plantaris muscle from rats that were exercised for 5 weeks (Burniston, 2009). This analysis found 15 proteins that were differentially expressed between exercise-trained and untrained animals. Proteins that increased with exercise included myoglobin, which facilitates diffusion of molecular oxygen

within the myofiber and the mitochondrial enzyme succinyl-CoA dehydrogenase (SDH). This study also found an increase in slow isoform of myosin light chain (MLC), which is known to be preferentially expressed in slow-twitch oxidative muscle fibers. These findings agree with and expand upon histochemical and morphometric measures, showing that aerobic exercise training produces a slower contracting and more fatigue-resistant fiber type in muscle. The list of downregulated proteins appear to confirm this interpretation, with decreases in metabolic enzymes most commonly associated with type II (glycolytic) muscle fibers such as lactate dehydrogenase A (LDHA) and phosphoglucomutase 1 (PGM1) (Burniston, 2009). Another interesting finding from this study was that mitochondrial aconitase, an important tricarboxylic acid (TCA) cycle enzyme (Burniston, 2009), was represented by five distinct spots with identical molecular weight but separated by charge on a 2D gel. Further, while one of these spots decreased with exercise, another one increased. Using a combination of 2DGE and MS, the authors were able to determine that the differences in protein charge on the 2D gel corresponded to varying amounts of protein carbonylation, the most frequent type of protein modification in response to oxidative stress. Chemically, oxidative carbonylation occurs at the amino acids proline, threonine, lysine, and arginine, presumably through the metal-catalyzed activation of hydrogen peroxide to a reactive intermediate. Originally, it was assumed that protein carbonylation is irreversible and destined to induce protein degradation in a nonspecific manner; however, recent evidence suggests that it may play a role in signaling cellular redox state (Anderson et al., 2009). The fact that this and other PTMs could not be predicted from mRNA abundance or DNA sequence strengthens our argument that proteomics is a practical approach for discovering novel adaptations to exercise.

It is a well-established fact that exercise training produces functional improvements in performance and maintenance of cardiac function under stress (Boluyt et al., 2006). In fact, rats subjected to an exercise protocol simulating moderate aerobic exercise show significant improvements in both aerobic capacity ($\dot{V}O_2$) and cardiac function (Boluyt et al., 2006). In this study, a classic proteomics experimental workflow (Figure 28.4) was able to identify 624 protein spots from isolated rat hearts, 23 of which were differentially expressed with exercise training. A number of energy metabolism proteins were upregulated with exercise training, including heart-type fatty acid-binding protein (HFABP), short- and long-chain acyl-CoA dehydrogenase (SCAD, LCAD), malate dehydrogenase (MDH), and aspartate aminotransferase (AAT). These adaptations are not entirely surprising, given the increased metabolic demands being placed on the exercising heart. Both MDH and AAT are components of the malate–aspartate shuttle that couples glycolysis with the TCA cycle. In a previous exercise study using DNA microarrays, we also found increased expression of AAT, MDH, and SCAD in skeletal muscle, presumably to improve the economy of glucose metabolism (Hittel et al., 2005). This finding suggests that conserved mechanisms

regulate metabolic adaptations to exercise in both skeletal muscle and the heart.

In addition to a number of contractile proteins, the expression and phosphorylation of heat shock protein 20 (HSP20) is also increased in rat heart with exercise training (Burniston, 2009). It is worth noting that the phosphorylation of HSP20 was detected because of an "acidic shift" in the position of the protein spot on a 2D gel (Spot B → A in Figure 28.4) indicating the presence of a phosphate group. HSP20 is expressed in cardiac, smooth, and skeletal muscle, where it protects cells from thermal and ischemic insult. Further, HSP20 phosphorylation increases the rate of myocardial shortening and prevents catecholamine-induced cardiomyocyte death (Burniston, 2009). Collectively, these findings provide novel information about the molecular adjustments underlying the exercise-induced improvements in myocardial contractility, stress adaptation, and metabolism.

Emerging Proteomics Technologies

In the previous section, we provided examples of how a classic proteomics workflow (Figure 28.4) could be used to discover novel protein adaptations with exercise training. However, this approach is limited by the relatively small proportion of proteome that can be identified by gel-based staining techniques. This may explain, in part, why most of the publications describing the skeletal muscle proteome are dominated by contractile and metabolic proteins (Flueck, 2009; Hittel et al., 2005).

Mass spectrometry-based proteomic analyses are carried out at the peptide level where the peptides are ionized and transferred into a gas phase that the instrument can analyze (see Chapter 5). The mass spectrometer instrument consists of an ion source, a mass analyzer that measures the mass to charge ratio (m/z), and a detector that registers the number or intensity of ions at each m/z value. Electrospray ionization (ESI) and matrix-assisted laser desorption/ionization (MALDI) are the two most common techniques to ionize peptides for MS analysis (Smith & Figeys, 2006). Whereas MALDI is carried out on individual proteins, ESI-MS lends itself to automation since it relies on a steady liquid phase, which can be coupled to a liquid-chromatography (LC) column to separate complex mixtures of proteins and peptides prior to analysis. The quantitative aspects of classic 2DGE/MS-based proteomics rely on densitometric measurements of spot intensity to ascertain differences in protein abundance between samples (Figure 28.4). Though densitometry scanners and image-processing software have improved significantly over the years, they still deal poorly with estimating relative changes in protein abundance, particularly for spots which are saturated or poorly focused (Flueck, 2009; Hittel et al., 2007; Smith & Figeys, 2006). To enhance the quantitative aspects of MS-based proteomics, techniques have been developed that exploit the ability of MS instruments to differentiate chemically identical proteins that contain different ratios of stable carbon isotopes (carbon 13). The practical relevance is that the ratio of MS signal intensities for (labeled and unlabeled) peptide pairs can be used to calculate the ratio of their corresponding protein abundance in a complex mixture (Figure 28.5). Interestingly, the stable isotope analysis of urine metabolites has also been used to detect exogenous testosterone doping in sport (Aguilera et al., 1999). The three most common ways to label protein mixtures are stable isotope labeling by amino acids in culture (SILAC) using ^{13}C-labeled amino acids, proteolytic or end-labeling with ^{18}O-water, and chemical labeling using isotope-coded affinity tags (ICAT) which are covalently attached to amine residues in peptides. Although we have found SILAC as the most satisfactory strategy for quantitative proteomics using cells in culture (Hathout, 2007; Hittel et al., 2007, 2008), most exercise studies use human or animal models; therefore, we will limit this review to post-isolation isotope tagging.

A stable isotope proteomics workflow differs from a 2D gel-based approach (Figure 28.4) in that mixtures of proteins are first digested into peptides

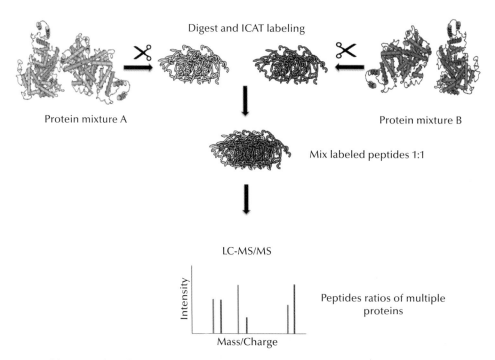

Figure 28.5 A stable isotope-based proteomics experimental workflow. Two protein pools are first digested into peptides, which are then isotopically coded by covalent modification. Samples are then mixed and separated by liquid chromatography coupled directly to an ESI-MS/MS instrument. MS/MS analysis allows for both the calculation of peptide ratios and for the identification of the peptide based on its amino acid sequence.

and then separated using multidimensional liquid chromatography prior to MS (LC-MS/MS) (Figure 28.5). Since one measure of success in proteomics is the number of proteins identified, a current challenge to high-throughput proteomics is identifying large numbers of proteins from complex peptide mixtures. Presently, the interpretation of large spectra such as those generated by LC-MS/MS must go through several rounds of software-driven data analysis to identify and quantify the many thousands of peptide pairs in each sample. Even with these drawbacks, the potential benefits of a high-throughput proteomics platform vastly outweigh the current technical limitations; in fact, many of these same issues plagued the early days of genome sequencing and microarray analysis prior to improvements in equipment and data analysis. These technologies and tools are currently being developed to create robust platforms for quantitative high-throughput proteomics. In fact, commercial kits and software packages are now available for ICAT, SILAC, and ^{18}O labeling of protein samples, and many companies currently offer quantitative analysis on protein or whole tissue samples. This is often the more prudent choice for those labs without access to core facilities, or which cannot devote the time to retool their lab for proteomic analysis. Regardless, the number of publications utilizing stable isotope-based quantitative proteomics has dramatically increased since the technology first appeared in 2003.

To overcome the complexity of large proteomics data sets and their analysis, a strategy used by researchers has been to analyze subcellular proteomes rather than entire homogenates of a cell, tissue, or organ (Vitorino et al., 2007). One of the most studied and well-characterized subcellular proteomes is the mitochondria, the organelle that produces more than 90% of cellular adenosine triphosphate (ATP) by oxidative phosphorylation. The mitochondrial proteome has long been considered the ideal subproteome to study in the context of physiological adaptation for a variety of reasons: it is highly abundant in metabolically active tissues such as skeletal muscle, heart, and liver; it is easy to isolate using differential centrifugation; and its purity and intactness can be accurately assessed using a combination of Western blotting and respirometry (measuring O_2 uptake). The protein composition, structure, and function of mitochondria vary significantly depending on cellular location, disease, and physiological state. Furthermore, the mitochondrial proteome has been well studied and annotated (Kurland & Andersson, 2000; Mootha et al., 2003; Schon, 2007; Taylor et al., 2003) in a number of tissues and species. To date, studies of the mitochondrial proteome have identified almost 750 proteins, many of which were not predicted by sequence analysis of the human genome for mitochondrial targeting sequences (Jullig et al., 2008; Taylor et al., 2003; Yates et al., 2005). These studies have provided important insights into mitochondrial function, adaptation to exercise, and maladaptation in disease states such as the mitochondrial myopathies and type 2 diabetes.

A recent exercise study used stable isotope-based proteomics to study changes in subsarcolemmal (SS) and intermyofibrillar (IMF) mitochondria from the hearts of rats before and after endurance exercise training (Kavazis et al., 2009). The purpose of this study was to investigate exercise-induced changes in mitochondrial proteins by comparing the proteome of these functionally distinct mitochondrial populations from the myocardium of sedentary control and exercise-trained animals. To achieve this goal, the authors utilized a combination of subcellular fractionation and isobaric tags for relative and absolute quantitation (iTRAQ), which allows for the identification and quantification of proteins between multiple samples. Techniques such as iTRAQ allow for labeling of free amine groups; all proteins are labeled without any specificity. Further, this strategy allows for multiplexed quantitation of multiple samples in one MS/MS experiment (Figure 28.5). To demonstrate that the exercise training protocol used in this study had a measurable effect, the authors conducted cardiac functional measurements prior to mitochondrial isolation. This method identified 222 total cardiac mitochondrial proteins, seven of which had never been characterized as mitochondrial proteins when compared against a gene and protein ontology database compiled from previous surveys of human and rodent mitochondria.

Of these, 11 proteins were significantly altered by repeated bouts of exercise. Differentially expressed proteins were divided into functional categories including beta-oxidation, respiratory chain, TCA cycle, redox regulation, and amino acid metabolism. Interestingly, there were differences in the response of IMF versus SS mitochondria that are thought to supply ATP to different cellular structures in heart and skeletal muscle. Notably, many respiratory chain proteins were present in lower abundance in SS versus IMF mitochondria, agreeing with previously published data suggesting that SS mitochondria have lower rates of oxidative phosphorylation compared to IMF mitochondria (Kavazis et al., 2009). While there were many interesting differences between IMF and SS mitochondria, the primary findings of this study showed that both populations of mitochondrial from exercised hearts had lower levels of monoamine oxidase A (MAO-A) and increased thioredoxin-dependent peroxide reductase (PRDXIII) relative to sedentary controls. Given the respective roles of these proteins as oxygen radical generators and scavengers, the authors interpreted these novel findings to mean that mitochondria from exercise-trained hearts are positively adapted to produce fewer oxygen free radicals under increased workloads. As with the other studies described in this chapter, this finding has important implications for our understanding of the molecular events that underlie improved cardiac function with exercise training. This study provided novel and significant information about the exercise-induced adaptations of the heart mitochondrial proteome that would probably not have been discovered using either mRNA profiling or a classic experimental design.

Serum Proteomics, the Secretome, and Doping

The use of proteomics techniques to screen for circulating biomarkers is increasingly common in exercise training and health research. In fact, the circulating levels of a wide variety of proteins are routinely used to evaluate pathophysiological changes in organs such as skeletal muscle (CK),

the liver (AAT), and heart (FABP). Furthermore, circulating proteins such as interleukin 6 (IL-6), myostatin (MSTN), and insulin-like growth factor 1 (IGF-1) are now recognized for their important role in the systemic coordination of exercise training adaptation. As such, there is significant interest in finding novel biological markers to gauge human health and performance. Furthermore, since many performance-enhancing substances are themselves proteins or peptide hormones, proteomic screening techniques have become another valuable tool for detecting new doping agents in blood and urine of athletes.

A "top–down" approach toward identifying novel protein biomarkers in blood is hampered by an abundant background of serum proteins, wherein a protein of interest may be diluted several orders of magnitude. To achieve this goal, the protein detection technique must either be highly sensitive to detect all or a group of proteins in a complex mixture or highly specific for a protein of interest. As described previously, recombinant EPO and CERA can be directly detected in both serum and urine because differences in glycosylation patterns relative to endogenous EPO can be readily detected using a combination of IEF and highly specific immunoblotting (Figure 28.3). In some cases, however, when a compound of interest cannot be reliably detected, a surrogate or indirect biomarker can be used. For instance, it has been shown that chronic abuse of anabolic steroids increases circulating levels of IGF-1 and alters lipoprotein profiles. Presumably, a coincident change in both of these protein biomarkers in an individual could indicate past steroid abuse, warranting more scrutiny of the athlete in future tests (Riedmaier et al., 2009). Another highly sensitive but unbiased method to screen for circulating protein biomarkers is surface-enhanced laser desorption/ionization (SELDI), a variant of MALDI MS that utilizes a biochemical affinity matrix to reduce the complexity of protein mixtures. In effect, binding to the SELDI surface acts as a separation step to reduce the complexity of the serum, making the subset of proteins that bind to the surface easier to analyze. Further, this surface can be functionalized with antibodies, other proteins, or DNA depending on the type of biomarker being sought. SELDI, which is normally used to identify novel disease biomarkers, has recently been applied to detect performance-enhancing substances. One such study found that the administration of recombinant human growth hormone (rHGH; a potent but difficult to detect anabolic agent) resulted in the sustained elevation of alpha-hemoglobin long after rHGH could be detected in the blood (Baoutina et al., 2008).

While endocrine organs specialize in the secretion of proteins into circulation, there is mounting evidence that adipose tissue and skeletal muscle constitutively or intermittently secrete bioactive proteins, particularly in response to overnutrition (e.g., leptin) and exercise (e.g., IL-6). However, a top–down approach toward identifying novel protein biomarkers in blood is hampered by an abundant background of serum proteins, wherein a protein of interest may be diluted several orders of magnitude making it difficult to detect even using the most sensitive techniques. To overcome these limitations, an approach adopted by our lab and others is a "bottom–up" experimental design to characterize the secreted protein complement or "secretome" of a variety of cell types, specifically primary human muscle cells. Secreted proteins constitute an important class of biologically active molecules that are released into circulation by a variety of cells, tissues, and organ (Hathout, 2007; Sarkissian et al., 2008). In fact, many secreted proteins function as either anabolic (IGF-1) or anti-anabolic (MSTN) agents that coordinate the systemic adaptations to exercise among several organ systems. As such, there is significant interest in mining the secretome for novel diagnostic biological markers or targets for pharmacological intervention (Hathout, 2007). In a recent study, we quantified differences in the muscle secretome of cells derived from lean and extremely obese patient populations. However, this model could easily be extended to characterize cell secretomes with exposure to anabolic steroids or with simulated exercise (cell contraction or stretching). From this study, we found that the secreted protein complement was of relatively low complexity and, as such,

ideal for using stable isotope-based proteomic experimental design (Figure 28.5) for the identification and quantification of secreted proteins from muscle (Hittel et al., 2008). Using this technique, we identified a number of novel proteins not previously identified as being secreted from muscle. Unexpectedly, we also discovered that MSTN, a potent anti-anabolic regulator of muscle mass, is highly expressed and secreted from insulin-resistant skeletal muscle. Because of this discovery, we are presently evaluating circulating MSTN levels as a surrogate biological marker for skeletal muscle insulin resistance.

Proteomics and Systems Biology

One of the more ambitious goals of systems biology is the integration of genomic, proteomic, and metabolic data into computer models that will predict how a cell, tissue, or organism adapts to physiological stress (see chapters in Part 4). The practical application of this approach will broaden all aspects of biology and medicine, including the exercise and sports sciences. For instance, the practice of following changes in the expression of individual genes in response to exercise training is being abandoned in favor of gene regulatory networks (Keller et al., 2007). Proteomics has become essential to achieving these goals, which involves the systematic study of all concurrent physiological processes in a cell, or tissue by global molecular measurements (RNA, protein, and metabolite levels). Born from the technological advances that accelerated the sequencing of whole genomes, the field of systems biology has been greatly enhanced and refined by the vast numbers of published genome-wide expression studies. However, the large amount of data generated by molecular profiling studies requires sophisticated software tools and novel statistical analysis for a meaningful interpretation. Considerable progress has been made on this front with the appearance of many free (www.GenMAPP.org) and commercially available (www.ingenuity.com) network analysis tools. Although there have been many gene and protein expression studies of muscle in disease, exercise, and obesity, it continues to be a challenge to produce a comprehensive molecular model that links gene expression patterns with their corresponding physiological effects. Instead, these studies have traditionally focused on clusters of temporally or coordinately expressed genes with loosely defined gene ontologies. Network analysis software has recently evolved to a point where it has become useful for modeling complex gene expression data sets. In a recent study, we described the differential secretion of proteins from obese versus lean human-derived muscle cells (Hittel et al., 2008). Using this list of secreted proteins, we generated a network of interactions consisting of differentially expressed proteins in a graphical representation of their molecular relationships (Figure 28.6). Proteins are represented

as nodes, and the biological relationship between two nodes is represented as an edge (line); nodes are displayed using various shapes that represent the functional class of the gene product (www.ingenuity.com). All edges are supported by at least one reference from the literature or textbook. This analysis revealed significant associations between secreted proteins and components of the extracellular matrix (MSTN and bone morphogenic protein 1 to be specific) that we would have otherwise not have noticed using a classic approach. A systems biology approach toward complex biological models will increasingly be used to interpret transcriptional and proteomic data sets and help generate novel hypothesis that would not otherwise arise from a linear approach.

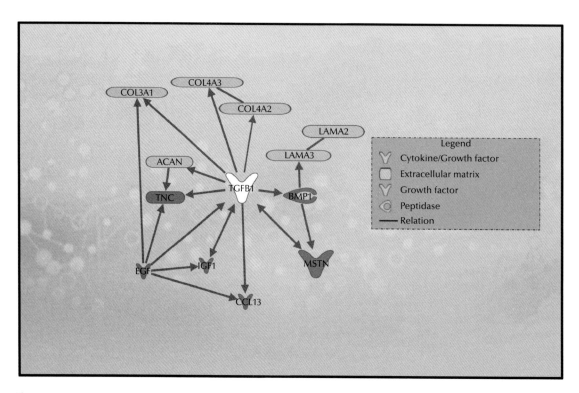

Figure 28.6 Interaction network generated (www.ingenuity.com) from a list of proteins secreted from primary human muscle cells derived from lean and obese individuals. Interactions are represented as an edge (line); nodes are displayed using various shapes that represent the functional class of the gene product proteins. All edges are supported by at least one reference from the literature or textbook.

Summary

The goal of quantitative proteomics is to describe and compare all proteins from a cell, tissue, or organism in a certain physiological state or in response to a treatment. In the past decade, proteomic profiling using MS has transformed from a highly specialized subdiscipline to a much more accessible technique for the nonspecialist (Hittel et al., 2007). The high-quality protein expression data generated by these advances has contributed greatly to the nascent field of systems biology as a new approach toward biological research that seeks to model and, ultimately, to predict concurrent processes in the cell, tissue, or organism. With the advent of new proteomic tools and their enabling technologies, exercise physiologists are rediscovering proteins that, after all, are the ultimate physical interface between our genome and the environment. However, for proteomics to solve biologically meaningful questions in exercise training research, the data generated must be of high quality, the analysis semi-quantitative and statistically robust. Proteomics is proving to be highly complementary to other functional genomics approaches such as microarray analysis (Yi et al., 2008). The integration of these measurements could be used to build a powerful database of gene–environment interactions, which will be useful to sports science researcher to both build and test hypotheses.

References

Aguilera, R., Catlin, D.H., Becchi, M., et al. (1999). Screening urine for exogenous testosterone by isotope ratio mass spectrometric analysis of one pregnanediol and two androstanediols. *Journal of Chromatography B, Biomedical Sciences and Applications* 727, 95–105.

Anderson, E.J., Lustig, M.E., Boyle, K.E., et al. (2009). Mitochondrial H2O2 emission and cellular redox state link excess fat intake to insulin resistance in both rodents and humans. *The Journal of Clinical Investigation* 119, 573–581.

Baoutina, A., Alexander, I.E., Rasko, J.E. & Emslie, K.R. (2008). Developing strategies for detection of gene doping. *The Journal of Gene Medicine* 10, 3–20.

Boluyt, M.O., Brevick, J.L., Rogers, D.S., Randall, M.J., Scalia, A.F. & Li, Z.B. (2006). Changes in the rat heart proteome induced by exercise training: Increased abundance of heat shock protein hsp20. *Proteomics* 6, 3154–3169.

Burniston, J.G. (2009). Adaptation of the rat cardiac proteome in response to intensity-controlled endurance exercise. *Proteomics* 9, 106–115.

Flueck, M. (2009). Plasticity of the muscle proteome to exercise at altitude. *High Altitude Medicine & Biology* 10, 183–193.

Hathout, Y. (2007). Approaches to the study of the cell secretome. *Expert Review of Proteomics* 4, 239–248.

Hittel, D.S., Hathout, Y., Hoffman, E.P. & Houmard, J.A. (2005). Proteome analysis of skeletal muscle from obese and morbidly obese women. *Diabetes* 54, 1283–1288.

Hittel, D.S., Hathout, Y. & Hoffman, E.P. (2007). Proteomics and systems biology in exercise and sport sciences research. *Exercise and Sport Sciences Reviews* 35, 5–11.

Hittel, D.S., Berggren, J.R., Shearer, J., Boyle, K. & Houmard, J.A. (2008). Increased secretion and expression of myostatin in skeletal muscle from extremely obese women. *Diabetes* 58, 30–38.

Jullig, M., Hickey, A.J., Chai, C.C., et al. (2008). Is the failing heart out of fuel or a worn engine running rich? A study of mitochondria in old spontaneously hypertensive rats. *Proteomics* 8, 2556–2572.

Kavazis, A.N., Alvarez, S., Talbert, E., Lee, Y. & Powers, S.K. (2009). Exercise training induces a cardioprotective phenotype and alterations in cardiac subsarcolemmal and intermyofibrillar mitochondrial proteins. *American Journal of Physiology. Heart and Circulatory Physiology* 297, H144–H152.

Keller, P., Vollaard, N., Babraj, J., Ball, D., Sewell, D.A. & Timmons, J.A. (2007). Using systems biology to define the essential biological networks responsible for adaptation to endurance exercise training. *Biochemical Society Transactions* 35, 1306–1309.

Kurland, C.G. & Andersson, S.G. (2000). Origin and evolution of the mitochondrial proteome. *Microbiology and Molecular Biology Reviews* 64, 786–820.

Lasne, F. & de Ceaurriz, J. (2000). Recombinant erythropoietin in urine. *Nature* 405, 635.

Mootha, V.K., Bunkenborg, J., Olsen, J.V., et al. (2003). Integrated analysis of protein composition, tissue diversity, and gene regulation in mouse mitochondria. *Cell* 115, 629–640.

Riedmaier, I., Becker, C., Pfaffl, M.W. & Meyer, H.H. (2009). The use of omic technologies for biomarker development to trace functions of anabolic agents. *Journal of Chromatography A* 1216, 8192–8199.

Sarkissian, G., Fergelot, P., Lamy, P.J., et al. (2008). Identification of pro-MMP-7 as a serum marker for renal cell carcinoma by use of proteomic analysis. *Clinical Chemistry* 54, 574–581.

Scheele, C., Nielsen, S. & Pedersen, B.K. (2009). ROS and myokines promote muscle adaptation to exercise. *Trends in Endocrinology and Metabolism* 20, 95–99.

Schon, E.A. (2007). Appendix 6. Changes in the mitochondrial transcriptome and proteome under various stresses and growth conditions. *Methods in Cell Biology* 80, 877–887.

Sheehan, D. (2006). Detection of redox-based modification in two-dimensional electrophoresis proteomic separations. *Biochemical and Biophysical Research Communications* 349, 455–462.

Smith, J.C. & Figeys, D. (2006). Proteomics technology in systems biology. *Molecular Biosystems* 2, 364–370.

Taylor, S.W., Fahy, E., Zhang, B., et al. (2003). Characterization of the human heart mitochondrial proteome. *Nature Biotechnology* 21, 281–286.

Vitorino, R., Ferreira, R., Neuparth, M., et al. (2007). Subcellular proteomics of mice gastrocnemius and soleus muscles. *Analytical Biochemistry* 366, 156–169.

Yates, J.R., III, Gilchrist, A., Howell, K.E. & Bergeron, J.J. (2005). Proteomics of organelles and large cellular structures.

Nature Reviews Molecular Cell Biology 6, 702–714.

Yi, Z., Bowen, B.P., Hwang, H., et al. (2008). Global relationship between the proteome and transcriptome of human skeletal muscle. *Journal of Proteome Research* 7, 3230–3241.

Chapter 29

The Influence of Physical Exercise on Adult Stem Cells

FAWZI KADI

School of Health and Medical Sciences, Örebro University, Örebro, Sweden

Repeated bouts of structured bodily exertion requiring energy expenditure above resting levels result in the occurrence of multiple molecular and cellular events leading to several functional changes and providing countless health-related benefits. The disruption of the dynamic equilibrium of body homeostasis is the *sine qua non* of the exercise-induced adaptations at the level of the cardiovascular and neuromuscular systems.

Exercise is one of the most powerful non-pharmacological strategies, which is able to affect nearly all cells and organs in the body. In this context, a new research avenue focusing on the action of exercise on adult stem cells has emerged during the last decade. Changes in the behavior of adult stem cells from different regions including skeletal muscle and the cardiovascular system have been shown to occur in response to exercise training. Through its action on adult stem cells, exercise may act on the regenerative potential of tissues by altering the ability to generate new stem cells and differentiated cells that are able to carry out tissue-specific functions. The present review, which is not meant to be exhaustive, mainly focuses on the behavior of non-embryonic adult stem cells under the influence of physical exercise.

Adult Stem Cells

Mammalian organisms are characterized by the existence of a machinery devoted to tissue and organ regeneration and repair throughout the life span. This important property is due to the presence of undifferentiated cells able to proliferate and generate differentiated cells capable of performing tissue-specific functions. The undifferentiated cells allowing the repair of tissues are called stem cells. According to their origin, stem cells can be broadly classified into embryonic stem cells, derived from early mammalian embryos and capable of differentiating into several cell types and non-embryonic stem cells also known as adult stem cells (Young et al., 2004). Under the influence of specific cellular microenvironments known as niches, adult stem cells contribute to the generation of a wide range of differentiated progenies.

The essential properties of adult stem cells are their ability to renew their own pool and their capacity to generate new tissue-specific cells through cell division. Given these characteristics, adult stem cells are important for the maintenance of tissue integrity and the restoration of tissue function. Under homeostatic conditions, adult stem cells can repopulate the tissue in which they reside and serve as an internal repair system. Evidence also exists supporting the participation of specific stem cells with a broad plasticity and differentiation potential in the regeneration of more than one tissue (Wright et al., 2001).

Genetic and Molecular Aspects of Sport Performance, 1st edition.
Edited by Claude Bouchard and Eric P. Hoffman.
Published 2011 by Blackwell Publishing Ltd.

The presence of tissue-specific stem cells in adult tissues has been discovered in several tissues including bone marrow, skeletal muscle, heart, brain, blood vessels, gut, liver, and peripheral blood. The fact that most adult tissues contain a reservoir of adult stem cells has been especially striking in organs, which were typically known for their poor self-renewal capacity such as the heart or the brain.

New data suggest that physical exercise can exert powerful effects on different stem cell niches by altering their microenvironment. Currently, the mechanisms behind the maintenance of a quiescent state within each stem cell niche as well as the exact signals leading to the proliferation of stem cells following exercise are not fully understood. As the interaction between the adult stem cells and their microenvironment influences their behavior (Moore & Lemischka, 2006), mechanical, hormonal, and growth factor-related signals are thought to be the major actors in the chain of events leading to the exercise-induced transition from dormant to activated cells.

Stem Cells of Skeletal Muscle

Skeletal muscle is a dynamic tissue able to adapt to various physiological conditions. The ability of skeletal muscle to regenerate is mainly due to small mononucleated cells, called the satellite cells, located between the basal lamina and the sarcolemma of muscle fibers (Mauro, 1961). Satellite cells are considered as skeletal muscle stem cells as they can reenter the cell cycle to generate differentiated cells and new undifferentiated myogenic precursor cells, allowing the renewal of their own population (Kadi et al., 2005). In vastus lateralis of healthy active young individuals, satellite cells are equally distributed in slow oxidative type I and fast glycolytic type II muscle fibers (Kadi et al., 2006). However, the satellite cell pool, which accounts for a mean of 2–7% of sublaminal nuclei, can increase or decrease in response to several physiological and pathological conditions (Ambrosio et al., 2009; Kadi, 2008; Kadi et al., 2005).

The CD56, which is an isoform of the neural cell adhesion molecule (NCAM) reacting with the Leu19-antigen and the transcription factor Pax-7, represent two markers delineating the majority of satellite cells in human skeletal muscle (Kadi, 2000; Kadi et al., 2005; Mackey et al., 2009). Although the satellite cell is considered the major source of muscle regeneration, other adult multilineage progenitor cell populations with a myogenic potential have been isolated in skeletal muscle. Although rare, side population cells (Jackson et al., 1999), CD133$^+$ progenitors (Torrente et al., 2004), and muscle-derived stem cells (Qu-Petersen et al., 2002) can be detected in skeletal muscle. However, the extent to which these cells contribute to the regeneration of human skeletal muscle is currently unknown. In the context of healthy skeletal muscle, satellite cells would carry out muscle regeneration without the involvement of other myogenic stem cells. Interestingly, non-satellite cell populations isolated in skeletal muscle can also progress to spontaneously beating cardiomyocytes (Winitsky et al., 2005), indicating that within skeletal muscle might exist a small population of stem cells having myogenic and non-myogenic potentials.

Effects of Exercise on Skeletal Muscle Satellite Cells

The number of satellite cells within skeletal muscle is dynamic. The study of the influence of aging on skeletal muscle revealed an age-related decline in satellite cell number in humans (Kadi et al., 2004a; Renault et al., 2002; Sajko et al., 2004) and rodents (Gibson & Schultz, 1983; Snow, 1977). In humans, it is hypothesized that the decline in satellite cell pool becomes prominent after the age of 70 and that this phenomenon is influenced by the exercise history of individuals (Kadi et al., 2004a). Interestingly, it has recently been shown that fast type II muscle fibers in lower limb muscles are the specific sites of a reduced number of satellite cells in elderly subjects (Verdijk et al., 2007; Verney et al., 2008). The reason behind the lower satellite cell content in type II fibers is unknown. It is possible that satellite cells are recruited to rescue the muscle fibers, which leads to their depletion. To test this hypothesis, further experiments should examine the relationship between the area of type II fibers and their satellite cell content.

If satellite cells are implicated in the etiology of skeletal muscle atrophy, would exercise training counterbalance these age-related events? Both longitudinal and cross-sectional studies addressed this question in humans and revealed that physical exercise can induce the proliferation of satellite cells allowing for the renewal of their pool (Kadi, 2008; Kadi & Ponsot, 2009; Kadi et al., 2005; Snijders et al., 2009). Several exercise modalities including maximal eccentric contractions, non-damaging resistance training, and endurance training have been associated with the enhancement of satellite cell pool in human skeletal muscle (Charifi et al., 2003; Crameri et al., 2004; Dreyer et al., 2006; Kadi, 2000; Kadi et al., 2004b; Mackey et al., 2007a, b; Murphy et al., 2008; O'Reilly et al., 2008; Olsen et al., 2006; Petrella et al., 2008; Roth et al., 2001; Verdijk et al., 2009; Verney et al., 2008). Interestingly, enhancement of the satellite cell pool can occur specifically in fast type II muscle fibers of elderly individuals (Verdijk et al., 2009; Verney et al., 2008), and the duration of the exercise stimulus is an important factor able to modulate the magnitude of satellite cell self-renewal in human skeletal muscle (Kadi et al., 2004b). Thus, existing data strongly indicate that the expansion of the satellite cell pool can be achieved in human skeletal muscle through exercise. However, it remains important to understand the usefulness of satellite cell pool expansion, especially in pathological conditions. In this respect, it has been shown that resistance training in patients with single, large-scale deletions of mitochondrial DNA induces satellite cell expansion (Murphy et al., 2008). The newly generated satellite cells have low or undetectable levels of deleted mitochondrial DNA (Murphy et al., 2008). At present, the exact chain of events mediating the exercise-induced expansion of satellite cells is not known. This being said, several circulating and local growth factors are known to influence the satellite cell and its microenvironment (Kadi, 2005, 2008). Specific exercise modalities can create microenvironments promoting the enhancement of satellite cell pool and favoring the recruitment of muscle precursor cells (MPCs). This has been shown in exercised *mdx* mice (a model of Duchenne muscular dystrophy), where the recruitment and the fusion of transplanted MPCs in dystrophic tibialis anterior muscle were significantly improved (Bouchentouf et al., 2006).

The Vascular System and Endothelial Progenitor Cells

Circulating endothelial progenitor cells (EPCs) represent a heterogeneous population of circulating cells emanating from different territories and having in common the ability to differentiate into endothelial cells (Asahara et al., 1997). In this context, some bone marrow-derived cell populations such as the $CD133^{+}/VEGFR2^{+}$ cells and non-bone marrow-derived cells have been found to have endothelial progenitor capacities (Urbich & Dimmeler, 2004). The EPCs are important for vascular homeostasis and neovascularization of tissues as they can promote angiogenesis and vasculogenesis in ischemic tissue (Kalka et al., 2000).

The EPCs are normally quiescent cells that are able to enter the cell cycle and to differentiate into functional endothelial cells under the influence of several stimuli such as increased blood flow and changes in the concentrations of VEGF (vascular endothelial growth factor), angiopoietin-1, and several cytokines (Aicher et al., 2003; Hattori et al., 2001).

The capacity to repair the endothelial monolayer can be important in preventing the development of atherosclerotic lesions (Dzau et al., 2005). Therefore, it can be hypothesized that a higher number of circulating EPCs would ensure an efficient repair of the endothelium. In agreement with this statement, a significant relationship was found between the number of circulating EPCs and the endothelial function in men with various degrees of cardiovascular risk (Hill et al., 2003).

Effects of Exercise on EPCs

Unless the physiological context at the time of the analysis is well defined, the baseline levels of circulating EPCs in humans is difficult to interpret.

Nevertheless, existing data suggest an age-related reduction in survival rate, proliferation (Heiss et al., 2005), and mobilization of circulating EPCs (Scheubel et al., 2003). It has also been shown that the number of EPC colony-forming units and the migratory activity of EPCs are higher in middle-aged women compared to men (Hoetzer et al., 2007). These gender differences in EPC characteristics might reflect the gender-specific differences in the prevalence of cardiovascular morbidity (Hoetzer et al., 2007).

It is well established that physical inactivity is an important risk factor for the development of cardiovascular diseases and that endurance exercises are associated with an improvement of the vascular endothelial function (Hambrecht et al., 2000; Van Guilder et al., 2005). In this context, it has been shown that the EPC response to training is influenced by the exercise modality. A prospective crossover study showed that 30 min of intensive (82% $\dot{V}O_2$ max) and moderate (68% $\dot{V}O_2$ max), but not 10 min of moderate (68% $\dot{V}O_2$ max), running induced a significant increase in circulating EPCs quantified by flow cytometry and cell culture (Laufs et al., 2005). It has also been shown that 45 min of daily intensified and supervised school sports was successful in increasing the amount of EPCs, but failed to increase the migratory capacity of the cells in high-school students (Walther et al., 2008). The stimulation of the proliferation of EPCs as well as the enhancement of their pool in response to exercise have been suggested to be mediated by erythropoictin (EPO), VEGF, and shear stress-induced endothelial nitric oxide synthase (eNOS) (Asahara et al., 1999; Bahlmann et al., 2003; Prior et al., 2003). Interestingly, it has been suggested that the level of tissue ischemia is an important trigger of the exercise-induced EPC response as daily ischemic, but not non-ischemic, endurance training can lead to an increase in circulating progenitor cells in patients with cardiovascular disease (Sandri et al., 2005). Moreover, intensive exercise leading to severe inflammatory events can downregulate circulating hematopoietic stem cells (Adams et al., 2008).

Although the physiological significance of increased circulating EPCs is not fully understood, enhanced EPC pool is viewed as an important exercise-induced cardiovascular benefit in physically active humans. However, it is important to note that the expansion of EPC pool has not been reported in some other studies (Palange et al., 2006; Thijssen et al., 2006). Instead, it was suggested that exercise might exert its effects on the functional properties of EPCs rather than modulating their number.

Cardiac Muscle and Cardiac Stem Cells

While the cardiac muscle has long been considered as a post-mitotic tissue lacking regenerative potential, the discovery of adult cardiac stem cells (CSCs) opened new avenues in the biology of this tissue (Beltrami et al., 2003). Although the origin of CSCs is not well defined, the c-Kit$^+$-CSCs have been identified as multipotent cells able to differentiate into cardiomyocytes, smooth muscle cells, and endothelial cells (Thijssen et al., 2009). The number of CSCs in the myocardium is low and increases upon myocardial damage. However, it has previously been suggested that the activation of CSCs might not occur at the site of injury (Urbanek et al., 2005), although more recent data suggest that CSCs are resistant to the noxious effects of an acute hyperadrenergic state (Ellison et al., 2007).

Interestingly, the study of the effects of aging on the cardiac tissue revealed the occurrence of functionally impaired CSCs evidenced by an increase in the expression of markers for cell senescence (Chimenti et al., 2003). This finding suggests that aging might affect the functional status of CSCs and thereby the regenerative potential of the heart.

Effects of Exercise on CSCs

The importance of exercise as a physiological tool to modulate the number of CSCs and their functional properties in healthy organisms remains to be elucidated.

As for skeletal muscle, it can be hypothesized that under the influence of exercise, the CSCs would generate differentiated progenies contributing to the improvement of the myogenic and vascular characteristics of the cardiac tissue. Indeed, similar to what occurs during the hypertrophy of skeletal muscle, increased cardiac muscle mass after human aortic stenosis is associated with myocyte hypertrophy as well as with the formation of new myocytes (Urbanek et al., 2003). In agreement with these results, preliminary data suggest that a period of 4 weeks treadmill exercise at two different intensities (55–60% and 85–90% $\dot{V}O_2$ max) induces an increase in c-kit$^+$-CSCs together with myocyte hypertrophy and the formation of new myocytes in rats (Waring et al., 2009). It is important to note that in physiological conditions, the formation of new muscle cells in both skeletal and cardiac muscle would lead to fiber hyperplasia only if the exercise-induced new fiber formation exceeds fiber loss.

Muscle Stem Cells in Therapeutics or Doping

Muscle stem cells were an early focus of experimental therapeutics in muscle disease (muscular dystrophies) (Partridge et al., 1989). Since then, hundreds of reports have been published on attempts to use muscle stem cells to improve muscle function in preclinical models and human patients. There have been three hurdles limiting the efficacy of this approach: production of adequate numbers of appropriate stem cells, challenges in sufficient delivery of cells to muscles throughout the body, and immune barriers. Of these three hurdles, the delivery problems may be the most insurmountable. As noted earlier, muscle stem cells are nestled against each myofiber, within the basal lamina, and the effective function of the stem cell appears to require this specific geographic niche. Therapeutic stem cells, when injected intravenously, must navigate through the blood stream, pass through the capillaries, and transverse the basal lamina to reach this specific location in the tissue. This extent of "homing" has been difficult to achieve, particularly in a large proportion of all myofibers throughout the body. To date, there is no evidence that muscle stem cell transplants have improved the quality of life of a patient with neuromuscular disease.

Until these three hurdles are navigated, it is unlikely that stem cells can be effectively utilized as a form of doping to improve muscle performance.

Conclusion

The identification of tissue-specific adult stem cells has generated a great interest and enthusiasm in the field of regenerative medicine where the main focus is the development of novel cellular therapies aiming the restoration of normal function. Currently, the *in vivo* modulation of tissue-resident adult stem cells is not fully understood. At present, data suggest that exercise can be viewed as an important physiological strategy modulating the stem cell itself and its environment leading to the improvement of the adaptive and regenerative potential of healthy tissues. In the context of regenerative medicine, exercise might be viewed as a physiological complement to other therapies aiming the restoration of damaged tissues. For these reasons, more efforts should be made in order to better understand the complex cellular regulations occurring in the environment of stem cells during the transition from rest to physical exertion.

References

Adams, V., Linke, A., Breuckmann, F., et al. (2008). Circulating progenitor cells decrease immediately after marathon race in advanced-age marathon runners. *European Journal of Cardiovascular Prevention and Rehabilitation* 15, 602–607.

Aicher, A., Heeschen, C., Mildner-Rihm, C., et al. (2003). Essential role of endothelial nitric oxide synthase for mobilization of stem and progenitor cells. *Nature Medicine* 9, 1370–1376.

Ambrosio, F., Kadi, F., Lexell, J., Fitzgerald, G.K., Boninger, M.L. & Huard, J. (2009). The effect of muscle loading on skeletal muscle regenerative potential: An update of current research findings relating to aging and neuromuscular pathology. *American Journal of Physical Medicine & Rehabilitation* 88, 145–155.

Asahara, T., Murohara, T., Sullivan, A., et al. (1997). Isolation of putative progenitor endothelial cells for angiogenesis. *Science* 275, 964–967.

Asahara, T., Takahashi, T., Masuda, H., et al. (1999). VEGF contributes to postnatal neovascularization by mobilizing bone marrow-derived endothelial progenitor cells. *The EMBO Journal* 18, 3964–3972.

Bahlmann, F.H., DeGroot, K., Duckert, T., et al. (2003). Endothelial progenitor cell proliferation and differentiation is regulated by erythropoietin. *Kidney International* 64, 1648–1652.

Beltrami, A.P., Barlucchi, L., Torella, D., et al. (2003). Adult cardiac stem cells are multipotent and support myocardial regeneration. *Cell* 114, 763–776.

Bouchentouf, M., Benabdallah, B.F., Mills, P. & Tremblay, J.P. (2006). Exercise improves the success of myoblast transplantation in mdx mice. *Neuromuscular Disorders* 16, 518–529.

Charifi, N., Kadi, F., Feasson, L., Denis, C. (2003). Effects of endurance training on satellite cell frequency in skeletal muscle of old men. *Muscle & Nerve* 28, 87–92.

Chimenti, C., Kajstura, J., Torella, D., et al. (2003). Senescence and death of primitive cells and myocytes lead to premature cardiac aging and heart failure. *Circulation Research* 93, 604–613.

Crameri, R.M., Langberg, H., Magnusson, P., et al. (2004). Changes in satellite cells in human skeletal muscle after a single bout of high intensity exercise. *The Journal of Physiology* 558, 333–340.

Dreyer, H.C., Blanco, C.E., Sattler, F.R., Schroeder, E.T. & Wiswell, R.A. (2006). Satellite cell numbers in young and older men 24 hours after eccentric exercise. *Muscle & Nerve* 33, 242–253.

Dzau, V.J., Gnecchi, M., Pachori, A.S., Morello, F. & Melo, L.G. (2005). Therapeutic potential of endothelial progenitor cells in cardiovascular diseases. *Hypertension* 46, 7–18.

Ellison, G.M., Torella, D., Karakikes, I., et al. (2007). Acute beta-adrenergic overload produces myocyte damage through calcium leakage from the ryanodine receptor 2 but spares cardiac stem cells. *The Journal of Biological Chemistry* 282, 11397–11409.

Gibson, M.C. & Schultz, E. (1983). Age-related differences in absolute numbers of skeletal muscle satellite cells. *Muscle & Nerve* 6, 574–580.

Hambrecht, R., Wolf, A., Gielen, S., et al. (2000). Effect of exercise on coronary endothelial function in patients with coronary artery disease. *The New England Journal of Medicine* 342, 454–460.

Hattori, K., Dias, S., Heissig, B., et al. (2001). Vascular endothelial growth factor and angiopoietin-1 stimulate postnatal hematopoiesis by recruitment of vasculogenic and hematopoietic stem cells. *The Journal of Experimental Medicine* 193, 1005–1014.

Heiss, C., Keymel, S., Niesler, U., Ziemann, J., Kelm, M. & Kalka, C. (2005). Impaired progenitor cell activity in age-related endothelial dysfunction. *Journal of the American College of Cardiology* 45, 1441–1448.

Hill, J.M., Zalos, G., Halcox, J.P., et al. (2003). Circulating endothelial progenitor cells, vascular function, and cardiovascular risk. *The New England Journal of Medicine* 348, 593–600.

Hoetzer, G.L., MacEneaney, O.J., Irmiger, H.M., et al. (2007). Gender differences in circulating endothelial progenitor cell colony-forming capacity and migratory activity in middle-aged adults. *The American Journal of Cardiology* 99, 46–48.

Jackson, K.A., Mi, T. & Goodell, M.A. (1999). Hematopoietic potential of stem cells isolated from murine skeletal muscle. *Proceedings of the National Academy of Sciences of the United States of America* 96, 14482–14486.

Kadi, F. (2000). Adaptation of human skeletal muscle to training and anabolic steroids. *Acta Physiologica Scandinavica* 646, 1–52.

Kadi, F. (2005). Hormonal and growth factor-related mechanisms involved in the adaptation of skeletal muscle to exercise. In: W. Kraemer & A. Rogol (eds.) *Encyclopaedia of Sports Medicine An IOC Medical Commission Publication, The Endocrine System in Sports and Exercise,* Vol. 11, pp. 306–318. Blackwell, Malden, MA, USA.

Kadi, F. (2008). Cellular and molecular mechanisms responsible for the action of testosterone on human skeletal muscle. A basis for illegal performance enhancement. *British Journal of Pharmacology* 154, 522–528.

Kadi, F. & Ponsot, E. (2009). The biology of satellite cells and telomeres in human skeletal muscle. Effects of aging and physical activity. *Scandinavian Journal of Medicine & Science in Sports* 20(1), 39–48.

Kadi, F., Charifi, N., Denis, C. & Lexell, J. (2004a). Satellite cells and myonuclei in young and elderly women and men. *Muscle & Nerve* 29, 120–127.

Kadi, F., Schjerling, P., Andersen, L.L., et al. (2004b). The effects of heavy resistance training and detraining on satellite cells in human skeletal muscles. *The Journal of Physiology* 558, 1005–1012.

Kadi, F., Charifi, N., Denis, C., et al. (2005). The behavior of satellite cells in response to exercise: What have we learned from human studies? *Pflugers Archiv* 451, 319–327.

Kadi, F., Charifi, N. & Henriksson, J. (2006). The number of satellite cells in slow and fast fibers from human vastus lateralis muscle. *Histochemistry and Cell Biology* 126, 83–87.

Kalka, C., Masuda, H., Takahashi, T., et al. (2000). Transplantation of ex vivo expanded endothelial progenitor cells for therapeutic neovascularization. *Proceedings of the National Academy of Sciences of the United States of America* 97, 3422–3427.

Laufs, U., Urhausen, A., Werner, N., et al. (2005). Running exercise of different duration and intensity: Effect on endothelial progenitor cells in healthy subjects. *European Journal of Cardiovascular Prevention and Rehabilitation* 12, 407–414.

Mackey, A.L., Esmarck, B., Kadi, F., et al. (2007a). Enhanced satellite cell proliferation with resistance training in elderly men and women. *Scandinavian Journal of Medicine & Science in Sports* 17, 34–42.

Mackey, A.L., Kjaer, M., Dandanell, S., et al. (2007b). The influence of anti-inflammatory medication on exercise-induced myogenic precursor cell responses in humans. *Journal of Applied Physiology* 103, 425–431.

Mackey, A.L., Kjaer, M., Charifi, N., et al. (2009). Assessment of satellite cell number and activity status in human skeletal muscle biopsies. *Muscle & Nerve* 40, 455–465.

Mauro, A. (1961). Satellite cell of skeletal muscle fibers. *The Journal of Biophysical and Biochemical Cytology* 9, 493–495.

Moore, K.A. & Lemischka, I.R. (2006). Stem cells and their niches. *Science* 311, 1880–1885.

Murphy, J.L., Blakely, E.L., Schaefer, A.M., et al. (2008). Resistance training in patients with single, large-scale deletions of mitochondrial DNA. *Brain* 131, 2832–2840.

O'Reilly, C., McKay, B., Phillips, S., Tarnopolsky, M., Parise, G. (2008). Hepatocyte growth factor (HGF) and the satellite cell response following muscle lengthening contractions in humans. *Muscle & Nerve* 38, 1434–1442.

Olsen, S., Aagaard, P., Kadi, F., et al. (2006). Creatine supplementation augments the increase in satellite cell and myonuclei number in human skeletal muscle induced by strength training. *The Journal of Physiology* 573, 525–534.

Palange, P., Testa, U., Huertas, A., et al. (2006). Circulating haemopoietic and endothelial progenitor cells are decreased in COPD. *The European Respiratory Journal* 27, 529–541.

Partridge, T.A., Morgan, J.E., Coulton, G.R., Hoffman, E.P. & Kunkel, L.M. (1989). Conversion of mdx myofibres from dystrophin-negative to dystrophin-positive by injection of normal myoblasts. *Nature* 337, 176–179.

Petrella, J.K., Kim, J.S., Mayhew, D.L., Cross, J.M. & Bamman, M.M. (2008). Potent myofiber hypertrophy during resistance training in humans is associated with satellite cell-mediated myonuclear addition: A cluster analysis. *Journal of Applied Physiology* 104, 1736–1742.

Prior, B.M., Lloyd, P.G., Yang, H.T. & Terjung, R.L. (2003). Exercise-induced vascular remodeling. *Exercise and Sport Sciences Reviews* 31, 26–33.

Qu-Petersen, Z., Deasy, B., Jankowski, R., et al. (2002). Identification of a novel population of muscle stem cells in mice: Potential for muscle regeneration. *The Journal of Cell Biology* 157, 851–864.

Renault, V., Thornell, L.E., Eriksson, P.O., Butler-Browne, G. & Mouly, V. (2002). Regenerative potential of human skeletal muscle during aging. *Aging Cell* 1, 132–139.

Roth, S.M., Martel, G.F., Ivey, F.M., et al. (2001). Skeletal muscle satellite cell characteristics in young and older men and women after heavy resistance strength training. *Journals of Gerontology Series A: Biological Sciences and Medical Sciences* 56, 240–247.

Sajko, S., Kubinova, L., Cvetko, E., Kreft, M., Wernig, A. & Erzen, I. (2004). Frequency of M-cadherin-stained satellite cells declines in human muscles during aging. *Journal of Histochemistry and Cytochemistry* 52, 179–185.

Sandri, M., Adams, V., Gielen, S., et al. (2005). Effects of exercise and ischemia on mobilization and functional activation of blood-derived progenitor cells in patients with ischemic syndromes: Results of 3 randomized studies. *Circulation* 111, 3391–3399.

Scheubel, R.J., Zorn, H., Silber, R.E., et al. (2003). Age-dependent depression in circulating endothelial progenitor cells in patients undergoing coronary artery bypass grafting. *Journal of the American College of Cardiology* 42, 2073–2080.

Snijders, T., Verdijk, L.B. & van Loon, L.J. (2009). The impact of sarcopenia and exercise training on skeletal muscle satellite cells. *Ageing Research Reviews* 8, 328–338.

Snow, M.H. (1977). The effects of aging on satellite cells in skeletal muscles of mice and rats. *Cell and Tissue Research* 185, 399–408.

Thijssen, D.H., Vos, J.B., Verseyden, C., et al. (2006). Haematopoietic stem cells and endothelial progenitor cells in healthy men: Effect of aging and training. *Aging Cell* 5, 495–503.

Thijssen, D.H., Torella, D., Hopman, M.T. & Ellison, G.M. (2009). The role of endothelial progenitor and cardiac stem cells in the cardiovascular adaptations to age and exercise. *Frontiers in Bioscience* 14, 4685–4702.

Torrente, Y., Belicchi, M., Sampaolesi, M., et al. (2004). Human circulating AC133(+) stem cells restore dystrophin expression and ameliorate function in dystrophic skeletal muscle. *The Journal of Clinical Investigation* 114, 182–195.

Urbanek, K., Quaini, F., Tasca, G., et al. (2003). Intense myocyte formation from cardiac stem cells in human cardiac hypertrophy. *Proceedings of the National Academy of Sciences of the United States of America* 100, 10440–10445.

Urbanek, K., Torella, D., Sheikh, F., et al. (2005). Myocardial regeneration by activation of multipotent cardiac stem cells in ischemic heart failure. *Proceedings of the National Academy of Sciences of the United States of America* 102, 8692–8697.

Urbich, C. & Dimmeler, S. (2004). Endothelial progenitor cells: Characterization and role in vascular biology. *Circulation Research* 95, 343–353.

Van Guilder, G.P., Hoetzer, G.L., Smith, D.T., et al. (2005). Endothelial t-PA release is impaired in overweight and obese adults but can be improved with regular aerobic exercise. *American Journal of Physiology—Endocrinology and Metabolism* 289, E807–E813.

Verdijk, L.B., Koopman, R., Schaart, G., Meijer, K., Savelberg, H.H. & van Loon, L.J. (2007). Satellite cell content is specifically reduced in type II skeletal muscle fibers in the elderly. *American Journal of Physiology—Endocrinology and Metabolism* 292, E151–E157.

Verdijk, L.B., Gleeson, B.G., Jonkers, R.A., et al. (2009). Skeletal muscle hypertrophy following resistance training is accompanied by a fiber type-specific increase in satellite cell content in elderly men. *Journals of Gerontology Series A: Biological Sciences and Medical Sciences* 64(3), 332–339.

Verney, J., Kadi, F., Charifi, N., et al. (2008). Effects of combined lower body endurance and upper body resistance training on the satellite cell pool in elderly subjects. *Muscle & Nerve* 38, 1147–1154.

Walther, C., Adams, V., Bothur, I., et al. (2008). Increasing physical education in high school students: Effects on concentration of circulating endothelial progenitor cells. *European Journal of Cardiovascular Prevention and Rehabilitation* 15, 416–422.

Waring, C., Torella, D. & Ellison, G. (2009). Cardiac stem cell activation and ensuing myogenesis and angiogenesis contribute in cardiac adaptation to intensity-controlled exercise training. *14th Annual ECSS Congress*, Oslo, Norway, abstract p. 124.

Winitsky, S.O., Gopal, T.V., Hassanzadeh, S., et al. (2005). Adult murine skeletal muscle contains cells that can differentiate into beating cardiomyocytes in vitro. *PLoS Biology* 3, e87.

Wright, D.E., Wagers, A.J., Gulati, A.P., Johnson, F.L., Weissman, I.L. (2001). Physiological migration of hematopoietic stem and progenitor cells. *Science* 294, 1933–1936.

Young, H.E., Duplaa, C., Romero-Ramos, M., et al. (2004). Adult reserve stem cells and their potential for tissue engineering. *Cell Biochemistry and Biophysics* 40, 1–80.

PART 5

ETHICAL AND SOCIETAL IMPLICATIONS

Chapter 30

Genetics and Ethics in Elite Sport

SIGMUND LOLAND

Department of Social and Cultural Studies, The Norwegian School of Sport Sciences, Oslo, Norway

Human variation, exemplified by inherent talent for sport, has been driven by evolution, the selection for beneficial traits, and selection against detrimental traits, based on environmental pressures. Good genetic predispositions for socially and culturally valued abilities and skills, or what is often referred to as talent, have been considered the fortunate outcome of the so-called *natural lottery*. New knowledge from molecular biology and genetics and its associated technologies have the potential of changing this picture.

I will take a closer look at some of the main ethical challenges of genetics to sport in this respect. Firstly, I will discuss performance-enhancing strategies based on genetic science that do not require direct modification of genetic material. The most relevant example is genetic testing for athletic talent. Secondly, I will examine interventions that require active modification of the human genome. Examples can be somatic gene transfer techniques and the more radical possibility of germline modification.

Before proceeding, there is need for a few clarifications. The distinction between therapy and performance enhancement is a difficult one (Juengst, 1998). I will not engage in the broader issues here. My discussion is limited to the specific use of genetic knowledge and technology to enhance in healthy athletes what are considered relevant phenotypes for particular athletic performances. Moreover, genetic

technology is in an early stage of development. As of today, few, if any, therapeutic techniques have been shown to be successful and safe (Friedman & Hoffman, 2009). My discussion is based both on what is currently possible and on what seems possible, at least from a theoretical point of view, in the future. Finally, my focus is on competitive sport (and not on exercise and fitness activities), as I believe this is where the most challenging ethical dilemmas arise.

Genetic Testing for and Selection of Athletic Talent

Knowledge of the role of genes and genetic individuality on the performance and trainability of the human organism is rapidly increasing (Bouchard et al., 1997; Bray et al., 2009). This knowledge is of relevance to the development of athletic performances in several ways. There are many constructive applications here. For example, relevant genetic information about individual athletes can make training and competition schemes both safer and more efficient.

Other implications can be more problematic. The so-called pharmacogenetics refers to the design of drugs made by recombinant DNA technology according to an individual's genetic profile. The use of genetic information to make drugs more efficient raises similar ethical questions as the use of traditional performance-enhancing drugs and somatic gene transfer techniques. These will be discussed later. In the following sections, I will examine the use of genetic knowledge in the development of tests for athletic potential. One example is the finding

Genetic and Molecular Aspects of Sport Performance, 1st edition.
Edited by Claude Bouchard and Eric P. Hoffman.
Published 2011 by Blackwell Publishing Ltd.

of a variant of the *ACTN3* gene and its expression in fast- and/or slow-twitch muscle fibers (Yang et al., 2003) (see Chapter 18 for a full description of research to date). Commercial companies offer a test for this gene which again can indicate predispositions for developing performance in speed/power or endurance sports. In the time to come, more extensive descriptions of favorable genetic predispositions are possible. One assumption is that more complete tests can be developed upon which athletic genotypes can be detected and selected.

Genetic testing for talent is a scientific development of the traditional practice of talent detection and selection that has been a part of sport since its very beginning. Such tests can be performed on children and youth by simple means. A one-cell sample is sufficient and there is essentially no risk of harm. Tests can also take more radical forms and actually be undertaken before the individuals from which to select exist, for example, in terms of samples from potential sperm or ova donors in assisted reproduction. We may also imagine testing for good athletic predispositions in fetuses or early embryos in *in vitro* fertilization.

Let us first look at genetic tests for athletic predispositions in children and youth. This may have valuable potential as one among many elements in an ethically responsible and mild paternalist scheme of sport education. Reflective parent and coaches could give good advice when it comes to choosing a sport that fits a child's capacities. This would presumably reduce the possibilities of experiences of loss and failure and increase the chances for mastery and joy. In the larger schemes of elite sport, information about genetic predispositions could assist in sound talent selection and career planning and the efficient use of resources. This could cohere with perfectionist ideals prescribing the realization in rational ways of "human nature" (Hurka, 1993).

However, genetic testing for sport talent can also be practiced in ethically problematic ways (McNamee et al., 2009; Miah & Rich, 2006; Savulescu & Foddy, 2005; van Hilvoorde, 2005). One main challenge is that of early specialization. In complex societies, individual genetic predispositions can be developed in many directions.

An individual with good predispositions for speed may end up with track and field sports but may just as well become a dancer. An individual with a talent for endurance events may become a marathon runner but may also be attracted to long-distance hiking or mountain climbing. Both individuals may also have completely other interests and not explore their sport talents at all.

Feinberg (1980) articulates a widely shared ethical ideal in his prescription of a child's right to an open future. Education is seen as an empowering process in which children develop into autonomous agents with both freedom and responsibility to explore their own possibilities in life. Taken in the context of sport, we can say that children ought to be given the opportunity to explore their talents in a variety of activities that again will enable them to make their own choices.

The social logic of testing for sport talent may easily threaten this ideal. Parents, coaches, and sport systems look for information to determine a child's direct path to sport success. Trial and error processes in which the child is exposed to alternative activities are easily overruled. The social dimension of children's lives in which they enjoy the activities of their peers is often restricted. Genetic tests of children for talent may lead to early specialization that limits a child's development in ethically problematic ways.

A second main critique of genetic tests for talent focuses on questions of their validity, both in terms of lack of predictive power and overinterpretation of results. There is very active debate on this issue due to the availability of very large testing panels marketed directly to consumers (direct-to-consumer; now referred to as DTC testing). Consider the following statement from the leading society of genetics professionals, "Claims made regarding DTC genetic tests may in some cases be exaggerated or unsupported by scientific evidence. Exaggerated or unsupported claims may lead consumers to get tested inappropriately or to have false expectations regarding the benefits of testing" (http://ashg.org/pdf/dtc_statement.pdf). In events in which basic bio-motor phenotypes such as speed, strength, and endurance play a significant role, genetic tests may contribute with information about one specific condition to succeed, but there are certainly many other conditions that are additively important to the individual's success. Describing, in precise terms, genetic components of more complex technical and tactical skills, as in team games such as football or ice hockey, is far more problematic. But even in speed, strength, and endurance sports, genetic information can only give a small part of the story. Athletic performances are the products of an almost infinite number of gene–gene–environment interactions from the moment of conception to the moment of performance (Loland, 2002). The chances of becoming an Olympic champion of the 100-m sprint event depends upon being born with a predisposition for developing fast-twitch muscle fibers, but are also more or less dependent upon growing up in a family that supports sport participation and ambition, living in a society in which competitions are appreciated, being close to a track and field facility with a good coach and with a good social climate, preparing well and avoiding injuries, etc. These interactions have outcomes based on chance, luck, and merit and can never be fully controlled. Given these considerations, one can raise the question whether genetic testing for athletic skill empowers or instead endangers the person being tested (http://www.dnapolicy.org/policy.issue.php?action=detail&issuebrief_id=32).

A moderate conclusion is that genetic tests in sport can provide some information that, if used wisely, might be of some limited value in the detection and selection of talent in some sports. The radical conclusion is that due to the chances of misuse and their relatively limited predictive power, there is no need for genetic tests, as currently practiced for talent in sport at all. The radical conclusion is strengthened by the existence of what seems to be far more reliable strategies. In sport, we evaluate, compare, and rank participants according to the actual athletic performance. If a child runs fast and enjoys it, the child has a short-distance running talent. If a child develops well in the technical and tactical skills of soccer and has a passion for the game, the child has a soccer talent. A well-educated and experienced coach with good

judgment evaluates not only physically observable phenotypes but also children's motivations, attitudes, and passions. Based on concern for both ethical challenges and validity, observation of phenotypes more than genotypes in talent detection and selection is the rational way to go.

So far, I have discussed genetic tests for talent in children and have not touched upon testing at the embryonic or fetal stage. The reason is that the arguments above solve much of the dilemma. Selection of fetuses implies the controversial practice of abortion and a significant risk of harm. Laboratory procedures with testing and selection of embryos are also ethically ambiguous and have significant financial costs. If the lack of predictive force is as apparent as it seems, testing embryos and fetuses for sport talent is not just ethically problematic but irrational.

Interventions Involving Active Manipulation of Human Genes

I now turn to genetic technologies that involve the active manipulation of human genes. One obvious possibility is somatic gene transfer techniques as used in gene therapy (Friedman & Hoffman, 2009; Scherling, 2007). This involves injection of genetic material into the target mechanism or tissue that can turn on and off protein expressions with short- or long-term, local or systemic effect. Examples are genes that stimulate the production of erythropoietin and hence hematocrit and the oxygen-carrying capacity of the blood, or genes that stimulate the production of insulin-like growth factor-I (IGF-I) that leads to increase in muscle growth and strength. Gene transfer techniques can also be used before fertilization or at the embryonic stage. Germline modifications become part of the reproductive cells and are inheritable.

Although in an early stage of development and potentially hazardous, the therapeutic use of gene transfer techniques is promising and may revolutionize clinical medicine in the future. However, if used in healthy individuals, with the primary intention of enhancing sport performance, these techniques become more problematic. Actually, the World Anti-Doping

Agency (WADA) has imposed a ban on the performance-enhancing use of genetic technologies. The reference is to gene or cell doping: "…the non-therapeutic use of genes, genetic elements and/or cells that have the capacity to enhance athletic performance" (WADA, 2009). Why is the use of performance-enhancing somatic gene transfer banned? What are the ethical issues involved?

Fairness

Performance-enhancing use of gene transfer techniques raises ethical questions similar to the use of performance-enhancing drugs. A predominant argument is that gene doping involves rule violations and is unfair. Fairness is seen as a moral obligation on rule adherence that arises when we voluntarily engage in rule-governed practices (Rawls, 1971). Athletes taking part in organized sports have an obligation to obey their rules. In this context, gene transfer is cheating. For gene transfer to be acceptable, all competitors ought to be given equal access to the technology. Cheaters are free riders of the sport system and treat other competitors as a mere means to their own success.

But this understanding of fairness does not really help in a more principled discussion. The justification of a ban on gene doping by reference to the wrongness of breaking the rule is circular and invalid. References to fairness can also be used to support a liberal policy toward performance enhancement. If a significant number of athletes break the rules without being caught, rule-adhering athletes have a disadvantage. The situation is unjust and the obligation of fairness becomes problematic. Because of the challenges of detecting the use of gene transfer techniques (Scherling, 2007), this argument has gained momentum. To restore fairness, one option could be to invest in medical control and nonharmful initiatives and make genetic technologies open and accessible to all (Kayser & Smith, 2008).

This is, however, a simplistic understanding of fairness. Rawls (1971) defines fairness not just as an individual obligation on rule adherence but as an institutional norm on equal opportunity. Within a more extensive normative theory of sport,

arguments based on fairness may provide more critical force. I will return to this later.

Health

Another standard argument in the debate over performance-enhancing means and methods is linked to athletes' health. Gene transfer techniques imply health risks (Friedman & Hoffman, 2009; Scherling, 2007). Increased production of erythropoietin increases the production of red blood cells and the viscosity of the blood which may lead to cardiovascular and heart problems. Increased production of human growth hormone may cause muscle and ligament ruptures and cancer tumors. Gene transfer techniques also carry unknown risks. Introducing foreign proteins from virus or bacteria may activate harmful responses from the immune system. Nontherapeutic and nonlaboratory settings increase the chance of contamination and the possible production of pyrogens and virulent new viruses. People in close contact with the athletes might be infected by viral vectors as well and develop nonintended traits.

The potential harm of gene transfer techniques is used by WADA as one main reason for the current ban. To be included in the WADA Prohibited List, a method or a substance has to meet the requirements of (1) scientific evidence or experience which demonstrates that the method or substance has the potential to enhance, or enhances, sport performance; (2) medical evidence or experience suggests that the use of the substance or method represents an actual or a potential health risk to the athlete (WADA, 2009).

But as with the fairness argument, the health argument is not simple and straightforward. Competitive sport is about performance enhancement. Many means and methods integral to sport imply a significant risk of harm. Long-term hard training implies constant balancing of the anabolic and catabolic processes of the body. Imbalances can result in overtraining and possibly injuries. Acute injuries may arise both in training and competition.

Moreover, in sport, the risk of harm is not always considered in negative terms. In fact, in some sports, risk is a constitutive and valuable element.

In parachute jumping and downhill skiing, there is an inherent possibility for serious harm and even death. In boxing, avoiding pain and harm to oneself while imposing pain and harm to opponents are key technical and tactical challenges. If banning genetic technologies in sport due to health risks became common practice, we might end up banning elite sport as a whole.

Still, this conclusion is unreasonable as no distinctions are made on the relevance of health risks to the nature of sport. Different social practices have different goals and values. In medicine, the overriding goal is to prevent and treat illness and to maintain and restore health. As indicated in the Olympic motto, *citius, altius, fortius*, competitive sport is characterized by a strong drive to improve performance, to realize athletic potential, and to test the possibilities of individual and team talent. The challenge of training and competition is to put in the necessary effort to succeed while avoiding injuries. One of the fascinating challenges in both downhill skiing and boxing is the proper calculation of how much risk to take.

Health risks linked to the use of genetic technologies seem to be of a different kind. There seems to be a significant difference between stimulating the production of erythropoietin in a patient with a pathological lack of it, and to do exactly the same thing in a professional and perfectly healthy bicycle racer. An idea often expressed by sport leaders and athletes is that biotechnologically enhanced performance comes about without training and individual effort (Houlihan, 2002). Such enhancements are seen as "artificial" and somehow undeserved. As with the fairness argument, the health argument does not really make sense without a clearer interpretation of the meaning and value of sport.

"The Spirit of Sport"

WADA has recognized this and refers to a third criterion in defining the list of prohibited means and methods: the substance or method should not violate "the spirit of sport." In the so-called fundamental rationale for the WADA Code (WADA, 2009, p. 14), the spirit of sport is defined as "…the celebration of the human spirit, body, and mind, and is characterized by the following values:

- Ethics, fair play, and honesty
- Health
- Excellence in performance
- Character and education
- Fun and joy
- Teamwork
- Dedication and commitment
- Respect for rules and laws
- Respect for self and other participants
- Courage
- Community and solidarity"

These are general references, however, and hard to apply when it comes line drawing in concrete cases. How can the idea of the spirit of sport be interpreted and operationalized? I will sketch two possibilities and their implications for the use of genetic science and technology.

The Thin Interpretation

One alternative can be found in what can be called the thin interpretation. The only restrictions on performance are to be found in the constitutive rules of a sport and valid only within competitions themselves. In football, there is a rule against touching the ball with the hands; in the 100-m sprint running race, athletes have only one starting attempt and must stay within their lane throughout the race. Outside of competition, athletes are free to choose whatever performance-enhancing means and methods they find appropriate.

There are two branches of the thin interpretation. Some see sport as a sphere for objective progress and athletic improvement (Savulescu et al., 2004; Tamburrini, 2000). The Olympic motto, *citius, altius, fortius*, is taken literally. The interpretation echoes the Coubertanian quote of the record as having the same status in Olympic sport as the law of gravity in Newtonian mechanics: it is the eternal axiom (Loland, 1995). Others emphasize sport as spectacle; as dramatic but amoral entertainment with primarily esthetic values (Barthes, 1972; Møller, 2010). Sport can be more or less fun and spectacular but has little or no moral significance.

The implications for the use of performance-enhancing technologies are clear. Within competitions, fairness is crucial. Without rule adherence, evaluation of performance and improvement would not be possible at all. Outside of competition, however, there is no need for regulations. As long as requirements are met for athlete autonomy along with informed and free choice, athletes may use whatever means and methods they find appropriate to reach their goals.

Within sporting communities, the thin interpretation is politically incorrect and has few public defenders. Still, experiences from subcultures within cross-country skiing, track and field, and professional cycling indicate that it exerts a certain impact among athletes and supporter systems (Waddington & Smith, 2009). Ideas of "the natural" and of biological limits are rejected as prejudices and irrational traditionalism. Nature is considered the raw material to be shaped and developed according to human interests and goals. Genetic technology is seen as a particularly powerful tool in this process. Traditional distinctions between biological and technological and between organic and mechanical are no longer relevant (Miah, 2004). Biotechnological innovations are seen as opening valuable possibilities, for instance by equalizing unfair advantages inherited in the natural lottery (Mehlman, 2009). Tamburrini (2000) does not see any particular problems with germline modification as long as the techniques are efficient and safe. What could be better than enhancing our children so that they can become more successful than their parents? Savulescu (2007) argues that as long as it increases the potential for human well-being, genetic enhancement is not just a possibility but a moral obligation.

The thin interpretation is clear and consistent. It challenges traditional sport views and seems close to the social logic of competitive sport and its strong quest for improvement. However, this view of sport has received several objections. One objection is that the position seems sociologically naive. The social position of an athlete in competitive sport poses significant challenges to autonomy and the notion of unforced and informed choices. Athletes are embedded in complex networks of power relations (Waddington & Smith, 2009). In early stages of their career in particular, young athletes depend more or less totally upon guidance from coaches and supporting systems whose own

survival depends upon sport success. In liberal biotechnological regimes, athletes end up being in vulnerable positions for exploitation. In worst-case scenarios, sport could turn into something like a grand experiment of human performance with athletes as the guinea pigs (Hoberman, 1992).

Even if requirements on athlete autonomy were met, we can still question whether the use of performance-enhancing genetic technologies is of value to sport. Imagine a 100-m sprint race with the fastest runners of the time, with a winning time of 10 s but with all competitors performing without the use of means and methods on WADA's Prohibited List. On the other hand, imagine a 100-m race with a winning time of 9 s with the same athletes but after they have received an identical genetic enhancement protocol resulting in a 10% improvement of performance. Is the latter 100-m race the more valuable form of the sport event?

With everything else being equal, the athletes of both races provided the same sporting challenge, and the relative performance differences between them remained the same. Whether the race lasts 9 or 10 s should not matter to the public as long as they know that the sprinters are the fastest of their time and that the race is tight and uncertain in terms of outcome. There are, however, differences between the two races when it comes to costs. One cost of the genetically enhanced race is an increased risk of harm. And even if the risk of harmful effects can be reduced, there will be technological and financial costs. Why should athletes and sport systems spend time and resources on genetic enhancement if it is legal and does not give an exclusive, competitive advantage? What ideals would a sport with a liberal enhancement policy promote? Why should societies put emphasis on such practices and provide support in terms of resources and attention? Why should communities engage in recruiting and developing young people and sporting talent in such a sport system?

Quantitative improvements of sport performance are not necessarily improvements of sport values. The thin interpretation offers no strong value theory of sport in the wider sociocultural context of perfectionism and human flourishing. I now turn to a more extensive interpretation of the spirit of sport and "thicker" theory of value.

The Thick Interpretation

Thick interpretations build on the premise that sport has meaning that extends progress and performance. Sport is a sphere for the realization of more general sociocultural and moral values. Rule-defined ends of sport are not the only guideline in this respect. Whether an athlete wins the 100-m sprint race with the use of genetic technologies or without it is morally significant. Means, including those used outside of competition, matter to the outcome.

The thick interpretation has diversified roots: classic ancient thinking about athletics, British amateur ideology, and Olympic ideology in its full version. Thick interpretations are expressed in more critical and systematic ways in works of, among others, Loland (2002), Morgan (2006), McNamee (2008), and Simon (2004). Murray (2009) sums up with the idea of sport as "the virtuous development of natural talents" and as a particular expression of human excellence.

What is meant by "virtuous" and "natural talent"? In most accounts of the thick interpretation, virtue is linked to perfectionist ideals. Athletes are seen to flourish only and insofar as they realize their innate and trained abilities into increased levels of complexity. "Natural talent" refers to genetic predispositions for sport performance. In contradiction to the thin interpretation in which nature is seen as the raw material to be shaped by human interests and goals, talent is understood as morally relevant contours of the given and as a field of moral deliberation (Murray, 2007). Genetic inequalities are not necessarily unjust but have to be dealt with in morally reflective ways. Listening to nature makes sense.

What are the implications of the thick interpretation of the spirit of sport for the use of genetic technology? The perspective shares with the thin interpretation the view that competitions have to be fair. Rules have to be followed. But this is not enough. The rule systems of sport express more general ideals with relevance outside of sport. One example is the classification of athletes. Women do not compete with men in sports in which biological sex exerts significant impact on performance such as in sprint running. Tall and heavy athletes do not compete with short and light athletes in sports in which body size exerts a significant impact on performance such as in combat sports.

Why is this so? One reason could be a quest for drama, excitement, and uncertainty of outcome. Such values, however, can be realized to an even larger extent through handicap competitions in which men and women compete together, or in spectacular competitions between humans and animals. A closer look indicates that classification is designed to eliminate or compensate for genetic inequalities for which individuals cannot be held responsible and which they cannot influence in any significant way. Typically, classification concerns not talent that can be developed with training and effort but absolute biological differences such as sex and body size. This resonates with a general principle of equal opportunity to be found in most ethical theories, the so-called fair opportunity principle (Beauchamp, 1991):

> "We should eliminate or compensate for inequalities in significant matters that individuals can not influence or control in any significant way and for which they cannot be held responsible."

The implication of the equal opportunity principle for the use of somatic gene transfer techniques seems clear. As in the case of traditional doping, thick interpretations imply skepticism to such use as it tends to reduce athlete control of and responsibility for performance. In this way, sport loses its potential for the virtuous perfection of natural talents and for cultivating athletes as free and responsible moral agents. More specifically, the view is that the use of somatic gene transfer does not add value to sport and represents an unnecessary risk of harm, such that the method should be avoided and if possible banned.

What can be said of germline modifications in which potential athletes are enhanced at the fetal or embryonic stage? At first sight, the thick interpretation may not offer a clear answer. After all, if these methods work and are safe the result will be individuals with a more fortunate genetic makeup than what inheritance can offer them. These individuals can then flourish in both sport and life.

Still, although thin interpreters such as Savulescu (2007) and Tamburrini (2000) generally embrace enhancement of these kinds, others, such as Murray (2007) and Sandel (2007), are genuinely skeptical.

This discussion extends sport ethics and belongs to the more general ethical discourse of the possibilities and limitations of genetic enhancement in the wider context of human life and society. One response relevant to the sport setting, however, can be found by returning to the arguments about the limits of genetic tests. An athletic performance involves an almost infinite number of gene–gene–environment interactions. At present, knowledge of genetic predispositions alone offers low prediction for sport success. Spending resources and intervening genetically in ethically controversial ways at the fetal or embryonic state, with the sole purpose of "designing" sport talent, seems irrational.

There are several possible objections to the thick interpretation. Why should this particular interpretation of sport be accepted globally? What about other interpretations of the meaning and value of sport, for example Asian ones, with more emphasis on collective values than on individualism and autonomy (Brownell, 2008). As common ground in a stand against the use of genetic technologies in sport, the thick interpretation is simply too thick.

The critique is relevant. But even if we distance the interpretation from ethnocentric ideas such as those of amateurism and Olympism, sport is still not completely open to interpretation. As Suits (1978) demonstrates, games are built on a particular playful logic that defines them as social practices. Simply said, and in contrast to the instrumental pursuits of everyday life, games are voluntary attempts to overcome unnecessary obstacles. There is a ban on hitting the ball more than once in volleyball; there are restrictions on particular styles and techniques in swimming; and in the 3000-m hurdle event, athletes have to jump over (and not run around) a series of obstacles. A good sporting game seems to strike the right balance between challenge and mastery and is open to more or less infinite possibilities for improvement of relevant abilities and skills.

To proponents of the thick interpretation, this logic of moderation provides sport with moral potential (Loland, 2002; Morgan, 2006). When practiced according to the ideal of equal opportunity and without the use of performance-enhancing genetic technology, sports enable in particularly strong ways the virtuous development of natural talent. Athletic performances both in relative and absolute terms can be an admirable form of human excellence that seems to strike important chords in social and cultural value systems all over the world.

Concluding Comments

In this chapter, I have discussed some of the main ethical challenges of genetic knowledge and genetic technologies to sport. I first looked at the use of genetic science without direct genetic modification such as in the case of genetic testing for talent. I raised ethical doubts about the use of these methods in sport, and argued that probably the most ethically sound and valid talent detection and selection is based on the systematic study of functional phenotypes by well-educated and experienced coaches. I then examined possibilities of gene transfer techniques to enhance sport performance. I argued that just as in the case of traditional performance-enhancing drugs, an ethically informed viewpoint has to include a normative interpretation of what WADA calls the spirit of sport. I sketched two such interpretations—a thin and a thick one—and examined critically their implications.

The thick interpretation implies a view of sport as a sphere for the virtuous development of natural talent to develop human excellence in accordance with sociocultural and moral ideals. The view expresses skepticism toward biotechnological performance-enhancement technologies as they seem to have significant costs without adding value to sport. The thick interpretation provides a moral justification for a ban. The thin interpretation implies strict requirements on fairness in competition but no restrictions on performance-enhancing means and methods outside of competition. Whether improvement in performance is due to talent and hard training or genetic modification does not morally matter. For the strongest proponents of this interpretation, the use of genetic technology to enhance performance is not only a possibility but also a moral obligation.

The thin interpretation has merits. Why should we be skeptical toward performance-enhancing technologies if they can enhance the possibilities for human flourishing both inside and outside of sport? Thin interpretations challenge our preconceptions and attitudes in direct and relevant ways. From the thick interpretation perspective, however, the use of technology in sport cannot be reduced to a question of objective progress and records alone. In terms of explanatory and moral power, I believe thick interpretations still have the upper hand. It remains to be seen whether such ideas can survive in a world in which the performance-enhancing use of biotechnological means and methods is on the rise.

References

Barthes, R. (1972). *Mythologies*. Hill and Wang, New York.

Beauchamp, T.L. (1991). *Philosophical Ethics: An Introduction to Moral Philosophy*, 2nd edn. McGraw Hill, New York.

Bouchard, C., Malina, R.M. & Perusse, L. (1997). *Genetics of Fitness and Physical Performance*. Human Kinetics, Champaign, IL.

Bray, M.S., Hagberg, J.M., Pérusse, L., et al. (2009). The human gene map for performance and health-related fitness phenotypes: The 2006-2007 update. *Medicine and Science in Sports and Exercise* 41(1), 35–73.

Brownell, S. (2008). *Beijing's Games. What the Olympics Means to China*. Rowham & Littlefield, Lanhan, MD.

Feinberg, J. (1980). A child's right to an open future. In: W. Aiken and H. LaFollette. *Whose Child? Parental Rights, Parental Authority and State Power*. Totowa, NJ: Littlefield, Adams, and Co., 124–153.

Friedman, T. & Hoffman, E. (2009). Genetic doping in sport: Applying the concepts and tolls of gene therapy. In: T.L. Murray, K.J. Maschke & A.A. Wasunna (eds.) *Performance-Enhancing Technologies in Sport. Ethical, Conceptual and Scientific Issues*, pp. 241–253. The Johns Hopkins University Press, Baltimore.

Hoberman, J. (1992). *Mortal Engines. The Science of Performance and the Dehumanization of Sports*. The Free Press, New York.

Houlihan, B. (2002). *Dying to Win. Doping in Sport and the Development of Anti-Doping Policy*. Council of Europe Publication, Strasbourg.

Hurka, T. (1993). *Perfectionism*. Oxford University Press, Oxford.

Juengst, E. (1998). The meaning of enhancement. In: E. Parens (ed.) *Enhancing Human Traits: Ethical and Social Implications*, pp. 29–47. Georgetown University Press, Washington.

Kayser, B. & Smith, A.C.T. (2008). Globalisation of anti-doping: The reverse side of the medal. *British Medical Journal* 337, a584.

Loland, S. (1995). Coubertin's ideology of Olympism from the perspective of the history of ideas. *Olympika* 1, 55–77.

Loland, S. (2002). *Fair Play. A Moral Norm System*. Routledge, London.

McNamee, M. (2008). *Sport, Vices and Virtues*. Routledge, London.

McNamee, M.J., Müller, A., van Hilvoorde, I. & Holm, S. (2009). Genetic testing and sports medicine ethics. *Sport Medicine* 39, 339–344.

Mehlman, M. (2009). Genetic enhancement in sport: Ethical, legal, and policy concerns. In: T.L. Murray, K.J. Maschke & A.A. Wasunna (eds.) *Performance-Enhancing Technologies in Sport. Ethical, Conceptual and Scientific Issues*, pp. 205–224. The Johns Hopkins University Press, Baltimore.

Miah, A. (2004). *Genetically Modified Athletes: Biomedical Ethics, Gene Doping and Sport*. Routledge, London.

Miah, A. & Rich, E. (2006). Genetic tests for ability? Talent identification and the value of an open future. *Sport, Education and Society* 11(3), 259–273.

Møller, V. (2010). *The Ethics of Doping and Anti-Doping*. Routledge, London.

Morgan, W. (2006). *Why Sport Morally Matters*. Routledge, London.

Murray, T.H. (2007). Enhancement. In: B. Steinbock (ed.) *The Oxford Handbook of Bioethics*, pp. 491–515. Oxford University Press, Oxford.

Murray, T.H. (2009). In search for the ethics of sport: Genetic hierarchies, handicappers general, and embodied excellence. In: T.L. Murray, K.J. Maschke & A.A. Wasunna (eds.) *Performance-Enhancing Technologies in Sport. Ethical, Conceptual and Scientific Issues*, pp. 225–238. The Johns Hopkins University Press, Baltimore.

Rawls, J. (1971). *A Theory of Justice*. Harvard University Press, Cambridge.

Sandel, M.J. (2007). *The Case Against Perfection. Ethics in the Age of Genetic Engineering*. Belknap Press, Cambridge, MA.

Savulescu, J. (2007). Genetic interventions and the ethics of enhancement of human beings. In: B. Steinbock (ed.) *The Oxford Handbook of Bioethics*, pp. 516–535. Oxford University Press, Oxford.

Savulescu, J. & Foddy, B. (2005). Comment: Genetic test available for sports performance. *British Journal of Sports Medicine* 39(8), 472.

Savulescu, J., Foddy, B. & Clayton, M. (2004). Why we should allow performance enhancing drugs in sport. *British Journal of Sports Medicine* 38, 666–670.

Scherling, P. (2007). The basics of gene doping. In: T. Tännsjö & C. Tamburrini (eds.) *Values in Sport. Elitism, Nationalism, Gender Equality and the Scientific Manufacture of Winners*, pp. 19–31. Routledge, London.

Simon, R.L. (2004). *Fair Play. The Ethics of Sport*, 2nd edn. Westview, Boulder, CO.

Suits, B. (1978). *The Grasshopper. Games, Life and Utopia*. University of Toronto Press, Toronto.

Tamburrini, C. (2000). *The 'Hand of God'? Essays in the Philosophy of Sport*. Acta Universitatis Gothoburgensis, Gothenburg.

van Hilvoorde, I. (2005). Sport and genetics: Moral and educational considerations regarding athletes' "predestination." In: C. Tamburrini & T. Tännsjö (eds.) *Genetic Technology in Sport. Ethical Questions*, pp. 91–103. Routledge, London.

WADA. (2009). *World Anti-Doping Code*. WADA, Toronto. (http://www.wada-ama.org/en/dynamic.ch2?pageCategory.id=250).

Waddington, I. & Smith, A. (2009). *An Introduction to Drugs in Sport. Addicted to Winning?* Routledge, London.

Yang, N., MacArthur, D.G., Gulpin, J.P., et al. (2003). ACTN3 genotype is associated with human elite athletic performance. *American Journal of Human Genetics* 73, 627–631.

Chapter 31

Genes and Talent Selection

STEPHEN M. ROTH

Department of Kinesiology, School of Public Health, University of Maryland, College Park, MD, USA

A role for genetics in athletic potential and performance has long been recognized. Only recently have both scientific knowledge and technological advancements emerged to begin to provide an opportunity for the direct genetic selection of athletic talent. Talent selection is notoriously difficult, as selection procedures are performed most often in children in order to provide early exposure to sport-specific training. Because athletic performance is thought to be partly due to genetic factors, genetic testing offers a way to identify adult performance traits prior to their development, with the goal of improving the talent selection process. But genetic contributions to athletic performance are remarkably complex, including impacts on physiological, motor, and psychological traits, with unique gene–environment interactions unique to each athlete. We discuss the complexities and the serious ethical dilemmas that confront athletes and their sponsors as part of genetic profiling. Today, only a very limited number of robust performance-related genetic loci are known, such as the *ACTN3* (alpha-actinin-3) polymorphism. However, the use of genome-wide association studies (GWAS) in large cohorts of normal volunteers and athletes is likely to quickly expand the genetic report card repertoire. Current genetic technologies may not meaningfully improve upon the talent selection strategies already used, yet the rapid pace of applications of genome-enabled technologies to sports-relevant traits may change the landscape in the near future.

Nature Versus Nurture: An Introduction

"Can you tell me, Socrates, whether virtue is acquired by teaching or by practice; or if neither by teaching nor by practice, then whether it comes to man by nature, or in what other way?" (Plato's *Meno*, ~380 B.C.E.)

Though Meno queried Socrates about the source of virtue, no doubt the concept of talent could have easily been substituted. Thus, for thousands of years, humans have debated the roles of innate character and external forces in the development of any number of traits. Francis Galton is attributed with adding the phrase "nature versus nurture" to our lexicon in the late 1800s, and, since then, the debate has shifted back and forth with competing views and arguments from these two camps. A given today is that the vast majority of traits are necessarily derived from some balance of both innate genetic and external environmental factors, but that balance appears to differ depending on the trait in question, which provides reason to continue to debate and study this remarkably complex issue.

The importance of dedicated training is recognized as a necessity to elite sport performance, but we also recognize the importance of inherent abilities that appear to contribute to athletic success. Athletes with "good genes" or some perceived inherent talent have long been recognized and multigeneration families of successful athletes can

Genetic and Molecular Aspects of Sport Performance, 1st edition. Edited by Claude Bouchard and Eric P. Hoffman. Published 2011 by Blackwell Publishing Ltd.

be identified across a number of sports. In 1869, Galton published *Hereditary Genius* and provided the first evidence-based argument for the inherited nature of extreme talent across a variety of domains, including athletic performance in rowing and wrestling. Heritability studies beginning in the mid-20th century provided strong evidence for genetic contributions to a variety of performance- and fitness-related traits, and beginning in the late 1980s, specific genes were investigated as potential contributors (Bouchard et al., 1997). In 1993, de la Chapelle and colleagues (de la Chapelle et al., 1993) provided the first direct evidence that genetic mutation facilitated athletic performance by identifying a unique, rare mutation in the erythropoietin receptor of the Finnish cross-country skier Eero Mäntyranta, who had dominated that sport throughout the 1960s. Research advancements and improvements in technology have allowed more studies of genetic association with performance-related traits, with more than 230 genes identified in the recently published Human Gene Map for Performance and Health-Related Fitness Phenotypes (Bray et al., 2009). Thus, evidence both anecdotal and empirical has emerged over many years arguing for the importance of genetic factors to athletic talent, providing a strong rationale for parents, coaches, and sports organizations to push for the use of genetic technologies in identifying and selecting future athletes for sport-specific training.

Strategies for talent selection have a long history in many sports, including varying combinations of physiological, mental, and performance-based tests (and at times family history as well) that have some relevance or perceived correlation to future athletic prowess. So the addition of genetic information to the talent selection toolkit of coaches and talent scouts is logical and on the surface does not appear to drastically alter the complexion of athletic recruitment and talent identification strategies. But on closer inspection, several areas of concern are raised, perhaps the most important being whether or not the remarkable complexity of genetic contributions to performance can be identified in a way that will appreciably improve on the nongenetic selection techniques already used. And when considerations of athlete autonomy, genetic privacy, and discrimination are considered, genetic screening for athlete selection becomes a more challenging issue. Whether the ethical dilemmas raised by the addition of genetic testing to the toolbox of talent selection are outweighed by the potential benefits of improvements in athlete selection are uncertain and discussed in detail in this chapter.

Principles of Talent Selection

Defining "talent" is itself challenging, let alone developing tools to identify it. Howe et al. (1998) argued that talent is exceptional performance that is partly innate, relatively domain-specific, found only in a limited minority of individuals, and partly identifiable at an early stage of development. Importantly, a key aspect of talent is the idea that early indications of talent provide a basis for predicting future success. If talent is identifiable at some level in children, then talent identification programs could yield improvements in future performance by identifying potential athletes sooner, providing guidance on the particular domains best suited to their innate abilities, and allowing for more time for dedicated training. A clear theme in discussions of talent and athletic performance is the critical importance of dedicated, sport-specific training, without which innate physical, mental, or motor abilities are unlikely to yield successful performance in a particular sport. Multiple lines of evidence point to the idea that many years of dedicated training are required to reach elite levels in most sports for all athletes, with 10,000 h considered a rough threshold for mastery (Ericsson et al., 1993b).

Early talent identification and selection is institutionalized for many sports around the world. With the limited resources available to train future athletes paired with the importance of athletic performance to national and/or cultural pride, most sport governing bodies, sponsors, and coaches are compelled to identify top potential from the masses of youngsters with an interest in a particular sport. From the Australian National Talent Identification and Development program to the Olympic Development programs of the U.S. Olympic Committee, nearly all sports have a means of identifying and developing young athletes in the hopes of producing future champions.

For the most part, talent identification is based on morphological and motor characteristics, psychological factors, and performance in sport-specific activities. Numerous papers have reported on various strategies across a range of sport domains, though authors acknowledge the imperfect nature of the discipline (Vaeyens et al., 2008). Talent selection at an early age is thought to be necessary to provide future athletes adequate training time in sport-specific activities, but the nonlinear process of physical and psychological development of children is a significant detriment to the success of talent identification programs (Abbott et al., 2005). When studied over various ages, the traits of an individual relative to age-matched peers will differ according to nonlinear developmental pathways, which will be particular to each trait. Thus, body proportions, psychological behaviors, motor skills, etc., will develop along independent and inconsistent trajectories that will result in different talent selection outcomes depending on when traits are measured in a particular individual, especially near puberty when many such identification programs are implemented. Thus, the risk for "false negative" and "false positive" decisions in talent identification programs are quite high, especially when dictated by single measures of discrete traits. The best programs thus monitor youth across a broad span of time and focus on psychological factors as much as physical factors, with an emphasis on determining *potential* for future performance rather than predicting performance itself (Abbott et al., 2005).

These difficulties in typical talent identification programs provide an opening for the addition of genetic technologies into the selection process. If at least some of the traits of a successful adult athlete are genetically determined (and perhaps not obvious within the developing child), then identification of these key genetic factors in otherwise unremarkable youngsters could enable the reduction of false negative decisions in the identification process. Selection by morphology or performance on agility tests, etc., is arguably no different from selection with genetic technologies: both attempt to predict the likelihood for future success based on small bits of information. Athletes who meet some performance threshold (or genetic profile) are selected to participate in special training opportunities that will further develop their skills until the most elite of the athletes becomes clear over multiple, high-level selections and associated competitions. Crude genetic information in the form of family history is already taken into account when identifying potential athletes, and newer technologies are simply attempting to make more measurable the "art" of athlete selection based on family history, morphology, etc. (Munthe, 2000). The critical issue then is the extent to which elite athletic performance is influenced by identifiable genetic factors.

Heritability and the Complexity of Genetic Contributions to Performance

Ultimately, talent scouts are trying to identify in children traits that are likely to result in success as young adult athletes, traits which are very often developed and refined over the course of years of dedicated training. Because many traits in the adult elite athlete cannot be directly observed in the untrained child, prediction of future traits and their potential for development are the key ideas behind a genetic selection program.

The standard measure of genetic contribution to a trait of interest is heritability, which provides a quantitative estimate of the importance of genetic factors to a trait. Through twin and other family studies, researchers can estimate the importance of genetic and shared environmental factors in comparison to unique environmental factors for a particular trait measure. Heritability in this context should not be confused with the idea of direct transmissibility of a trait from parent to offspring. Rather, heritability is a statistic that estimates the general importance of genetic and shared (familial) environmental factors to the trait in a population, with the remaining influences coming from non-shared (unique) environmental factors. Heritability values can theoretically range from 0% to 100%, though the vast majority of performance-related traits fall in the range of 15–60%, which would indicate small to moderate genetic and shared environmental influences (Bouchard & Malina, 1983; Bouchard et al., 1997; Roth, 2007).

Importantly, while moderate heritability values have been observed for a number of motor skill and performance-related traits (e.g., muscle mass, maximal oxygen consumption, and so on), little is known about the heritability of performance *per se*, which is a much more challenging trait to investigate (Bouchard et al., 1997). Elite-level performance is by definition limited to only a small number of superior athletes (making studies difficult), and performance requires success across a variety of physiological, motor, and psychological tasks (making measurement difficult). One of the few comprehensive surveys of sport performance and genetics was performed by Luigi Gedda in the 1950s, in which he reported on the characteristics of more than 350 athlete twin pairs. Though rudimentary, the studies examined differences between monozygotic and dizygotic twins with regard to sport participation and specialization within sports, concluding a genetic component was important to "sports activity in man" (Gedda, 1960). Klissouras and colleagues (2001) studied an identical twin pair of Olympic-level athletes with matched training and competition histories, but differing levels of athletic success. Interestingly, the twins exhibited nearly identical physiological characteristics that were likely critical to their reaching an elite level, but showed differences in key psychological traits that correlated with the differences in their competitive successes. Thus, although psychological traits are under partial genetic control, the authors argued for the importance of both genetic and unique environmental contributions to elite athletic performance, again emphasizing the complex contributions of physical, motor, and mental traits. Though the vast majority of genetic studies have focused on individual contributing traits rather than athletic performance itself, the assumption in the field is that a meaningful component of overall athletic performance is based on genetic factors.

By definition, a complex trait is one that is governed by many genes and multiple environmental factors, and athletic performance is arguably one of the most complex traits given the multiple body systems that contribute to a successful outcome. This means that many (likely hundreds of) genes of differing levels of influence will contribute to

the traits underlying performance and thus in an additive way to performance itself. Such complexity is a limitation to the identification of specific genes for performance, but not necessarily an impossible obstacle. Specific traits may be more important for certain performance domains and the genes important to those heritable traits could be targets for genetic screening in a talent identification program (e.g., lactate threshold or maximal oxygen consumption and aerobic performance).

Williams and Folland (2008) used available data on gene polymorphisms associated with physiological aspects of endurance performance and calculated the number of people who would be carriers of various fractions of the 23 identified genotypes. If those particular genotypes could be conclusively identified as predisposing someone to success, then certainly genetic screening could be used to identify carriers. But, the authors' calculations show that remarkably few individuals carry even a fraction of the 23 genetic factors they targeted (Table 31.1), and the odds ratio of carrying the "perfect" profile was greater than 1 in 1000 trillion. While these calculations do argue that future elite athletes will occasionally emerge with remarkable genetic profiles, they also show how futile genetic screening would be if sponsors were seeking athletes with even modest combinations of the preferred genetic factors.

The studies to date have mostly focused on identifying the genes underlying physiological traits that make up athletic performance, but there is no question that psychological traits such as attitude, motivation, strategic thinking, and others contribute to performance as well. Defining the role of psychological factors in performance is an ongoing challenge for sport science researchers and the extent to which these factors are necessary is not clear, owing to their complexity (Davids & Baker, 2007). And while many of these psychological traits are themselves heritable, a lack of clear recognition of how these traits manifest in successful performance in different contexts makes their predictive value in genetic profiling limited. Thus, while genetic profiling of psychological traits that predict motivation and dedication is more likely to yield success in a talent identification program

Table 31.1 Data from Williams and Folland (2008) showing the probability (percent chance and odds ratio) of any one individual carrying a specific number of the 23 optimal genotypes identified as being important for endurance performance

Number of genotypes carried of the 23 total identified as important for endurance performance	Percent chance of carrying the combination of genotypes	Odds ratio of carrying the combination of genotypes
1	21.0	1:5
2	3.8	1:25
3	2.3	1:40
4	0.8	1:120
5	0.7	1:150
6	0.2	1:600
7	0.1	1:800
8	6.4×10^{-3}	1:16,000
9	9.6×10^{-4}	1:100,000
10	4.7×10^{-4}	1:200,000
11	1.5×10^{-4}	1:600,000
12	2.9×10^{-5}	1:3 million
13	1.2×10^{-6}	1:85 million
14	9.1×10^{-7}	1:110 million
15	1.8×10^{-8}	1:5.5 billion
16	1.7×10^{-8}	1:6 billion
17	1.2×10^{-9}	1:85 billion
18	8.3×10^{-11}	1:1.2 trillion
19	5.8×10^{-12}	1:17 trillion
20	4.0×10^{-12}	1:25 trillion
21	1.6×10^{-12}	1:62 trillion
22	2.7×10^{-13}	1:364 trillion
23	8.2×10^{-14}	1:1212 trillion

With a current world population approaching 7 billion in early 2010, no more than two individuals on the planet are predicted to carry 15 of the 23 optimal genotypes.

(Abbott et al., 2005; Ericsson et al., 1993a), the genes involved are more likely to be acting in complex gene–environment interactions that make their use in profiling considerably more challenging than genes associated with physiological traits.

Further complexity is contributed by the inherent randomness or stochasticity of development. Genes are regulated within complex networks interacting with signaling pathways and biochemical processes, each of which has an inherent unpredictability to its outcome. While the processes of gene transcription and translation are predictable outcomes to a particular cell signaling event, the specific timing and final outcome of such an event are not. This makes biological systems inherently "noisy" and susceptible to minor fluctuations in the variety of cellular processes that interact in response to deviations in homeostasis (Davids & Baker, 2007; Kaern et al., 2005). Thus, even the "right genes" and the right environment may not be enough to result in optimal development owing to the stochasticity of the cellular enterprise. Anecdotally, this may speak to those children of athletes who do not go on to greatness themselves.

Ultimately, genetic association studies revealing specific genetic factors that contribute to a physiological, motor, or psychological trait are performed in large samples of individuals and speak to the probability that a particular factor is contributing to that trait in a population. The magnitude of importance for an individual athlete will be more variable than that observed across a sample of multiple athletes. Thus, while a larger proportion of individuals with genotype X might be expected in a group of elite endurance athletes, athletes with genotype Y would not necessarily be absent from that group owing to the complexity of genetic influences. This argues against any certainty of genetic profiling, as profiling is necessarily about selecting individuals while the profiled gene targets are chosen based on population-based genetic association studies. Genetic profiling is likely to add more confusion to the already challenging and imperfect process of talent identification rather than providing a clear improvement.

The Current State of Genetic Screening

In athletics, genetic technologies are envisioned as a way to identify individuals carrying particular combinations of genotypes suited to success in particular sports. In the extreme, genetic screening for athlete identification could take place prior to birth, either through preimplantation genetic diagnosis in *in vitro* fertilization, screening of potential parents, or testing fetuses themselves (Munthe, 2000; Wackerhage et al., 2009). While such extreme situations are possible, the more likely scenario is the early identification (i.e., in childhood) of individuals predisposed to (or, alternatively, unlikely to attain) success in a particular performance domain. Once identified, selected children would be targeted for early sport-specific training with individualized environments matched to the genome of each individual and designed for attaining maximum performance during adolescence and beyond.

The idea of pre-birth selection is not new, but has previously relied on the old-fashioned technique of "matchmaking." Gifted athletes will necessarily have opportunities to form relationships as part of their training and competitive interactions, and even forced marriage is not unprecedented. In the controversial biography *Operation Yao Ming* by Brook Larmer, Chinese basketball star Yao Ming's birth is traced both to his grandparents who were first identified for their remarkable height and then to the arguably forced marriage of his parents, both successful professional basketball players. That said, even when gifted athletes have offspring, there is no guarantee that those offspring will be endowed with the same genetic combinations that provided an advantage for either parent. The processes of reproductive biology in many ways stand in direct opposition to such a "clone" outcome, hence the interest in direct genetic profiling of individuals.

Direct genetic testing (or genetic screening) has an established place in modern health care. Around the world, newborn screening for a variety of genetic disorders is a common practice, although this is typically done using biochemical methods and not by DNA testing. In adults, genetic testing is used in many situations, including diagnostic testing to confirm a diagnosis and assist with disease treatment; carrier testing for parents prior to pregnancy (or preimplantation testing of embryos prior to *in vitro* fertilization); or predictive testing, especially in cases in which a family member has a late-onset genetic disorder but the patient is as yet asymptomatic (e.g., *BRCA1* testing and breast cancer risk). While many tests are available only through established medical sources, many more tests are being targeted directly to consumers. A large number of direct-to-consumer (DTC) genetic tests are becoming available in which individuals submit a DNA sample (most often a cheek swab) to a company that then returns genetic test results directly to the consumer often with no involvement of healthcare providers. Such DTC genetic tests in the United States have little regulatory oversight and are frequently available via the Internet. The advent of microarray-based genotyping technologies has opened the possibility of extensive DTC genetic screening for genes across the genome. Companies such as 23andMe Inc. offer whole-genome testing of thousands of genotypes with information provided about susceptibility to hundreds of diseases, along with information about maternal and paternal ancestry.

Athletes, parents, coaches, and sports teams are beginning to pursue such testing despite the uncertainties of the underlying science. As early as 2005, some professional sports teams were testing their athletes and modifying training programs (Davids & Baker, 2007). In the realm of athletic performance, genetic screening technology is considerably more advanced than the understanding of genetic factors underlying performance traits. While many genes have been tentatively associated with performance-related traits (see Bray et al., 2009 and many chapters in this volume), few if any have risen to a level that would be called conclusive.

Despite these limitations, multiple companies have marketed genetic tests purporting to provide important information about the likelihood of future athletic success. Testing for the alpha-actinin-3 (*ACTN3*) R577X nonsense polymorphism has been available for several years from multiple companies, and CyGene Laboratories Inc. markets a test called "CyGene Optimum Athletic Performance" that profiles genotypes across multiple genes (e.g., *ACE, APOE,* and *VDR*). Consumers are provided with personal results across all tested genes along with descriptions about what the results indicate about their predisposition to succeed in endurance or power sports, susceptibility to injury, etc. The scientific validity of such recommendations is questionable given the early stages of the underlying science, but there is little doubt that such DTC genetic tests for athletic potential will only become more numerous moving into the future.

The likelihood that genes will be identified that predict injury susceptibility seems far more likely than genes that explain elite sport performance. As these cases of genetic profiling fall outside the realm of talent selection and could be strictly limited to protecting athletes from harm (thereby sidestepping some key ethical concerns), they will not be reviewed here but have been discussed elsewhere (McNamee et al., 2009; Wackerhage et al., 2009).

Ethical Concerns Regarding Genetic Screening

The inclusion of genetic information in talent identification programs is problematic for a variety of reasons as discussed earlier, including the remarkable complexity of athletic performance, the numerous genes that contribute only small fractions to various aspects of performance, and the intricate gene–environment interactions thought to modify developmental and performance outcomes. The attraction of genetic testing, however, and the ease and availability with which athletes and sponsors

can gain access to such information suggests that the early evidence of its inclusion in talent identification is only the beginning of a broader expansion of genetic profiling. Such an expansion raises several ethical concerns, including issues of athlete autonomy, privacy, and discrimination.

Primary among the ethical concerns is the issue of athlete autonomy in the genetic screening process. The principle of individual autonomy suggests that if a person does not want to know about a particular disease susceptibility or condition, then he or she should not be forced into learning that information. The same holds for athletic potential or lack thereof. The typical age of talent selection programs comes prior to adulthood and sometimes occurs in young children. A full understanding of the consequences of genetic test results is very likely beyond the capacity of these young people, and their parents, coaches, and sponsors are unlikely to have a full grasp of the meaning of the results themselves. Because of the enormous pressure placed on different members of the athlete's support group, a very real possibility exists that a young child will be subjected to screening without full understanding or consent (Williams & Wackerhage, 2009). The likelihood is that a child will not understand that his or her future sport participation might be affected by the outcomes of the test (whether or not justified by the science). Parents may feel compelled to push children into certain sports that are considered more matched to the resulting genetic profiles in order to maximize chances of success despite the interests of the child. Coaches may be inclined to limit participation of those children with poorly matched genetic profiles. These same arguments can be made for the results of more typical talent selection processes as well. Genetic screening, however, can occur much earlier than when a child is able to perform typical skill tests (e.g., embryonic screening), making the issue of autonomy and consent more salient for genetic profiling (Wackerhage et al., 2009).

All genetic screening should be performed under the established guidelines of informed consent, such that the individual understands the purposes of the test, can refuse or abandon testing at any point without adverse consequences, and will receive full information regarding the results of the test. With the misunderstandings about genetic determinism that are common for so many people, such conditions are unlikely to be met by most parents and coaches who may be most involved in the test initiation and interpretation. In medical genetic testing, genetic counseling is a standard of care that provides patients with both general and specific information about the benefits and risks of genetic testing and assists with the interpretation of testing outcomes; such counseling is unlikely to be satisfactory in the area of talent selection. Scientists would do well to discourage the myth of genetic determinism by emphasizing the complexity not only of genetic influences but also of performance itself.

For established athletes for whom concerns of autonomy may not be present, the voluntary nature of their consent to undergo genetic testing is a concern. The power differential between an athlete and a sponsor (e.g., coach, franchise, and so on) can result in conflicts of interest and cloud the voluntary consent process (McNamee et al., 2009). The concern of coercion is very real given the high costs to franchises of signing top-caliber athletes and benefiting from their sporting success. What is the degree of comprehension and thus informed consent of the athletes undergoing testing? Who is interpreting the results and is genetic counseling available? Does the athlete have a say in the implementation of changes to the training program, etc. following the test results? These issues are especially important in the case of injury susceptibility profiling, but also have a place in genetic selection processes as well. Because human interpretation is required following genetic profiling, the processes by which genetic screening results are conveyed to the athlete are subject to abuse and may not conform to the best interests of the athlete (McNamee et al., 2009). Genetic testing may also have the unintended consequence of identifying potentially harmful disease predispositions.

Privacy of genetic information is also a critical issue. Especially for established athletes, genetic information is remarkably sensitive information that could impact several aspects of an athlete's life and will be of critical interest to the athlete's sponsors (Wackerhage et al., 2009). To whom will the genetic information be provided and can such information remain confidential? DNA sequence is shared in a predictable manner with family members, so information from one person necessarily provides information about that person's relatives (e.g., twins, parents, and siblings). If genetic information is used to select athletes, then predictions can also be made about the likelihood that the athlete's parents, siblings, and extended family carry these same genetic profiles. Again, the power differential and potential conflicts of interest between athletes and sponsors could result in abuse of an athlete's genetic information with consequences for the athlete's relatives as well (Munthe, 2000).

A primary concern in previous discussions of athlete genetic screening has been the privacy and discrimination concerns of athletes as employees. Athlete sponsors and especially franchises with which athletes are under contract could be anticipated to require testing for a range of reasons, from injury susceptibility to individualized training to determination of psychological traits that may contribute to future success. Concerns about employment issues have been diminished in the United States by the recent passage of the Genetic Information Nondiscrimination Act (GINA), which prohibits discrimination in health coverage and employment based on genetic information. Similar laws in Europe (Council of Europe Bioethics Convention) may also prevent required genetic testing for athlete selection or the abuse of such testing if the athlete pursues it independently (McNamee et al., 2009). These laws also appear to protect athletes from a sponsor that might seek genetic information from familial relatives (e.g., parents and siblings). An interesting conflict arises in the case of determining the willingness of an athlete to adhere to the difficulties of prolonged and difficult training regimens: the franchise is justified in determining the likelihood of an athlete fully committing to such hardship prior to offering a contract, but can genetic screening of psychological factors be included in this predetermination? Several employment-related ethical concerns such as these have been reviewed recently (McNamee et al., 2009).

Genetic testing for athlete selection has the potential to be misleading and thus discriminatory, as the complexities of genetic influence will necessarily mean that some "causal" genes are more or less important in different individuals. The consequence will be that some individuals will be selected for sport greatness based on genetic information but not actually achieve it (i.e., false-positives), while at the same time other individuals will be selectively removed from sport participation when in fact the unmatched genetic profile would not have prevented elite status (Munthe, 2000). These risks are not different from the inclusions and exclusions based on current talent selection programs, but are unique in that the current programs require performance of various tasks by the individual while a genetic profile is beyond individual control.

Some ethical concerns focus at the broader societal level. Fairness is a common argument when considering the role of gene selection in sport, but the issue is complex. On the one hand, fairness in athletics is defined as a level playing field, with no undue advantage for any one player over another in terms of equipment, preparation, etc. As such, genetic selection might be viewed as further leveling the field by minimizing genetic differences that might exist through selection of those most likely to excel in a particular sport. Some argue that the use of genetic technology is desirable for sport by promoting enhanced, optimal performance, which speaks to the "sport as entertainment" context (Munthe, 2000). This assumes that all athletes will have access to testing, which seems reasonable given the remarkable advancements and cost reductions in genetic testing technology. On the other hand, genetic selection will necessarily eliminate individuals who may otherwise have been able to compete; even if considered genetically inferior, the inability to participate runs counter to the tradition of many sports (Munthe, 2000). And will the performances of some athletes who are identified as exceptionally unique with regard to their genetic profile (e.g., Eero Mäntyranta) be labeled as inherently unfair, even though it likely was these same types of individuals crowned as champions in the past? Broad societal discussion

about these issues is lacking, owing in part to the misconceptions regarding genetic determinism and the current low profile of genetic screening in sport.

Discussion and Conclusions

Since the time of Galton, debate has raged about the importance of innate talent versus motivation and persistence in success. Though some have argued that determination and dedicated effort over the many years and thousands of hours of training are all that are required to produce elite talent (Ericsson et al., 1993a), the heritability studies and images of multigeneration athlete families provide compelling evidence that innate qualities provide certain individuals an advantage in pursuing athletic endeavors. The question is whether or not genetic screening techniques are likely to identify that innate advantage as part of talent identification programs. While the genes for particular physiological, motor, and psychological traits are likely to be identified in the coming years, will identification of these factors in childhood contribute to successful athlete identification?

Certainly, those athletes with a favorable genetic profile who interact with a favorable, matched training environment are more likely to achieve higher levels of performance. But the likelihood remains that the possible unique combinations of genetic and environmental factors resulting in elite-level performance are enormous and generally unpredictable. Such complexity will limit the usefulness of genetic screening as part of a talent identification program. As noted in Williams and Folland (2008), the prediction that only a small number of individuals will carry any particular combination of important genetic factors indicates that genetic profiling may never overcome the risk of false-negatives. While it may be that only 1 in 10,000 individuals is destined for high-level sport success, the likelihood that genetic screening will identify that person any better than our current strategies for talent identification seems highly optimistic.

The promise of genetic screening is the ability to separate core talent from the usual interconnections of innate talent, parental support, and opportunities for training and performance (Gardner, 1993). Parents, coaches, sponsors, and governing bodies naturally seek to identify those likely to succeed as early as possible, as success is ultimately tied to access to quality training and related resources (Davids & Baker, 2007). But athletes and their sponsors make the incorrect assumption that earlier access to specialized, sport-specific training will necessarily improve the athlete's chance of future success. Not only are there negative consequences of such early specialization, but whether such specialization is necessary for future success has also been called into question (Baker, 2003; Davids & Baker, 2007; Wiersma, 2000).

Perhaps the most likely use of genetic testing will be the identification of individuals with some inherent general capacities in psychological or motor skills that are transferrable across multiple domains (Abernethy et al., 2005; Millet et al., 2002). Encouraging these individuals to pursue athletic activities that are enjoyable and motivating will provide key opportunities to develop both physically and emotionally; without pleasurable involvement, athletes risk burnout and are ultimately more likely to leave a sport (Davids & Baker, 2007; Ericsson et al., 1993a). Ultimately, personal interest and enjoyment can provide important contributions to one's effort and motivation, and presumably issues of personal satisfaction and enjoyment will be sport specific and subject to the unique environmental exposures of childhood. The inclination to particular psychological behaviors or temperaments that are conducive to sport-related training and competition activities are thus the factors that may prove to have the most use for genetic screening in talent identification programs.

So many extraneous factors come together on the occasion of a sporting event that the outcome is nearly always in some doubt, even when one competitor is recognized as superior over another. Even when athletes are perfectly matched in skill level and preparation, such environmental inconsistencies and random events result many times in deciding the outcome. We recognize that "good genes" likely propel an athlete to greatness, but also that such good genes don't guarantee an outcome either; training, motivation, environment, etc. must still be accounted for when explaining excellence. Because of the remarkable complexity of athletic performance and the requirement of years of intensive, sport-specific training, genetic information can at best only explain a fraction of an individual's likelihood of future performance success. The complexity of genetic and environmental influences on physiological, motor, and psychological traits suggests that successful genetic profiling will be extremely limited in scope or will be targeted to such general performance characteristics as to be of little use beyond typical talent selection approaches.

References

Abbott, A., Button, C., Pepping, G.J. & Collins, D. (2005). Unnatural selection: Talent identification and development in sport. *Nonlinear Dynamics, Psychology, and Life Sciences* 9, 61–88.

Abernethy, B., Baker, J. & Cote, J. (2005). Transfer of pattern recall skills may contribute to the development of sport expertise. *Applied Cognitive Psychology* 19, 705–718.

Baker, J. (2003). Early specialization in youth sport: A requirement for adult expertise? *High Ability Studies* 14, 85–94.

Bouchard, C. & Malina, R.M. (1983). Genetics of physiological fitness and motor performance. *Exercise and Sport Sciences Reviews* 11, 306–339.

Bouchard, C., Malina, R.M. & Perusse, L. (1997). *Genetics of Fitness and Physical Performance*. Human Kinetics, Champaign, IL.

Bray, M.S., Hagberg, J.M., Perusse, L., et al. (2009). The human gene map for performance and health-related fitness phenotypes: The 2006–2007 update. *Medicine and Science in Sports and Exercise* 41, 35–73.

Davids, K. & Baker, J. (2007). Genes, environment and sport performance: Why the nature-nurture dualism is no longer relevant. *Sports Medicine* 37, 961–980.

de la Chapelle, A., Traskelin, A.L. & Juvonen, E. (1993). Truncated erythropoietin receptor causes dominantly inherited benign human erythrocytosis. *Proceedings of the National Academy of Sciences of the United States of America* 90, 4495–4499.

Ericsson, K.A., Krampe, R.T. & Heizmann, S. (1993a). Can we create gifted people? *Ciba Foundation Symposium* 178, 222–231.

Ericsson, K.A., Krampe, R.T. & Tesch-Romer, C. (1993b). The role of deliberate practice in the acquisition of expert performance. *Psychological Review* 100, 363–406.

Gardner, H. (1993). The relationship between early giftedness and later achievement. *Ciba Foundation Symposium* 178, 175–182.

Gedda, L. (1960). Sports and genetics: A study on twins (351 pairs). *Acta Geneticae Medicae et Gemellologiae* 9, 387–406.

Howe, M.J., Davidson, J.W. & Sloboda, J.A. (1998). Innate talents: Reality or myth? *Behavioral and Brain Sciences* 21, 399–407.

Kaern, M., Elston, T.C., Blake, W.J. & Collins, J.J. (2005). Stochasticity in gene expression: From theories to phenotypes. *Nature Reviews Genetics* 6, 451–464.

Klissouras, V., Casini, B., Di Salvo, V., et al. (2001). Genes and Olympic performance: A co-twin study. *International Journal of Sports Medicine* 22, 250–255.

McNamee, M.J., Muller, A., van Hilvoorde, I. & Holm, S. (2009). Genetic testing and sports medicine ethics. *Sports Medicine* 39, 339–444.

Millet, G.P., Candau, R.B., Barbier, B., Busso, T., Rouillon, J.D. & Chatard, J.C. (2002). Modelling the transfers of training effects on performance in elite triathletes. *International Journal of Sports Medicine* 23, 55–63.

Munthe, C. (2000). Selected champions. Making winners in the age of genetic technology. In: T. Tännsjö & C. Tamburrini (eds.) *Values in Sport. Elitism, Nationalism, Gender Equality and the Scientific Manufacture of Winners*, pp. 217–231. E&FN Spon, London.

Roth, S.M. (2007). *Genetics Primer for Exercise Science and Health*. Human Kinetics, Champaign, IL.

Vaeyens, R., Lenoir, M., Williams, A.M. & Philippaerts, R.M. (2008). Talent identification and development programmes in sport: Current models and future directions. *Sports Medicine* 38, 703–714.

Wackerhage, H., Miah, A., Harris, R.C., Montgomery, H.E. & Williams, A.G. (2009). Genetic research and testing in sport and exercise science: A review of the issues. *Journal of Sports Sciences* 27, 1109–1116.

Wiersma, L.D. (2000). Risks and benefits of youth sport specialization: Perspectives and recommendations. *Pediatric Exercise Science* 12, 13–22.

Williams, A.G. & Folland, J.P. (2008). Similarity of polygenic profiles limits the potential for elite human physical performance. *The Journal of Physiology* 586, 113–121.

Williams, A.G. & Wackerhage, H. (2009). Genetic testing of athletes. *Medicine and Sport Science* 54, 176–186.

Chapter 32

Performance Enhancement by Gene Doping

ANNA BAOUTINA

National Measurement Institute, Lindfield, NSW, Australia

Gene doping refers to enhancement of athletic performance via nontherapeutic transfer or manipulation of endogenous expression of performance-related genes. It is feared that gene doping would become a new form of doping in sport and may be preferred over "traditional" drug-based doping because of the difficulty of its detection. As gene doping is an illegitimate deviation of gene therapy, in this chapter the principles and technologies of gene transfer developed for therapy of human disease are firstly described. Recent progress in gene therapy is also examined with focus on safety of gene transfer technologies in humans, especially when used outside a strictly controlled clinical setting. Candidate genes for performance enhancement and their use for therapeutic purposes are then reviewed. The likelihood of achieving the desired goal of performance enhancement using current technologies for gene manipulation and the approaches for gene doping detection are finally discussed. Because of space constraints of this publication, a large number of high-quality original articles to support certain statements could not be cited and, where possible, the reader is referred to a suitable review.

Genetic and Molecular Aspects of Sport Performance, 1st edition.
Edited by Claude Bouchard and Eric P. Hoffman.
Published 2011 by Blackwell Publishing Ltd.

Gene Transfer for Therapeutic Purposes

Gene Therapy

Gene therapy is the treatment of human diseases using the transfer of genetic material. *Ex vivo* gene therapy involves harvesting cells from a patient or a donor, genetically modifying the cells outside the body, and implanting them back to the patient. *In vivo* gene therapy involves delivery of the gene straight into tissues in the body. While *ex vivo* gene therapy is more complicated, costly, and time consuming, it is potentially safer and results are more consistent since modified cells are screened and selected for optimal expression of the target protein. In the most common form, gene therapy is in essence gene addition therapy, where a functional copy of a gene is added into cells containing a defective copy of that gene. Recently, gene transfer strategies have extended to the use of small nucleic acids that downregulate an endogenous gene that contributes to a disease. Examples of small nucleic acids are synthetic antisense oligonucleotides, short interfering RNA (siRNA), ribozymes, and aptamers. A novel approach involves modulation of expression of endogenous disease-related gene using genes for transgenic zinc finger protein transcription factors. Strategies are being developed to correct a defective gene; these include the use of zinc-finger nucleases, targeted genomic editing using synthetic oligonucleotides, and a transcriptional repair through exon skipping using antisense sequences.

Gene Therapy Vectors and Their Delivery of Potential Relevance to Gene Doping

The success of gene therapy relies on efficient gene delivery technologies. The required gene with appropriate regulatory elements (referred to as a transgene) is delivered to target cells by a vector. The two vector categories are (1) viral, including retroviral (rRV or rLV, derived from gammaretrovirus or lentivirus) and DNA viral (including rAd and rAAV, derived from adenovirus or adeno-associated virus) and (2) nonviral, including plasmid DNA, naked or with a chemical carrier. Nonviral vectors are safe and easy to produce yet inefficient compared to viral vectors, although their efficiency can be enhanced by physical methods, for example electroporation. The characteristics and applications of vectors have been reviewed and their use in clinical trials of relevance to gene doping was also examined (Baoutina et al., 2007).

Successful gene therapy also depends on the administration mode and its selection is influenced by several factors (Baoutina et al., 2007). Skeletal muscle is an attractive target tissue for gene therapy because it is abundant, easily accessible, has good vascularity, and its postmitotic state allows for prolonged expression of episomal DNA. For these reasons and also because many candidate genes for gene doping encode either muscle-related or secreted proteins, muscle would be the most convenient tissue for vector delivery in gene doping. Muscle-directed gene therapy using various vectors has been investigated for muscle disorders such as muscular dystrophy and for expression of secreted proteins, some with relevance to physical performance (e.g., growth hormone [GH], erythropoietin [EPO], and vascular endothelial growth factor [VEGF]) (Baoutina et al., 2007). Among viral vectors, gene transfer into muscle using rAAV in animals is safe and leads to long-term transgene expression. On the other hand, rAd, which also efficiently transduces muscle, often causes inflammation and minimal gene expression due to induction of strong immunity, although approaches to reduce these limitations are being developed. Successful *ex vivo* gene transfer into muscle in animals for delivery of VEGF, EPO, GH, and insulin-like growth factor 1 (IGF-1) have been reported. In humans, examples of transgenes delivered into skeletal muscle include coagulation factor-IX, α-1-antitrypsin, or lipoprotein lipase using rAAV in patients with deficiencies of these proteins and VEGF, fibroblast, or hepatocyte growth factors using plasmid or rAd in peripheral arterial disease (Baoutina et al., 2007; Mueller & Flotte, 2008). The high efficiency for transgene delivery to muscle of pseudotyped rAAV1 serotype has further broadened applications of rAAV (Mueller & Flotte, 2008), although reported rAAV immunogenicity in clinical trials (Mingozzi & High, 2007) requires further evaluation of this vector for human applications. Several approaches for systemic administration of vectors targeted to muscle have been developed in animals (Baoutina et al., 2007). However, while some of them are unlikely to be acceptable for use in humans, the safety and applicability of the others, for example hydrodynamic limb vein delivery, in man is yet to be established. Another approach to achieve gene expression specifically in muscle is the use of muscle-specific promoters, for example derived from muscle creatine kinase.

Regulation of Transgene Expression

Timing, level, and duration of transgene expression can be regulated by the incorporation into the expression construct of sequence elements that respond to the local environment or an administered inducer. An example of potential relevance to gene doping involves regulated expression of transgenes (e.g., *EPO* or angiogenic genes) in response to hypoxia by inclusion of a hypoxia-responsive element within the promoter. The recently developed double oxygen-sensing vector system may provide a robust and effective "sensor" for physiologically relevant induction of transgene expression during hypoxia (Fomicheva et al., 2008). Ligand-inducible promoters allow control of transgene expression by small molecular drugs, for example some antibiotics. Other regulatory elements of the expression cassette and bacterial sequences in plasmid DNA, including immunostimulatory unmethylated CpG motifs, can also affect the kinetics and amplitude of gene expression (Baoutina et al., 2007). Of relevance, minicircle DNA, a novel minimal expression cassette free of plasmid bacterial sequences, provided increased expression efficiency of a transgene, for example *VEGF* in muscle.

In summary, despite significant advances and ongoing extensive research, technical challenges remain considerable with respect to vector design and administration techniques, gene delivery efficiency, and tissue-specific transgene expression for human applications.

Human Gene Therapy and Its Safety

Since early 1990s, when the first gene therapy trials began, more than 1000 trials have been undertaken worldwide (for details see http://www.wiley. co. uk/genmed/clinical or http://www.clinicaltrials. gov). Yet, in this long history, gene therapy has been short on clinical success. The first clearly successful gene therapy (*ex vivo*) was reported in 2000 and 2002 to treat children with two forms of severe combined immunodeficiency (SCID). Other advances include the partial correction of chronic granulomatous disease in two patients (also *ex vivo*), improvement of visual sensitivity in a childhood inherited blindness, and some cancer therapies.

Although recent clinical trials have provided some exciting and encouraging news, they have also demonstrated potential risks associated with gene therapy (Baoutina et al., 2007). These have been linked to the oncogenic property of integrating rRV used in *ex vivo* trials or immune responses to viral vectors. Because of insertion of rRV into an oncogene that led to its activation, five children with X-linked SCID treated to date (about 25%) developed leukemia, leading to death in one case. An earlier tragic setback occurred in another gene therapy trial when an 18-year-old patient died of multiorgan failure as a result of an acute and uncontrollable inflammatory reaction to rAd administered through the hepatic artery. Apart from these severe adverse effects of human gene therapy, other trials have reported toxicity mediated by the viral vectors. In a trial for severe hemophilia B with coagulation factor-IX delivered by rAAV through the hepatic artery, liver toxicity was

reported in two subjects, while rAAV administered intramuscular (i.m.) to treat lipoprotein lipase deficiency also caused low toxicity (Mingozzi & High, 2007). In a subject with hemophilia A, rAd elicited inflammatory responses leading to hematologic and liver abnormalities (Chuah et al., 2004). These adverse events in human trials could not have been anticipated from preclinical studies, indicating that results from animal models do not always predict the outcome of similar protocols in humans and that many hurdles exist to translating strategies for vector design and delivery from animal studies to humans (Baoutina et al., 2007).

The history of gene therapy has demonstrated that, despite its potential to correct serious human diseases, it may also have many dangers to a patient, even under a highly controlled and regulated clinical environment. The misuse of this technology for performance enhancement may carry the risk of inducing serious health problems or even death when used in healthy people outside a clinical setting. Some of the adverse effects associated with the transfer of specific genes are discussed below. Other risks may include induction of immune responses and development of autoimmunity to the endogenous protein, the target for manipulation. Oncogenic effects as a result of transgene integration into the genome, generation of replication competent virus, and side effects associated with uncontrolled transgene expression, the quality of the vector preparation or the procedure of its administration in an uncontrolled athletic setting are also among potential dangers. There is also a possibility of vector integration into germ cells, transmitting the changes and associated risks to future generations. Importantly, long-term effects of gene transfer to healthy people may not be predicted from animal models or trials on patients with debilitating diseases and thus, if they occur, may remain unnoticed, uncontrolled, or untreated.

Potential Application of Gene Transfer in Sport

For almost a decade now, concerns have been raised over the possibility that gene transfer technology may be used to enhance athletic performance.

Potential gene doping applications will be similar to those for which drugs are used and may include enhancing endurance and increasing muscle size and power. The genes discussed earlier are categorized by their likely effect on physical performance; other less likely candidates for gene doping have been reviewed elsewhere (Baoutina et al., 2007).

Endurance Genes

EPO and Erythropoiesis Stimulators

Performance in endurance sports could be improved via enhanced oxygen delivery to tissues by stimulating production of erythrocytes (erythropoiesis). Naturally, erythropoiesis is stimulated by hypoxia as a result of elevated serum EPO and its interaction with EPO receptor on erythroid precursor cells in the bone marrow.

The therapeutic potential of EPO lies in the treatment of severe anemia associated with renal failure, HIV-AIDS, or cancer chemotherapy. Delivery of the *EPO* gene rather than repeated injection of recombinant EPO protein has been investigated for a number of clinical and economical reasons. Since abuse with recombinant EPO by endurance athletes has already been reported, it would not be surprising that once *EPO* gene therapy is developed for human application, it will attract interest from some sportspeople. In fact, the first indication of the imminent reality of gene doping was an inquiry by a German coach about the *EPO* gene therapy construct Repoxygen on the black market.

Several approaches for *EPO* gene transfer have been developed in animal models. These include i.m. injection of plasmid with or without electroporation, hydrodynamic limb vein plasmid delivery, injection of rAAV-*EPO* or rLV-*EPO* i.m. or subcutaneously or injection of rAAV-*EPO* into adipose tissue or salivary glands, and implantation of various cell types engineered to produce EPO (Baoutina et al., 2007; Campeau et al., 2009; Sebestyen et al., 2007). In humans, only *ex vivo* strategy has been tested to date. The first *EPO* gene therapy trial for patients with chronic renal failure-associated anemia demonstrated limited success following implantation of autologous rAd-*EPO*-transduced dermal cores.

Another trial proposes to test rRV-*EPO*-transduced smooth muscle cells seeded into vascular grafts (Baoutina et al., 2007).

As excessive EPO production may cause severe adverse effects, such as polycythemia and associated thrombosis, hypertension, hyperviscosity, stroke, and heart failure, regulated expression systems have been developed (Baoutina et al., 2007). For example, for physiological regulation of EPO expression under hypoxia, a promoter containing a hypoxia-responsive element is used. Another adverse effect of *EPO* gene therapy—autoimmune anemia—has been reported in nonhuman primates and in mice due to the appearance of neutralizing antibodies against the endogenous and transgene-derived EPO (Baoutina et al., 2007; Campeau et al., 2009).

Besides transfer of additional copies of *EPO* gene, the use of pharmacological agents that stimulate endogenous *EPO* gene expression may be considered for gene doping with EPO (Jelkmann, 2007). Hypoxia-inducible factor (HIF) stabilizers promote EPO expression in humans, but their use in clinical trials was suspended due to side effects. GATA transcription factor inhibitors that prevent natural negative regulation of EPO expression are also under investigation to treat anemia, and one such agent, K-11706, increased endurance in mice following oral administration.

Small synthetic drugs "chemical inducers of dimerization" increase the level of erythrocytes independent of EPO by inducing proliferation of *ex vivo* genetically manipulated hematopoietic cells (Miller & Blau, 2008). This approach, however, requires further development before being used in humans, and its application for gene doping is highly unlikely due to vector safety concerns and the requirement for bone marrow transplantation.

In summary, the lack of encouraging results from human gene therapy trials with EPO and reported severe adverse effects associated with elevated serum EPO may deter interest in EPO gene doping. Pharmacological stimulation of erythropoiesis increases risks of death, serious cardiovascular and thromboembolic events, and tumor progression in patients with cancer- and renal failure-associated anemia. Moreover, production of neutralizing anti-EPO antibodies and the development of pure red cell aplasia have been associated with the use of erythropoiesis-stimulating agents in patients with renal anemia (Pollock et al., 2008), while EPO over-expression in several animal studies also resulted in autoimmune responses (Baoutina et al., 2007).

Angiogenic Genes

Endurance can potentially be also increased using genes for angiogenic factors which, by promoting generation of new blood vessels, increase tissue blood perfusion and oxygen delivery.

Therapeutic angiogenesis using gene therapy approaches has been investigated to treat periph-eral or coronary arterial disease (Baoutina et al., 2007; Tongers et al., 2008). Although it has been val-idated in animals, results from a large number of clinical trials have been inconsistent. One promis-ing strategy has been the plasmid-based i.m. deliv-ery of fibroblast growth factor-1 in patients with critical limb ischemia. An alternative approach utilizes transcription factors upstream of the ang-iogenesis pathway such as HIF-1α, and a study with constitutively active hybrid form of the HIF-1α subunit showed clinical improvement in some patients with peripheral arterial disease. A trial was approved to investigate an engineered transgenic zinc finger protein transcription factors for the induction of VEGF-A and improvement of blood flow in patients with intermittent claudication.

As available approaches have shown little sus-tained benefit to patients with arterial disease, new strategies for therapeutic neovasculariza-tion have been proposed and tested in animals in recent years. Some of these strategies are given here: expression of a gene for urokinase plasmino-gen activator (a natural activator of angiogenesis) or of different VEGF isoforms or use of comple-mentary factors, for example angiopoietin- 1. These could partly counteract one limitation of gene therapy with VEGF that is vascular per-meability and insufficient maturity of the new blood vessels. Short hairpin RNA targeting prolyl hydroxylase-2 gene to suppress natural HIF-1α degradation was recently tested in myocardial ischemia in mice to increase angiogenic effect of HIF-1α (Huang et al., 2008).

The major concerns to health with angiogenic gene transfer are an increased risk of neoplastic disease and malignancy, and worsening of athero-sclerosis or retinopathy; flu-like symptoms with the use of rAd and local edema have also been reported (Baoutina et al., 2007; Tongers et al., 2008). As for novel angiogenic strategies, more work will be required to optimize formulations, dose, and delivery modality before they will be considered in human trials. Thus, despite a long history, ang-iogenic gene therapy has not fulfilled expectations, giving cause to doubt the efficacy of gene doping using proangiogenic approaches to increase oxygen supply to tissues.

Metabolic Energy Regulator Genes

Other genes of potential relevance to endurance encode proteins that are involved in production and use of metabolic energy. Among them are the perox-isome proliferator-activated receptor-delta (*PPARD*), peroxisome proliferator-activated receptor-gamma co-activator 1α (*PPARGC1A*) and 1β (*PPARGC1B*) genes (Handschin, 2009).

Overexpression or activation of these genes increases muscle fiber mitochondrial content and promotes a shift to oxidative "slow" fatigue-resistant fiber type, a conversion naturally pro-moted by exercise (reviewed in Baoutina et al., 2007; Wells, 2008). The role of *PPARD* in endur-ance exercise was best demonstrated by Wang et al. (ref. 66 in Baoutina et al., 2007), who generated mice constitutively overexpressing *PPARD* in skel-etal muscles. These "marathon" mice ran twice the distance compared with wild-type littermates and were resistant to high-fat diet-induced obesity and glucose intolerance. PPARGC1B overexpression also led to increased capacity for aerobic exercise and, when placed on a high-fat diet, *PPARGC1B*-transgenic mice consumed more food but remained leaner than controls (reviewed in Baoutina et al., 2007). Studies involving *PPARGC1A* have pro-duced some conflicting results. Its *ex vivo* or *in vivo* expression in muscles resulted, respectively, in greater resistance to fatigue or enhanced ability to exercise (refs. 14 and 45 in Handschin, 2009); how-ever, in an inducible *PPARGC1A* transgenic mouse model, a diminished exercise performance during a high-intensity exercise was reported (Wende et al., 2007). Controversy also surrounds the effect of this gene on muscle atrophy. Muscle-specific PPARGC1A overexpression was protective against muscular atrophy promoted by inactivity, fasting, or activation of atrophy-related genes; however, in another study, it resulted in muscle wasting (refs. 52 and 87 in Handschin, 2009). Importantly, another adverse effect of PPARGC1A overexpression was impaired insulin sensitivity (ref. 50 in Handschin, 2009).

Activation of the endogenous PPARD, PPARGC1A, or PPARGC1B could be achieved with pharmacological agents. For example, PPARGC1A is upregulated with β2-adrenergic receptor agonists or by activation of adenosine monophosphate (AMP)-activated protein kinase with 5-aminoimidazole-4-carboxamide ribonucleoside (Miura et al., 2007 and ref. 17 therein). PPARD is activated with GW501516, which in combination with exercise also increased running endurance in adult mice (Narkar et al., 2008); although in another study, administration of a different PPARD agonist initiated in skeletal muscle the activation of a muscle atrophy program (Constantin et al., 2007).

Thus, because some animal models have demonstrated performance-enhancing properties of PPARD, PPARGC1A, or PPARGC1B, these genes may be attractive for endurance athletes. However, currently it is not really known whether an increase of these proteins in humans is beneficial or harmful and whether doping with these genes will bring the desired outcome of improved performance.

Recently, transgenic mice were created with specific and very high (over 100 times higher than in controls) expression of enzyme phosphoenolpyruvate carboxykinase (PEPCK-C) in muscle (Hanson & Hakimi, 2008). The PEPCK-C^{mus} mice had exceptionally high running endurance and, during strenuous exercise, used fatty acids rather than glycogen for energy and produced less lactate in their blood compared to control animals. Whether or not the remarkable physical activity of the PEPCK-C^{mus} mice has direct application to human performance was answered by the investigators themselves: "there is currently no way to overexpress PEPCK-C in all of the skeletal muscles without introducing the gene into the germ line of humans, as we do with the transgenic mice; this is neither ethical nor possible."

Genes for Muscle Growth and Repair

Positive effectors of muscle growth IGF-1 and follistatin (FST), and negative growth regulator myostatin (MSTN) can also become targets for gene doping.

IGF-1 Isoforms

Postnatally, IGF-1 plays a critical role in growth, repair, and maintenance of skeletal muscle and is a major mediator of the actions of GH. The ability of the muscle-expressed isoforms IGF-1Ea and mechano growth factor (MGF, or IGF-1Ec in humans) to promote skeletal muscle hypertrophy has been demonstrated by several methods (reviewed in Baoutina et al., 2007). Most convincingly, transgenic mice with muscle-specific IGF-1Ea expression were characterized by marked muscle growth and increased force generation compared with wild-type siblings. They had little body fat and escaped age-related muscle atrophy. In rodents, i.m. injection of rAAV-*IGF-1Ea* also resulted in muscle hypertrophy, improved strength, and decline in muscle atrophy due to aging or inactivity, while i.m. nonviral transgene delivery promoted regeneration in a muscle injury model. Significant increase in muscle size was also observed following i.m. injection of a plasmid carrying a MGF-transgene in normal mice or of rAAV-MGF in young (but not mature) animals.

Because of anabolic properties, muscle-specific IGF-1 isoforms have been investigated for therapy of muscle-wasting diseases and are likely to be of interest to power athletes. Their recombinant proteins have a short half-life and will need to be used at high concentrations to produce a substantial effect. Also, systemic IGF-1 protein administration may have serious safety concerns because of its glucose lowering properties and potential effects on other muscles and organs, the likelihood of causing cardiac problems, and increasing the risk of cancer progression. Gene transfer, on the other hand, may be an effective method to deliver stable high concentrations of IGF-1Ea or IGF-1Ec to a muscle and could be relatively safe given that the effects seem to be localized to the targeted muscle. Athletes may be further attracted to this approach by the observation that a combination of resistance training and IGF-1 overexpression resulted in a more significant gain in muscle mass and strength than was observed with either alone (ref. 80 in Baoutina et al., 2007). However, such changes have not yet been demonstrated for humans, and the success of this strategy for performance enhancement remains doubtful. One can only speculate whether a hypothetical gain in muscle mass and power output in response to gene doping with IGF-1 will be sufficient to outperform competitors and will outweigh associated risks.

MSTN and FST

MSTN is a potent negative regulator of skeletal muscle growth and regeneration. This role is supported by distinctive excessive muscle growth associated with its inhibition or absence due to natural mutations in cattle, sheep, dogs, and even in a human child, or targeted disruption in *MSTN* gene or transgenic expression of MSTN inhibitors in mice (reviewed in Baoutina et al., 2007; Tsuchida, 2008). Importantly, MSTN signaling contributes to muscle mass and strength not only during development, but also postnatally and even in adults. The best evidence of this is the recent study in which *MSTN* gene knockout in mature muscle led to significant increase in mouse muscle mass (ref. 16 in Tsuchida, 2008). In contrast to MSTN, FST is a positive regulator of muscle growth and a potent antagonist of MSTN. Transgenic mice overexpressing FST in skeletal muscles showed a dramatic increase in muscle growth, which was attributed at least partially to FST inhibitory effect on MSTN (reviewed in Baoutina et al., 2007).

The function of MSTN on muscle growth and regeneration has triggered interest in the MSTN-related pathway as a promising strategy for treatment of muscle-wasting disorders and muscle injury. The potential usefulness of MSTN blockade for therapy of muscle atrophies has been confirmed in murine studies using various pharmacological agents or with *MSTN* gene manipulation strategies

(Baoutina et al., 2007; Tsuchida, 2008). The latter included transgenic or rAAV-mediated overexpression of MSTN-propeptide (that naturally maintains MSTN in the circulation in its inactive form) or FST, targeted deletion of the *MSTN* gene, using antisense RNA oligonucleotides or siRNA to MSTN. In addition to muscle hypertrophy, the MSTN blockade led to reduction of fibrosis and improvement of muscle healing in dystrophic muscle or after acute or chronic injury, and increased muscle strength in aged mice (Li et al., 2008).

Importantly and more relevant to the possibility of anti-MSTN doping, MSTN suppression was also effective in promoting anabolic effects in nonpathological adult animal models (Foster et al., 2009; Qiao et al., 2009; Tsuchida, 2008). Among the gene transfer approaches, MSTN DNA vaccine and antisense RNA targeted to MSTN caused an increase in muscle mass and, in vaccinated mice, also in grip endurance. Muscle hypertrophy was also observed after rAAV-mediated transfer of a gene for MSTN-propeptide in mice and dogs or transfer of siRNA targeting MSTN in rodents. *FST* gene delivered with rAAV or a nonviral vector in rodents led to enhanced muscle mass and, in one study, increased strength (Gilson et al., 2009 and ref. 11 therein).

Understandably, there is a growing concern among anti-doping authorities that genetic manipulation of the MSTN-related pathway may be among the targets for gene doping. However, MSTN blockade may not bring the desired results to performance enhancement. The lack of MSTN in mice in one study, though promoting muscle growth, resulted in impaired force generation, a shift to a fast fiber type, and mitochondrial depletion indicative of decrease of muscle oxidative capacity (ref. 106 in Baoutina et al., 2007). In another study, postnatal MSTN inhibition with rAAV-MSTN-propeptide did not change the force output in fast muscle despite hypertrophy (Foster et al., 2009). Moreover, no improvements in muscle strength or function were observed in the human trial on MSTN blocking with its neutralizing antibody in adults with muscular dystrophy. Also, not surprisingly, marketed muscle-enhancing supplements with MSTN-neutralizing properties did not increase muscle mass and strength or inhibit serum MSTN (Baoutina et al., 2007). All other MSTN-targeting gene transfer strategies have been evaluated only in animal models and many of them are not applicable to human application. Thus, the likelihood that MSTN inhibition in athletes can increase muscle power is questionable.

GH and GH-Releasing Hormone

GH is synthesized and released in a pulsatile pattern from the pituitary gland and its secretion is stimulated by a hypothalamic hormone, GH-releasing hormone (GHRH). It is essential for growth, homeostasis of carbohydrates, proteins, and lipids and its effects are mediated by IGF-1, the downstream effector in the GHRH-GH-IGF-1 axis.

As was recently reviewed (Baoutina et al., 2007), there is insufficient evidence regarding the "enhancing" effects of GH on body composition, muscle strength and cardiac and pulmonary function in highly trained healthy subjects. In contrast, health risks, including risk of diabetes and cancer, and detrimental effects on the cardiovascular and pulmonary systems, have become increasingly evident. Despite this, recombinant GH is widely abused by athletes.

Several strategies for *GH* gene delivery, both *in vivo* (with rAd or rAAV) and *ex vivo*, have been developed in animal models, with some systems allowing regulated GH expression by pharmacological compounds (Baoutina et al., 2007). The use of a muscle-regulatory cassette in one recent study resulted in selective GH expression in skeletal muscle following i.m. rAAV-transgene delivery (Martari et al., 2009). Although this system may avoid some potential deleterious effects associated with ubiquitous GH expression, GH gene therapy is thought to be still far from a clinical application.

An alternative method to increase GH production is administration of GHRH or the corresponding transgene. As GHRH stimulates pulsatile synthesis of GH and is capable of feedback regulation of GH release, such an approach should avoid adverse effects of the recombinant GH therapies, in which feedback regulation is lost. Gene therapy using a plasmid containing *GHRH* with a muscle-specific promoter has been applied in animal models

leading to stimulation of the GHRH-GH-IGF-1 axis and overall improvement in health, accompanied by anabolic responses and improvement of exercise tolerance in some models. Although a trial to test this system in patients with cancer-associated cachexia has been proposed, it is important to note that GHRH may potentially promote detrimental effects on human health similar to GH and IGF-1. For example, GHRH may be involved in pathogenesis and growth of cancers (Kovacs et al., 2008) and, although in patients the benefit of GHRH gene therapy may outweigh the risks, in healthy individuals the shift may occur toward the risks.

Gene Doping Detection

The World Anti-Doping Agency banned gene doping and established a research program aimed at gene doping detection. The latter, however, remains a major challenge. The approaches to detect gene doping and their challenges were recently reviewed (Baoutina et al., 2008; Wells, 2008) and a number of them are being pursued (http://www.wada-ama .org/en/Science-Medicine/Research/Funded-Research-Projects, Baoutina et al., 2010).

Several research groups are developing direct testing approaches targeted at a transgene or transgenic protein. The approaches focus on subtle differences between a doping agent and its endogenous counterpart in a human body, for example unique transgene sequences or a distinctive posttranslational modification of a transgenic protein due to its ectopic expression. The methods intended to be applicable to blood have the potential to detect gene doping if sampling is performed within a relatively short period after the doping event. This could be achieved with "out-of-competition" testing regimes already in practice in doping control.

An alternative strategy pursued by several research groups is to use indirect detection methodologies to detect responses of the body to a doping agent. These include changes at the level of other genes expression (transcriptomics), proteins (proteomics), or biochemical pathways and their metabolites (metabolomics). These changes may provide identification of specific biomarkers that could form the basis for detection methodologies of doping with not only genes, but also with the equivalent recombinant proteins, although implementation of this approach will require biomarkers' reference ranges for each athlete. Immune responses to viral vectors could also be measured. Drugs used for regulated gene expression systems may also provide an indirect target for gene doping detection. In addition, *in vivo* imaging can be utilized for detecting gene doping and it was recently tested in a "proof-of-principle" study to detect EPO ectopically expressed in muscle in mice. However, currently available imaging technologies restrict their application to animals and patients with life-threatening diseases.

In summary, even though a routine method for gene doping detection in sport does not exist yet, recent progress in research in this area indicates that the time may not be far away when either a direct detection technique or a combination of indirect approaches may provide sufficient information to detect and confirm gene enhancement in athletes.

Conclusion

To date, apart from a couple of reports of sports representatives inquiring about "gene enhancement" products, no known case of gene doping exists. The views about the danger of gene doping to sport vary. Some scientists believe it is imminent and an opinion exists that even low gene transfer efficiency may produce an improved performance. Others suggest that the danger of gene doping is exaggerated, at least based on the current "state of the art" of human gene therapy and the fact that many most impressive phenotypes of performance enhancement in animals were achieved in transgenic models through germ line genetic manipulation rather than somatic gene transfer (Wells, 2008). The potential for abuse also relates to limitations associated with production of sufficient amount of doping material which vary depending on the vector to be utilized. Nevertheless, it is apparent that concern over misuse of the gene transfer technology to enhance athletic performance is growing among the scientific, anti-doping and sports communities. The potential risks gene doping presents to the health and even lives of athletes and to a broader community together with the uncertainty whether the desired effect of performance enhancement will be achieved may discourage athletes from considering this form of abuse. Furthermore, the interest in this banned practice may also be deterred by progress in research toward development of methods for gene doping testing and education of sports community about the true scope and implications of gene doping.

Lovingly dedicated to the memory of Peter de Montfort with gratitude for his love, and for his interest, encouragement, and support in all my professional endeavors

References

Baoutina, A., Alexander, I.E., Rasko, J.E. & Emslie, K.R. (2007). Potential use of gene transfer in athletic performance enhancement. *Molecular Therapy* 15, 1751–1766.

Baoutina, A., Alexander, I.E., Rasko, J.E. & Emslie, K.R. (2008). Developing strategies for detection of gene doping. *The Journal of Gene Medicine* 10, 3–20.

Baoutina, A., Coldham, T., Bains, G.S. & Emslie, K.R. (2010). Gene doping detection: evaluation of approach for direct detection of gene transfer using erythropoietin as a model system.

Gene Therapy May 13, Epub ahead of print.

Campeau, P.M., Rafei, M., Francois, M., Birman, E., Forner, K.A. & Galipeau, J. (2009). Mesenchymal stromal cells engineered to express erythropoietin induce anti-erythropoietin antibodies and anemia in allorecipients. *Molecular Therapy* 17, 369–372.

Chuah, M.K., Collen, D. & Vandendriessche, T. (2004). Clinical gene transfer studies for hemophilia A. *Seminars in Thrombosis and Hemostasis* 30, 249–256.

Constantin, D., Constantin-Teodosiu, D., Layfield, R., Tsintzas, K., Bennett, A.J. & Greenhaff, P.L. (2007). PPARdelta agonism induces a change in fuel metabolism and activation of an atrophy programme, but does not impair mitochondrial function in rat skeletal muscle. *The Journal of Physiology* 583, 381–390.

Fomicheva, E.V., Turner, I.I., Edwards, T.G., et al. (2008). Double oxygen-sensing vector system for robust hypoxia/ ischemia-regulated gene induction in cardiac muscle in vitro and in vivo. *Molecular Therapy* 16, 1594–1601.

Foster, K., Graham, I.R., Otto, A., et al. (2009). Adeno-associated virus-8-mediated intravenous transfer of myostatin propeptide leads to systemic functional improvements of slow but not fast muscle. *Rejuvenation Research* 12, 85–94.

Gilson, H., Schakman, O., Kalista, S., Lause, P., Tsuchida, K. & Thissen, J.P. (2009). Follistatin induces muscle hypertrophy through satellite cell proliferation and inhibition of both myostatin and activin. *American Journal of Physiology—Endocrinology and Metabolism* 297, E157–E164.

Handschin, C. (2009). The biology of PGC-1alpha and its therapeutic potential. *Trends in Pharmacological Sciences* 30, 322–329.

Hanson, R.W. & Hakimi, P. (2008). Born to run: The story of the PEPCK-Cmus mouse. *Biochimie* 90, 838–842.

Huang, M., Chan, D.A., Jia, F., et al. (2008). Short hairpin RNA interference therapy for ischemic heart disease. *Circulation* 118, S226–S233.

Jelkmann, W. (2007). Control of erythropoietin gene expression and its use in medicine. *Methods in Enzymology* 435, 179–197.

Kovacs, M., Schally, A.V., Varga, J.L. & Zarandi, M. (2008). Endocrine and antineoplastic actions of growth hormone-releasing hormone antagonists. *Current Medicinal Chemistry* 15, 314–321.

Li, Z.B., Kollias, H.D. & Wagner, K.R. (2008). Myostatin directly regulates skeletal muscle fibrosis. *The Journal of Biological Chemistry* 283, 19371–19378.

Martari, M., Sagazio, A., Mohamadi, A., et al. (2009). Partial rescue of growth failure in growth hormone (GH)-deficient mice by a single injection of a double-stranded adeno-associated viral vector expressing the GH gene driven by a muscle-specific regulatory cassette. *Human Gene Therapy* 20, 759–766.

Miller, C.P. & Blau, C.A. (2008). Using gene transfer to circumvent off-target effects. *Gene Therapy* 15, 759–764.

Mingozzi, F. & High, K.A. (2007). Immune responses to AAV in clinical trials. *Current Gene Therapy* 7, 316–324.

Miura, S., Kawanaka, K., Kai, Y., et al. (2007). An increase in murine skeletal muscle peroxisome proliferator-activated receptor-gamma coactivator-1alpha (PGC-1alpha) mRNA in response to exercise is mediated by beta-adrenergic receptor activation. *Endocrinology* 148, 3441–3448.

Mueller, C. & Flotte, T.R. (2008). Clinical gene therapy using recombinant adeno-associated virus vectors. *Gene Therapy* 15, 858–863.

Narkar, V.A., Downes, M., Yu, R.T., et al. (2008). AMPK and PPARdelta agonists are exercise mimetics. *Cell* 134, 405–415.

Pollock, C., Johnson, D.W., Horl, W.H., et al. (2008). Pure red cell aplasia induced by erythropoiesis-stimulating agents. *Clinical Journal of the American Society of Nephrology* 3, 193–199.

Qiao, C., Li, J., Zheng, H., et al. (2009). Hydrodynamic limb vein injection of AAV8 canine myostatin propeptide gene in normal dogs enhances muscle growth. *Human Gene Therapy* 20, 1–10.

Sebestyen, M.G., Hegge, J.O., Noble, M.A., et al. (2007). Progress toward a nonviral gene therapy protocol for the treatment of anemia. *Human Gene Therapy* 18, 269–285.

Tongers, J., Roncalli, J.G. & Losordo, D.W. (2008). Therapeutic angiogenesis for critical limb ischemia: Microvascular therapies coming of age. *Circulation* 118, 9–16.

Tsuchida, K. (2008). Targeting myostatin for therapies against muscle-wasting disorders. *Current Opinion in Drug Discovery & Development* 11, 487–494.

Wells, D.J. (2008). Gene doping: The hype and the reality. *British Journal of Pharmacology* 154, 623–631.

Wende, A.R., Schaeffer, P.J., Parker, G.J., et al. (2007). A role for the transcriptional coactivator PGC-1alpha in muscle refueling. *The Journal of Biological Chemistry* 282, 36642–36651.

Chapter 33

Bioethical Concerns in a Culture of Human Enhancement

ANDY MIAH

Faculty of Business & Creative Industries, University of the West of Scotland, Ayr Campus, UK

Identifying the moment when genetic science became an ethical concern for the world of sport is difficult. When Watson and Crick first characterized DNA in 1953, with the assistance of Rosalind Franklin (Hubbard, 2003), this initiated a whole series of discoveries and questions about the future of humanity, the implications of which are only beginning to become manifest within social practices, such as sports. Yet, genetics in sport certainly gained special attention around the publication of the initial sequence of the human genome in 2001. At this time, athleticism was one of many characteristics that the popular science press represented as being genetically determined, via headlines about "performance genes" (Coghlan, 1998). Yet, many of the scientific studies on which such claims were based were far more modest in what they claimed the findings revealed. Nevertheless, this period gave rise to numerous debates about how genetic science would have a major impact on sport. Moreover, public concern over the direction of science required institutions to take into account the ways in which science may be applied to the detriment of individuals or society more generally. This anxiety extended to the world of sport, either directly via the International Olympic Committee (IOC) or the World Anti-Doping

Agency (WADA), or indirectly through agencies that dealt with the ethics of emerging technology, many of which identified sport as a likely area of application.

Since the early 2000s, debates about the ethics of genetics in sport have been dominated by the prospect of gene doping, the use of gene transfer technology for nontherapeutic or enhancing purposes. This technology promises a new era of performance enhancement in sport, which may call into question the possibility of detecting and catching users. Despite this possibility, WADA began to investigate the prospect in 2001 and responded in 2003 by prohibiting "gene" or "cell" doping within the 2004 World Anti-Doping Code (see WADA, 2003) which read: "M3. Gene Doping. Gene or cell doping is defined as the non-therapeutic use of genes, genetic elements and/or cells that have the capacity to enhance athletic performance" (p. 14). Additionally, WADA's funding of scientific research into establishing indirect testing methods for gene doping places the anti-doping authorities in a unique position, in which they may be ahead of the users. In particular, by establishing lines of cooperation between the world of sport and the developing biotechnology industries, whose products might be used by athletes, there is an opportunity to affect the use of gene doping before it is readily available.

In this context, this chapter provides an overview of the ethical and policy issues that surround the development and use of genetic and molecular

Genetic and Molecular Aspects of Sport Performance, 1st edition.
Edited by Claude Bouchard and Eric P. Hoffman.
Published 2011 by Blackwell Publishing Ltd.

science in sport. It begins by exploring the characteristics of sports and identifying why ethical concerns about technology generally and genetics specifically should be of special concern to sports medicine. It then considers the main purposes to which genetic science may be put within sport—specifically, research, genetic information, and gene transfer—in order to outline the ethical and policy considerations that they provoke.

An integral part of this analysis involves taking into account the vast range of cultural values that surround sports practice and genetic research, which differ remarkably throughout the world. Arguably, what distinguishes the ethics of genetic science in sport from other forms of scientific application is the way that it interfaces with a range of fundamental moral concern about the sanctity of life, human dignity, and what it means to be human (Miah, 2004). When developing policy guidelines on the use of genetic science, sport's challenge is to take into account these complex philosophical concepts where diverse cultural differences also exist.

Universal Rules and Particular Cultures

A distinguishing feature of sports is their reliance on universally accepted rules, which permit a reasonable, meritocratic basis on which to evaluate and compare performances. For example, in sprinting, athletes—via their federation—agree on the means by which they will run, the distance they will cover, when they will start, and so on. Of course, this is not a perfect system, and sports federations are sometimes subject to external pressures to modify their rules for political reasons, such as, with regard to the timetabling of sports during mega-events like the Olympics. To this end, it is misleading to suggest that the rules are set entirely by the sports community, absent of any political influence. One can distinguish between various types of such rules: constitutive, regulative, and auxiliary. Thus, *constitutive* rules "stipulate a goal and the means…by which this goal can be attained" (Loland, 2002), while *regulative* rules are "logically independent of the process of competing," and include dimensions such as fixing the "appearance, size, and weight of the golf ball." *Auxiliary* rules are extraneous to the competitive situation, but nevertheless govern

what takes place. As Loland describes, this includes rules such as separating male and female into different competitions.

These rules describe the formal mechanisms through which sports promote fairness of competition. Additionally, there are tacit, unwritten rules that shape the ethical environment of sports or their *ethos*. For example, often in soccer an attacking team will deliberately put the ball out to the touchline, if a member of the opponent's team has become injured. Subsequently, despite the rules indicating that the opponents will then have the advantage—which they may rightfully use to attack—it is commonplace that they will voluntarily return the advantage to their opponents as a reciprocal courtesy. Sports are full of such examples of ethically praiseworthy behavior, indicating that what the rules indicate formally does not encompass the richness of moral action that exists in competition. Arguably, it is through such gestures where the strongest examples of fair play take place within sports, thus challenging the idea that elite sports results are won at all costs. Moreover, there is often consensus over the acceptance of such rules, in order to preserve the possibility of meaningful competition.

When expanding our range of ethical concern to the training field, rather than just competition, the ethos is much more ambiguous. Moreover, the way that science is used to affect sporting potential is a clear exemplar of this ambiguity. In part, this is because sports have changed over time from being practices largely constituted by technology, but only rarely modified by it, to practices that are wholly immersed within a scientific culture. Thus, today, the development of technology and technique is an integral part of what we understand as sports performance, and this shift coincides with how the value system of sports has changed over time. As Guttmann (1978) puts it, sports have shifted "from ritual to record" and, consequently, have become practices that are especially interested in quantification and measurement, rather than ceremony and ritual. The breadth of this technologization is considerable and includes modifications in technique, transformations to equipment, the use of sophisticated biomonitoring devices, and greater knowledge about biological systems. Such knowledge informs a number of behavioral patterns that alter athletic performance, such as hydration, nutrition, rest, and so on.

While it is uncontroversial to embrace such changes when they are clearly therapeutic, unease about these transformations is evident in the concomitant concern that technologization may lead to the dehumanization and mechanization of athletic performance (Hoberman, 1992). Additionally, various authors have identified the unintended consequences of sport technologies, which may have been designed to make competition safer, but which have had the opposite effect (Gelberg, 1995). While the immediate historical context of this concern is the 1980s, when East Germany undertook programs to enhance their athletes via illegal doping methods, there are broader sociological roots that inform such perspectives. Specifically, heightened concern about automation within modern society as a deeply alienating process leads some to conclude that society may be, overall, worse off from such changes.

The mechanization thesis is made even more complex—technically and ethically—in an era of genetic and molecular science. The prospect of selecting or engineering humans creates frightening images often portrayed through totalitarian societies, where humanity collapses under the hyper-scientization of society or misguided attempts to build a better human race, which has appeared more than once in human history. As well, cultural texts have often served as a warning about the prospect of a technologically dependent future. The kind of surveillance described in Huxley's *Brave New World* and movies such as *GATTACA* or *Bladerunner* offer an imagined future where technology undermines what is valuable about being human, often to its destruction. Thus, the science of molecular mechanization implies a more invasive form of transformation, even if the actual complexity of the procedure is nothing more than an injection with a needle. This is because there are both cultural and moral distinctions between, say, adapting the body via an exercise machine such as a treadmill and transforming the genetic composition of an athlete. The relevant differences may not reside in the actual effect upon performance; both may allow an athlete to run longer. For example, consider the recurrent discussions about swimming costumes that purport to

provide a performance advantage. Such technology might similarly improve performance, as might a biological enhancement. Yet, the critical moral concerns involve issues such as reversibility or the capacity to affect subsequent generations. Such matters concern sports federations for a number of reasons, but principally because the use of such technology is often associated with affecting the merit of an athlete's performance. Thus, attempts at enhancement may affect the conditions of competition in such a way as to invalidate the kind of praise and admiration many people feel athletes deserve.

To this end, ethical considerations of genetics and molecular science in sport are fundamentally concerns about justice in sport. However, they also imply broader questions about social justice; since the conversation is often about access to sophisticated medical technology, the privatization of which could contradict broader societal aspiration to distribute primary goods fairly. Yet, there are also important moral considerations that arise from *withholding* the use of science and technology. Thus, while their use might disturb equality within sport, withholding their use may also preserve existing inequalities. Recall that sports try to promote equality, not maintain it. For example, one might claim that some people have a genetic constitution that is more suited than others for particular kinds of sport. A simple exemplar of this is the genetic predisposition toward height, which can be helpful for a number of sports. Thus, if it was possible to genetically alter the eventual height of an individual, one might claim that this is reasonable on the grounds that failing to do so permits an unfair inequality to exist among an athlete population. This pro-enhancement perspective has gained prominence over the last 10 years and will be revisited later in this chapter.

Bioethics in Sport

Since the popularization of genetic science in the 1990s, scientific research in this area has been matched by the expansion of bioethical investigations into the subject. These inquiries include the consideration of issues such as the patenting of human DNA, the use of genetic selection, and the possible use of gene transfer for nonthera-peutic applications. In addition to this, genetic and molecular science are part of the converg-ing sciences, which encompass nanotechnologi-cal, biotechnological, informational, and cognitive sciences, thus making the realm of ethical debate even more contested and subject to various, shift-ing political interests (Roco & Bainbridge, 2002). A concomitant research agenda has emerged over this period to evaluate how science reaches the public (see various essays in the SAGE journal *Public Understanding of Science*). Thus, studies in the *Public Understanding of Science* and evalua-tions of media narratives on genetic science have become an important feature of the biopolitical landscape. (As an indication of this, the nanote-chnology debates 10 years later have endeavored to undertake more sophisticated approaches to public engagement via NanoJurys and the like.) Prominent controversies surround the publication of research that often relies on small-scale popula-tions, though generating large-scale media cover-age. Such reporting often involves dramatic claims about the potential to transform humanity via such research findings. Sport is one area that has been a focal point for such propositions and is generally a practice that is subject to regular discussions about the future of humanity and its ongoing transforma-tion by technology (Miah & Eassom, 2002).

To this extent, the cultural context of genetic and molecular science plays an important role in shap-ing how sports organizations react to its utiliza-tion. For example, consider the story that indicated five British footballers planned to store the stem cells of their children to protect themselves (and their children), should they become injured during competition (Templeton, 2006). In this situation, a number of complex issues arise that reflect the cul-tural context within which genetic technology has developed. First, the footballers' intentions imply an acceptance of the legitimacy to harvest stem cells, which is not something that is shared in all countries. Second, the fathers presume an entitle-ment to utilize the cells of their children for their own means, rather than consider that these cells belong only to the child. Third, the possibility of undertaking such a decision exists in a country

where the industry of commercial stem cell storage exists, which is a possibility that is not available to all nations.

How should the sports world react to such pos-sibilities? Is it fair that these footballers have access to a technology that could assist them in recovering more effectively and quicker, when others do not? Should the sports world be obliged to promote the utilization of such technology to optimize condi-tions of equality? Alternatively, does the utilization of such technology fall outside of what the sports world can assume as its regulatory responsibility? Must sports organizations ultimately accept what society considers to be an acceptable use of tech-nology, rather than seek to prohibit or control it? These questions reflect the complexity of the ethi-cal debates both for individual users who may seek more technology to enhance their performance or their resilience and for the sports industries, which attempt to promste fair competition. They also reinforce the claim that policy development on genetic science in sport cannot exist in isolation from broader social policies on genetics.

However, the challenges presented by genetics are not exceptional. For many years, sports have negotiated such ethical terrain where, at times, it has sought to strictly restrict the use of technol-ogy and, on other occasions, has embraced it fully (Miah, 2005). Nevertheless, in subsequent years, this challenge will become even greater, as new applications emerge and as opinions differ on their legitimacy to alter humanity. Moreover, it is not just humans that are affected by this science. For exam-ple, controversies have already arisen about the cloning of racehorses (Coghlan, 2005).

To what extent can sports rely on established ethical codes of conduct to address this emerging science? Is it reasonable to expect the direct appli-cation of bioethical principles to meet the range of ways that genetic science will be applied to sport? In the mid-1990s, Murray (1997) argued *against* the idea of "genetic exceptionalism," a concept that supports the treatment of genetic science as ethi-cally distinct from other sciences. Instead, Murray explains that there are some new challenges that genetics faces, but that the way we have addressed ethics in science and medicine previously should not be dismissed, as if it is no longer relevant or useful. Yet, cultural theorists Nelkin and Lindee (1995) argue that the symbolic status of the gene does require that we treat it differently, if not ethically, than at least sociologically. They argue that genes have a special status and this requires us to distinguish them from other units of our biology, if only because people hold such beliefs. This need not mean that societies devise new ethical codes to take into account such beliefs, but it might require taking into account special sensitivities that arise because of the importance people give to genes. It might also imply revising ethical protocols, such as the consent process, to account for the different ways in which people make sense of genetic information. Arguably, this is why the importance of genetic counseling has been emphasized in a clinical context, as these sensitivities give rise to new obligations for the world of sport to address.

Understanding how best to approach these questions involves looking back on how ethical theory has changed in a postgenomic era. In the context of medicine and technology, early approaches to bioethics and medical ethics focused on developing principles that could govern good practice, notably through the work of Beauchamp and Childress (2008), which dominated the rise of bioethics in the postwar period. Their four principles—autonomy, beneficence, nonmaleficence, and justice—have shaped the development of ethical codes within modern medicine and science. In recent times, scholars have critiqued this top–down principalism, arguing that the lived reality of ethical conduct is much more complex and that ethical codes must be informed by these circumstances. Thus, a bottom–up approach to deriving ethical guidelines has also emerged, so-called casuistry (the study of cases). Today, a method of *reflective equilibrium* has become commonplace within applied ethical settings, which relies on a combination of the two approaches, while increasingly empirical research is informing the characterization of ethical dilemmas within the medical setting (Haimes, 2002). To some extent, the world of sport has yet to fully adopt these methods of ethical reasoning, though some studies do emphasize their importance (see Butryn, 2003; Butryn & Masucci, 2009).

Categories of Ethical Concern

The ethical debates surrounding genetics and molecular science in sport can be separated into three distinct territories: research and development, using genetic information, and gene transfer. Each of these areas is closely related in a variety of ways, and the latter two are reliant heavily on the first. This is why many of the struggles fought by genetic scientists have aimed to dampen the claims over controversial, radical applications, such as designer babies, and instead emphasize how research will be instrumental to our treatment of seriously debilitating illnesses. While other chapters in this volume deal with each of these issues in detail, it is useful to identify a range of ethical and policy issues that are common themes within all three and address the ways in which the policy world has attempted to deal with them.

A particular controversy that surrounds the use or development of genetic science in sport is the distinction between research and application. Unfortunately, discourse about the two often blurs together, and the development of science can suffer due to a misunderstanding within the public imagination made manifest through public policy. Indeed, an established complexity of scientific research that might have a bearing on the world of sport is its tendency to develop within a medical context. For example, Lee Sweeney's research on insulin-like growth factor 1 (IGF-1) aspires to treat muscle-wasting diseases. Yet, Sweeney's work has also been at the forefront of the gene doping debate—much more than the research might otherwise have been—in part due to his willingness to engage publicly on how the future of his work could be utilized. Sweeney's 2004 publication on IGF-1 (Sweeney, 2004) was also one of the earliest studies to indicate the potential for sports:

> "However, this work raises a number of ethical considerations about the use of IGF-I in gene therapy, as its beneficial effects could be used in humans for athletic or cosmetic purposes, rather than for disease treatment (Barton-Davis et al., 1998)."

There are many other genetic scientists whose work holds similar implications and this presents challenges for anti-doping policy makers. This is because the kind of developmental work that goes into medical research is highly protected until commercialization. Yet the capacity of the sports world to address illegal uses of such technology relies on early indications of the products that are likely to emerge on the market, to ensure it can develop robust anti-doping tests. In addition, the controversy surrounding gene doping has meant that experimental research surrounding genetics and exercise science has also met with skepticism within the policy community. As a result, the British Association for Exercise and Sport Science (Wackerhage et al., 2009) recently published a position statement arguing on behalf of genetic research in sport. The authors assert that there are novel challenges presented by genetic science in sport, but that there should be encouragement to continue. For example, one of the difficulties with separating research from application is the unexpected knowledge that might derive from research, as time goes on. For example, they discuss how

"…a polymorphism in the gene encoding the human bradykinin receptor B2 was initially shown to be associated not only with exercise-induced cardiac hypertrophy (Brull, et al., 2001) but later that it also predicted coronary risk (Dhamrait, et al., 2003)."

Outside of sport, the use of genetic information is also taken very seriously. Hendriks (1997) argues that "the unrestricted use of genetic information poses a number of threats to the exercise and enjoyment of human rights" (p. 557). The apparent desire of employers or insurance companies to have access to genetic information, as a tool for limiting economic risk, invites scrutiny due to its potential for discrimination by the Australia Law Reform Commission (ALRC) (2003). Moreover, the possibility of identifying specific genetic characteristics, coupled with the possibility of selecting in or out certain traits, may lead to considerable social pressure to undertake such decisions. Indeed, one might envisage that such choices become an integral part

of what is deemed to be responsible practice in sports talent identification.

Again, these circumstances reveal the broad societal concerns that orbit the use of genetic science in sport. Genetic testing in sport has arisen in two prominent cases. The first is outlined by the ALRC (2003), which was the first major investigation into the use of genetic information in sport. Here, they outline the case of the Professional Boxing and Combat Sports Board of Victoria, which discussed whether genetic testing could be used to help physicians advise (or better inform) the athlete about the level of risk in their competition. McCrory (2001) mentions a similar concern, explaining how "delayed cerebral edema after minor head trauma" has been linked to "an abnormality with the CACNA1A calcium channel subunit gene" (p. 142). McCrory argues that these findings are important enough to require physicians to offer advice against participation in sports and even to require athletes to take genetic tests, where such risk exists. Moreover, the ALRC note that a "milder form of this condition can occur in players of rugby, soccer, and other sports associated with repetitive blows to the head" (Section 38.29, p. 964). Yet, the Boxing Board decided against the use of such tests. A second case was of Ed Curry, basketball player for the Chicago Bulls team, which required him to undertake a genetic test after he had missed games due to an irregular heartbeat. Curry refused the test and transferred to the New York Knicks in 2005, a club that did not require that he take such a test.

The two examples highlight the complexity of maintaining ethical principles—such as confidentiality—in what are often high-profile cases. Moreover, they highlight the legal uncertainty within the sports world over how to address claims from a range of parties over access to genetic information. Alongside these debates, a number of conversations have also taken place about the use of genetic information to make selection decisions on the basis of *athletic potential*, rather than liability of health risk. For some time, there was ambivalence about the legitimacy of such selection decisions when even the IOC President Jacques Rogge reportedly indicated that there is nothing obviously wrong about refining talent identification techniques using genetic information: As Clarey (2001) notes,

> "Though it is too soon for sports legislation, Rogge favors a declaration of principle now: 'Yes' to genetic screening, which is the examination of a prospective athlete's genes to see which sport would be the perfect fit (perhaps less than five years away). 'No' to genetic manipulation."

However, the lived reality of such testing has led sports leaders to revise their position on such use. In particular, in 2004, the first commercial test for a performance gene reached the market, called the ACTN3 Sports Performance Test™. It claimed to identify whether the user may be "naturally geared towards sprint/power events or towards endurance sporting ability." (Genetic Technologies Ltd, 2004). Around the same time, *Nature* reported that an Australian rugby league team would experiment with genetic tests to improve their ability to train athletes and direct them toward success within competition (Dennis, 2005). Soon after, WADA responded with its *Stockholm Declaration* (WADA, 2005), which "strongly discouraged" the use of such tests by sports organizations, especially to make selection decisions. The declaration emerged as part of WADA's second landmark meeting on gene doping.

The use of genetic information reinforces the broader societal implications of the genetics issue in sport. Such concerns can involve extending the realm of parental autonomy, though in dramatically different ways. For prenatal selection decisions, it would involve presenting parents with decisions about what kind of embryo they bring into existence. Alternatively, in a postnatal setting, it can imply using a mouth swab to identify what sports children might excel in. The ethical and moral objections to such technology being used range from a concern about engendering an endless spiral of biotechnological competitiveness to anxieties that such selection decisions exhibit unjustified and inappropriate prejudices toward certain kind of people over others (Miah, 2007). Yet, it may also be argued that these freedoms fall well within the realm of parental liberty and so do not present such great harms as to require prohibition.

A number of practical questions also emerge from such prospects, which have yet to be resolved. For example, would athletes be required to undergo genetic screening to establish their genetic constitution before being allowed to compete in sports? In what ways could sports authorities and their stakeholders have access to the information derived from genetic tests to identify doping practices? How would genetic testing influence an athlete's enjoyment of sport? If an athlete has an unusually favorable phenotype, would this lead to their disqualification from competition?

Where genetic testing is used to identify talent, concerns over discrimination are of a different character. Here, the concern is that the testing method may not be an adequate indicator of performance potential. Indeed, the complexity of sports is such that making absolute judgments about what characteristics will ensure or even increase the probability of success is difficult. For example, one may reasonably argue that extreme shortness (or extreme height) prevents an individual from performing the required skills of many sports. As such, by claiming that height is a relevant characteristic of sporting performance, one may then claim that genetic tests could be used to justify why short children are not selected in an elite training program. Yet, this conclusion is insufficient, since there are many people who may welcome the chance to become an elite athlete, but for whom there is no opportunity due to the tests. Indeed, such arguments could be made in relation to a number of disabilities, for which it would be unreasonable to claim that such aspirations do not deserve support. Thus, clearly there is a sense in which sport depends on providing opportunities for different kinds of people. Moreover, it seems preferable to adapt the structures in sport to allow the possibility of such people to pursue elite competition, rather than to endorse a system, which excludes certain kinds of people from participating. If this additional commitment requires creating greater divisions within sports, then this should be the responsibility of sports federations, since the value of sport depends on inclusivity.

An additional complication in the context of sport is taking into account the life course, where participation in sport often starts at a very young age.

This has a specific bearing upon the use of such testing and selection in children since a child may enjoy many years as a competitive athlete, before reaching a point where genetic factors limit competitiveness at an adult level. Consider a young basketball player who is destined to be 165 cm tall, who may enjoy being an excellent player until his peers have undergone their final growth spurts. The value a child may accrue from these experiences is clearly sacrificed, if a genetic test to be used early in life revealed that his eventual height would likely limit competitiveness in adulthood.

If discussions about genetic information reflect the present-day use of genetic information, conversations about gene transfer in sport are its future. These debates have been dominated by the prospect of *gene doping*, the use of gene transfer for nontherapeutic or enhancing purposes. A range of institutions take this prospect seriously and include WADA, the American Association for the Advancement of Science, the United States President's Council on Bioethics, and the British Government (House of Commons, 2007). A number of philosophical and ethical issues surround the debate on gene transfer in sport. On a philosophical level, there is a need to distinguish between types of therapy and nontherapy (Miah, 2008). For example, while the WADA Code accepts the use of gene transfer for therapeutic use, it is unclear whether the distinction between therapy and enhancement can be sustained in the long term. Thus, insofar as genetic disorders are often linked to age-related diseases—such as muscular depletion—it might be medically desirable to "enhance" people in order to maintain a reasonable level of health. Moreover, it might appear that individuals must be treated with gene transfer well before they are symptomatic, that is, when they are considered healthy. These prospects are receiving careful consideration from a range of governments around the world. As noted in the introduction, the United States President's Council on Bioethics (2003) discussed this prospect in the context of sport and identified no clear consensus on what should follow in policy terms. In addition, the British House of Commons (2007) investigation also reinforced the likely expansion of such modifications in society. More recently,

the European Parliament committee Scientific Technology Options Assessment (STOA) (Coenen et al., 2009) identified a series of key priorities around human enhancement, which included gene doping.

Ethically, the recurrent questions are about how such technology would affect equality in sport and broader notions of justice. WADA's approach to any new technology is to identify whether it engages two of the three of its criteria: performance enhancing, risky to health, and against the spirit of sport. If two conditions are engaged, then it will *consider* prohibition. While the use of gene transfer remains highly experimental, it may give rise to forms of performance enhancement that are safer than current methods that rely on synthetic substances—often from an illegal black market. On this basis, there might be a good reason to promote these healthier modes of human enhancement in order to diminish the illegal trade of substances that currently overshadows elite sports. Moreover, through the utilization of biomarkers and DNA passports, there might be a greater potential to monitor the detrimental health risks that an athlete could face through such modifications. These arguments are part of broader perspectives that argue on behalf of human enhancement in sport, which have gained prominence in recent years (Miah, 2004).

Conclusion

The ethics of performance enhancement in sport are operationalized through WADA as a principle of "strict liability," which deems that any positive anti-doping test means immediate suspension pending an inquiry. Yet, there are many biotechnological modifications that the sports world does not address, such as functional elective surgery. To this extent, questions remain about how genetic and molecular modifications or knowledge should be treated in the long term. Arguably, as humanity's continued pursuit of health progresses, it will become apparent that the use of such science implies seeking to alter those biological processes that are a part of the aging process, and our intervention ultimately will ensure a collapse of the distinction between therapy and enhancement. If societies accept such continued pursuit, then the attempts to maintain sport as an environment free from enhancement will not simply be impractical or undesirable, they would also contravene fundamental human rights.

To this end, as the sports world races ahead to criminalize doping practices and treat the widespread use of performance enhancement as a broad public health issue, it will need to consider the interface between the local, national, and international policy debates. Arguably, the political history of sport in the postwar period ensured that genetic science would be treated as a questionable technology for sports, where gene doping would become an integral part of the war on drugs. Yet, as the American Academy of Pediatrics (2005) noted, young people are not using steroids just for competitive sport. Rather, there is a broad culture of enhancement that underpins the use of technology. In time, genetic modification may become a part of this culture, though its integration within society will emerge first through applications that are medically justified, and sports authorities will have to resolve how they will address the genetically modified athlete that society deems to be medically permissible.

References

American Academy of Pediatrics (2005). Guidelines for pediatricians: Performance-enhancing substances. *Sport Shorts* 12, 1–2.

Australia Law Reform Commission (2003). *Essentially Yours: The Protection of Human Genetic Information in Australia.* Southwood Press, Sydney.

Barton-Davis, E.R., Shoturma, D.I., Musaro, A., Rosenthal, N. & Sweeney, H.L. (1998). Viral mediated expression of insulin-like growth factor I blocks the aging-related loss of skeletal muscle function. *Proceedings of the National Academy of Sciences of the United States of America* 95, 15603–15607.

Beauchamp, T.L. & Childress, J.F. (2008). *Principles of Biomedical Ethics*, 6th edn. Oxford University Press, New York.

Brull, D., Dhamrait, S., Myerson, S., et al. (2001). Bradykinin B2BKR receptor polymorphism and left-ventricular growth response. *Lancet* 358, 1155–1156.

Butryn, T.M. (2003). Posthuman podiums: Cyborg narratives of elite track and field athletes. *Sociology of Sport Journal* 20, 17–39.

Butryn, T.M. & Masucci, M.A. (2009). Traversing the matrix: Cyborg athletes, technology, and the environment. *Journal of Sport and Social Issues* 33(3), 285–307.

Clarey, C. (2001). Chilling new world: Sports and genetics. *New York Times*, January 26, 2001. Available at http://www.nytimes.com/2001/01/26/sports/26iht-arena.2.t.html.

Coenen, C., Schuijff, M., Smits, M., et al. (2009). Human enhancement: Study. Brussels, science and technology options assessment, European Parliament, ref: (IP/A/STOA/FWC/2005-28/SC35, 41 & 45) PE 417.483. Available at http://www.europarl.europa.eu/stoa/events/workshop/20090224/background_en.pdf.

Coghlan, A. (1998). Sporty types. *New Scientist*, May 23, 1998. Available at http://www.newscientist.com/article/mg15821351.000-sporty-types.html.

Coghlan, A. (2005). First clone of champion racehorse revealed. *New Scientist*, April 14, 2005. Available at http://www.newscientist.com/article.ns?id=dn7265.

Dennis, C. (2005). Rugby team converts to give gene tests a try. *Nature* 34, 260.

Dhamrait, S.S., Payne, J.R., Li, P., et al. (2003). Variation in bradykinin receptor genes increases the cardiovascular risk associated with hypertension. *European Heart Journal* 24, 1672–1680.

Gelberg, J.N. (1995). The lethal weapon: How the plastic football helmet transformed the game of football, 1939-1994. *Bulletin of Science, Technology, and Society* 15(5–6), 302–309.

Genetic Technologies Limited (2004). *Your Genetic Sports Advantage™*. Genetic Technologies Limited, Fitzroy, Victoria, Australia.

Guttmann, A. (1978). *From Ritual to Record: The Nature of Modern Sports*. Columbia University Press, New York.

Haimes, E. (2002). What can the social sciences contribute to the study of ethics? Theoretical, empirical and substantive considerations. *Bioethics* 16(2), 89–113.

Hendriks, A. (1997). Genetics, human rights and employment: American and European perspectives. *Medicine and Law* 16, 557–565.

Hoberman, J.M. (1992). *Mortal Engines: The Science of Performance and the Dehumanization of Sport*. The Free Press, New York (reprinted 2001, The Blackburn Press).

House of Commons (2007). *Human Enhancement Technologies in Sport: Second Report of Session 2006-7 (22 February)*. Science and Technology Select Committee, The Stationary Office, London, HC67.

Hubbard, R. (2003). Science, power, gender: How DNA became the book of life. *Signs: Journal of Women in Culture and Society* 28(3), 791–799.

Loland, S. (2002). *Fair Play in Sport: A Moral Norm System*. Routledge, London.

McCrory, P. (2001). Ethics, molecular biology, and sports medicine. *British Journal of Sports Medicine* 35(3), 142–143.

Miah, A. (2004). *Genetically Modified Athletes: Biomedical Ethics, Gene Doping and Sport*. Routledge, London.

Miah, A. (2005). From anti-doping to a 'performance policy': Sport technology, being human, and doing ethics. *European Journal of Sport Science* 5(1), 51–57.

Miah, A. (2007). Genetic selection for enhanced health characteristics. *Journal of International Biotechnology Law* 4(6), 239–264.

Miah, A. (2008). Engineering greater resilience or radical transhuman enhancement? *Studies in Ethics, Law and Technology* 2(1). Available at http://www.bepress.com/selt/vol2/iss1/art5.

Miah, A. & Eassom, S.B. (eds) (2002). *Sport Technology, Vol. 21: History, Philosophy & Policy. Research in Philosophy and Technology*. Elsevier Science, Oxford.

Murray, T.H. (1997). Genetic exceptionalism and 'future diaries': Is genetic information different from other medical information? In: M.A. Rothstein (ed.) *Genetic Secrets: Protecting Privacy and Confidentiality in the Genetic Era*, pp. 60–73. Yale University Press, New Haven.

Nelkin, D. & Lindee, M.S. (1995). *The DNA Mystique: The Gene as a Cultural Icon*. W.H. Freeman & Co, New York.

Roco, M.C. & Bainbridge, W.S. (eds.) (2002). *Converging Technologies for Improving Human Performance: Nanotechnology, Biotechnology, Information Technology and Cognitive Science*. National Science Foundation, Kluwer, Dordrecht.

Sweeney, H.L. (2004). Gene doping. *Scientific American*, 291, 62–69.

Templeton, S.K. (2006). Footballers use babies for 'repair kits'. *The Sunday Times*, August 27, 2006. London. Available at http://www.timesonline.co.uk/tol/news/uk/article620835.ece.

The President's Council on Bioethics (2003). *Beyond Therapy: Biotechnology and the Pursuit of Happiness*. The President's Council on Bioethics, Washington, DC.

Wackerhage, H., Miah, A., Harris, R.C., Montgomery, H.E. & Williams, A.G. (2009). BASES position stand on 'Genetic Research and Testing in Sport and Exercise Science': A review of the issues. *Journal of Sports Science and Medicine* 31, 1–8.

World Anti-Doping Agency (2003). International Standard for the Prohibited List 2004. Available at http://www.wada-ama.org/docs/web/standards_harmonization/code/list_standard_2004.pdf.

World Anti-Doping Agency (2005). The Stockholm Declaration, World Anti-Doping Agency. Available at http://www.wada-ama.org/Documents/Science_Medicine/Scientific%20Events/WADA_Stockholm_Declaration_2005.pdf.

Index

393